Proceedings of GeoShanghai 2018 International Conference: Transportation Geotechnics and Pavement Engineering

Xianming Shi · Zhen Liu
Jenny Liu
Editors

Proceedings of GeoShanghai 2018 International Conference: Transportation Geotechnics and Pavement Engineering

Springer

Editors
Xianming Shi
Washington State University
Washington, WA
USA

Jenny Liu
Missouri University of Science and Technology
Missouri, MO
USA

Zhen Liu
Department of Civil and Environmental Engineering
Michigan Technological University
Houghton, MI
USA

ISBN 978-981-13-0010-3 ISBN 978-981-13-0011-0 (eBook)
https://doi.org/10.1007/978-981-13-0011-0

Library of Congress Control Number: 2018939621

Printed on acid-free paper

This Springer imprint is published by the registered company Springer Nature Singapore Pte Ltd.
part of Springer Nature
The registered company address is: 152 Beach Road, #21-01/04 Gateway East, Singapore 189721, Singapore

Preface

The 4th GeoShanghai International Conference was held on May 27–30, 2018, in Shanghai, China. GeoShanghai is a series of international conferences on geotechnical engineering held in Shanghai every four years. The conference was inaugurated in 2006 and was successfully held in 2010 and 2014, with more than 1200 participants in total. The conference offers a platform of sharing recent developments of the state-of-the-art and state-of-the-practice in geotechnical and geoenvironmental engineering. It has been organized by Tongji University in cooperation with the ASCE Geo-Institute, Transportation Research Board, and other cooperating organizations.

The proceedings of the 4th GeoShanghai International Conference include eight volumes of over 560 papers; all were peer-reviewed by at least two reviewers. The proceedings include Volumes 1: Fundamentals of Soil Behavior edited by Dr. Annan Zhou, Dr. Junliang Tao, Dr. Xiaoqiang Gu, and Dr. Liangbo Hu; Volume 2: Multi-physics Processes in Soil Mechanics and Advances in Geotechnical Testing edited by Dr. Liangbo Hu, Dr. Xiaoqiang Gu, Dr. Junliang Tao, and Dr. Annan Zhou; Volume 3: Rock Mechanics and Rock Engineering edited by Dr. Lianyang Zhang, Dr. Bruno Goncalves da Silva, and Dr. Cheng Zhao; Volume 4: Transportation Geotechnics and Pavement Engineering edited by Dr. Xianming Shi, Dr. Zhen Liu, and Dr. Jenny Liu; Volume 5: Tunneling and Underground Construction edited by Dr. Dongmei Zhang and Dr. Xin Huang; Volume 6: Advances in Soil Dynamics and Foundation Engineering edited by Dr. Tong Qiu, Dr. Binod Tiwari, and Dr. Zhen Zhang; Volume 7: Geoenvironment and Geohazards edited by Dr. Arvin Farid and Dr. Hongxin Chen; and Volume 8: Ground Improvement and Geosynthetics edited by Dr. Lin Li, Dr. Bora Cetin, and Dr. Xiaoming Yang. The proceedings also include six keynote papers presented at the conference, including "Tensile Strains in Geomembrane Landfill Liners" by Prof. Kerry Rowe, "Constitutive Modeling of the Cyclic Loading Response of Low Plasticity Fine-Grained Soils" by Prof. Ross Boulanger, "Induced Seismicity and Permeability Evolution in Gas Shales, CO_2 Storage and Deep Geothermal Energy" by Prof. Derek Elsworth, "Effects of Tunneling on Underground Infrastructures" by Prof. Maosong Huang, "Geotechnical Data Visualization and Modeling of Civil

Infrastructure Projects" by Prof. Anand Puppala, and "Probabilistic Assessment and Mapping of Liquefaction Hazard: from Site-specific Analysis to Regional Mapping" by Prof. Hsein Juang. The Technical Committee Chairs, Prof. Wenqi Ding and Prof. Xiong Zhang, the Conference General Secretary, Dr. Xiaoqiang Gu, the 20 editors of the 8 volumes and 422 reviewers, and all the authors contributed to the value and quality of the publications.

The Conference Organizing Committee thanks the members of the host organizations, Tongji University, Chinese Institution of Soil Mechanics and Geotechnical Engineering, and Shanghai Society of Civil Engineering, for their hard work and the members of International Advisory Committee, Conference Steering Committee, Technical Committee, Organizing Committee, and Local Organizing Committee for their strong support. We hope the proceedings will be valuable references to the geotechnical engineering community.

Shijin Feng
Conference Chair
Ming Xiao
Conference Co-chair

Organization

International Advisory Committee

Herve di Benedetto	University of Lyon, France
Antonio Bobet	Purdue University, USA
Jean-Louis Briaud	Texas A&M University, USA
Patrick Fox	Penn State University, USA
Edward Kavazanjian	Arizona State University, USA
Dov Leshchinsky	University of Illinois, USA
Wenhao Liang	China Railway Construction Corporation Limited, China
Robert L. Lytton	Texas A&M University, USA
Louay Mohammad	Louisiana State University, USA
Manfred Partle	KTH Royal Institute of Technology, Switzerland
Anand Puppala	University of Texas at Arlington, USA
Mark Randolph	University of Western Australia, Australia
Kenneth H. Stokoe	University of Texas at Austin, USA
Gioacchino (Cino) Viggiani	Université Joseph Fourier, France
Dennis T. Bergado	Asian Institute of Technology, Thailand
Malcolm Bolton	Cambridge University, UK
Yunmin Chen	Zhejiang University, China
Zuyu Chen	Tsinghua University, China
Jincai Gu	PLA, China
Yaoru Lu	Tongji University, China
Herbert Mang	Vienna University of Technology, Austria
Paul Mayne	Georgia Institute of Technology, USA
Stan Pietruszczak	McMaster University, Canada
Tom Papagiannakis	Washington State University, USA
Jun Sun	Tongji University, China

Scott Sloan	University of Newcastle, Australia
Hywel R. Thomas	Cardiff University, UK
Atsashi Yashima	Gifu University, Japan

Conference Steering Committee

Jie Han	University of Kansas, USA
Baoshan Huang	University of Tennessee, USA
Maosong Huang	Tongji University, China
Yongsheng Li	Tongji University, China
Linbin Wang	Virginia Tech, USA
Lianyang Zhang	University of Arizona, USA
Hehua Zhu	Tongji University, China

Technical Committee

Wenqi Ding (Chair)	Tongji University, China
Charles Aubeny	Texas A&M University, USA
Rifat Bulut	Oklahoma State University, USA
Geoff Chao	Asian Institute of Technology, Thailand
Jian Chu	Nanyang Technological University, Singapore
Eric Drumm	University of Tennessee, USA
Wen Deng	Missouri University of Science and Technology, USA
Arvin Farid	Boise State University, Idaho, USA
Xiaoming Huang	Southeast University, China
Woody Ju	University of California, Los Angeles, USA
Ben Leshchinsky	Oregon State University, Oregon, USA
Robert Liang	University of Dayton, Ohio, USA
Hoe I. Ling	Columbia University, USA
Guowei Ma	Hebei University of Technology, China
Roger W. Meier	University of Memphis, USA
Catherine O'Sullivan	Imperial College London, UK
Massimo Losa	University of Pisa, Italy
Angel Palomino	University of Tennessee, USA
Krishna Reddy	University of Illinois at Chicago, USA
Zhenyu Yin	Tongji University, China
ZhongqiYue	University of Hong Kong, China
Jianfu Shao	Université des Sciences et Technologies de Lille 1, France
Jonathan Stewart	University of California, Los Angeles, USA

Wei Wu	University of Natural Resources and Life Sciences, Austria
Jianhua Yin	The Hong Kong Polytechnic University, China
Guoping Zhang	University of Massachusetts, USA
Jianmin Zhang	Tsinghua University, China
Xiong Zhang (Co-chair)	Missouri University of Science and Technology, USA
Yun Bai	Tongji University, China
Jinchun Chai	Saga University, Japan
Cheng Chen	San Francisco State University, USA
Shengli Chen	Louisiana State University, USA
Yujun Cui	École Nationale des Ponts et Chaussees (ENPC), France
Mohammed Gabr	North Carolina State University, USA
Haiying Huang	Georgia Institute of Technology, USA
Laureano R. Hoyos	University of Texas at Arlington, USA
Liangbo Hu	University of Toledo, USA
Yang Hong	University of Oklahoma, USA
Minjing Jiang	Tongji University, China
Richard Kim	North Carolina State University, USA
Juanyu Liu	University of Alaska Fairbanks, USA
Matthew Mauldon	Virginia Tech., USA
Jianming Ling	Tongji University, China
Jorge Prozzi	University of Texas at Austin, USA
Daichao Sheng	University of Newcastle, Australia
Joseph Wartman	University of Washington, USA
Zhong Wu	Louisiana State University, USA
Dimitrios Zekkos	University of Michigan, USA
Feng Zhang	Nagoya Institute of Technology, Japan
Limin Zhang	Hong Kong University of Science and Technology, China
Zhongjie Zhang	Louisiana State University, USA
Annan Zhou	RMIT University, Australia
Fengshou Zhang	Tongji University, China

Organizing Committee

Shijin Feng (Chair)	Tongji University, China
Xiaojqiang Gu (Secretary General)	Tongji University, China
Wenqi Ding	Tongji University, China
Xiongyao Xie	Tongji University, China

Yujun Cui	École Nationale des Ponts et Chaussees (ENPC), France
Daichao Sheng	University of Newcastle, Australia
Kenichi Soga	University of California, Berkeley, USA
Weidong Wang	Shanghai Xian Dai Architectural Design (Group) Co., Ltd., China
Feng Zhang	Nagoya Institute of Technology, Japan
Yong Yuan	Tongji University, China
Weimin Ye	Tongji University, China
Ming Xiao (Co-chair)	Penn State University, USA
Yu Huang	Tongji University, China
Xiaojun Li	Tongji University, China
Xiong Zhang	Missouri University of Science and Technology, USA
Guenther Meschke	Ruhr-Universität Bochum, Germany
Erol Tutumluer	University of Illinois, Urbana—Champaign, USA
Jianming Zhang	Tsinghua University, China
Jianming Ling	Tongji University, China
Guowei Ma	Hebei University of Technology, Australia
Hongwei Huang	Tongji University, China

Local Organizing Committee

Shijin Feng (Chair)	Tongji University, China
Zixin Zhang	Tongji University, China
Jiangu Qian	Tongji University, China
Jianfeng Chen	Tongji University, China
Bao Chen	Tongji University, China
Yongchang Cai	Tongji University, China
Qianwei Xu	Tongji University, China
Qingzhao Zhang	Tongji University, China
Zhongyin Guo	Tongji University, China
Xin Huang	Tongji University, China
Fang Liu	Tongji University, China
Xiaoying Zhuang	Tongji University, China
Zhenming Shi	Tongji University, China
Zhiguo Yan	Tongji University, China
Dongming Zhang	Tongji University, China
Jie Zhang	Tongji University, China
Zhiyan Zhou	Tongji University, China
Xiaoqiang Gu (Secretary)	Tongji University, China
Lin Cong	Tongji University, China
Hongduo Zhao	Tongji University, China

Fayun Liang	Tongji University, China
Bin Ye	Tongji University, China
Zhen Zhang	Tongji University, China
Yong Tan	Tongji University, China
Liping Xu	Tongji University, China
Mengxi Zhang	Tongji University, China
Haitao Yu	Tongji University, China
Xian Liu	Tongji University, China
Shuilong Shen	Tongji University, China
Dongmei Zhang	Tongji University, China
Cheng Zhao	Tongji University, China
Hongxin Chen	Tongji University, China
Xilin Lu	Tongji University, China
Jie Zhou	Tongji University, China

Contents

Pavement Structures and Materials

About the Editors

Xianming Shi, P.E. received his PhD in Chemistry from the Institute of Chemistry, Chinese Academy of Sciences. He was a Research Professor of Civil Engineering at Montana State University before joining Washington State University as an Associate Professor of Civil and Environmental Engineering in August 2014. He is currently the Geotechnical and Transportation Engineering Group Coordinator at WSU and an Associate Director for the Center for Environmentally Sustainable Transportation in Cold Climates. He is a Control Member of the American Society of Civil Engineers (ASCE) Bituminous Materials Committee.

Zhen Liu completed his PhD in Geotechnical Engineering at Case Western Reserve University. He is an Assistant Professor of Civil and Environmental Engineering and an affiliated Assistant Professor of Geological and Mining Engineering and Sciences at Michigan Technological University. He is a member of American Society of Civil Engineers.

Jenny Liu received her PhD from Texas A&M University, USA. She is currently an Associate Professor in the Department of Civil, Architectural and Environmental Engineering, Missouri University of Science and Technology.

Geotechnical Data Visualization and Modeling of Civil Infrastructure Projects

Anand J. Puppala(✉), Surya S. C. Congress, Tejo V. Bheemasetti, and Santiago Caballero

The University of Texas at Arlington, Arlington, TX 76019, USA

Abstract. Geomaterial data analysis and visualization are important tools as they can provide geotechnical information in a visualized way that will help design and practice engineers understand the site to materials conditions in a holistic manner so they can make prudent decisions on foundation infrastructure to ground improvement options. With the advancements made in the computational arena, the use of material property sets including those of large data sets, one can expect a nicer presentation of site to material characterization if they are presented in a visual manner. GIS-based visualization tools are often used in many areas of the geotechnical engineering including terrain digital elevation modeling, subsurface material characterization, infrastructure monitoring and hazard mapping. Tests including laboratory and field studies on geomaterials, sensor instrumentation, remote sensing and photogrammetry studies, and others have provided geomaterial characterization to performance details in the visualization models. This paper highlights a few of these highlights and advancements made for better understanding of construction and performance of infrastructure using geospatial and visualization tools. Case studies ranging from data captured from photogrammetry-based measurements using unmanned aerial vehicles (UAVs) platforms to material characterizations to in-situ cone penetration tests are described to illustrate the significance of 2-Dimensional and 3-Dimensional based visualization analyses. Future applications are also briefly covered.

Keywords: Geotechnical data · Visualization · Civil infrastructure
Unmanned aerial vehicles · Cone penetration test

1 Introduction

Geotechnical data visualization is valuable for many application areas including subsurface characterization, performance monitoring, hazard mitigation and disaster response (Olsen et al. 2013; Montgomery 2014; Ellis and Vessely 2015), and it can include a wide list of available tools to collect, interpret, analyze, and represent geotechnical features to address wide variety of applications including laboratory to field measurements, geospatial mapping contours, presentation of field captured data in rich visualization dense point cloud plots, 3D GIS presentations, and many others. Many of the existing research works focus on the interpretation of collected data using enhanced visualization data presentations, which could range from a simple graphical plot to an interactive three-dimensional view of subsurface conditions to hazard

X. Shi et al. (Eds.): GSIC 2018, *Proceedings of GeoShanghai 2018 International Conference: Transportation Geotechnics and Pavement Engineering*, pp. 1–12, 2018.
https://doi.org/10.1007/978-981-13-0011-0_1

mappings. With the advancements in high-speed computing systems to data analysis tools for processing large data sets, there appears to be opportunities in civil engineering in particular geotechnical engineering field to present geotechnical results in a rich and visual manner. Such presentation would enhance practical aspects of geotechnical and urban infrastructure design, construction and rehabilitation in normal to hazardous conditions. Figure 1 presents the potential roles of visualization analysis during different stages of an infrastructure construction project.

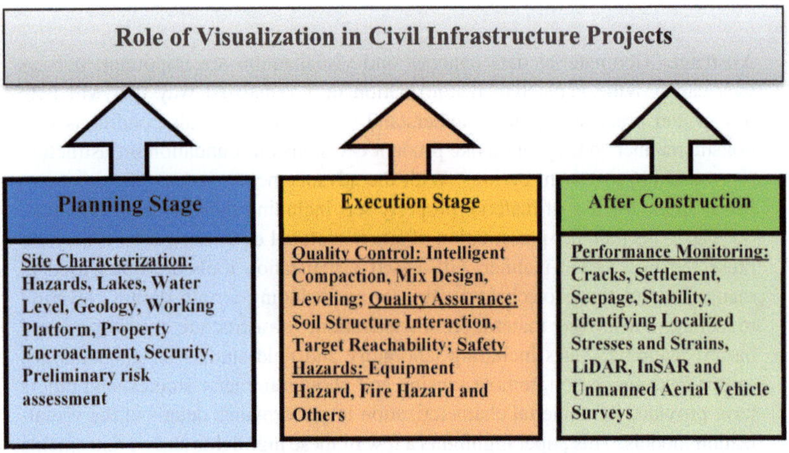

Fig. 1. Role of visualization during different stages of civil infrastructure

Many characterization methods including in situ to geophysical tools are used for characterizing subsurface geotechnical conditions. Other methodologies such as remote sensing and photogrammetry are also used to develop visualization models that depict any object closer to its original form in the field condition (Puppala and Congress 2018). In recent years, the Unmanned Aerial Vehicles (UAVs) coupled with Close Range Photogrammetry (CRP) have received increasing attention due to the advancements in drone technologies and image analyses software (Fernández et al. 2015). Reliability and reproducibility of this UAV-CRP methodology further helped in finding its way into many applications in civil engineering fields (Puppala and Congress 2018).

The paper presents an overview of geotechnical visualization and geostatistical techniques traditionally used by researchers and practitioners to better understand subsurface to geomaterial properties and conditions before performing geotechnical hazard assessment to infrastructure design. Both two-dimensional and three-dimensional visualization models for various geotechnical site related information based on laboratory and field measurements as well as UAV based photogrammetry field studies are described and explained. The next section describes the background information on the visualization modeling in geotechnical engineering.

2 Background

Geotechnical engineering is associated with interpretation and interpolation of soil properties using available insufficient knowledge of true soils and their property variations, guided by judgment and experience, to estimate the probable ranges of soil properties for design. Recent years have seen rapidly growing research into better property characterization and comprehensive hazard analysis. Therefore, increased interests in applied advanced computational facilities have facilitated optimization of geotechnical properties, incorporation of probability concepts into the design, and analysis with advanced numerical and analytical models.

Often these concepts present paradigm approaches; however, the knowledge of certainty of the soils beneath the ground still eludes the engineers, leading to the adoption of conventional and conservative designs over designs with calculated risks. This gap has perplexed geotechnical engineers for many decades. However, the recent advancements in the interpolation techniques with the aid of visualization and statistical tools have made it easy to better study and perceive the ground truth variation of subsoils. Geotechnical data visualization includes a wide list of available tools to collect, interpret, analyze and represent various geotechnical features (Ellis and Vessely 2015). A report prepared by Ellis and Vessely (2015) includes a list of existing geotechnical visualization technologies: spreadsheets and boring log generators, fence diagrams, Geographical Information Systems (GIS), image analysis software and web-based imaging analysis and presentation systems.

The spreadsheets are often used to generate X-Y scatter plots to visualize and explore the extent of correlation between measured and/or observed soil properties. Since the geotechnical engineering mostly deals with the naturally available materials, engineers often use the spreadsheets to develop empirical models for property trends. The boring log generator is often referred to as software that incorporates user data input, stores the data electronically and provides a completed log interpretation (Ellis and Vessely 2015). It is considered as the simplest way to visualize collected field data in geotechnical engineering. However, the interpretation of single boring log is not the most adequate approach to understand the subsurface features before any geotechnical assessments. The traditional technique used to interpret boring logs is known as fence diagrams, which is the method that merges the information from several subsurface explorations to create a cross-sectional view of soil layering and material properties.

Visualization techniques for subsurface studies can be addressed based on high-quality data collected over a specific site in a given period. For example, geotechnical data can be collected, stored and accessed by different reconnaissance tools and methods. This data collecting devices include field intrusive testing (Cone Penetration Testing – CPT; Standard Penetration Testing or SPT; Dilatometer Testing or DMT and others), geophysical devices (Seismic Refraction and Seismic Logging methods), geotechnical instrumentation (Piezometers, Inclinometers, and others), remote sensing devices (field surveys, satellite-based data collection and photogrammetry technologies) and others. Irrespective of the type of collecting data method used, the vital step in analyzing and modeling geotechnical data is to use and apply available tools in an efficient and realistic manner.

The University of Texas at Arlington (UTA) research team has been involved in several research studies that explored the use of geotechnical data in both two-dimensional and three-dimensional visualization mode analysis in spatial coordinate system of real field projects. An attempt is made here to present some of these studies in this paper along with observations and findings of each study. The following sections cover these studies by presenting both two-dimensional and three-dimensional maps created on soil properties to performance of infrastructure obtained from the laboratory to field studies.

3 Geotechnical Visualization Modeling

3.1 2D Visualization Models

Integrated Pipeline (IPL) project, which is a joint effort between the Tarrant Regional Water District (TRWD) and Dallas Water Utilities (DWU) intends to bring additional water supplies to the Dallas/Fort Worth metropolis. As a part of the project, a few geotechnical research studies were conducted by the University of Texas at Arlington to evaluate the sulfate levels in the soils along the pipeline and reuse potential of the excavated materials in the form of Controlled Low Strength Material (CLSM) or flowable fills as a bedding and backfill material during pipeline construction.

Sulfate Concentration Contours. In a 240+ km (150-mile) long pipeline construction in North Texas, one of the main challenges is to address the highly sporadic sulfate concentrations in the ground within six different geological formations. Soluble sulfates in the subsurface are known to cause high heave distresses when in contact with water and/or calcium-based stabilizers (Puppala 2016). Hence, it is important to know the sulfate levels in the soils if the excavated soil is reused for pavement base layers and other infrastructure projects. Several soil samples were collected along the pipeline alignment, and these samples were tested for soluble sulfates in the laboratory using modified UTA method (Bheemasetti et al. 2016). Figure 2 presents the contour map developed for soluble sulfate concentrations using 2-D visualization tools, GIS, and spatial variability models.

The 2-D visualization models presented in Fig. 2 provide engineers in the design and selection of dosage content to stabilize medium to high sulfate soils (those with more than 2000 ppm to 5000 ppm and those more than 5000 ppm). It is also interesting to note that the sulfate levels varied along each geological formation, signifying the spatial variability of them, which will aid engineers in the appropriate selection of chemical treatments of the excavated soils if they are reused.

Moduli Variations with Time. In this research study, the excavated soil was subjected to a laboratory mix design to prepare it as Controlled Low Strength Material (CLSM) for supporting the pipeline infrastructure and mix design details can be found in Puppala et al. (2014).

A field prove out section was constructed to validate that the CLSM prepared in the field meet the laboratory mix design property requirements. A series of Spectral Analysis of Surface Waves (SASW) studies were performed inside the prove out

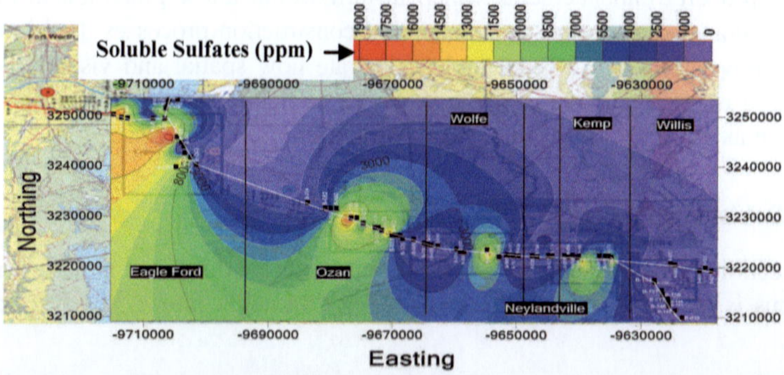

Fig. 2. 2D visualization model for spatial soluble sulfate concentrations in ppm along the pipeline alignment in north texas

section at selected points to study the CLSM property enhancements with the setting time. SASW measurements were collected at 1, 3, 7, 14 and 28 days after the CLSM placement in the pipe section. A total of 85 points corresponding to 17 sections along the pipeline were monitored. Kriging, a stochastic interpolation procedure was utilized to make stiffness contours along the pipeline using interpreted moduli data from various sections. Figure 3 presents the stiffness contours around the pipeline at different time intervals.

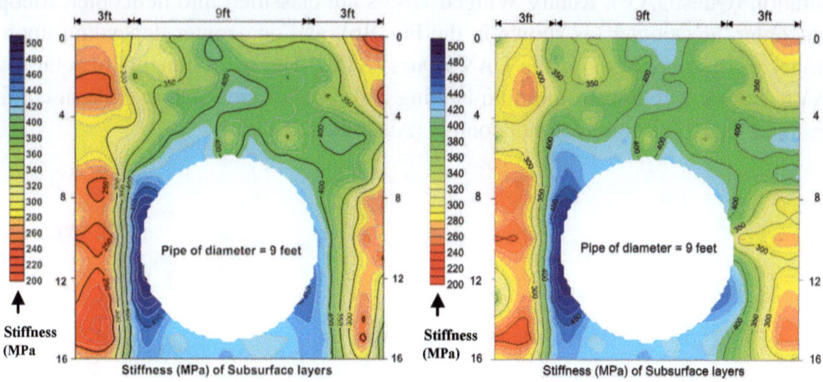

Fig. 3. Kriging analysis of stiffness of CLSMs around the pipeline at different time intervals

Kriging test results and analyses presented in Fig. 3 show that the 2-D visualization contours of stiffness data indicate that the material around the pipeline exhibits high stiffness value, corresponding to CLSM. The stiffness values decreases with an increase in the distance from the pipeline depicting the transition from CLSM to soil. The soil above the CLSM depicts high stiffness values which can be due to continuous compaction effort performed in the field. Visualization of 2-D analysis indeed explain the time rate property enhancements which in turn would allow practitioners to address quality assurance and control assessments of construction processes used in the field.

The above described 2-D examples illustrate how spatial and visualization modeling based on material characterization would be helpful in quality control works to decision-making abilities of the engineering team.

3.2 3D Visualization Modeling

In this section, geostatistical and visualization models are described in 3D environment. The focus is on spatial variability and property variation into x, y and z coordinate system.

Close Range Photogrammetry (CRP) Data Analysis Using UAV Platform. In the United States, the initial development of unmanned aerial vehicles or UAV has taken place in the early 90's with special interest in using them for long-term reconnaissance videos to assist in defense applications (Haulman 2003). The last decade has witnessed a growth and demand for UAVs owing to its capabilities and potential applications in numerous civilian missions with high social benefits (Pereira et al. 2009). Few of these areas include transportation infrastructure studies, traffic safety assessments, and precision agriculture applications (El Meouchea et al. 2016).

Small Unmanned Aerial Vehicles (sUAVs) are classified into two types namely: Fixed Wing and Rotary Wing depending upon the mechanism used for attaining the lift required to become airborne (Australian UAV). Fixed Winged UAVs, shown in the Fig. 4a, are classified into high, mid, and low wing types based upon the wing configuration (QuestUAV). Rotary Winged UAVs are classified into helicopter, tricopter, quadcopter, hexacopter (as shown in the Fig. 4b), and octocopter depending upon the number of rotors present on the UAV (QuestUAV). The main benefits of rotary wing UAVs are that it requires minimum landing and take-off area, can operate in confined areas and can offer good camera control (Australian UAV).

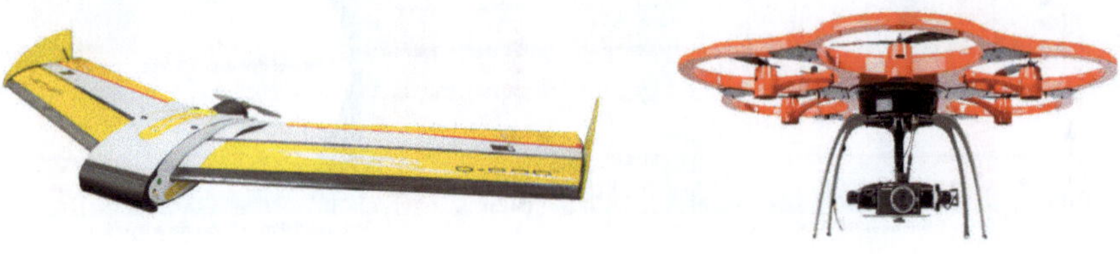

(a) (b)

Fig. 4. (a) Fixed wing UAV; (b) Rotary wing hexacopter (Source: OnestUAV)

Photogrammetry is a technique that utilizes this principle by extracting data from the images collected by a sensor mounted on either a stationed or mobile platform. American Society for Photogrammetry and Remote Sensing (ASPRS) referred it as the art, science, and technology of obtaining reliable information about physical objects and the surrounding environment through the process of recording, measuring and interpreting images and patterns of electromagnetic radiant energy and other phenomena (Udin and Ahmad 2014). Adoption of UAVs for photogrammetry comes under the category of Close-Range Photogrammetry (CRP) (Siebert and Teizer 2014).

In this paper, the CRP works were collected from UAVs, and hence the results are reported as UAV-CRP method results. As part of our UAV-CRP studies, researchers used the rotary winged hexacopter shown in Fig. 4b. Prior to any field operations, comprehensive calibration studies were conducted on UAV accessories and operation variables to address and ensure the required accuracy.

Captured images from UAV surveys are georeferenced using Global Navigation Satellite System (GNSS) data accessed either by Real Time Kinematic (RTK) or Post Processing Kinematic (PPK). Image analysis software are later used on the field images to build 3-Dimensional dense point cloud and orthorectified images that are exported in LAS and TIFF formats, respectively. The generated output files corresponding to the infrastructure asset under inspection are further analyzed using algorithms to identify major distress features such as cracks, deformations, failure planes, and others. Different image data sets pertaining to the condition of various infrastructure assets were collected using a visual range camera mounted on the UAV. Some of these visualized data are analyzed and discussed in the subsequent sections.

Pavement Distress. One important application of UAV-CRP studies is in the area of pavement infrastructure mapping. Pavement is the backbone of any community's infrastructure, and hence its health condition from CRP studies would provide an assessment of the infrastructure condition. Details shown in Fig. 5 consists of three pavement dense point cloud imaging views namely top view, 3-Dimensional view, and profile section view of the pavement. A closer examination of these images provides engineering performance of the pavement under consideration. Results in Fig. 6 indicate cracking distress pattern in the top view to differences in contours in 3-D view to considerable rutting or permanent deformation damage in profile view. All these images provide valuable pavement health condition at a given location and time.

Rock Cut Stability Assessments. Stability of rock cuts is warranted for safe operation and preservation of any infrastructure asset that passes through hilly terrains. A proper assessment of the stability of cuts would prevent potential landslides during hazard events including seismic and hurricane events. UAV-CRP methodology assists in collecting such data in a safe and efficient manner. UAV operations are performed on one of the rock-cut sections in Alpine, Texas region where a Texas Department of Transportation (TxDOT) owned railway track was located.

Multiple flight altitude mission was conducted manually due to the variations in the ground profile, wind speed, and bird threat in this remote area of Texas, USA. The copter was flown away from the wind tunnel effect created by the rock walls on either side of the cut. Maintaining the line of sight of the copter, data was collected while the area being covered was monitored real-time using live video display unit (DLVP). Servo gimbal was not only useful in damping the vibrations caused due to flying in windy conditions but also in collecting nadir and oblique images. Oblique images were helpful in accurate representation of uneven slope surface of the rock walls.

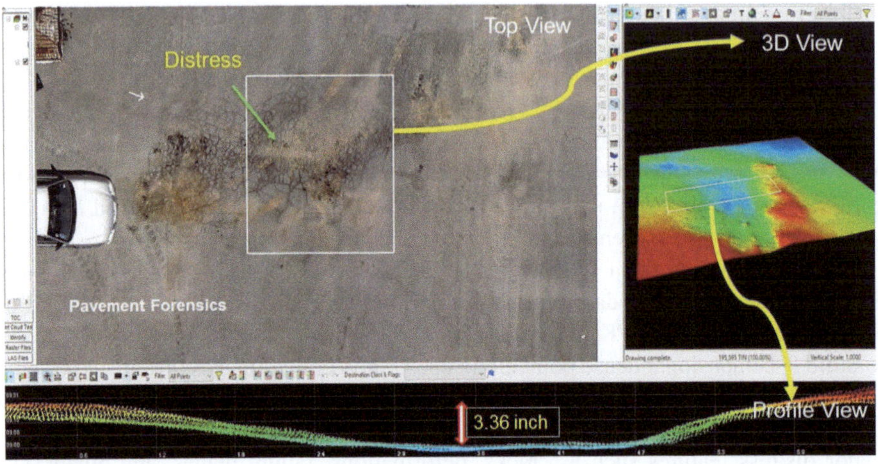

Fig. 5. Pavement distress assessment from UAV-CRP visualization assessments

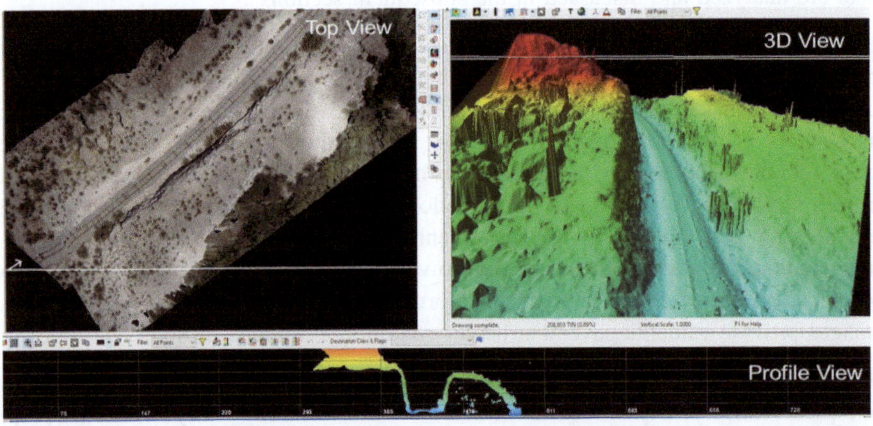

Fig. 6. Rock cuts – visualization images

Geotagged images were used to generate high quality 3-dimensional dense point cloud models and orthophotos.

UAV-CRP data from this site was analyzed using GeoCue software, and the results are presented in Fig. 6. All three views including top or plan view, 3D view, and profile section views of the railway infrastructure passing through a rock cut are presented in this figure. These different view perspectives will help in identifying the most critical rock slopes and plan for performing any stabilizing measures in case if the slopes appear to be close to a failure state. The debris on the track made up of the rocks that rolled down from the rock outcrops were represented, in the two- and three-dimensional models, close to its existence in the field. Also, potential rock debris volume estimation can be performed to help in planning for the debris removal post the rock slides. Hence, visualization plots from UAV-CRP can provide comprehensive information on the rock cuts and its management.

Visualization of Dam from CPT Studies

In this section, 3D visualization of a hydraulic fill dam using in situ cone penetration studies results is attempted. One of the many constraints in assessing the stability of these structures is identifying the localized critical zones within a dam or levee, where the initial failures could potentially trigger catastrophic failures. In order to assess the state of soils in these dams, the piezocone and seismic cone penetration soundings were performed along the crest, upstream and downstream sides of the dam. The visualization models are developed using Ordinary Kriging and by considering spatial anisotropy of data as well and present soil properties models based on interpolation of CPT data collected on the dam at different locations. These models were developed using the integration of spatial variability analyses and visual interface programs. Figure 7 presents the 3D visualization model developed for the dam using Soil Behavior Type (SBT) values determined from the Robertson (2010) charts.

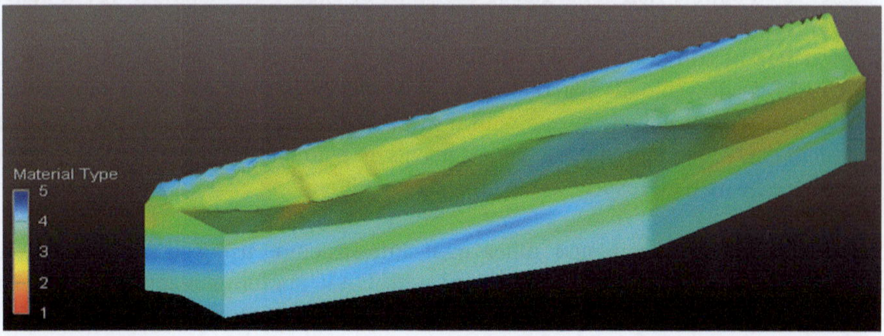

Fig. 7. Three-dimensional visualization model of Soil Behavior Type

Several insights are drawn on this dam based on the spatial variability analyses and developed visualization models. The visualization model depicted the presence of dense sand layers at top 3 m (10 ft) followed by mixtures of silt and clay layers. Several sand lenses were identified which may be susceptible to liquefaction or slope stability problems. Foundation soils were primarily composed of compact and dense sands; therefore, the significant settlement is not expected to occur at the main dam. In both models, the clay core was illustrated by filtering data within ranges of $2 < SBT < 3$. A significant finding is the absence of clayey soils at the core was found at the extreme end of the dam.

Three-dimensional visualization modeling can also enhance evaluation of geotechnical soil properties and hazards including seismic hazard. Equipped with the data optimization and visualization tools, the geotechnical engineer now has a larger role in preserving the resilience of the levee system, which governs the sustainability of the community protected by it.

Overall, it can be safe to mention that in-situ studies including UAV platform based CRP to LIDAR studies as well as intrusive penetration and laboratory characterization tests would help in the establishment of visualization profiles with soil layering and properties. Such visual geotechnical data would provide valuable guidance in the construction, quality control/assessments and proper precautionary measures needed for safeguarding the construction activities.

4 Summary and Conclusions

The following provides some of the major conclusions derived from the research results presented in this paper:

1. Two-dimensional spatial modeling of sulfates based on 1-D kriging of laboratory measured soluble sulfates indicate that the levels varied considerably within the geomaterials. Though this information is reported for shallow depths of exploration, this data will be useful for screening chemical stabilizers that would best work with excavated sulfate rich soils. Reuse decisions of excavated materials would benefit from this information.
2. Modeling of stiffness property enhancements of Controlled Low Strength Material (CLSM) or flowable fill material placed along the pipeline segment has shown shear moduli property enhancements with respect to the elapsed time period. Visualization of data has captured these enhancements, explaining the importance of property enhancements with respect to the pipeline length and the elapsed curing time in the field.
3. Monitoring pavements and other civil infrastructure using Close-Range Photogrammetry (CRP) technique coupled with UAV platforms resulted in two- and three-dimensional visualization plots derived from the point clouds of the infrastructure footprints. These two- and three-dimensional plots include dense point clouds, orthophotos, and digital elevation models. These images have provided information of the infrastructure along any section within the infrastructure and have provided cracking imaging as well as any other structural failure patterns. For rock cut sections, the visualization plots from UAV-CRP provided volumes of stable and unstable slope portions of the rocky hills, and this DEM information can be input into slope stability software for further stability assessments.
4. A comprehensive 3-D visualization of a hydraulic fill dam was successfully attempted using cone penetration results on the dam. The visualization analyses showed soil profiling of the dam and then soil strength property characterization. Some of the features noted in the visualization analysis including soil characterization from Seismic Cone Penetration Test (SCPT) data have shown excellent agreement with borehole soil profiling data.

Overall, the analysis attempted here showed the importance of incorporating visualization elements for a comprehensive understanding of the infrastructure stability and performance. Visualization framework developed for liquefaction analysis can be applied to other mega infrastructure projects as well.

Acknowledgements. This work presented in the paper has been supported by several research grants from the Texas Department of Transportation (TxDOT), National Science Foundation (NSF) and Tarrant Regional Water District (TRWD). Authors would like to acknowledge the support provided by Mr. Joe Adams (RTI Manager - TxDOT), Matt Gaughan and Louie Verrault (TRWD) and several others for their involvement. Authors also would like to thank many UTA students, Bhaskar Chittoori, Justin Thomey, Aravind Pedarla, Sayanthan Chakraborty, Anil Raavi, Rathna Mothkuri and others.

References

Bheemasetti, T.V., Chittoori, B., Zou, H., Puppala, A.J., Thomey, J.: Spatial mapping of soluble sulfate concentrations present in natural soils using geostatistics. J. Geotech. Geoenviron. Eng. **143**(2), 04016090 (2016)

Ellis, H.L., Vessely, M.J.: Visualization of Geotechnical Data for Hazard Mitigation and Disaster Response (No. Project 20–05, Topic 45–17) (2015)

El Meouchea, R., Hijazib, I., Ponceta, P.A., Abunemeha, M., Rezouga, M.: UAV photogrammetry implementation to enhance land surveying, comparisons and possibilities. Int. Arch. Photogrammetry, Remote Sens. Spat. Inf. Sci. **XLII-2/W2**, 20–21 (2016)

Fernández-Hernandez, J., González-Aguilera, D., Rodríguez-Gonzálvez, P., Mancera-Taboada, J.: Image-based modelling from unmanned aerial vehicle (UAV) photogrammetry: an effective. low-cost tool archaeological applications. Archaeometry **57**, 128–145 (2015)

Haulman, D.L.: Unmanned Aerial Vehicles in Combat, 1991–2003. Air Force History Research Agency, June 9, 18 p. (2003)

Montgomery, D.: Map the Runout Risk for Landslides, The Seattle Times, April 11, p. A11 (2014)

Olsen, M., Raugust, J., Roe, G.: NCHRP Synthesis 446: Use of Advanced Geospatial Data, Tools, Technologies, and Information in Department of Transportation Projects. Transportation Research Board of the National Academies, Washington, D.C. (2013). 97 p.

Pereira, E., Bencatel, R., Correia, J., Felix, L., Goncalves, G., Morgado, J., Sousa, J.: Unmanned air vehicles for coastal environmental research. J. Coastal Res., SI 56 (Proceedings of the 10th International Coastal Symposium), Lisbon, Portugal, pp. 1557–1561 (2009). ISSN 0749-0258

Puppala, A.J., Chittoori, B., Raavi, A.: Flowability and density characteristics of controlled low-strength material using native high-plasticity clay. J. Mater. Civ. Eng. **27**(1), 06014026 (2014)

Puppala, A.J.: Advances in ground modification with chemical additives: from theory to practice. Transp. Geotech. **9**, 123–138 (2016)

Puppala, A.J., Congress, S.S.C.: Recent advances in design, construction, and maintenance of pavement infrastructure. In: Keynote Presentation, 2nd International Conference on Advances in Concrete, Structural, and Geotechnical Engineering (ACSGE 2018), 26–28 February, BITS Pilani, Pilani Campus (2018)

Robertson, P.K.: Soil behaviour type from the CPT: an update. In: 2nd International Symposium on Cone Penetration Testing, USA pp. 9–11 (2010)

Siebert, S., Teizer, J.: Mobile 3D mapping for surveying earthwork projects using an unmanned aerial vehicle system. Autom. Constr. **41**, 1–14 (2014)

Udin, W.S., Ahmad, A.: Assessment of photogrammetric mapping accuracy based on variation flying altitude using unmanned aerial vehicle. In: 8th International Symposium of the Digital Earth, Earth and Environmental Science, vol. 18, pp. 1–7 (2014)

https://www.auav.com.au/articles/drone-types/. Accessed 4 Dec 2017

https://www.questuav.com/media/case-study/fixed-wing-versus-rotary-wing-for-uav-mapping-applications/. Accessed 4 Dec 2017

Transportation Geotechnics and Geomechanics

Investigation of Ground Displacement Induced by Hydraulic Jetting Using Smoothed Particle Hydrodynamics

Pierre Guy Atangana Njock[1,2] and Shuilong Shen[1,2,3](✉)

[1] State Key Laboratory of Ocean Engineering, School of Naval Architecture, Ocean, and Civil Engineering, Shanghai Jiao Tong University, Minhang District, Shanghai 200240, China
slshen@sjtu.edu.cn

[2] Collaborative Innovation Center for Advanced Ship and Deep-Sea Exploration (CISSE), Shanghai Jiao Tong University, Shanghai 200240, China

[3] Department of Civil and Construction Engineering, Centre for Sustainable Infrastructure, Swinburne University of Technology, Hawthorn, Melbourne, VIC 3122, Australia

Abstract. The few methodologies available for estimating the displacement induced by the action of a fluid jet on the soil are almost all based on the cavity expansion theory, which the adaptiveness to model this phenomenon can be questioned due to some significant drawbacks. This paper investigates the suitability of the SPH approach to investigate the ground displacement induced by jet grouting. A simulation of a water jet impinging on a soil mass is performed using AUTODYN-2D. In this simulation, the soil is modelled as a granular elastoplastic material, whereas the water is assumed to perform as a Newtonian fluid. The interaction between these two bodies is therefore investigated in order to demonstrate the advantage of the SPH method over others. In particular, the strain and pressure variations in the soil are exploited to clearly expose the limitations of the cavity expansion based approaches. It is apparent that the SPH approach simulates the fluid-soil interaction more realistically and thus can be used to investigate the aforementioned mechanism. The results also highlight some computation instabilities that occurred during the simulation. It is therefore recommended to adopt coding as solving technique, notably to have a greater flexibility in dealing with issues of this type.

Keywords: Jet grouting · Smoothed particle hydrodynamics
Ground displacement · Water-soil interaction

1 Introduction

The ground displacement induced by high pressure fluid jet has become a central issue in jet grouting practice and has generated considerable recent research interest. Initially registered in the UK in 1950s [1], the jet grouting method consists of creating in-situ rigid elements in the soil, along with the successive processes of drilling, jetting and grouting. However, during the construction of jet grouting column, a large amount of

X. Shi et al. (Eds.): GSIC 2018, *Proceedings of GeoShanghai 2018 International Conference: Transportation Geotechnics and Pavement Engineering*, pp. 15–22, 2018.
https://doi.org/10.1007/978-981-13-0011-0_2

grout is injected into the ground via a small diameter nozzle. For the single fluid system for example, the injection pressure even varies from 40 to 70 MPa [2, 3]. Moreover, it was demonstrated that the injection of cement slurry at high pressure tends to cause detrimental effects to the surrounding subsoil. In densely urbanized areas, these effects adversely affect the constructed structures in the vicinity of the implementation zone [4]. Few approaches have been dedicated to the assessment of fluid jet induced displacement in soil, and were regularly found acceptable [4–10]. However, these methods are largely established on the basis of the cavity expansion theory, which the appositeness to describe the mechanism of relevance in this paper can be questioned owing to some intrinsic limitations. The exploration of a new approach such as Smoothed Particle Hydrodynamics (SPH), which incorporates numerous desirable advantages, thus seems legitimate.

This paper analyzes the effect of a water jet impinging on a soil mass modeled as granular material (single phase), using the SPH method. The main purpose underlying this approach is to investigate the suitability of the SPH approach to model and evaluate the ground displacement observed during the jet grouting procedure. The fundamentals of the Smoothed Particle Hydrodynamics formalism are first presented, followed by a detailed methodology of the analysis of water-soil interaction. The results of the simulation (obtained using the commercial software ANSYS-Autodyn) tend to confirm the aptitude of the SPH method to simulate jet grouting mechanism.

2 Fundamentals of Smoothed Particle Hydrodynamics (SPH)

2.1 Definition

The smoothed particle hydrodynamics (SPH) is an adaptive meshfree particle method that harmonically combines particle approximation and Lagrangian formulation. Its adaptive nature is a remarkable advantage that provides a certain ease in handling large deformations problems [11]. Originally, this meshless method was developed independently by [12, 13] for astrophysics purposes; and thenceforth, has been extended to a wide range of applications owing to a number of advantages detailed in [14]. In this approach, physical phenomena are translated into mathematical models (conventionally in the form of governing equations) and then expressed in SPH form.

2.2 SPH Formulation

The prominent idea behind the SPH formulation is that numerical results rely on an interpolation process to describe both the space and time evolution of the problem investigated. Basically, the domain of interest is discretized into a finite number of particles, each of which independently carries the material properties. The material properties at a given point x (particle i) is subsequently determined through an interpolation process with respect to its neighboring points x' (particles j), and within a support domain Ω (see Fig. 1). Noticeably, the particles have a spatial distance h (usually termed smoothing length), over which their properties are approximated by using a “Kernel” or smoothing function W. This smoothing function thus plays a crucial role in the computation process, as it not only determines the interpolation pattern of field variables, but also the stability, the accuracy and the efficiency of this interpolation. More details related to the definition or the choice of the smoothing function can be found in [14]. Usually, the aforementioned interpolation is based on the integral representation of a field function that can be a density, a velocity, an energy functions, etc., depending on the problem to be solved. Yet, one must understand that the solution scheme to the discretized governing equations (that define the problem) should be formulated following three principal steps:

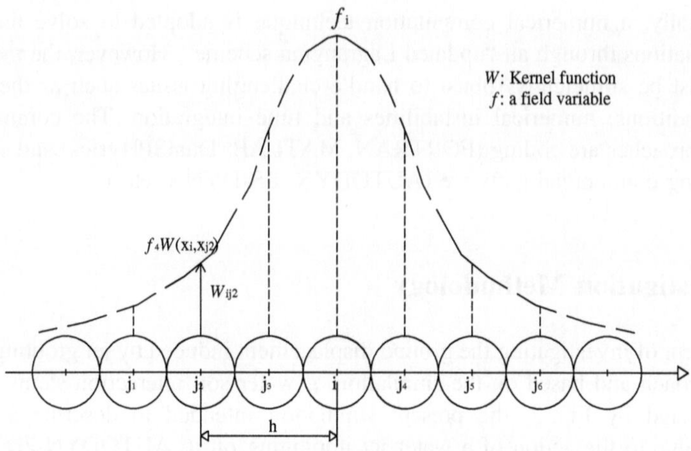

Fig. 1. Example of a field variable approximated at a given particle *i*.

(i) The first step also called the Kernel approximation consists of representing a given field function and its derivatives in a continuous integral representation as shown by the following Eqs. (1) and (2).

$$f(x) = \int_{\Omega} f(x')\, W(x - x', h)\, dx' \tag{1}$$

$$\frac{\partial f(x)}{\partial x} = -\int_{\Omega} f(x')\, \frac{\partial W(x - x', h)}{\partial x'}\, dx' \tag{2}$$

(ii) In the second step (particle approximation), the field variables at a given particles are approximated using the information of its neighboring particles, Eqs. 3, 4. This step is clearly illustrated by Fig. 1.

$$f(x_i) = \sum_{j=1}^{N} \frac{m_j}{\rho_j} f(x_j) W_{ij} \tag{3}$$

$$\frac{\partial f(x_i)}{\partial x} = \sum_{j=1}^{N} \frac{m_j}{\rho_j} f(x_j) \cdot \frac{\partial W_{ij}}{\partial x_i} \tag{4}$$

(iii) Finally, a numerical computation technique is adopted to solve the resulting equations through an "updated Lagrangian scheme". However, the method used must be sufficiently robust to handle challenging issues such as the boundary conditions, numerical instabilities and time integration. The commonly used approaches are coding (FORTRAN, MATLAB, DualSPHysics) and simulations using commercial software (AUTODYN, LS-DYNA, etc.)

3 Investigation Methodology

The concept of investigating the ground displacement induced by jet grouting using the SPH approach and based on the simulation of water-soil interaction stems from [15]. As illustrated by Fig. 2, the present simulation intended to describe a soil mass response due to the action of a water jet impinging on it. AUTODYN-2D SPH-code was used in this case to model the relevant mechanisms. AUTODYN-2D is an ANSYS software product that is accessible from ANSYS Workbench (under software license). This software possesses a rich materials library that also includes their constitutive models and equations of states (EOS). Taking advantage of this feature, the soil was modelled as a granular material with the following properties: reference density ρ = 1674 kg/m^3; bulk modulus K = 125 MPa and poisson's ratio υ = 0.3. Its behavior was assumed elastoplastic under the action of water jet. Additionally, the water was modelled as a Newtonian fluid with a virtual sound of speed C = 1483 m/s, and a density ρ_{fo} = 1000 kg/m^3. The initial velocity of the water jet was however taken as 400 m/s.

The water jet was discretized by 750 particles, while 56250 particles were used for the soil mass. The shape and packing of the particles in these two bodies are depicted in the Fig. 2. Moreover during the calculation, the relevant planes were modelled by "rigid wall" boundary conditions that are available in AUTODYN as velocity boundaries conditions.

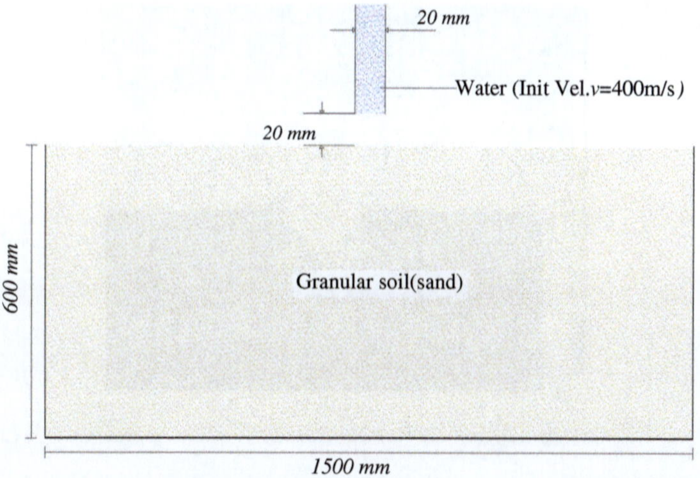

Fig. 2. Simulation arrangement

4 Simulation Results and Discussion

Three typical variables were investigated for the purposes of this study namely, velocity vectors, pressure and strains variations. These results are respectively presented in Figs. 3, 4 and 5. However, it is crucial to understand the limitations of the cavity expansion models while discussing these results.

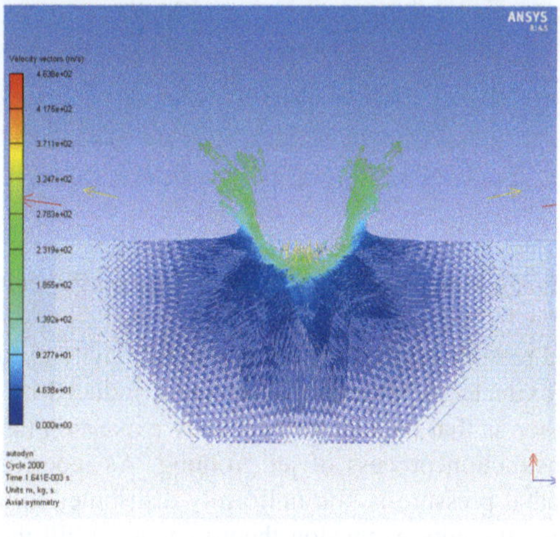

Fig. 3. Velocity vectors of SPH particles after the water jet impact

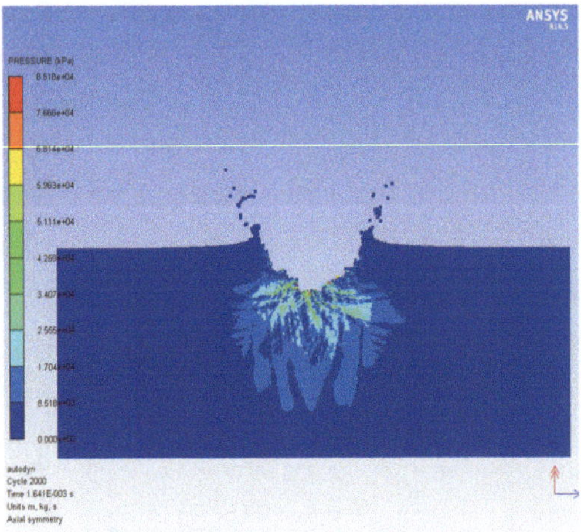

Fig. 4. Pressure distribution in the soil

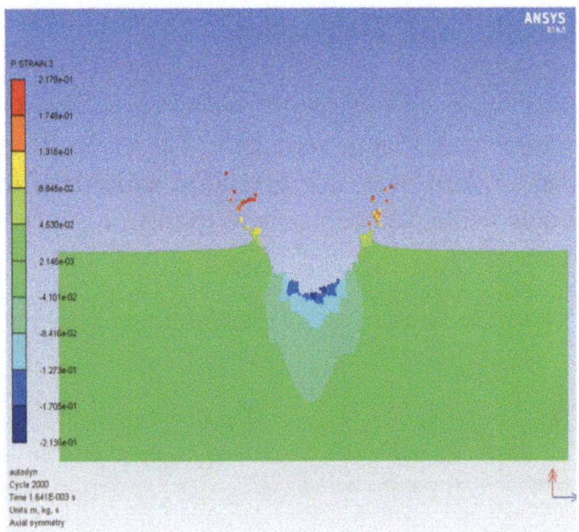

Fig. 5. Strain distribution in the soil during the jetting

There are several reasons why the pertinence of the cavity expansion based models to realistically approximate the ground displacement induced by jet grouting can be questioned. One reason is that the cylindrical cavity expansion theory cannot simulate the incremental construction process of jet grouting. As reported by [6], during the jetting action, the radial pressure is non-uniformly distributed to the surrounding soil; however, the cylindrical cavity expansion theory cannot fulfill the non-uniform strain state of the soil undergoing the action of the fluid jet. In other words, the deformation of the soil is solely assumed in one direction and the other relevant effects (e.g. the effect of upward movement on lateral displacement) are ignored. Another important reason is that the cavity expansion models treat the soil as a continuum; and consequently, they are unable to simulate some fundamental behaviors of the soil that arise from its particulate nature. Indeed, these models neglect the influence of the hydraulic fracturing of soil particles on the final ground displacement. Importantly, it has been demonstrated that the mechanical agitation in the ground generated by the fluid pressure is the principal mechanism inducing ground displacement [16].

Conversely, the SPH simulation allows picturing some crucial soil responses to fluid jet. As illustrated by the Fig. 3, which presents the particles velocity vectors after the water jet impact, it can be seen that the water jet affects the surrounding mass in a non-uniform fashion. This effect regularly fades as the distance of the particles from the immediate impact area becomes larger. It can therefore be assumed that the stress in the soil decrease as the particles get rearranged in a compact way. This assumption is confirmed by Fig. 4 that presents pressure distributions in the soil. It is obvious that the farther from the impact area, the smaller the pressure in the soil. This pressure is also develops non-uniformly in the soil. More to the point, it is important to pay attention to the erosion of the soil by fluid jet. This erosion is the result of hydraulic fractures of soil particles, and represents a crucial phenomenon in jet grouting as it determines the strains distribution in the soil (see Fig. 5). It can be seen that the most affected zones are the ones located in the vicinity of the water jet. Interestingly, the development of strain in the soil is non-uniform and strains evolution can be observed in all directions. Though, one should notice that unexpectedly in the Fig. 4, the contours present some irregularities that can be ascribed to computation instabilities, given that the soil is assumed isotropic in this case. Furthermore, although the present study only investigates the case of granular media, it is believed that in the case of saturated soil, the contribution of pore water pressure could be a decisive parameter.

5 Conclusion

The present study investigated the suitability of the smoothed particle hydrodynamics (SPH) technique to simulate the mechanism of ground displacement during the jet grouting procedure. Therefore the approach adopted in this paper was the simulation of water-soil interaction using the software AUTODYN-2D. The results obtained allow affirming this approach is more realistic, and might overcome the limitations of the cavity expansion based methods. In addition, this exploration has also permit to underline some particles instabilities issues during the computation. We therefore recommend implementing this approach using a coding solving technique, which offers more confidence in controlling the relevant parameters for both single and double phase materials.

References

1. Essler, R., Yoshida, H.: Jet grouting. In: Ground Improvement, 2nd edn., pp. 160–196. Taylor and Francis (2004)
2. Lunardi, P.: Ground improvement by means of jet grouting. Ground Improv. **1**, 65–85 (1997)

3. Burke, G.K.: Jet grouting systems: advantages and disadvantages. In: GeoSupport 2004: Drilled Shafts, Micropiling, Deep Mixing, Remedial Methods, and Specialty Foundation Systems, Orlando, pp. 875–886 (2004)
4. Wang, Z.F., Shen, S.L., Ho, E.C., Kim, Y.H.: Investigation of field installation effects of horizontal twin-jet grouting in Shanghai soft soil deposits. Can. Geotech. J. **50**(3), 288–297 (2013). https://doi.org/10.1139/cgj-2012-0199
5. Shen, S.L., Wang, Z.F., Cheng, W.C.: Estimation of lateral displacement induced by jet grouting in clayey soils. ICE Geotech. **67**(7), 621–630 (2017). https://doi.org/10.1680/geot./16-P-159
6. Chai, J.C., Miura, N., Koga, H.: Lateral displacement of ground caused by soil–cement column installation. J. Geotech. Geoenviron. Eng. **131**(5), 623–632 (2005)
7. Chai, J.C., Carter, J.P., Miura, N., Zhu, H.H.: Improved prediction of lateral deformations due to installation of soil–cement columns. J. Geotech. Geoenviron. Eng. **135**(12), 1836–1845 (2009)
8. Wang, Z.F., Bian, X., Wang, Y.Q.: Numerical approach to predict ground displacement caused by installing a horizontal jet grout column. Mar. Georesour. Geotechnol. (2016). http://dx.doi.org/10.1080/1064119X.2016.1273288
9. Wu, Y.D., Diao, H.G., Ng, C.W.W., Liu, J., Zeng, C.C.: Investigation of ground heave due to jet grouting in soft clay. J. Perform. Constr. Facil. **30**(6), 06016003 (2016)
10. Liu, H., Zhou, H., Kong, G., Qin, H., Zha, Y.: High pressure jet-grouting column installation effect in soft soil: theoretical model and field application. Comput. Geotech. **88**, 74–94 (2017)
11. Liu, G.R., Liu, M.B.: Smoothed Particle Hydrodynamics: A Meshfree Particle Method. World Scientific, Singapore (2003)
12. Lucy, L.: A numerical approach to the fission hypothesis. Astron. J. **82**, 1013–1024 (1977)
13. Gingold, R.A., Monaghan, J.J.: Smoothed particle hydrodynamics - theory and application to non-spherical stars. Mon. Not. Roy. Astron. Soc. **181**, 375–389 (1977)
14. Liu, M.B., Liu, G.R.: Smoothed particle hydrodynamics (SPH): an overview and recent developments. Arch. Comput. Meth. Eng. **17**, 25–76 (2010). https://doi.org/10.1007/s11831-010-9040-7
15. Njock, P.G.A., Shen, J.S., Modoni, G., Arulrajah, A.: Recent advances in horizontal jet grouting (HJG): an overview. Arab. J. Sci. Eng., 1–18 (2017). https://doi.org/10.1007/s13369-017-2752-3
16. Miura, N., Shen, S.-L., Koga, K., Nakamura, R.: Strength change of clay in the vicinity of soil-cement column. J. Geotech. Geoenviron. Eng. **596**(43), 209–221 (1998)

Service Life Prediction of Underground Concrete Pipeline in Chloride Contaminated Soils

Jingpei Li[1,2(✉)], Hui Wang[1,2], and Lin Li[1,2]

[1] Key Laboratory of Geotechnical and Underground Engineering of Ministry of Education, Tongji University, Shanghai, China
lijp2773@tongji.edu.cn

[2] Department of Geotechnical Engineering, Tongji University, Shanghai, China

Abstract. This paper develops a theoretical model to predict the service life of concrete pipelines in contaminated soils. The service life of concrete pipeline is divided into four stages: the chloride ion diffusion stage, the initial corrosive stress generation stage, the initial crack generation stage and the crack propagation stage. For the first stage, the equation for chloride diffusion is formulated and solved analytically by using the integral transform method. For the other stages, the concrete around a corroding steel reinforcing bar is modeled by a thick-walled cylinder with a wall thickness be equal to the thinnest concrete cover. The concrete pipeline is supposed to reach the end of service life when the crack width exceeds the maximum allowable value at the limit state. The proposed model is validated by comparing the predictions with the numerical simulation results. The analysis results demonstrate that the soil pressure and the concrete cover thickness have significant effects on the service life of concrete pipelines exposed to the chloride contaminated soils.

Keywords: Concrete pipeline · Chloride diffusion · Steel corrosion Service life · Crack

1 Introduction

Concrete pipelines, such as the water pipeline, the sewer pipeline, the natural gas pipeline, as well as the utility tunnels, are increasingly applied in municipal engineering. However, the steel reinforcing bar of the concrete pipelines would be corroded by the chloride ion transferred from the chloride contaminated soil, and hence the performance of the pipeline will be greatly deteriorated. If the concrete pipeline is exposed to chloride contaminated soils, the chloride ions in the soil will diffuse into the concrete cover. When the chloride ions accumulate to a threshold level, the passive film on the reinforcing steel surface will be disrupted and the steel will be rusted under the presence of oxygen and moisture. The corrosion products, which are about 2–6 times the volume of the origin steel, give rise to tensile stresses on the concrete protective cover. The tensile stresses will eventually cause cracking of the concrete cover and lead to the failure of the concrete pipeline.

X. Shi et al. (Eds.): GSIC 2018, *Proceedings of GeoShanghai 2018 International Conference: Transportation Geotechnics and Pavement Engineering*, pp. 23–34, 2018.
https://doi.org/10.1007/978-981-13-0011-0_3

A lot of models have been developed to predict the service life of the concrete structures exposed to chloride ion erosion environments [1–7]. These models can be broadly divided into three main categories, (i) analytical models [1–3], (ii) empirical models [4, 5], (iii) numerical models [6, 7]. All these efforts have made great contributions in predicting the service life of the chloride ion eroded structures. However, most of those previous studies only focused on the concrete structures above the ground, little effort has been extended to determine the service life of the underground concrete structures. Especially these approaches predicting the underground concrete pipelines in chloride contaminated soils are not currently available. In contrast to the upper concrete structures, the chloride ions diffuse radially in the concrete of the pipeline. Moreover, the deterioration of the underground concrete pipelines is greatly affected by the soil pressure. Hence, it makes sense for this paper to develop a theoretical model considering the effect of soil pressure to estimate the service life of underground concrete pipelines exposed to chloride contaminated soils. For this, based on the chloride-induced reinforcing steel corrosion mechanism, this paper divides the service life of underground concrete pipelines into four stages: chloride ion diffusion stage, initial stress generation stage, initial crack generation stage and crack propagation stage, as shown in Fig. 1. A method is presented based on the elasticity theory and the multiplicative rule of corrosion products to calculate the total time of corrosion stages. Furthermore, the analytical method is validated by comparing analytical results with numerical results.

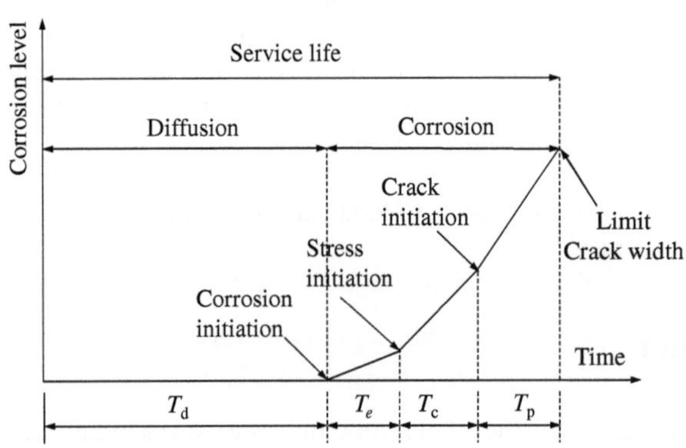

Fig. 1. Modified service life model of present study

2 Theoretical Model

For convenience, this paper merges the 2nd, 3rd and 4th stages into one stage, called corrosion stage. The service life of the concrete pipeline, T, is the sum of the time of the diffusion stage, T_d, and the time of the corrosion stage, T_{cor}.

2.1 Diffusion Stage

It's assumed that there is only the outer surface of the pipeline exposed to the chloride contaminated soil, the concrete pipeline is initially chloride-free and there won't be diffusion on the inner surface of the pipeline. To simplify the mathematical treatment of the problem, only the radial diffusion is considered. Hence, the chloride concentration in the concrete pipeline varies with time and radial radius during the diffusion stage. According to the Fick's second law of diffusion, the equation governing the mechanism of chloride diffusion into concrete pipeline can be given as

$$\frac{\partial c(Cl)}{\partial t} = D\left\{\frac{\partial c^2(Cl)}{\partial r^2} + \frac{1}{r}\frac{\partial c(Cl)}{\partial r}\right\} \tag{1}$$

where c (kg/m^3) is the chloride concentration at time t (s); r (m) is the radial radius, and D (m^2/s) is the chloride ions diffusion coefficient.

According to the previous assumption, the associated initial and boundary conditions of Eq. (1) can be written as

$$\begin{aligned} &c(r,t) = 0, \quad (t = 0, a \le r \le b) \\ &c(r,t) = c_s, \quad (t > 0, r = b) \\ &\frac{\partial c(r,t)}{\partial r} = 0, \quad (t > 0, r = a) \end{aligned} \tag{2}$$

where c_s (kg/m^3) is the chloride ion concentration outside the outer concrete cover; a (m) and b (m) are the inner and outer radii of the concrete pipeline.

The above equation can be solved by using the method of separation variables. The solution can be given as follows [8]:

$$c = c_s\left[1 - \pi\sum_{\mathrm{m}=1}^{\infty}\frac{J_0(b\beta_{\mathrm{m}})J_1(a\beta_{\mathrm{m}})u_0(\beta_{\mathrm{m}}r)}{J_1^2(a\beta_{\mathrm{m}}) - J_0^2(b\beta_{\mathrm{m}})}e^{-D\beta_{\mathrm{m}}^2 t}\right] \tag{3}$$

where $u_0(\beta_{\mathrm{m}}r) = J_0(r\beta_{\mathrm{m}})Y_1(\alpha\beta_{\mathrm{m}}) - Y_0(r\beta_{\mathrm{m}})J_1(\alpha\beta_{\mathrm{m}})$, β_{m} are the positive roots of characteristic equation $u_0(\beta_{\mathrm{m}}r) = 0$, m = 1, 2, 3…, J_0 and Y_0 are the first and second kind Bessel functions of order zero while J_1 and Y_1 of order one.

The total chloride ion content, Q_t, in concrete pipeline of unit length in time t can be obtained as follows:

$$Q_t = 2\pi\int_0^t\int_a^b\frac{\partial c(Cl)}{\partial t}rdrdt \tag{4}$$

To obtain the time of the diffusion stage, it is needed to investigate the threshold level. Generally, the ratio of total chloride ion content to the weight of cement is widely accepted to describe the threshold level [9]. Supposing that the cement weight in concrete pipeline of unit length is m_c and the threshold level of chloride ions is c_{cr}, one can obtain the following equation:

$$c_{cr} = \frac{Q_t}{m_c} \times 100\% \tag{5}$$

The time T_d of the diffusion stage can be obtained from Eqs. (3), (4) and (5).

2.2 Corrosion Stage

Modeling and Basic Assumptions. To obtain the time of the corrosion stage, a thick-walled double ring model is founded, as shown in Fig. 2. In the picture, D_0 is the initial diameter of the steel bar, C is the thickness of the inner and outer concrete cover, d_0 is the thickness of porous zone around the steel-concrete interface. d_{s1} and d_{s2} are the thickness of corrosion products to generate expansive pressure for inner and outer cover, respectively. To simplify the expression of the equation, the following denotations are defined: $R_1 = a$, $R_2 = a + C$, $R_3 = R_2 + 2d_0 + D_0$, and $R_4 = R_3 + C$.

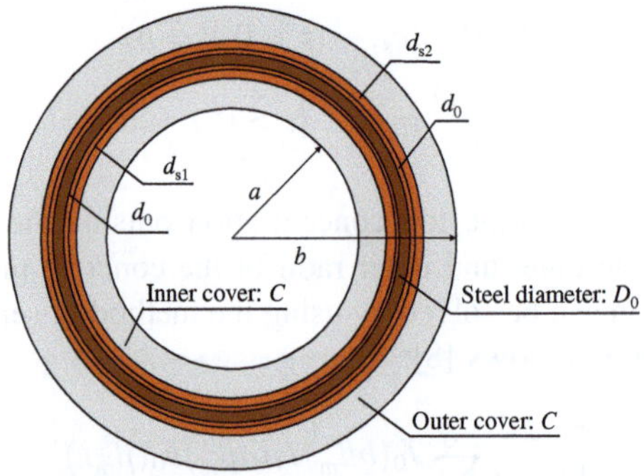

Fig. 2. Thick-walled double ring model

The assumptions made for the analytical model are as follows:

(a). The concrete of the pipeline is linear elastic and homogeneous. The material characteristics won't change over time. (b). The corrosion products are uniformly distributed on the surface of the steel bar and the expansive pressure induced by the corrosion products is uniformly distributed on the steel-concrete interface as well. (c). On account of the complexity of the practical problem, there is nothing but the expansive pressure induced by the corrosion products that leads to the cracking of the concrete cover. (d). After the crack initiation, the model keeps in the ring shape as the crack width is extremely small compared with the perimeter of both the inner and outer ring. (e). In the whole stage from corrosion initiation to structure failure, the soil is in elastic state.

Mechanical Analysis. Firstly, considering the outer concrete cover, the boundary problem of the thick-walled double ring model can be further reduced to a plane strain problem under the radially symmetric conditions, as shown in Fig. 3(a).

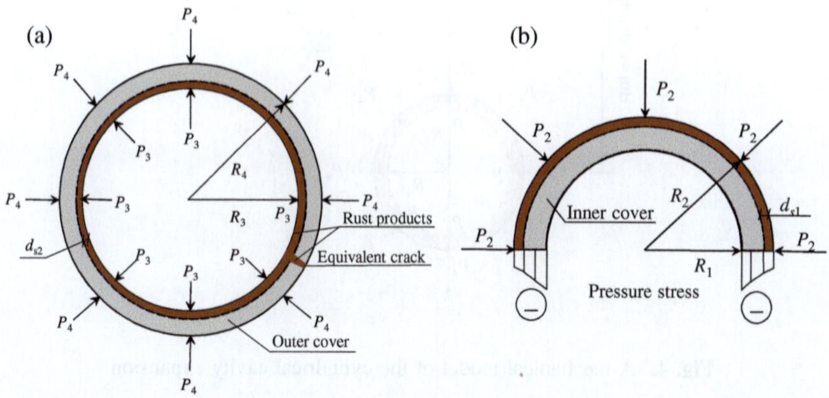

Fig. 3. Diagram of interaction for concrete cover: (a) Outer cover; (b) Inner cover

Based on the relationship between strain and displacement, the expression of the circumferential strain, $\varepsilon_{\theta 4}$, at R_4 can be obtained as follows:

$$\varepsilon_{\theta 4} = \frac{u_{r4}}{R_4} \tag{6}$$

where u_{r4} is the radial displacement at R_4.

Based on the previous assumption, the service life of concrete comes to end when the crack width on the outer surface exceeds the maximum allowable value. In this paper, we use $w_{\lim}$ to represent the limit crack width. Then one can obtain the equation:

$$w_{\lim} = 2\pi R_4(\varepsilon_{\theta 4} - f_t/E_{ef}) \tag{7}$$

where f_t/E_{ef} is the limit elastic strain, f_t is the tensile strength of the concrete, $E_{ef} = E_c/(1 + \varphi_{cr})$ is the effective elastic modulus, E_c is the elastic modulus of the concrete and φ_{cr} is creep coefficient of the concrete.

Substitution of Eq. (6) into Eq. (7) yields the radial displacement, u_{r4}, at R_4 given as:

$$u_{r4} = \frac{w_{\lim}}{2\pi} + R_4\frac{f_t}{E_{ef}} \tag{8}$$

Figure 4 presents a mechanical model of the cylindrical cavity expansion in the chloride contaminated soil. As the corrosion products grow, the radius of the initial round hole, R_4, increases to the final value, R_5, under the uniformly distributed internal pressure, P.

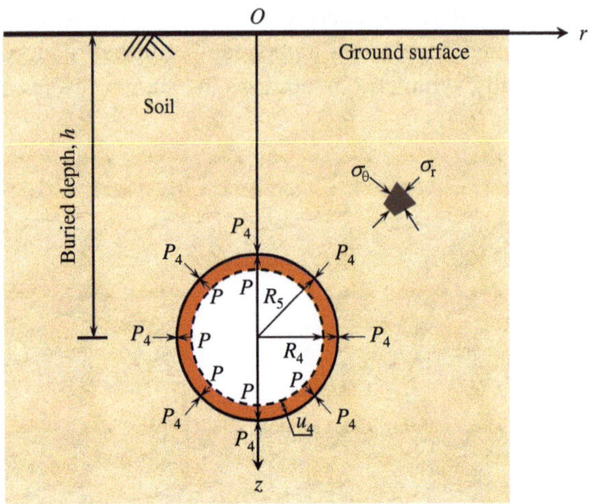

Fig. 4. A mechanical model of the cylindrical cavity expansion

Based on the elastic mechanics theory, the solution of the soil stress and displacement can be obtained as follows [10]:

$$\sigma_r = \sigma_0 + (P - \sigma_0)\left(\frac{R_4}{r}\right)^2 \tag{9}$$

$$\sigma_\theta = \sigma_0 - (P - \sigma_0)\left(\frac{R_4}{r}\right)^2 \tag{10}$$

$$u_r = \frac{P - \sigma_0}{2G}\left(\frac{R_4}{r}\right)^2 r \tag{11}$$

where G is the shear modulus of the soil; σ_r is the radial stress of the soil in elastic state; σ_θ is the circumferential stress of the soil in elastic state; u_r is the radial displacement of the soil in elastic state; σ_0 is the initial soil pressure at the buried depth, h, with $\sigma_0 = \gamma h$, in which γ is the unit weight of the soil.

According to the mechanical equilibrium conditions, when $r = R_5 = R_4 + u_{r4}$, combining Eqs. (9), (10) and (11) gives the uniformly distributed external pressure as follows:

$$P_4 = \gamma h + 2G\left(\frac{u_{r4}}{R_4 + u_{r4}}\right) \tag{12}$$

Given that the outer concrete cover under the external radial pressure and the internal expansive pressure, the radial displacement, u_{ro}, at r, can be obtained by using elasticity method:

$$u_{ro} = \frac{1+v}{E_{ef}r\left(R_4^2 - R_3^2\right)}\left[(1-2v)\left(P_3R_3^2 - P_4R_4^2\right)r^2 - (P_4 - P_3)R_3^2R_4^2\right] \tag{13}$$

where v is Possion's ratio of the concrete cover.

Substituting $r = R_4$ and $u_{ro} = u_{r4}$ into Eq. (13) yields the internal expansive pressure, P_3, as follows:

$$P_3 = \frac{1}{2R_3^2R_4^2(1-v^2)}\left[E_{ef}R_4\,u_{r4}\left(R_4^2 - R_3^2\right) + \left((1-2v)(1+v)R_4^4 + (1+v)R_3^2R_4^2\right)P_4\right] \tag{14}$$

When $r = R_3$, the thickness of corrosion products to generate expansive pressure for inner cover, d_{s2}, is equal to the radial displacement, u_3, at R_3. Hence, d_{s2} can be then obtained by substituting Eqs. (12) and (14) into Eq. (13).

The circumferential strain, $\varepsilon_{\theta 3}$, at R_3, can be derived as:

$$\varepsilon_{\theta 3} = \frac{u_{r3}}{R_3} \tag{15}$$

Therefore, the crack width, w_3, at R_3, can be expressed as follows:

$$w_3 = 2\pi R_3\left(\varepsilon_{\theta 3} - f_t/E_{ef}\right) \tag{16}$$

As shown in Fig. 3(b), for the inner concrete cover under the external expansive pressure, the radial displacement, u_{ri}, at r, can be derived from elastic mechanics theory as follows:

$$u_{ri} = -\frac{P_2R_2^2(1-v^2)}{E_{ef}r\left(R_2^2 - R_1^2\right)}\left[\left(r^2 + R_1^2\right) + \frac{v}{1-v}\left(R_1^2 - r^2\right)\right] \tag{17}$$

where P_2 is equal to P_3 according to the mechanical equilibrium conditions of the steel bar. When $r = R_2$, one can obtain the radial displacement, u_{r2}, at R_2, through Eq. (17).

It's easy to find that the radial displacement value is negative because there is only the external expansive pressure with the direction towards the center of the ring. As a result, the thickness of corrosion products to generate expansive pressure outer cover, d_{s1}, can be written as:

$$d_{s1} = -u_{r2} = \frac{P_3R_2}{E_{ef}\left(R_2^2 - R_1^2\right)}\left[\left(1-v^2\right)\left(R_2^2 + R_1^2\right) + v(1+v)\left(R_1^2 - R_2^2\right)\right] \tag{18}$$

Calculation of the Corrosion Products. The total weight of the corrosion products, m_t, consists of the following four parts:

(i) The weight of the corrosion products filled in the porous zone around the steel-concrete interface, m_0, which can be expressed as:

$$m_0 = \rho_r \cdot 2\pi[R_2 d_0 + (R_2 + d_0 + D_0)d_0] \quad (19)$$

where ρ_r is the density of the corrosion products;

(ii) The weight of the corrosion products to generate expansive pressure on both the inner and outer concrete cover, m_p, which can be expressed as:

$$m_p = \rho_r \cdot 2\pi[R_3 d_{s2} + (R_2 - d_{s1})d_{s1}] \quad (20)$$

(iii) The weight of the corrosion products filled in the crack of the outer concrete cover, m_{cr}, which can be expressed as:

$$m_{cr} = \rho_r \cdot \frac{c}{2}(w_3 + w_{lim}) \quad (21)$$

(iv) The weight of the corrosion products fill in the loss volume of the steel bar due to corrosion, m_s, which can be expressed as:

$$m_s = \rho_r \frac{\alpha m_t}{\rho_{st}} \quad (22)$$

where ρ_r is the density of the steel bar; α ranges from 0.523 to 0.622 by the main components of corrosion products from $Fe(OH)_3$ to $Fe(OH)_2$ [1].

Thus, one can obtain the total weight of the corrosion products as follows:

$$m_t = m_0 + m_p + m_{cr} + m_s \quad (23)$$

Substituting Eq. (22) into Eq. (23) gives the total weight of the corrosion products as follows:

$$m_t = \frac{\rho_{st}}{\rho_{st} - \rho_r \alpha}(m_0 + m_p + m_{cr}) \quad (24)$$

Growth of the Corrosion Products. As the increase of the corrosion products, the rust layer thickens. Consequently, the rust production decreases with the increase of the diffusion distance. In general, the rate of the rust production is inversely proportional to the total weight of the rust products. Hence, the rate of the rust production, J_{rust}, can be expressed as follows [4]:

$$J_{\text{rust}} = \frac{dm_t}{dt} = \frac{k_p}{m_t} \tag{25}$$

where k_p ($kg^2 \cdot m^{-2} \cdot a^{-1}$) is related to the rate of the steel corrosion and can be written as follows [11]:

$$k_p = \frac{0.098}{1.08}\left(\frac{1}{\alpha}\right)\pi D_0 i_{\text{cor}} \tag{26}$$

where D_0 (mm) is the diameter of the steel bar; i_{cor} ($\mu A \cdot cm^{-2}$) is the corrosion electric current density.

Therefore, the time of the corrosion stage, T_{cor}, can be obtained by substituting Eq. (26) into Eq. (25) as follows:

$$T_{\text{cor}} = \frac{1.08}{0.196} \cdot \frac{\alpha m_t^2}{\pi D_0 i_{\text{cor}}} \tag{27}$$

Combining the time of the corrosion stage, T_{cor}, and the time of the diffusion stage, T_d, the total time from the diffusion initiation to the point that the crack width reaches the limit value can be obtained. Then the service life of the concrete pipeline exposed to chloride contaminated soil can be estimated and predicted.

3 Verification and Discussion

3.1 Numerical Model Verification

To validate the proposed diffusion model, a 2D plane finite element model shown in Fig. 5 describing the chloride ion radial diffusion of unit length of concrete pipeline is developed and implemented in the commercial finite element software ABAQUS, using the heat transfer module.

Here, the following parameter of the pipeline model are used in the finite element analyses: the inner radius: $a = 0.3$ m, the outer radius: $b = 0.42$ m and the diffusion coefficient: $D = 1.8 \times 10^{-12}$ m^2/s. The boundary conditions are as follows: the chloride ion concentration on the outer surface, c_s, is 0.346 kg/m^3. The four-node linear heat transfer quadrilateral finite elements are adopted to represent the concrete pipeline. The model is divided into 1056 elements based on the sweeping method.

Figure 6 show the chloride ion concentration profiles obtained by the theoretical model and the finite element model at different locations and times. In Fig. 6, the chloride ion concentration, c_r, at arbitrary location r is normalized by the chloride ion concentration at the outer surface, c_s, and the normalized position, $r_k = (r - a)/(b - a)$ is adopted to describe the variations of the positions along the radial direction.

As seen in Fig. 6, the normalized chloride ion concentration increases with the position factor, which means the chloride ion concentration in the outer position is

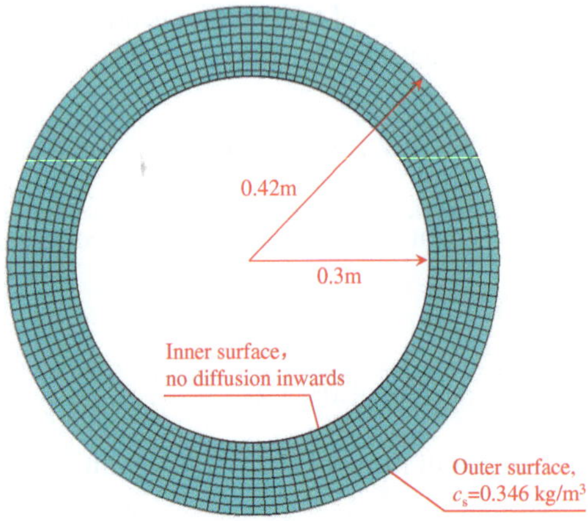

Fig. 5. A 2D plane finite element model for concrete pipeline exposed to chloride contaminated soils

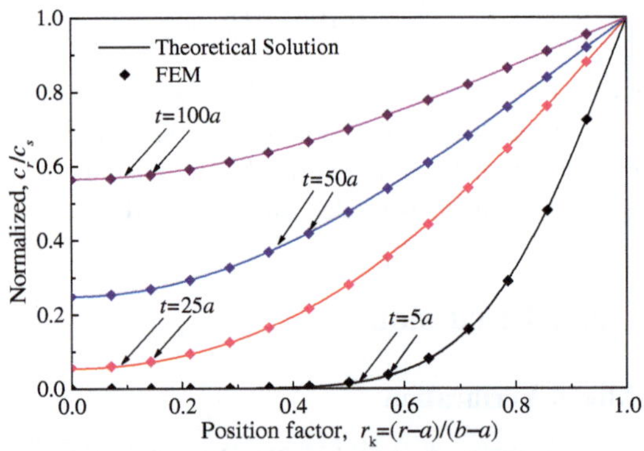

Fig. 6. Comparison between the theoretical solution and the FEM results of the chloride ion concentration profiles: Chloride concentration profiles with position

larger than that in the inner position. Moreover, the normalized chloride ion concentration increases with the time factor. All the above results show a good agreement with the actual condition. Furthermore, the theoretical results match very nicely with the FEM results, which provide a verification of the proposed methodology to predict the chloride ion profiles in concrete pipeline.

3.2 Parametric Analysis

In this study, a unit length of pipeline with 0.3 m inner radius and 0.42 m outer radius is selected to demonstrate an application of the proposed analytical method. The main components of the corrosion products is assumed to be $Fe(OH)_3$ and thus α is equal to 0.523 in this study [1]. In addition, the other parameters are assumed as follows:

$w_{\lim} = 3$ mm, $\gamma = 18$ kN/m^3, $d_0 = 12.5$ μm, $G = 75$ MPa, $E_{ef} = 9$ GPa, $f_t = 3.3$ MPa, $\nu = 0.18$, $\rho_r = 3.6$ g/cm^3, $\rho_{st} = 7.85$ g/cm^3. The effects of the soil pressure at different buried depth are conducted in the following section.

According to the proposed theoretical model, the relation curves that the service life of corrosion stage varies with the soil pressure are shown in Fig. 7. It can be seen obviously from Fig. 7 that the normalized service life increases linearly with the soil pressure at different normalized burial depth. Moreover, the normalized service life increases with the normalized concrete cover thickness. There are two reasons for this phenomenon. On one hand, the thicker the concrete cover is, the larger the critical expansive pressure to cause the initial crack is, which means that more corrosion products need to be generated. On the other hand, during the crack propagation stage, the corrosion products required to fill in the crack can rise as the concrete cover thickness increases. Therefore, a thicker concrete cover can not only protect the steel against corrosion, but also can prolong the service life of the concrete pipeline.

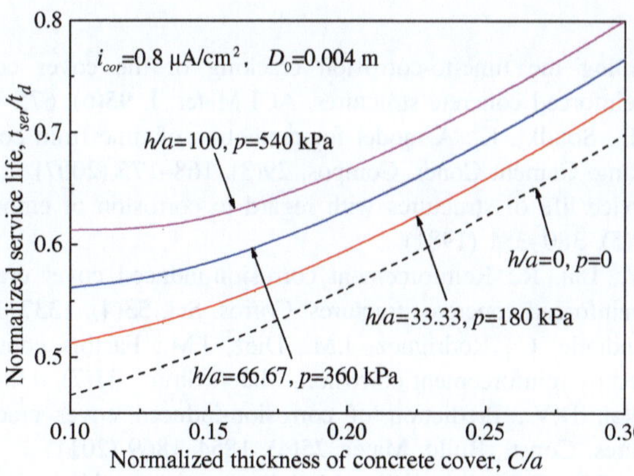

Fig. 7. Variations of the normalized service life of corrosion stage with the normalized thickness of concrete cover for different soil pressures at different burial depths

As for a series of cover thicknesses and inner diameters of the pipelines given by the Chinese Standard GB11836-2009, their normalized cover thicknesses are in the order of 0.2. According to the results shown in Fig. 7, their normalized lifetimes of corrosion stage vary from 0.56 to 0.67 with different burial depths. Since the time of corrosion stage is less than the time of diffusion stage, most concrete pipelines can meet the requirements of designed service life. Consequently, the stipulation of cover thicknesses and inner diameters of the pipelines given by the current Chinese standard is reasonable for most cases.

4 Conclusions

In this paper, a two-stage theoretical model is developed to predict the service life of the concrete pipeline exposed to chloride contaminated soil based on the deterioration mechanism. The proposed analytical model of diffusion stage is validated by comparing theoretical results with FEM results. The effects of the concrete cover thickness and the soil pressure at different buried depth on the service life of corrosion stage are analyzed. The analysis results indicate that the service life of the concrete pipeline exposed to chloride contaminated soil increases significantly with the growth of the soil pressure and the concrete cover thickness.

Acknowledgements. This study was financially supported by the National Natural Science Foundation of China (Grant No. 41772290).

References

1. Liu, Y.: Modeling the time-to-corrosion cracking of the cover concrete in chloride contaminated reinforced concrete structures. ACI Mater. J. **95**(6), 675–681 (1996)
2. Maaddawy, T.E., Soudki, K.: A model for prediction of time from corrosion initiation to corrosion cracking. Cement Concr. Compos. **29**(3), 168–175 (2007)
3. Tuutti, K.: Service life of structures with regard to corrosion of embedded steel. J. Am. Concr. Inst. **77**(5), 380–381 (1981)
4. Lu, C., Jin, W., Liu, R.: Reinforcement corrosion-induced cover cracking and its time prediction for reinforced concrete structures. Corros. Sci. **53**(4), 1337–1347 (2011)
5. Alonso, C., Andrade, C., Rodriguez, J.M., Diez, J.M.: Factors controlling cracking of concrete affected by reinforcement corrosion. Mater. Struct. **31**(7), 435–441 (1998)
6. Chernin, L., Val, D.V.: Prediction of corrosion-induced cover cracking in reinforced concrete structures. Constr. Build. Mater. **25**(4), 1854–1869 (2011)
7. Pantazopoulou, S.J., Papoulia, K.D.: Modeling cover-cracking due to reinforcement corrosion in RC structures. J. Eng. Mechanics **127**(4), 342–351 (2011)
8. Carslaw, H.S., Jaeger, J.C.: Conduction of Heat in Solids. The Clarendon Press, Oxford (1959)
9. Geng, C., Xu, Y., Weng, D., Wu, X.: A time-saving method to determine the chloride threshold level for depassivation of steel in concrete. Constr. Build. Mater. **24**(6), 903–909 (2010)
10. Zhou, H., Kong, G., Liu, H.: Study on pile sinking compaction effect of hydrostatic wedge pile using cavity expansion theory. China J. Highw. Transp. **27**(4), 24–30 (2014)
11. Reale, T., O'Connor, A.: A review and comparative analysis of corrosion-induced time to first crack models. Constr. Build. Mater. **36**(4), 475–483 (2012)

Layout Patterns of Underground Space Comprehensive Utilization Around Metro Station for Urban Commercial Area

Jian Peng[1,2,3] and Fang-Le Peng[1,2](✉)

[1] Department of Geotechnical Engineering, Tongji University, Shanghai, People's Republic of China
pj53129544@163.com, pengfangle@tongji.edu.cn
[2] Research Center for Underground Space, Tongji University, Shanghai, People's Republic of China
[3] Sustainable Energy and Environmental Engineering, Graduate School of Engineering, Osaka University, Suita, Japan

Abstract. In order to satisfy contemporary urban development requirements, it is essential to utilize urban underground space (UUS) effectively when planning new metropolitan subway (metro) stations, especially in commercial areas. Herein, two cases involving metro stations in commercial areas, the Xujiahui zone in Shanghai, China and the Namba zone in Osaka, Japan, are investigated with respect to land use and UUS utilization, after which their underground space distribution situations are summarized in terms of function and layout, and the differences between the two cases are discussed. Two kinds of comprehensive UUS utilization layout patterns are elucidated, i.e., station center layout and commercial street center layout. The station center layout emphasizes the smooth transfer of people between transportation modes, while the commercial street center layout emphasizes not only transportation transfer, but also more intensive commercial exploitation of space under municipal roads and higher floor area ratios. Here, it should be noted that the former can transform to the latter as the city develops. Our study also shows that the development of UUS around metro stations in urban commercial areas is based on urban "function and layout" interaction characteristics, but the layout pattern may also be different depending on different national context.

Keywords: Layout pattern · Urban underground space
Commercial area · Function · Interaction characteristics

1 Introduction

In recent years, serious traffic jams have become a significant issue in a number of large Chinese cities due to their rapidly increasing population levels. Rail transit offers many advantages in terms of urban traffic jam mitigation, and has been widely used. Correspondingly, urban underground space (UUS) developed in tandem with urban metro stations can be utilized to create commercial opportunities and transport space in ways that promote economic and social developments around the station. Thus, in order to maximize benefits from rail transit network improvements, it is essential to approach

X. Shi et al. (Eds.): GSIC 2018, *Proceedings of GeoShanghai 2018 International Conference: Transportation Geotechnics and Pavement Engineering*, pp. 35–43, 2018.
https://doi.org/10.1007/978-981-13-0011-0_4

metro station area planning from the standpoint of effective UUS utilization – and function and layout are key factors when planning UUS.

Such planning has been explored in a number of studies. For example, Boivin (1991) described the Montreal underground pedestrian network, which at that time linked 44 buildings and nine subway stations. Polzin (1999) analyzed the relationship between metro and land use in ways that challenged some conventional perceptions. Additionally, Hui (2002) studied the planning and construction of urban rail transit stations and classified them into four types. In addition, Bélanger (2007) discussed the underground network development patterns and their futures as important urban infrastructures while focusing on the origins and transformations of Toronto's underground pedestrian network in order to illustrate its multi-dimensionality. Furthermore, Li et al. (2016) studied the demand and driving factors of three metro station areas with respect to underground space. However, since none of the previous studies established a concrete relationship between the function and layout of UUS around metro stations, this paper will examine those issues and discuss classic UUS layout patterns around metro stations, with a particular focus on urban commercial areas.

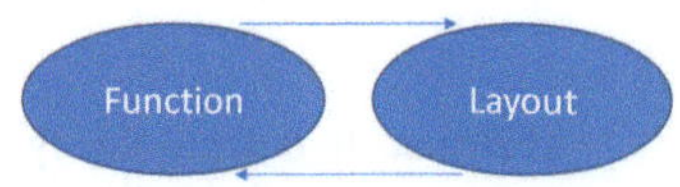

Fig. 1. Function and layout interaction

2 Methodology

City developments along metro lines tend to be directed by metro construction. In this process, the establishment of metro stations tips the balance of the original spatial structure and promotes regional accessibility. More specifically, the closer an establishment is to a metro station, the higher its accessibility and land value are, and vice versa (Bajic 1983; Du and Mulley 2007). Here, the environs of metro stations in commercial areas are defined. Choi and Asami (2002) defined close proximity to a station as within a distance of 200 m in a commercially dense area, and the first 500 m distance as a buffer zone, while Roukouni et al. (2012) presented a buffer zone with a radius of 500 m around metro stations. Thus, the scope of this study is defined as the area within 500 m of a commercial metro station area, within which commercial activities account for the biggest percentage of land use. Proper land use designations help fuse a supra-space city within a UUS city into one cohesive urban organism (Golany and Ojima 1996). However, UUS functions will vary depending on the land use. The concept of a proper UUS layout addresses how different functions should be considered, and what their plain and vertical form should be. To accomplish this, UUS function and layout must maintain relationships based on mutual coordination and mutual promotion. On one

Fig. 2. Study area – Xujiahui

hand, function changes almost always cause layout changes; but on the other hand, layout changes need new functions to fit those changes. A conceptual image of this interaction is shown as Fig. 1. Based on site investigations and data provided by local institutes, this paper will compare two cases of comprehensive UUS utilization in commercial metro station areas in terms of function and layout, and will then illustrate their relationships in order to clarify their interaction characteristics.

3 Data Collection

3.1 Xujiahui Zone, Shanghai Study Area

The Xujiahui zone, which is one of the city's commercial centers, is located in southwest part of Shanghai, China. It is also an area where substantial crowds aggregate and disperse, especially on weekends and holidays. Three metro lines cross the zone and three separate metro stations are situated there. The study area is defined as a radius of 500 m around each of the barycenter locations centered on the three stations. For the purposes of this research, due to the fact that the study area is surrounded by roads, only those blocks that are more than one half included in the buffer zone are considered (Fig. 2). Combining with site investigation and the datum from Shanghai Urban Planning and Design Research Institute, it is found that there are several land use function categories present in the zone, the specifics of which are presented in Fig. 4 and Table 1. Commerce and residential uses are dominant functions. The total area is approximately 80 ha. As shown in Fig. 4 and Table 2, the primary UUS function in the study area is parking, which accounts for more than 60% of the underground land use. Other underground commercial uses include shopping centers, retail outlets, clothing stores, entertainment centers, and catering establishments, while pedestrian passages provide access among the different facilities. Most commercial uses aggregate in the B1F of the buildings surrounding the Line 1 station, which was completed in the 1990s. Pedestrian passageways connecting different land parcels are almost always located under municipal roads. The total area utilized is approximately 445,000 m^2 from the B1F to B3F levels. As for layout, the underground space relatively close to the stations consists of an irregular assortment of basements and pedestrian passages, including an underground street on the B2F level. The transfer access passageways between the different stations on the B1F level are isolated from most of those basements, and underground space in areas relatively far from the stations do not appear to be connected to those stations.

Fig. 3. Study area – Namba

3.2 Namba Zone, Osaka Study Area

As one of the city's major commercial centers, the Namba zone is located in the downtown area of Osaka, Japan, in an area serviced by multiple metro lines. From the Xujiahui zone case study, it was found that UUS is mainly located in the blocks adjacent to roads, so the scope of the Osaka study area was narrowed and the Midosuji Line was chosen as the center, as shown in Fig. 3. In this study area, the specific land use categories are presented in Fig. 5 and Table 1 (Osaka City Planning Bureau 2017). It is obvious that the dominant land use is commerce. The total area is approximately 20.5 ha. Based on our site investigation, it is found that the primary UUS function of the study area is commerce, which accounts for 56.6% of the land use. Underground commercial uses also include shopping centers, retail outlets, clothing stores, restaurants, entertainment centers, and catering, and an extensive pedestrian passage system connects almost all of the underground areas. Most commercial area use is found below the municipal roads and in the B1F and B2F levels of the buildings surrounding the stations. Pedestrian passageways connecting the different land parcels almost always follow below the municipal roads. The total area utilized from the B1F to B3F levels is approximately 146,000 m^2 (Table 2). As for layout, the underground space consists of major four parts: Namba Walk, Namba Ekimo, Namba NanNan, and Namba City (Fig. 6), which together connect practically all the building basements and metro stations.

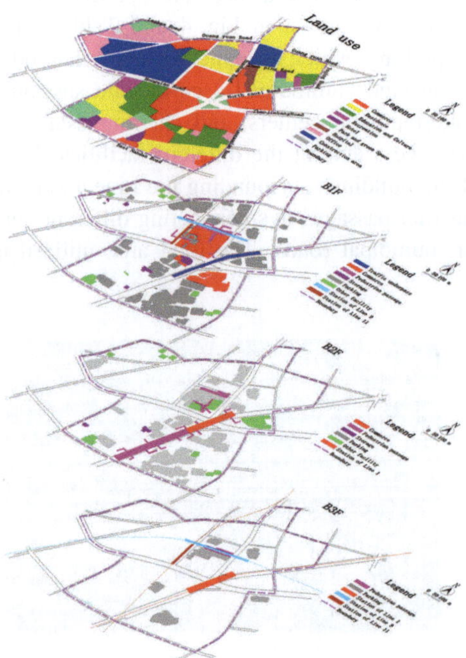

Fig. 4. Land use and UUS distribution of Xujiahui zone

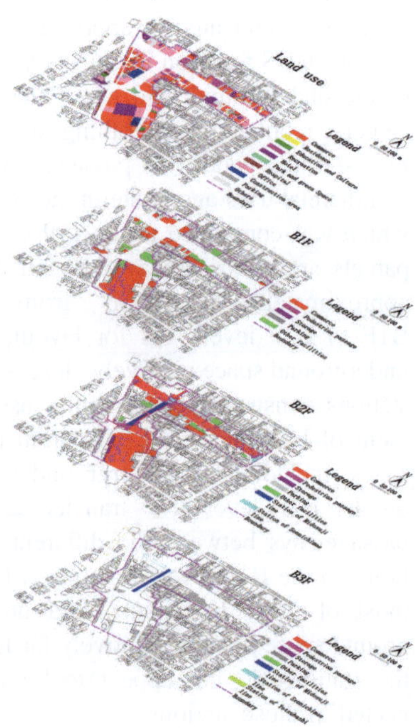

Fig. 5. Land use and UUS distribution of Namba zone

Table 1. Land use function categories in Xujiahui zone and Namba zone

Zone		C	R	E	Rn	H	G	Hl	O	Cn	P	Os	Rd	Sum
Xujiahui	Area (ha)	17.2	14.2	8.0	0.2	2.1	3.6	0.9	8.6	11.2	1.3	0.0	15.0	80.3
	Ratio (%)	21.4	17.7	10.0	0.2	2.6	4.5	1.1	10.7	13.9	1.6	0.0	18.7	100.0
Namba	Area (ha)	4.8	0.1	0.1	0.9	1.1	0.1	0.1	2.2	0.4	0.3	0.4	10.0	20.5
	Ratio (%)	23.4	0.5	0.5	4.4	5.4	0.5	0.5	10.7	2.0	1.5	2.0	48.8	100.0

Note: C-Commerce, R-Residence, E-Education and culture, Rn-Recreation, H-Hotel, G-Green space, Hl-Hospital, O-Office, Cn-Construction area, P-Parking, Os-Others, Rd-Road.

Table 2. UUS function categories in Xujiahui zone and Namba zone

Zone		F	S	Tu	Pp	C	P	Se	Os	Sum
Xujiahui	Area (m2)	B1F	8195	10025	8814	46672	130188	2892	10567	217353
		B2F	9753	0	25600	0	112996	0	26724	175073
		B3F	17400	0	1946	0	33374	0	0	52720
		Sum	35348	10025	36360	46672	276558	2892	37291	445146
	Ratio (%)		7.9	2.3	8.2	10.5	62.1	0.6	8.4	100.0
Namba	Area (m2)	B1F	0	0	5292	34771	1440	424	10515	52442
		B2F	4765	0	25000	47940	0	1934	5318	84957
		B3F	5270	0	0	0	3570	0	0	8840
		Sum	10035	0	30292	82711	5010	2358	15833	146239
	Ratio (%)		6.9	0.0	20.7	56.6	3.4	1.6	10.8	100.0

Note: F-Floor, S-Station, Tu-Traffic underpass, Pp-Pedestrian passage, C-Commerce, P-Parking, Se-Storage, Os-Others.

4 Discussion

Fig. 6. Layout of Namba zone

From the land use situation, it was found that commercial use occupies the main function in both study areas because the two zones are commercial centers servicing their entire cities. Commercial use is almost always found adjacent to the stations or roads, while other land uses do not seem to have any distinctive distribution characteristic with regard to distance to the station. For the Xujiahui zone, residential use and construction areas come in second and third, respectively. While for the Namba zone, office areas come in second, and there is little in the way of residential areas. The authors infer that this difference may be correlated to early stage urban renewal and residential district planning.

As for UUS utilization in the Xujiahui zone, it is found that underground parking occupies an overwhelming ratio of more than 60% of the land use, with the remainder widely distributed over almost all other usage types. Furthermore, it is found that, on the whole, this parking use was concentrated primarily in the B1F and B2F levels of the high-rise buildings completed after the year 2000. This can be explained by the fact that there were not so many cars in the years prior to that period, and that the impending scarcity of land resources had not yet attracted government attention. However, the situation in the Namba zone is quite different. There, underground parking is rare, accounting for just 3.4% of the study area. The authors infer that it is possible that the Japanese government's efforts to encourage public transportation use is more effective, and that the use of above-ground mechanical parking lots is more popular based on our investigation.

For both study zones, in addition to the underground commercial streets (called underground shopping malls in Japan, "*underground facility integrated with underground footpaths provided for public purposes and shops, offices and other similar facilities built on such footpaths that are linked to roads or station square areas provided for public purpose*", Kishii 2016), a large percentage of the underground commerce is located in the basements of buildings located within a 200-m distance from the metro stations. In the case of the Xujihui zone, almost all such commerce is located on the B1F level; while for Namba, it can be found on both the B1F and B2F levels. Here, the authors infer that it is influenced by the passenger flow rate as well as commercial land prices. In other words, it is plain that the UUS function distribution is strongly connected with metro stations located within a 200-m radius. Meanwhile, the vertical function of UUS distribution depends on different environmental and economic situations. As for layout, it was found that commercial underground space under the municipal roads in the Xujiahui zone is not exploited substantially, even though one pedestrian passage (or underground street), the Shanghai Metro Culture and Art Walkway, is located under the North Caoxi Road in the south of the Line 1 station (Fig. 4, B2F). However, there are almost no commercial uses along this passage. The access ways among the stations are located beneath the land parcel making up the study area, but are isolated from the building basements (Fig. 4, B1F), which means that passengers never go through those basements when they transfer from one station to another. As a result, the people flow tends to aggregate around the stations and transfer access areas, but not in the commercial areas. This utilization pattern can be called the station center layout pattern. In this layout pattern, commercial activities are almost always located in basements, and pedestrian passage between stations takes place below roads, separately from basements in the land parcel.

This pattern is significantly different from that seen in Japan. Although the Namba stations are transfer points for multiple subway lines, as shown in Fig. 3, it is found that commercial underground space is exploited substantially under the roads, with a variety of underground commercial streets present under main roads, and numerous commercial ventures alongside the underground streets. Transfer access ways among these multiple stations, i.e., underground streets, are also located beneath the roads. Passengers almost get off from one station, arrive at underground streets, and then either transfer to another line or go shopping in the commercial basement of the buildings. As a result, the people flow tends to aggregate in the commercial streets. This utilization form can be called the commercial street center layout pattern. For this layout pattern, large amounts of commercial activities are located in the commercial streets where pedestrian passage is amalgamated. Passengers can also transfer from one station to another through the commercial streets that are connected to the building basements in the area. For this pattern, UUS zoning (Fig. 6) is necessary for efficient wayfinding. Meanwhile, the flowrate of people along another access way, named the Chicago Gallery and located in the west side of Namba Walk, is much higher than that of Xujiahui. From this observation, the authors infer that the location of underground streets should be planned between two stations, not at one side of any station, so as to avoid decreased flowrates. Furthermore, the underground street exits should be amalgamated with the building basements in order to prevent flowrates from decreasing, especially when they are farther than 200 m from the metro stations. The importance of mobility and access in a contemporary metropolis brings to infrastructure the character of collective space. Based on these rough classifications, with respect to layout, UUS patterns around metro stations in urban commercial area can be divided into station center and commercial street center layouts. Their conceptual images are shown in Fig. 7.

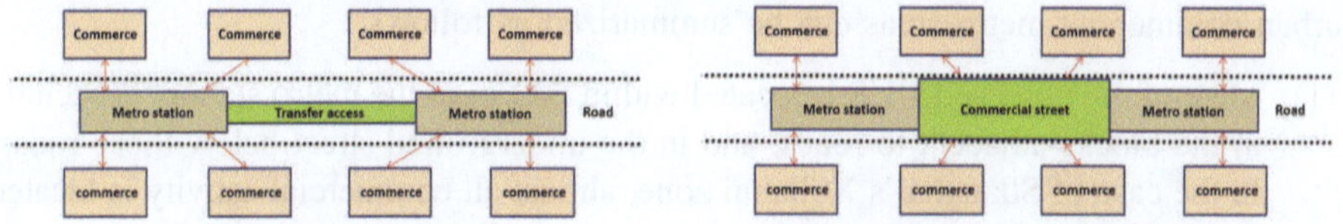

(1) Station center layout (2) Commercial street center layout

Fig. 7. Layout patterns of UUS around metro station in urban commercial area

Based on our site investigations, it is found that most of the UUS utilization around metro stations in the Shanghai commercial area, such as Zhongshan Park and People's Square Stations, follow the station center layout. In contrast, UUS around most of the metro stations in the commercial areas of Osaka, such as Shinsaibashi and Umeda Stations, follow the commercial street center layout. The authors infer that this is the result of differences in national context. The Shanghai underground pedestrian system development is largely based on metro stations and the connection layouts are determined by the principle of ensuring the quick movement of pedestrians to street level (Cui et al. 2012). However, while the Japanese government paid more attention to high intensive utilization of UUS several decades ago due to land scarcity, it is definitely the case that distances between adjacent metro stations in Osaka are shorter than those in Shanghai. Thus, there is more potential to exploit underground streets along municipal roads in order to connect adjacent stations in Japan. The authors contend that this consideration is extremely important when planning the location of an underground street in order to maximize benefits from rail transit networks, and is also why underground streets are developed only in relatively high-density commercial areas. For the Namba zone, the floor area ratio is 10.0 (Osaka City Planning Bureau 2017), although it is just 3–4 for Xujiahui zone (Mo et al. 2009). However, a station center layout can transform into a commercial street center layout over time. For example, Namba Station would have been considered to have a station center layout before 1970, when the Namba Walk and Sennichimae Line were still under construction. At that time, passengers transferred from the Midosuji to Yotsubashi Lines just through normal access ways, but underground streets connected all the underground facilities after they were completed. Thus, the layout pattern changed as the zone developed, which included the effects of traffic congestion due to population expansion. UUS function, e.g., commerce, cannot meet the needs of urban development on its own, so functions need to be improved, and such function changes almost always causes layout changes, thereby strengthening the presence of underground networks as dynamic public systems. Meanwhile, layout changes need new functions that take advantage of their features. This reflects the point that UUS function and layout should maintain a relationship of mutual coordination and promotion. Ultimately, upon adopting a commercial street center layout, the Namba Station area became a complex underground system with highly intensive space utilization.

5 Conclusion

The conclusions based on the research of UUS development around metro stations in urban commercial metro areas can be summarized as follows:

(1) Most of the observed UUS is located within 200 m of the metro station, especially in the blocks adjacent to roads, and in the underground street below those roads. In the case of Shanghai's Xujiahui zone, almost all commercial activity is located on the B1F level, and a substantial amount of underground parking scattered around the area. In the case of the Osaka's Namba zone, commercial use is located both on the B1F and B2F levels, and there is little underground parking. Thus, function and layout in the vicinity of the stations may vary according to national context, including city development history, construction regulation, property law, political issues, etc.

(2) Underground streets should be planned between two stations and the exits should be connected to the basements of surrounding buildings, in order to maximize commercial benefits. Meanwhile, commercial space should be mainly planned in B1F, and whether it can be planned in B2F, even B3F, should depend on the depth of underground street and other factors, such as passenger flow rate, etc.

(3) Two kinds of UUS comprehensive utilization layout patterns around metro station in commercial areas were elucidated: station center layout and commercial street center layout. Station center layout emphasizes smooth passenger transfers between transportation modes, while commercial street center layout emphasizes not only passenger transfer, but also higher intensive exploitation of commercial activities that are made possible by utilizing the available space under the municipal roads with higher floor area ratios. While the former can transform into the latter as the city develops, the layout pattern transformation process requires a greater level of integration between the metro station and surrounding streets. So we argue that exploitative space surrounding metro station should be reserved from a long view when it is in station center layout stage. This may also sound an alarm to the planners that utilizing underground space randomly runs counter to sustainable development.

(4) The development of UUS in a commercial metro station area is based on urban "function and layout" interaction characteristics. UUS function and layout are predicated on maintaining a relationship of mutual coordination and mutual promotion.

However, the patterns elucidated in this paper were developed based on ideal situations. In reality, commercial metro station areas may be a combination of both patterns, or something in between. Herein, a representative concept was introduced, which is critical to promoting further development, even though the classic UUS patterns discussed may not be applied directly. Meanwhile, our future work will focus on additional site investigations.

Acknowledgements. The authors gratefully acknowledge support provided by funds from the National Key Technology R&D Program grant 2012BAJ01B04, the National Basic Research Program (973 Program) grant 2015CB057806 and the China Scholarship Council grant 201606260162.

References

Bajic, V.: The effects of a new subway line on house prices in Metropolitan Toronto. Urban Stud. **20**, 147–158 (1983)

Bélanger, P.: Underground landscape: the urbanism and infrastructure of Toronto's downtown pedestrian network. Tunn. Undergr. Space Technol. **22**, 272–292 (2007)

Boivin, D.: Montreal's underground network: a study of the downtown pedestrian system. Tunn. Undergr. Space Technol. **6**(1), 83–91 (1991)

Choi, J., Asami, Y.: Analysis of distribution and attributes of redevelopment projects by the distance from railway stations. Jpn. J. Real Estate Sci. **16**, 91–101 (2002). (in Japanese)

Cui, J., Allan, A., Taylor, M.A., Lin, D.: Developing Shanghai underground pedestrian system under urbanization: mobility, functionality and equity. J. Archit. Urban. **36**(4), 283–297 (2012)

Du, H., Mulley, C.: The short-term land value impacts of urban rail transit: quantitative evidence from Sunderland, UK. Land Use Policy **24**, 223–233 (2007)

Golany, G., Ojima, T.: Geo-Space Urban Design. Wiley, Hoboken (1996)

Hui, Y.: Research for urban rail transit station area planning and construction. Urban Plan. Forum **138**(2), 30–33 (2002). (in Chinese)

Kishii, T.: Utilization of underground space in Japan. Tunn. Undergr. Space Technol. **55**, 320–323 (2016)

Li, X., Xu, H., Li, C., Sun, L., Wang, R.: Study on the demand and driving factors of urban underground space. Tunn. Undergr. Space Technol. **55**, 52–58 (2016)

Mo, Y., Deng, J., Wang, J.: A land use model for urban rail station area planning based on TOD strategy. Archit. Environ. Eng. **31**(2), 116–120 (2009). (in Chinese)

Osaka City Planning Bureau (2017). <https://www.mapnavi.city.osaka.lg.jp/webgis/>

Polzin, S.: Transportation/land-use relationship: public transit's impact on land use. J. Urban Plan. Dev. **125**(4), 135–151 (1999)

Roukouni, A., Basbas, S., Kokkalis, A.: Impacts of a metro station to the land use and transport system: the Thessaloniki Metro case. Soc. Behav. Sci. **48**, 1155–1163 (2012)

Evaluation of Two Thermo-TDR Probes for Soil Moisture, Density, and Thermal Conductivity

Xuelin Wang and Xinbao Yu(✉)

Department of Civil Engineering,
University of Texas at Arlington, Arlington, TX 76019, USA
xuelin.wang@mavs.uta.edu, xinbao@uta.edu

Abstract. Thermo-TDR probes can function both as a regular time domain reflectometry (TDR) probe, for measuring moisture and density, and a dual-heat probe for measuring thermal conductivity, thermal diffusivity, and volumetric heat capacity. The probe measurement sensitivity and accuracy are affected by probe length, diameter, and spacing. In this paper, two thermo-TDR probes were fabricated based on an improved design of a previous one. The two probes were first calibrated for measurement of dielectric constant and electrical conductivity using standard chemical solutions. Then, moisture, density, and thermal measurements were taken in custom designed compaction molds filled with ASTM fine-grade compacted sand using the thermo-TDR probes as well as KD2 probes. Calibrations are reported for the thermo-TDR probes measuring soil moisture, density, and thermal conductivity. The effect of sensor fabrication and sample preparation tests are evaluated. Recommendations are provided for using thermo-TDR probes for measuring in sand.

Keywords: Thermo-TDR probe · Thermal conductivity · Soil compaction
Fabrication effect

1 Introduction

Soil thermal properties along with soil moisture and density are the fundamental soil design parameters needed for design and construction of many engineering applications such as soil thermal energy storage, radioactive nuclear waste disposal, underground heat exchangers, and energy piles. Design of these geostructures requires better characterization of heat transfer in the soil, which includes heat conduction, convection, and radiation. Soil thermal properties include volumetric heat capacity, thermal diffusivity, and thermal conductivity. Needle-like heat probes are commonly used to measure soil thermal properties based on the line heat source theory. The thermal properties of the soil are obtained by back-calculation of the measured temperature response curves of heat probes subjected to a short heat pulse (Welty 1976).

The time domain reflectometry (TDR) technology is a quick, dependable, and safe method for soil moisture content measurement, both in the field and laboratory (Topp et al. 1980; Yu and Drnevich 2004; Yu et al. 2015). Liu et al. (2008) examined four probe designs for measurement of soil bulk density, volumetric water content, and

X. Shi et al. (Eds.): GSIC 2018, *Proceedings of GeoShanghai 2018 International Conference: Transportation Geotechnics and Pavement Engineering*, pp. 44–55, 2018.
https://doi.org/10.1007/978-981-13-0011-0_5

volumetric heat capacity. Yu et al. (2015) developed a thermo-TDR probe for laboratory measurement of soil moisture, dry density (bulk density), and thermal conductivity in compacted soil specimens.

TDR probe measurement sensitivity and accuracy are affected by probe length, diameter, and spacing. Previous studies have been conducted on probe design and performance evaluations. However, more studies are needed to optimize probe design for a comprehensive evaluation that can provide better measurements of soil moisture, dry density, and soil thermal properties of compacted soil specimens. In this study, two thermo-TDR probes were fabricated based on a modified probe design to comprehensively evaluate individual sensor performances and sample preparation methods for thermo-TDR measurements.

2 Sensor Design and Fabrication

The design of thermo-TDR probes shall satisfy the requirements of both TDR and dual probe heat pulse (DPHP) probes. The major design restraints of the TDR moisture probes include signal attenuation, sampling area, and sensor installation. The length of the probe is determined by the source voltage of the pulse generator, resolution of recorded signals, and signal loss, which is dominated by soil electrical conductivity (Ren et al. 1999). Ren et al. (1999) state that probe-to-probe spacing to probe diameter ratio should be less or equal to 10 as shown in Eq. 1. Design of a DPHP requires a heat source to be close to the infinite line heat source. The design constraints are presented by Eqs. 2–4 (Blackwell 1956; Ren et al. 1999).

$$s/d \leq 10 \tag{1}$$

$$L/2s > 2.2 \tag{2}$$

$$L/d > 25 \tag{3}$$

$$d/2s < 0.13 \tag{4}$$

where L is the probe length; s is the center-to-center spacing between the center probe and the outer probe, and d is the diameter of the probe. An ired design of the thermo-TDR probe by Yu et al. (2015) was developed as shown in Fig. 1. The probe has a longer length and wider spacing. The probes have a longer length extended into the probe head to increase the stability of probes during insertion.

Details of the fabrication process of thermo-TDR were introduced by Yu et al. (2015). Two thermo-TDR probes were fabricated based on the improved design just described. The two thermo-TDR probes are referred to as T1 and T2. The fabricated length including the tip of T1 and T2 is 54 mm and 53 mm, respectively.

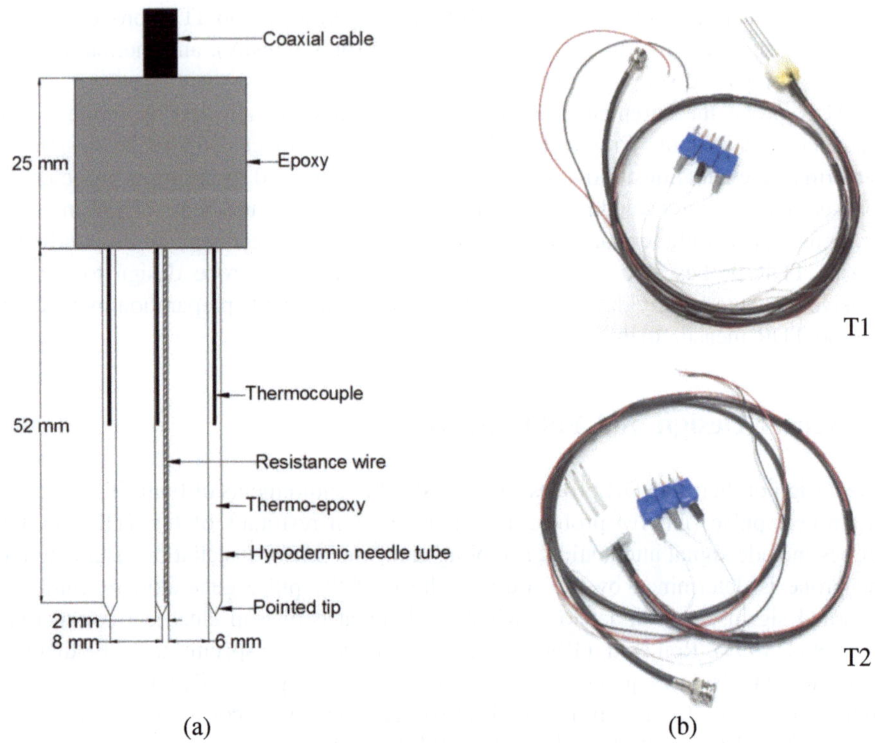

Fig. 1. (a) Design of thermo-TDR probe, (b) photos of T1 and T2

3 Calibration of Dielectric Constant and Electrical Conductivity

3.1 Dielectric Constant and Electrical Conductivity

Apparent dielectric constant and electrical conductivity of the measured soil can be directly determined by analyzing TDR waveforms. The apparent dielectric constant is determined by the Eq. 5 (Topp et al. 1980).

$$K_a = \left(L_a/L_p\right)^2 \tag{5}$$

where K_a is the apparent dielectric constant; L_p is the physical probe length, and L_a is the apparent length of the probe in the testing materials determined from TDR waveforms. The apparent length of the probe in the TDR waveform is determined by locating the electromagnetic wave reflection points at the probe head and end following the single tangent method proposed by Baker and Allmaras (1990).

Electrical conductivity measured by TDR can be determined using Eq. 6 (Giese and Tiemann 1975).

$$EC_b\left(Sm^{-1}\right) = \left(\varepsilon_o c Z_o / L_p Z_c\right)\left(2V_o/V_f - 1\right) \tag{6}$$

where EC_b is the electrical conductivity (Sm^{-1}); ε_o is the dielectric permittivity of free space; c is the speed of light in vacuum; Z_o is the characteristic probe impedance; L_p is the probe length; Z_c is the TDR cable tester output impedance; V_o is the source voltage; and V_f is the long-term voltage level of the reflected signal. Details of the calculations of K_a and EC_b can be found in Yu et al. (2015).

3.2 Calibration of Dielectric Constant (K_a) and Electrical Conductivity (EC_b)

Nine different chemicals with known dielectric constants were used to calibrate the dielectric constant K_a. Because the values of K_a for most testing soils are less than 20, only three different chemicals were selected for dielectric constant calibration. TDR waveforms of the chemicals and air are shown in Fig. 2. Calibrations of an apparent dielectric constant for T1 and T2 are shown in Fig. 3. The two probes have very similar calibration equations with high accuracy.

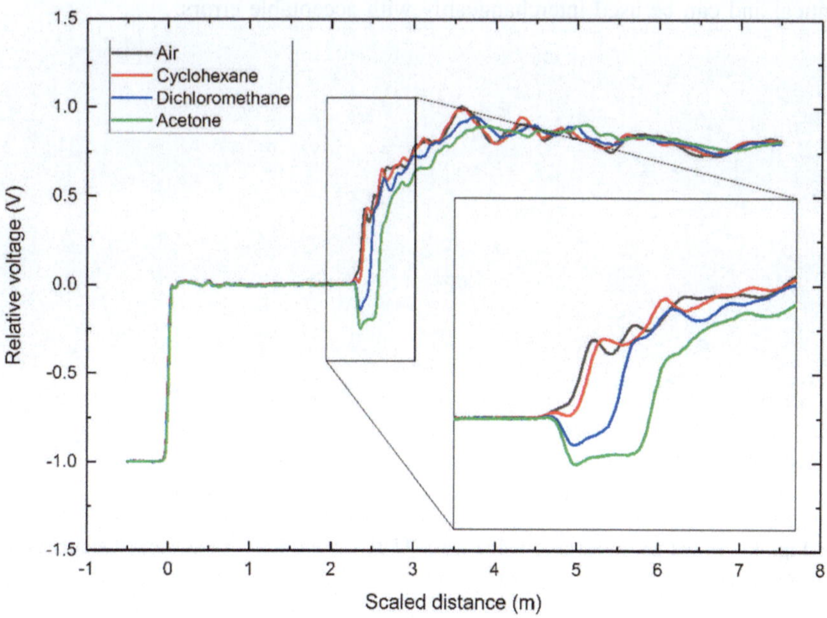

Fig. 2. TDR waveforms of four different chemicals for T1

The calibration of thermo-TDR for electrical conductivity measurement was performed by measuring NaCl solution of different concentration using the thermo-TDR probe as well as an EC meter. The electrical conductivity by the thermo-TDR probe is compared with the EC meter measurement as shown in Fig. 4. The calibration results

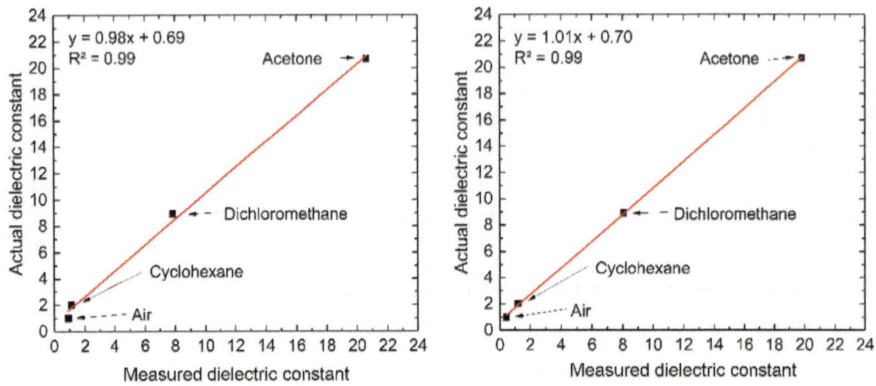

(a) T1 (b)T2

Fig. 3. Actual and measured value from TDR waveforms for dielectric constant

show that thermo-TDR can satisfactorily measure both dielectric constant and electrical conductivity with an R^2 of 0.99. The calibration equations of T1 and T2 are almost identical and can be used interchangeably with acceptable errors.

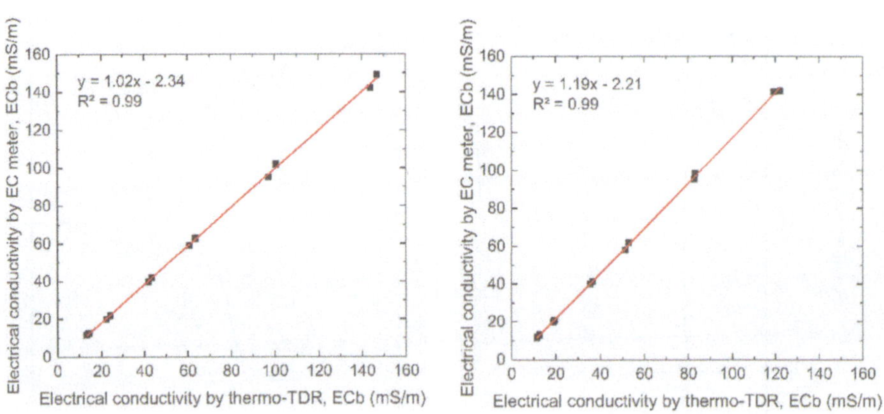

(a)T1 (b)T2

Fig. 4. Actual and measured value from TDR waveforms for electrical conductivity

4 Thermal Properties

A DC current source was connected to the resistance wire in the thermo-TDR center probe to generate a heat pulse in the center probe. The temperature variations of the three probes were recorded. Their temperature response can be predicted by the infinite line heat source theory as shown in Eq. 7 (De Vries 1952; Kluitenberg et al. 1993).

$$T(r,t) = (Q/4\pi\alpha)\left[Ei\left(-r^2/4\alpha(t-t_o)\right)-Ei\left(-r^2/4\alpha t\right)\right] \tag{7}$$

where T is the temperature change; Q is the strength of line heat source; α is the soil thermal diffusivity; $Ei(x)$ is the exponential integral; r is the radial distance; t is the time, and t_o is the duration of the heat pulse.

The volumetric heat capacity, thermal diffusivity, and thermal conductivity were calculated using Eqs. 8–10.

$$\rho_c = (q/4\pi\alpha T_m)\left[Ei\left(-r^2/4\alpha(t_m-t_o)\right)-Ei\left(-r^2/4\alpha t_m\right)\right] \tag{8}$$

$$\alpha = (r^2/4)\left[1/(t_m-t_o)\right]/ln\left[t_m/(t_m-t_o)\right] \tag{9}$$

$$k = \frac{q}{4\pi T_m}\left[Ei\left(\frac{-\ln\left[\frac{t_m}{t_m-t_o}\right]}{t_o/t_m}\right) - Ei\left(\frac{-\ln[t_m(t_m-t_o)]}{t_o/(t_m-t_o)}\right)\right] \tag{10}$$

In these equations, r is the radial distance between the center probe and outer probes; t_m is the time when the maximum temperature of the outer probes occurred; T_m is the maximum temperature increase of outer probes. Details of the analysis of heat pulse response curves can be found in Yu et al. (2015). Only thermal conductivity measurement is discussed in this paper. KD2 pro sensors include TR1 for measurement in soil and KS1 for measurement in liquids. Although KS1 is not recommended for soil, it was used to measure the thermal conductivity of ASTM fine-graded sand to evaluate the effect of probe size and heating pulse.

5 Test of Thermo-TDR Probes in ASTM Fine Graded Sand

Figure 5 shows the schematic of the experiment setup. The thermo-TDR probe inserted into the compaction mold is connected to the Campbell Scientific TDR100 through a coaxial cable to collect TDR waveforms. The resistance wire is connected to the DC power supply BK PRECISION 17850B to heat the soil specimens. Three thermal couples are connected to the thermocouple data logger Pico USB TC-08 to record temperature variations of the center probe and outer probes every second.

ASTM fine graded sand was selected to evaluate the two thermo-TDR probes for measurement of soil moisture, dry density, and thermal conductivity. The ASTM fine graded sand is Ottawa-type silica sand with a white color and round-to-subrounded shape. Total 4 group tests were performed to evaluate the effect of the sample preparation method and repeatability of test results. Each test group includes compaction of sand at various desired moisture content levels using custom-designed PVC compaction molds. The dimensions of the compaction mold and the extension collar are shown in Fig. 6.

For all test groups, dry sand samples were hand mixed with tap water at a desired mass ratio in a metal bowl and kept for 30 min before compaction. For the first test group, the prepared wet soil sample was poured into the compaction mold until the

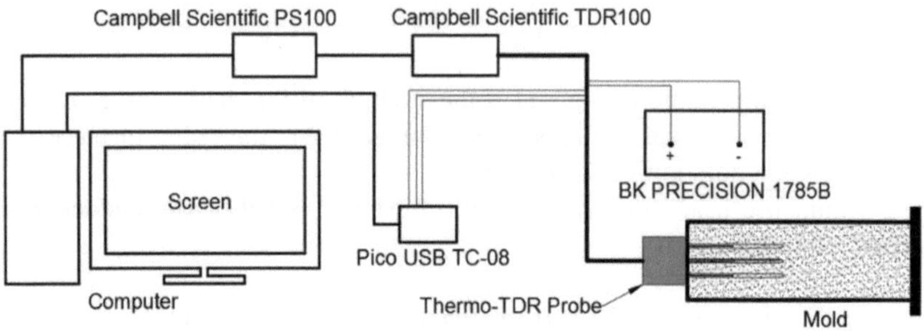

Fig. 5. Schematic of experiment setup

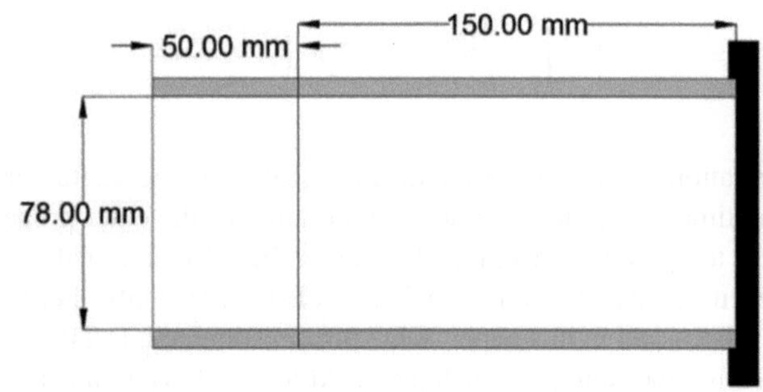

Fig. 6. Dimensions of the compaction mold

extension collar was full and then a beaker was used to compress the soil by hand compression. For the second test group, the prepared soil sample was poured into the compaction mold with extension collar until full and then a standard compaction rammer was used to compact the soil with the hammer raised to 10 cm above the soil surface. For Tests 3 and 4, the prepared soil samples were compacted at three equal lifts to reach the height of the compaction mold with the hammer at its full height of 30 cm. The soil compaction methods are illustrated in Fig. 7.

After the sand specimen was compacted by employing the procedure described in the above section, the thermo-TDR probe T1 was inserted into the sand with the probe head in close contact with the sand surface. During insertion, the probe should be kept vertical to maintain the probe in close contact with surrounding soils. Two TDR waveforms were acquired consecutively using PM-TDR SM Version 1.7. Then the DPHP test was performed by supplying a DC current of 0.15 A for 15 s. The thermocouple readings were acquired using a PicoLog Recorder. These above measurements were repeated one more time, and a total of four TDR waveforms and two heat pulse responses were recorded. The procedure was then repeated for probe T2. After the thermo-TDR probe tests, two heat probes, KD2 TR-1 and KS-1, were inserted into the specimens separately for heat pulse tests following the procedure recommended in the operation manual. Two readings were taken for each probe. After all the measurements were taken, two soil samples were collected from the top third of the

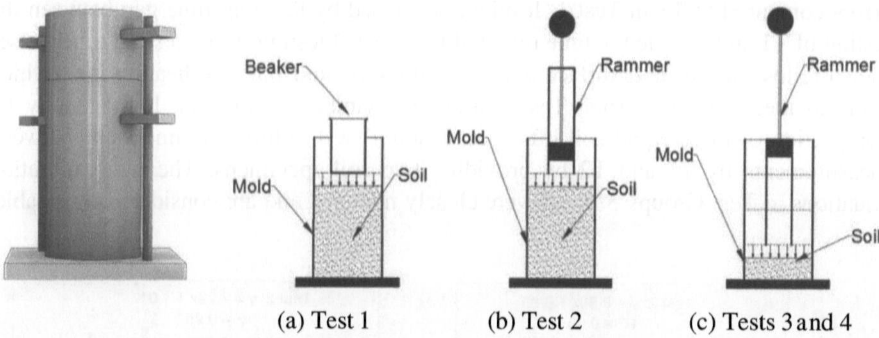

Fig. 7. Method of soil compression for four tests

compaction mold to determine oven-dry moisture content. Test 1 and Test 2 were performed with the compaction mold in a vertical position, while Tests 3 and 4 were conducted with the mold kept in a horizontal position to maintain uniform moisture along the height. For Tests 3 and 4, two compaction molds were used to prepare two identical compacted specimens for T1 and T2, respectively. Test 3 and 4 also reduced the time lapse between thermo-TDR and KD2 measurement.

6 Calibration of Thermo-TDR for Moisture and Dry Density

One step method proposed by Yu and Drnevich (2004) was used to develop calibration equations for measurement of the gravimetric moisture content and dry density. This method adopts two calibration equations shown in Eqs. 11 and 12 based on measured apparent dielectric constant and electrical conductivity of the measured soils.

$$\sqrt{K_a}\frac{\rho_w}{\rho_d} = a + bw \tag{11}$$

$$\sqrt{EC_b}\frac{\rho_w}{\rho_d} = c + dw \tag{12}$$

Here, a, b, c, and d are specific calibration constants; ρ_d is the dry density of soil, kg/m^3; ρ_w is the density of water, kg/m^3, and ω is the gravimetric moisture content, (based on percentage). Thus, gravimetric moisture content and dry density can be evaluated by solving Eqs. (11) and (12).

The calibration relationships for thermo-TDR probe T1 and T2 are presented in Figs. 8 and 9. Figures 8 and 9 present the calibration based on dielectric constant measurement and electrical conductivity respectively. Notably, the calibration equations from Test 1 and Test 2 significantly deviate from Test 3 and Test 4, especially calibration as per Eq. 11. Specimens were not compacted uniformly in Test 1 and Test 2 because the soil was compacted in one lift. This leads to large errors when average specimen moisture and dry density are used in the calibration. T2 has larger calibration errors compared to T1 in Test 1. It might be caused by the long-time gap between the testing of T1 and T2 due the time required for thermal testing. Long testing time causes moisture loss at the sand surface due to evaporation and possible heating loss, which leads to measurement errors. Test 3 improves sample compaction homogeneity by compaction of three equal soil lifts. Test 3 and Test 4 reduce the time lapse between measurements by T1 and T2 by providing two soil specimens. The two calibration equations in Test Groups 3 and 4 were closely matched and are considered repeatable.

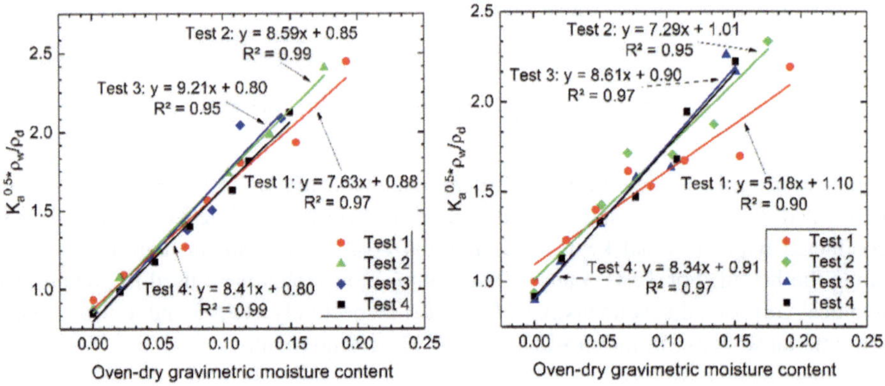

(a) T1 (b) T2

Fig. 8. Calibration of gravimetric moisture content and dry density measurement using measured dielectric constant. (a) probe T1 and (b) probe T2.

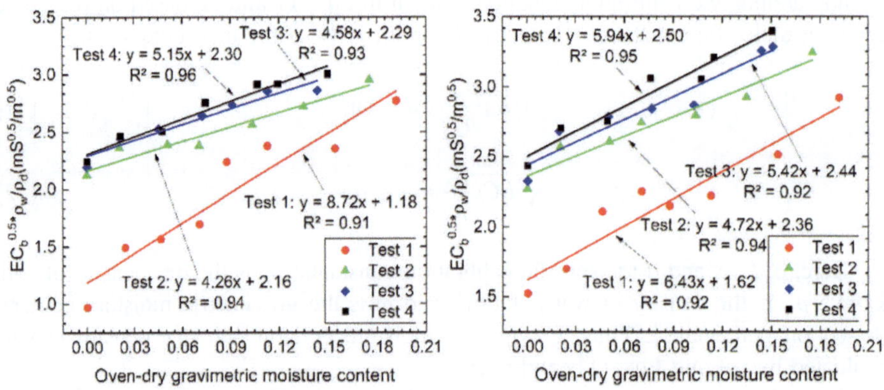

(a) T1 (b) T2

Fig. 9. Calibration of gravimetric moisture content and dry density measurement using measured electrical conductivity. (a) probe T1 and (b) probe T2.

7 Calibration of Thermal Conductivity Measurement

Comparison of measured thermal conductivity between thermo-TDR probes and KD2 TR-1 (long) probe for T1 and T2 are shown in Fig. 10(a) and (b), respectively. A comparison of measured thermal conductivity between thermo-TDR probes and the KD2 KS-1 (short) probe for T1 and T2 are shown in Fig. 11(a) and (b), respectively. Figures 10 and 11 show that TR-1 has a larger measurement range of thermal conductivity as compared with KS-1. Both Thermo-TDR probes yield a close match of TR-1 measurements. Sample preparation methods are much less effective when

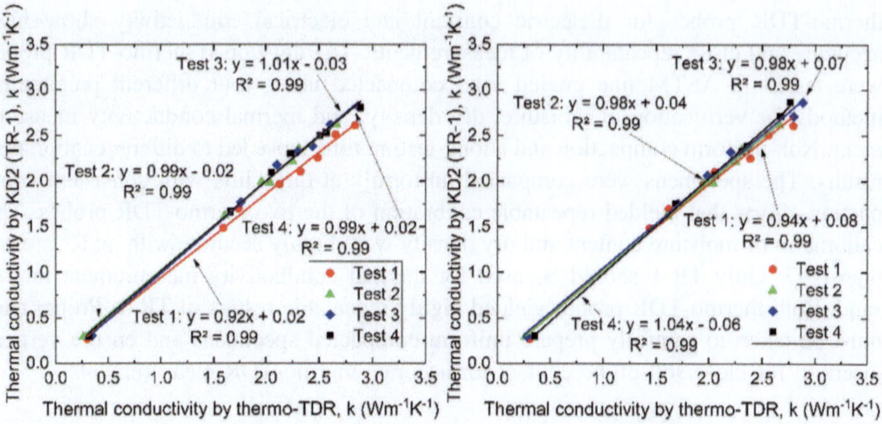

(a) T1 (b) T2

Fig. 10. Thermal conductivity comparison between thermo-TDR and KD2 (TR-1). (a) Thermo-TDR probe T1 and (b) thermo-TDR probe T2.

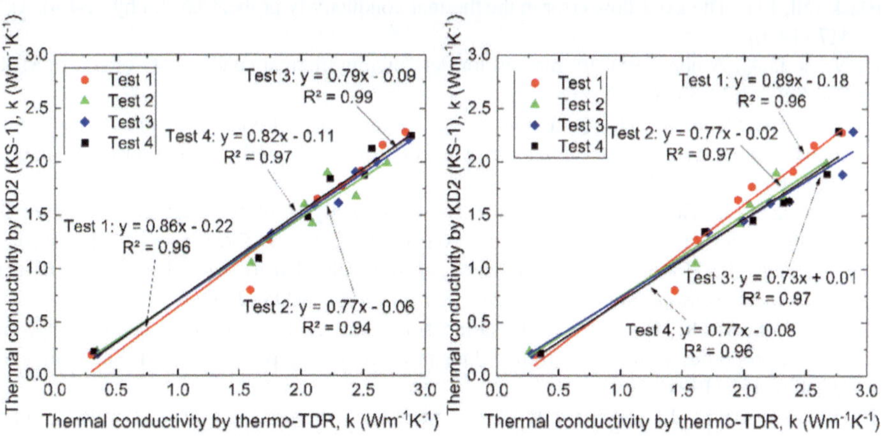

(a) T1 (b) T2

Fig. 11. Thermal conductivity comparison between thermos-TDR and KD2 (KS-1). (a) Thermo-TDR probe T1 and (b) thermo-TDR probe T2.

compared to moisture and density measurement. However, the loose compaction method in Test 1 still has a noticeable effect on thermal conductivity measurement. Figure 11 shows that the thermal conductivity measured by the KS-1 sensor is much smaller than that measured by the thermo-TDR probes. KS-1 sensors are designed for liquids and lose accuracy in sand.

8 Conclusions

In this paper, an improved thermo-TDR probe design was introduced, and two thermo-TDR probes were fabricated and tested comprehensively. Calibrations of thermo-TDR probes for dielectric constant and electrical conductivity show high accuracy and close repeatability of measurements. The developed thermo-TDR probes were tested in ASTM fine graded sand compacted using four different preparation methods for verification of moisture, dry density, and thermal conductivity measurement. Non-uniform compaction and a long-testing time lapse led to different calibration results. The specimens were compacted uniformly at three lifts with consistent compaction efforts that yielded repeatable calibration of the two thermo-TDR probes. The calibration of moisture content and dry density were highly accurate with an R^2 greater than 0.97. Only TR-1 should be used for thermal conductivity measurement in wet sand. Both thermo-TDR probes yielded highly repeatable results of TR-1. Proper care must be taken to carefully prepare uniform compacted specimens and ensure vertical insertion for close soil-probe contact for accurate thermo-TDR measurements.

References

Baker, J.M., Allmaras, R.R.: System for automating and multiplexing soil moisture measurement by time-domain reflectometry. Soil Sci. Soc. Am. J. **54**(1), 1–6 (1990)

Blackwell, J.H.: The axial-flow error in the thermal-conductivity probe. Can. J. Phys. **34**(4), 412–417 (1956)

De Vries, D.A.: A non-stationary method for determining thermal conductivity of soil in SITU. Soil Sci. **73**(2), 83–89 (1952)

Giese, K., Tiemann, R.: Determination of the complex permittivity from thins sample time domain reflectometry: implications for twin rod probes with and without dielectric coatings. Water Resour. Res. **32**(2), 271–279 (1975)

Kluitenberg, G.J., Ham, J.M., Bristow, K.L.: Error analysis of the heat pulse method for measuring soil volumetric heat capacity. Soil Sci. Soc. Am. J. **57**(6), 1444–1451 (1993)

Liu, X., Ren, T., Horton, R.: Determination of soil bulk density with thermo-time domain reflectometry sensors. Soil Sci. Soc. Am. J. **72**(4), 1000–1005 (2008)

Ren, T., Noborio, K., Horton, R.: Measuring soil water content, electrical conductivity, and thermal properties with a thermo-time domain reflectometry probe. Soil Sci. Soc. Am. J. **63** (3), 450–457 (1999)

Topp, G.C., Davis, J.L., Annan, P.: Electromagnetic determination of soil water content: measurements in coaxial transmission lines. Water Resour. Res. **16**(3), 574–582 (1980)

Welty, J.R., Wicks, C.E., Wilson, R.E.: Fundamentals of Momentum, Heat, and Mass Transfer, 2nd edn. Wiley, New York (1976). OCLC 2213384. ISBN 978-0-471-93354-0

Yu, X., Drnevich, V.P.: Soil water content and dry density by time domain reflectometry. J. Geotech. Geoenviron. Eng. **130**(9), 922–934 (2004)

Yu, X., Zhang, N., Pradhan, A., Thapa, B., Tjuatja, S.: Design and evaluation of a thermo-TDR probe for geothermal applications. Geotech. Test. J. **38**(6), 864–877 (2015)

Numerical Modeling of Temperature-Dependent Behavior of Saturated Clay

Qi-Yin Zhu[1(✉)], Zhen-Yu Yin[2,3], and Lian-Fei Kuang[1]

[1] State Key Laboratory for Geomechanics and Deep Underground Engineering, China University of Mining and Technology, Xuzhou 221116, China
qiyin.zhu@gmail.com

[2] Key Laboratory of Geotechnical and Underground Engineering of Ministry of Education, Department of Geotechnical Engineering, College of Civil Engineering, Tongji University, Shanghai 200092, China

[3] Research Institute of Civil Engineering and Mechanics (GeM), UMR CNRS 6183, Ecole Centrale de Nantes, 44321 Nantes, France

Abstract. The effect of temperature on the mechanical behavior of the soils around the special geo-structures, such as nuclear waste containers and energy piles can't be neglected. In this paper, some temperature-controlled laboratory test results are selected and analyzed, based on which and the thermal induced changes of main mechanical features (e.g., shear strength, critical state line, preconsolidation pressure) are identified. A constitutive model is then developed to interpret the temperature dependent behavior of saturated clays under the framework of critical state theory. Only one temperature related parameter is added in the present model comparing with the Modified Cam Clay model, and this parameter can be directly obtained from the yield loci of the soil. The performance of the proposed model is demonstrated with simulating temperature tests under drained condition.

Keywords: Constitutive model · Temperature effect · Critical state

1 Introduction

In the last few decades, there are a growing number of temperature related underground structures, such as nuclear waste isolation, buried high-voltage cables, geothermal structures, heat energy storage. The heat flow diffused from these high temperature sources will be dissipated in the surrounding clay. Thus, the effect of temperature on soil properties have become more interesting in the field of geotechnical engineering. Even though the temperature considered are below 100 °C in the above applications, this can also significantly affect the mechanical properties of soils in a way that can lead to failure of foundation. Thus, constitutive studies of the effects of temperature on clay behavior are of great importance.

After the pioneering work by [1], a variety of temperature-controlled laboratory tests have been conducted to study the effects of temperature on (1) volume change behavior, (2) permeability, and (3) stress-strain characteristics in drained heating or

X. Shi et al. (Eds.): GSIC 2018, *Proceedings of GeoShanghai 2018 International Conference: Transportation Geotechnics and Pavement Engineering*, pp. 56–62, 2018.
https://doi.org/10.1007/978-981-13-0011-0_6

undrained heating. Based on these experiments, some thermo-mechanical models have also been developed to interpret the temperature dependent soil behavior [2, 3]. However, each model has its own advantages and limitations from the theoretical point of view.

This paper attempts to propose a temperature-dependent elastoplastic model under the framework of critical state soil mechanics. The additional temperature-related parameter is analyzed. The predictive ability of the model is verified using datasets extracted from the literature.

2 Experimental Evidence

2.1 Temperature Dependent Preconsolidation Pressure

The experiments show that the elastic domain are sensitive to the temperature changes. As shown in Fig. 1, the elastic domain is largest at temperature T_0 with a preconsolidation pressure p_c'. Along with the increasing of temperature, the elastic domain is becoming smaller with a decreasing p_c'. However, the decrease of p_c' does not appear to be linear with the temperature. Based on many experimental data the following expression can be adopted

$$\frac{p_c'(T)}{p_c'(T_0)} = \left(\frac{T_0}{T}\right)^{\theta} \tag{1}$$

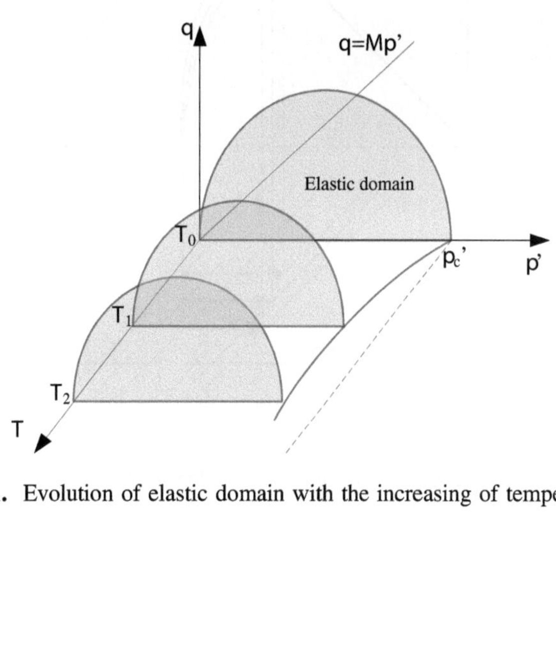

Fig. 1. Evolution of elastic domain with the increasing of temperature

2.2 Temperature Dependent Shear Strength

Based on the current available experimental observations, the effects of temperature on the shear strength behavior of saturated clays are confusing. On one hand, many experimental studies show that the increase in temperature causes strength degradation of clay [4, 5]. On the other hand, there are also many test results showing the reverse, that increasing temperature causes strengthening of clay [6–8]. Even though great efforts have been made, the current state of knowledge seems unable to explain the observed conflicting results.

2.3 Temperature Dependent Critical State Line

A clearer understanding of the critical state line (CSL) is needed when soils are undergoing an elevated temperature. The temperature dependent behaviors of CSL are studied on both the e-log p' and the q-p' planes.

In e-log p' plane, the slope of CSLs are usually considered as constant as observed in laboratory tests. As temperature increases, the CSLs shift downwards to a lower location. The CSLs at normal and elevated temperature usually parallel with each other [5, 6]. The NCL become closer to CSL with temperature increasing as shown by [5] and modelled by [2] as shown in Fig. 2.

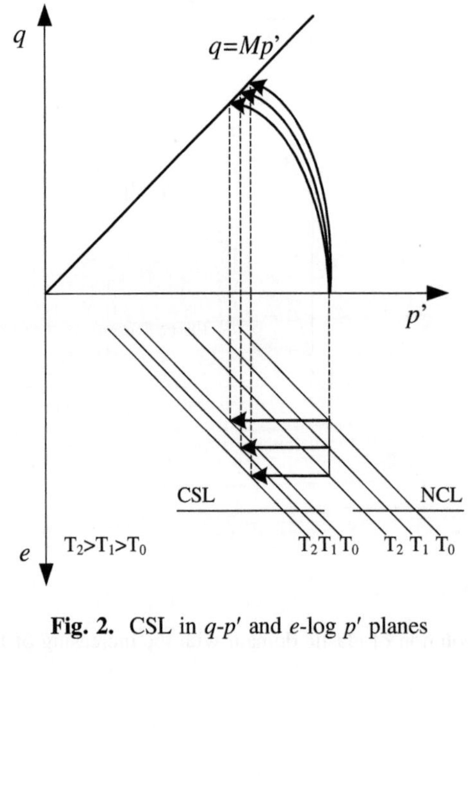

Fig. 2. CSL in q-p' and e-log p' planes

In p-q' plane, the CSL is written as follow:

$$q = Mp' \tag{2}$$

where q is deviatoric stress, p' is the mean effective stress and M is the slope of CSL. The evolution of M with temperature can be used to investigate the effect of temperature on the critical state line in p-q' plane. According to experiments, the evolution of the value of M with the temperature is uncertain, which can increase, stay constant or decrease. In this study, a constant M is assumed as shown in Fig. 2.

3 Temperature Dependent Mechanical Model

Based on the summaries discussed above, a temperature dependent shaped and sized yield surface was proposed

$$\frac{1}{n}\left(\frac{q}{Mp'}\right)^{n+1} + 1 - \frac{p_c'(T)}{p'} = 0 \tag{3}$$

where n is the parameter relating to the effect of temperature on the CSL (implicitly controlling the position of CSL in e-log p' plane [9–11]) and controlling the evolution of yield surface. It should be noted that the yield surface can be reduced to the MCC yield surface for $n = 1$. The parameter n at different temperature levels can be determined by curve-fitting the normalized effective stress path of a normally consolidated specimen. The other parameters are the same with the MCC model. Adopting the usual associated flow rule and the same hardening rule of MCC, the temperature dependent mechanical model can be used to simulate the thermal behavior of saturated clay. Figure 3 illustrate the change in the geometry of the yield surface with the parameter n.

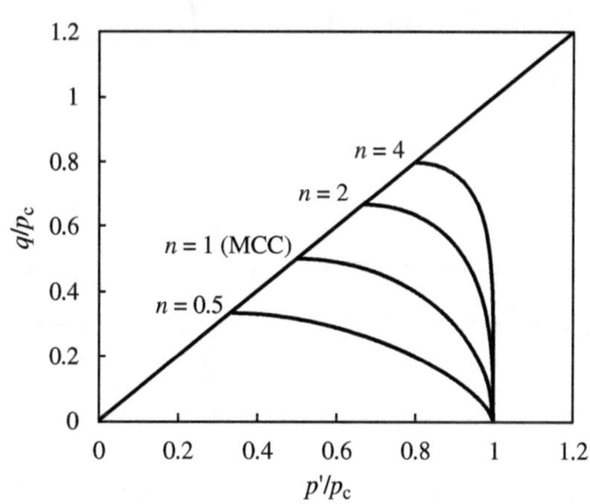

Fig. 3. Change in geometry of yield surface with fabric parameter n

4 Experimental Validation

The triaxial tests conducted by [6] were adopted to verify the predictive ability of the proposed model. The experiment was performed on the commercial Kaolin with triaxial apparatus with controlled temperatures. The samples were obtained from slurry with a water content two times the liquid limits of the soil. For the tests conducted at T = 22°, the samples were firstly consolidated with an initial confining pressure of 600 kPa. Then, the shear procedures were carried out with a strain rate of 0.3%/h in drained conditions. However, for the tests conducted at T = 90°, there is one more consolidation step that is heating the sample from 22 to 90 °C about 4 days. From the experimental results, the values of model parameters are $M = 0.8$, $\lambda = 0.11$, $\kappa = 0.025$. The temperature related parameter n were correlated, and $n = 1$ and $n = 1.5$ are obtained corresponding to $T = 22°$ and 90°, respectively.

Figure 4 shows a comparison between model predictions and test results in two temperatures for normal consolidated and slightly overconsolidated sample (OCR = 1.5) in drained condition. As can be observed from experimental results in Fig. 4a and b, the samples for OCR = 1 give a high strength than that OCR = 1.5. The model can capture the trend of the stress-strain curve, but slightly overestimates the strain which may due to the assumption of a constant M. Figure 4c and d also shows model predictions for volume

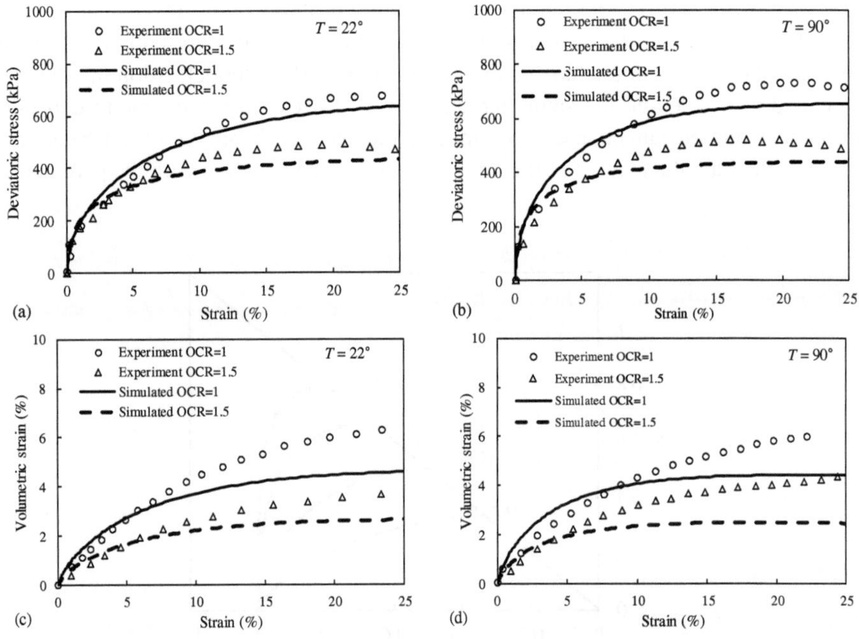

Fig. 4. Model predictions for drained triaxial tests on samples of Kaolin with OCR = 1 and OCR = 1.5: (a) deviatoric stress versus strain at $T = 22°$; (b) deviatoric stress versus strain at $T = 90°$; (c) volumetric strain versus strain at $T = 22°$; (d) volumetric strain versus strain at $T = 90°$

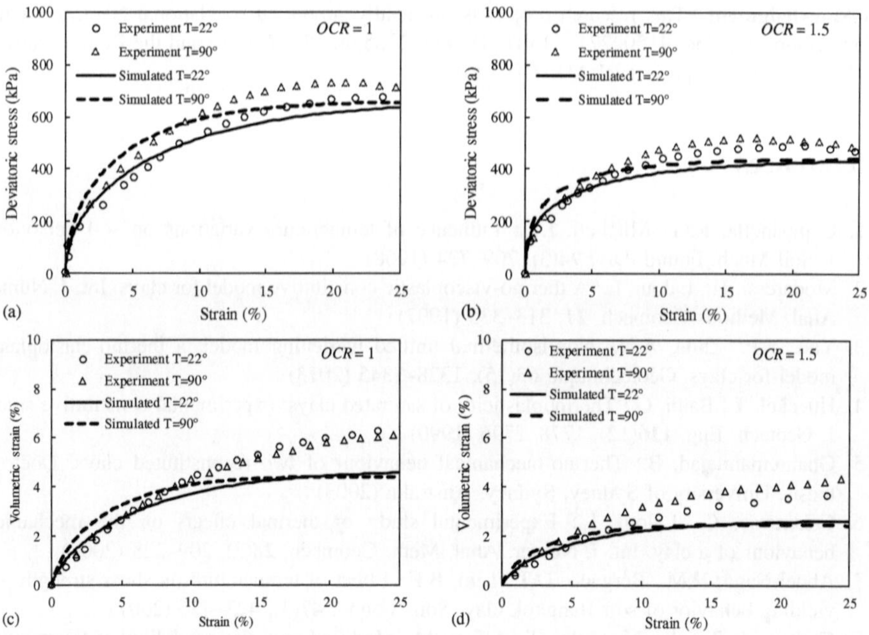

Fig. 5. Model predictions for drained triaxial tests on samples of Kaolin at different temperature: (a) deviatoric stress versus strain at $OCR = 1$; (b) deviatoric stress versus strain at $OCR = 1.5$; (c) volumetric strain versus strain at $OCR = 1$; (d) volumetric strain versus strain at $OCR = 1.5$.

changes. Lightly consolidated samples (OCR = 1.5) show smaller contraction comparing with the NC samples. The simulations estimate a lower volumetric strain.

The above results were replotted by comparing the difference for a varying of temperatures. A clear increase in strength can be observed by increasing the temperature as shown in Fig. 5a for NC samples and increasing of OCR presents the similar results (Fig. 5b). Heating of the normally consolidated samples produced a slight change of the volumetric strain (Fig. 5c). However, heating of the slightly consolidated samples (OCR = 1.5) shows larger volumetric strain in Fig. 5d.

5 Conclusion

In this paper, a constitutive model was proposed for predicting the temperature dependent behavior of saturated clays. It was formulated based on critical-state soil mechanics and the Modified Cam-Clay model. One temperature related parameter was added which can be correlated by the effective strain path. To verify the model, simulations have been compared with drained triaxial tests conducted on kaolin clay at different temperatures.

For ongoing research, it is desired to extend the current framework for predicting the mechanical behavior of stiff clays and partial saturated soil in elevated temperatures.

Acknowledgment. This research project is financially supported by National Natural Science Foundation of China (41502271, 51504245, 41372285 and 51579179), and the Region Pays de la Loire of France (project RI-ADAPTCLIM).

References

1. Campanella, R.G., Mitchell, J.K.: Influence of temperature variations on soil behaviour. J. Soil Mech. Found. Div. **94**(3), 709–734 (1968)
2. Modaressi, H., Laloui, L.: A thermo-viscoplastic constitutive model for clays. Int. J. Numer. Anal. Methods Geomech. **21**, 313–335 (1997)
3. Yao, Y.P., Zhou, A.N.: Non-isothermal unified hardening model: a thermo-elastoplastic model for clays. Géotechnique **63**(15), 1328–1345 (2013)
4. Hueckel, T., Baldi, G.: Thermoplasticity of saturated clays: experimental constitutive study. J. Geotech. Eng. **116**(12), 1778–1796 (1990)
5. Ghahremannejad, B.: Thermo-mechanical behaviour of two reconstituted clays. Doctoral thesis, University of Sydney, Sydney, Australia (2003)
6. Cekerevac, C., Laloui, L.: Experimental study of thermal effects on the mechanical behaviour of a clay. Int. J. Numer. Anal. Meth. Geomech. **28**(3), 209–228 (2004)
7. Abuel-Naga, H.M., Bergado, D.T., Lim, B.F.: Effect of temperature on shear strength and yielding behavior of soft Bangkok clay. Soils Found. **47**(3), 423–436 (2007)
8. Graham, J., Tanaka, N., Crilly, T., Alfaro, M.: Modified cam-clay modelling of temperature effects in clays. Can. Geotech. J. **38**, 608–621 (2001)
9. Yin, Z.-Y., Chang, C.S.: Non-uniqueness of critical state line in compression and extension conditions. Int. J. Numer. Anal. Meth. Geomech. **33**(10), 1315–1338 (2009)
10. Yin, Z.-Y., Xu, Q., Hicher, P.-Y.: A simple critical state based double-yield-surface model for clay behavior under complex loading. Acta Geotech. **8**(5), 509–523 (2013)
11. Yin, Z.-Y., Hicher, P.Y., Dano, C., Jin, Y.F.: Modeling the mechanical behavior of very coarse granular materials. J. Eng. Mech. ASCE **143**(1), C401600 (2017)

A Modified Method for Estimating the Thermal Conductivity of Sands

Aaron J. Rubin(✉) and Carlton L. Ho

University of Massachusetts Amherst, Amherst, USA
arubin@umass.edu

Abstract. Estimation of soil thermal conductivity is important for many practical geotechnical and civil engineering applications including the thermal performance of buried pipelines and geothermal heat pumps. A significant number of empirical-fit methods to estimate the thermal conductivity of sands exist in the existing literature. This paper provides a modification to an existing semi-analytical model developed by S.K. Haigh published in 2012. Haigh (2012) proposed an analytical 2D model developed based on unidirectional heat flow through a three-phase soil element. Haigh (2012) added an empirical fit based on a database of 155 test measurements to the analytical model to improve performance. This paper provides an alternative empirical fit to the analytical method to better retain the original physical origin of the model. The new model is validated with an additional 124 test measurements.

Keywords: Theoretical analysis · Thermal conductivity · Sand properties

1 Introduction

Thermal conductivity is an important soil property for a number of applications such as building insulation, geothermal HVAC applications, ground freezing, seabed pipelines carrying hot oil or gas and buried electric cables. Unfortunately, estimation of the thermal conductivity of soils remains challenging. There is a history of empirically derived correlations for thermal conductivity dating back to the late 1940s. Kersten (1949) proposed thermal conductivity correlations based on the soil dry density and water content for silt-clay and sandy soil mixtures. De Vries (1952) proposed a correlation for course grained soils between 10% and 20% saturation. Gemant (1952) proposed a method for estimating thermal conductivity based on the water content, thermal conductivity of the solids, and thermal conductivity of water. Van Rooyen and Winterkorn (1957) investigated the relationship between thermal conductivity of sands and gravels with saturation between 1.5% and 10%. Johansen (1975) created a predictive method for soils at any saturation between 20% and 100% if the conductivity was known at the fully saturated and dry states. More recently, similar work has been done by researchers suchs as Côté and Konrad (2005); Lu et al. (2007); Chen (2008); Lu and Dong (2015).

A recent model based on the micro-structure of soil was proposed by Haigh (2012) that presented a theoretical analysis based on thermal conduction through a simple soil element containing a soil particle, water and air. Validation of the model was done with

X. Shi et al. (Eds.): GSIC 2018, *Proceedings of GeoShanghai 2018 International Conference: Transportation Geotechnics and Pavement Engineering*, pp. 63–70, 2018.
https://doi.org/10.1007/978-981-13-0011-0_7

data collected previously by others. In particular, the development was heavily reliant on the dataset produced by Chen (2008). One shortcoming of Haigh (2012) was that after a significant amount of effort to establish an analytical model, the translation from 2D to 3D was done by empirical tuning of the model using the Chen (2008) dataset. Despite this shortcoming, Haigh (2012) is better than a purely empirical model since the core is a simple physical idealization. In this paper, a modification is made to the translation from 2D to 3D that improves the overall performance of the model. The model is then validated using a new thermal conductivity dataset.

2 Review of Haigh (2012) Model

Haigh (2012) is based on a highly simplified geometrical idealization of soil microstructure as shown in Fig. 1. The model assumes one-dimensional heat flow between two equally sized spherical particles of radius, R. In order to describe the fraction of water and air in the geometry, the parameters β and ξ are introduced. These parameters may be calculated based on saturation ratio and void ratio.

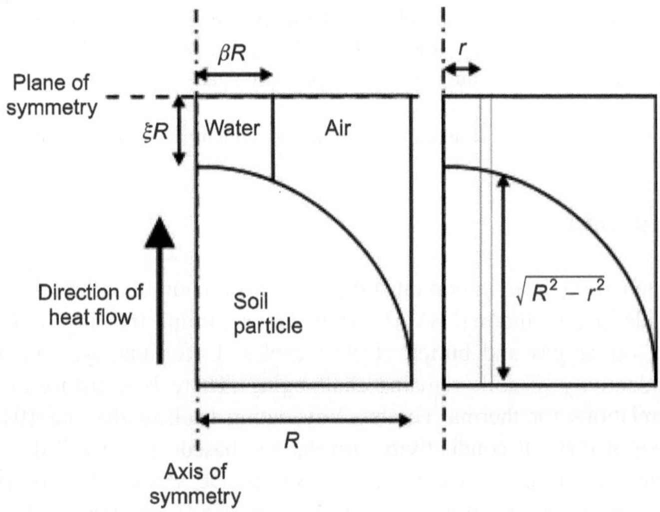

Fig. 1. Geometry of axisymmetric contact model from Haigh (2012).

For mathematical simplicity, the model assumes heat flow is confined to a specific, single, direction parallel to the axis of a cylindrical cell. The overall thermal conductivity of the cell, k, can be calculated by integrating over a series of parallel cylindrical shells. After integrating from 0 to βR, the following expression was developed:

$$\frac{k}{k_{solid}} = 2(1+\xi)^2\left\{\frac{a_w}{(1-a_w)^2}ln\left[\frac{(1+\xi)+(a_w-1)x}{\xi+a_w}\right]+\frac{a_a}{(1-a_a)^2}ln\left[\frac{(1+\xi)}{(1+\xi)+(a_a-1)x}\right]\right\} + \frac{2(1+\xi)}{(1-a_w)(1-a_a)}[(a_w-a_a)x-(1-a_a)a_w] \tag{1}$$

Where:

$$\xi = \frac{2e-1}{3} \tag{2}$$

and

$$x = \left(\frac{1+\xi}{2}\right)\left(1+\cos\theta-\sqrt{3}\sin\theta\right) \tag{3}$$

and

$$\cos 3\theta = \frac{2(1+3\xi)(1-S_r)-(1+\xi)^3}{(1+\xi)^3} \tag{4}$$

k = thermal conductivity of the cell
e = void ratio
a_a = thermal conductivity of air normalized by k_{solid}
a_w = thermal conductivity of water normalized by k_{solid}

Of important note, is that k is not the bulk thermal conductivity of the soil, only the thermal conductivity result of the prescribed geometry. Additionally, the prescribed heatflow is uni-directional and assumes no preferential heat flow paths. Haigh (2012) plotted estimated thermal conductivity (from 2D geometry), k, versus the measured thermal conductivity, it was found that they did not align 1-to-1. Haigh (2012) estimated that k (from the 2D geometry) underestimated the bulk thermal conductivity by a factor of 1.58. This factor is empirical in nature related to the Chen (2008) data set that was used to develop the model. Haigh (2012) also indicated that based on the geometry of idealized 3D particle orientation, one could calculate this factor as shown in Eq. 5.

$$Factor = 3\frac{\sqrt{\frac{8}{3}}R}{2R}\frac{\pi R^2}{2\sqrt{3}R^2} = \frac{\pi}{\sqrt{2}} \approx 2.2 \tag{5}$$

Haigh (2012) commented that this related favorably to the 1.58 calculated empirically. The bulk thermal conductivity was calculated by multiplying the result of Eq. 1 by k_{solid} and 1.58. Figure 2 shows the Chen 2008 data set and lines representing the 1.58 factor and 2.2 factor above the 1-to-1 line. It is important to recognize that

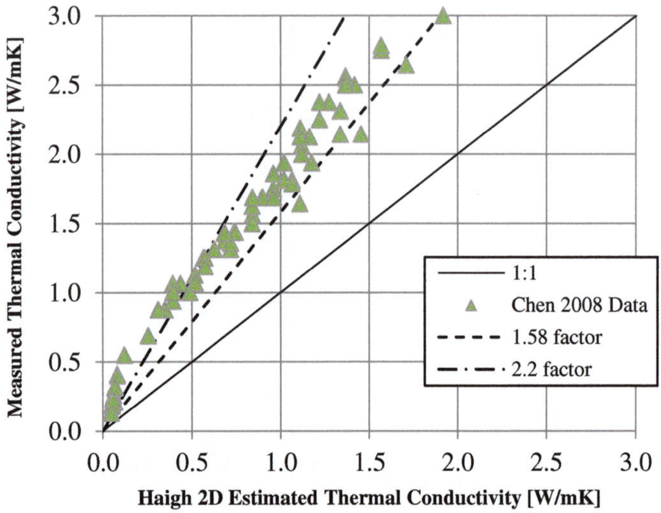

Fig. 2. Comparison of estimated to measured thermal conductivity from Haigh (2012) based on data from Chen (2008).

although Haigh (2012) concluded a constant value factor was appropriate to adjust the model results, by inspection, it is clear that the data do not closely mirror either the 2.2 or 1.58 Factor lines.

3 New Model Factor

3.1 Methodology for the New Factor

As previously stated, the analytical model proposed by Haigh (2012) based on the 2D geometry does not fit to the 1-to-1 line and a correction factor is needed. For each of the Chen (2008) data points, an individual correction factor can be determined by dividing the measured thermal conductivity by the 2D estimated thermal conductivity. These newly calculated correction factors can be plotted against the 2D estimated thermal conductivity as shown in Fig. 3. Note that there appears to be a difference in behavior from the dry data to the data with saturation greater than 10%. For 2D thermal conductivity less than 0.1 W/mK the correction factor for dry soil should be used otherwise the correction factor for saturated soil is appropriate.

3.2 Validation of New Factor

In order to validate that the new factor does a better job of estimating actual measurements, a dataset of 124 new thermal conductivity tests were conducted on five test soils.

Sample Preparation. Test specimens of the selected soils were prepared in a plastic cylindrical mold 15.24 cm (6.0 in.) in diameter by 19.05 cm (7.5 in.) tall. Dry specimens were placed using a funnel in general accordance with ASTM D4254-14

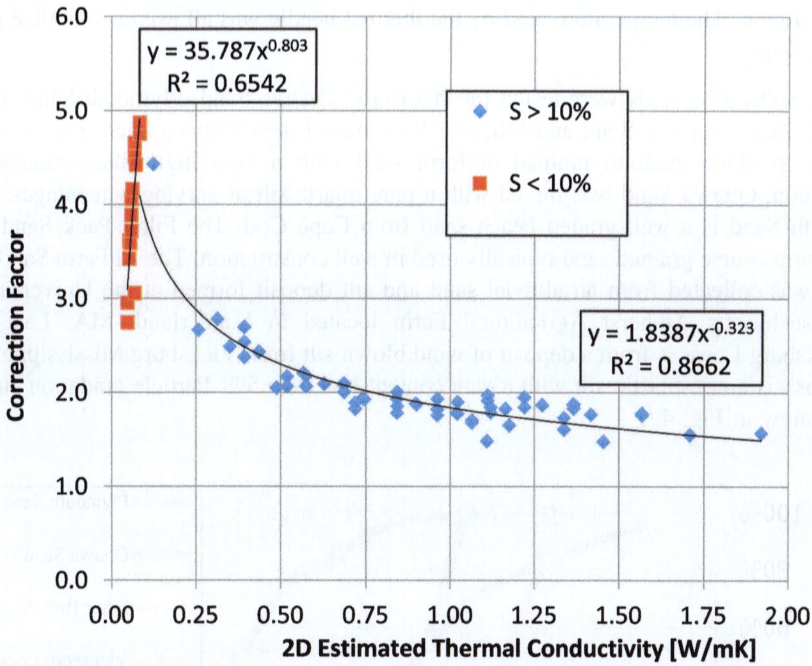

Fig. 3. New correction factor to estimate the thermal conductivity of sands.

Standard Test Methods for Minimum Index Density and Unit Weight of Soils and Calculation of Relative Density to achieve low-density samples. In order to provide samples of increasing density, samples were tamped and/or vibrated. Using these methods a range of specimens roughly ranging from the minimum dry density to the maximum dry density were prepared.

Moist specimens, that were not fully saturated, were prepared by pre-mixing soil to a target water contnet. The soil was then placed by shaking it through a No. 4 sieve (4.75 mm). Dense samples were created by tamping in layers using Standard Proctor Compaction equipment. The number of blows applied to each layer was increased or decreased to achieve the target density.

Wet specimens, that were fully saturated, were prepared by first placing soil dry (as described above) and then saturated from the bottom up. A porous stone separated incoming flow from the prepared specimen. Saturation was done slowly so as to prevent damage to the specimen particle structure. The water added to the specimen was measured by volume and mass so that the water content and saturation could be calculated.

Thermal Conductivity Measurements. Thermal conductivity was measured using a 15 cm thermal needle in accordance with ASTM 5334 Standard Test Method for Determination of Thermal Conductivity of Soil and Soft Rock by Thermal Needle Probe Procedure. The thermal needle was inserted into the prepared specimens after preparation. The temperature read by the thermal needle was allowed to stabilize prior to testing.

Test Soils. Five soils were tested for this study; Ottawa Sand, Plymouth Sand, Filter Pack Sand, AgFarm Sand and Silt, and Vicksburg Loess. Ottawa Sand is a commercially produce medium grained uniform sand with a very high silica content. In addition, Ottawa Sand was mixed with a pure quartz silt at varying percentages. Plymouth Sand is a well graded beach sand from Cape Cod. The Filter Pack Sand is a uniform course grained sand typically used in well construction. The AgFarm Sand and Silt was collected from an alluvial sand and silt deposit formed at the University of Massachusetts Amherst Agricultural Farm located in Sunderland, MA. Last, the Vicksburg Loess is from a deposit of wind blown silt from Vicksburg Mississippi. The deposit is a non-plastic silt with a clay content less than 5%. Particle gradation curves are show in Fig. 4.

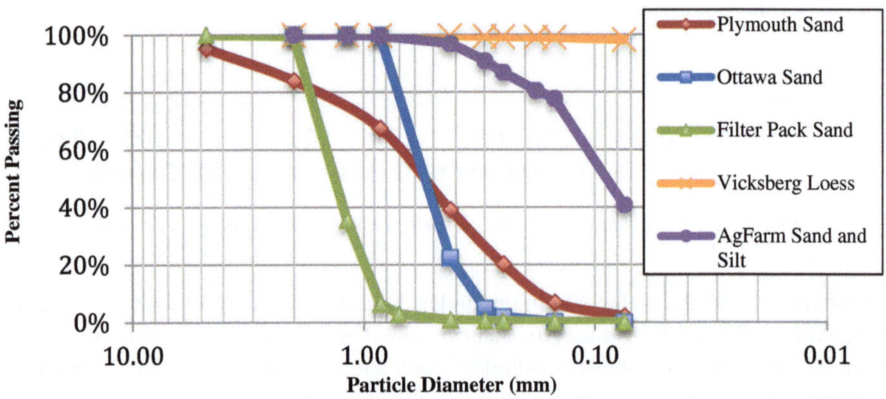

Fig. 4. Particle gradation of test soils.

Results. As can be seen in Figs. 5 and 6 the use of a new correction factor produces better results than the original Haigh (2012) methodology. As can be seen in Fig. 5, the estimated thermal conductivity appears too conservative at values less than 1 W/mK and are overly aggressive in predicting thermal conductivity values greater than 2.5 W/mK. The trend lines of both the Chen (2008) data and the new data in Fig. 5 present a general over estimation of thermal conductivity. Comparing Figs. 5 and 6 it is clear the new correction factor methodology produces a marginally better R^2 value and a markedly better gradient that closely matches the 1-to-1 line for the full range of values. Although there is more scatter in the new dataset, this is to be expected because the soils are more diverse both by mineral type, gradation, and in particle size.

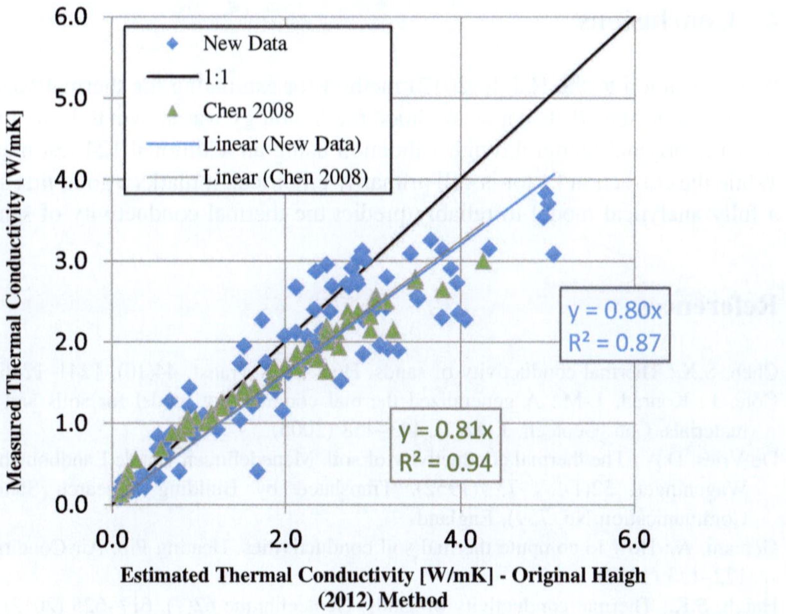

Fig. 5. Estimated versus measured thermal conductivity using original Haigh (2012).

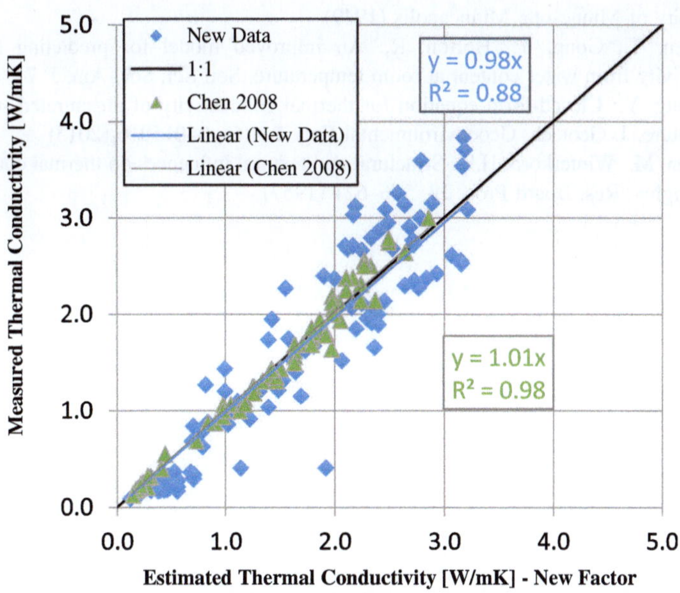

Fig. 6. Estimated versus measured thermal conductivity using modified Haigh (2012).

4 Conclusions

A modification to the Haigh (2012) method for estimating the thermal conductivity of sands was presented. The new modified methodology was shown to be an improvement over the original model through validation using an additional 124 test measurements. While the correction factor is still primarily empirical, it marks a good first step towards a fully analytical model to reliably predict the thermal conductivity of sands.

References

Chen, S.X.: Thermal conductivity of sands. Heat Mass Transf. **44**(10), 1241–1246 (2008)

Côté, J., Konrad, J.-M.: A generalized thermal conductivity model for soils and construction materials. Can. Geotech. J. **42**(2), 443–458 (2005)

De Vries, D.A.: The thermal conductivity of soil. Mededelingen van de Landbouwhogeschool te Wageningen **52**(1), 1–73 (1952). Translated by Building Research Station (Library Communication No. 759), England

Gemant, A.: How to compute thermal soil conductivities. Heating Pip. Air Conditioning **24**(1), 122–123 (1952)

Haigh, S.K.: Thermal conductivity of sands. Geotechnique **62**(7), 617–625 (2012)

Johansen, O.: Thermal conductivity of soils. Ph.D. thesis. Trondheim, Norway (CRREL Draft Translation 637, 1977) ADA 044002 (1975)

Kersten, M.S.: Thermal properties of soils. Bulletin 28. Engineering Experiment Station, University of Minnesota, Minneapolis (1949)

Lu, S., Ren, T., Gong, Y., Horton, R.: An improved model for predicting soil thermal conductivity from water content at room temperature. Soil Sci. Soc. Am. J. **71**, 8–14 (2007)

Lu, N., Dong, Y.: Closed-form equation for thermal conductivity of unsaturated soils at room temperature. J. Geotech. Geoenvironmental Eng. **141**(6), 04015016 (2015)

Van Rooyen, M., Winterkorn, H.F.: Structural and textural influences on thermal conductivity of soils. Highw. Res. Board Proc. **39**, 576–621 (1957)

Assessment of Direct Use of Geothermal Hot Water for Snow-Melting Pavements in Western North Dakota Using Finite Element Method

I-Hsuan Ho(✉)

Harold Hamm School of Geology and Geological Engineering,
University of North Dakota, Grand Forks, ND 58202, USA
ihsuan.ho@engr.und.edu

Abstract. Many advanced snow-melting methods for heating pavements have been studied, and the deep direct use (DDU) of geothermal hot water is considered to be a clean and economical resource for developing snow-melting pavement systems for cold regions. In western North Dakota, the cold climate hinders many economic activities due to the uncertainties associated with transportation. The presence of snow on roads and airport runways is a major problem, and thus, improving snow removal methods and transportation infrastructure in winter is vital to the economy and road safety. The use of geothermal hot water for snow-melting purposes provides a solution to these problems and has many advantages. In western North Dakota, the temperatures of geothermal hot water range between 30 °C to 130 °C, and water at these temperatures is available between 700 m and 3,500 m deep in several aquifers. These geothermal aquifers have been identified and are found to be able to supply good quality and abundant hot water. This paper focuses on two populated locations above these easily accessible aquifers. The paper considers the feasibility of a snow-melting system that uses DDU geothermal hot water based on heat transfer and the available heat supply. The heat requirements for melting snow are computed based on the climatic conditions in the studied areas using the Chapman and Katunich equations. Finite element analysis using COMSOL is employed to model the pavement heating processes, from pipes embedded in a concrete panel to the pavement surface according to the computed heat requirements. The heat supply also takes into account the cascading use of geothermal hot water for different temperatures, pipe properties, mass flow rates, reasonable water pressures, snow-melting efficiency, and the properties of concrete pavement panels. The feasibility of the proposed snow-melting pavement panels is assessed and an optimized design is recommended.

Keywords: Deep direct use · Snow-melting · Geothermal aquifers
Finite element analysis · COMSOL

1 Introduction

Although many advanced snow-melting techniques have been studied for pavement surfaces, many of them are either too energy consuming or not practical. Currently, advanced snow-removal systems used in pavements include electric heating, hydronic

X. Shi et al. (Eds.): GSIC 2018, *Proceedings of GeoShanghai 2018 International Conference: Transportation Geotechnics and Pavement Engineering*, pp. 71–80, 2018.
https://doi.org/10.1007/978-981-13-0011-0_8

systems, and shallow geothermal energy extraction [3–5]. These snow-melting methods for pavements avoid the traditional mechanical or chemical de-icing processes. However, these systems have disadvantages; some are not efficient and some are not economical.

Direct-use geothermal hot water has drawn significant attention and may be the most economical, clean, and efficient method for melting snow on pavements according to Lund [6]. The purpose of deep direct-use (DDU) geothermal energy is to achieve the goal of net-zero energy consumption for the nation's infrastructure. In cold regions, travel becomes especially difficult in winter, especially in states such as North Dakota. Without doubt, the major problem related to traveling in winter in cold regions is the presence of snow and ice on roadways and airport runways. Hence, if an efficient snow removal system could be developed, especially pavement panels that use renewable energy, such a pavement could revolutionize travel in cold regions.

In western North Dakota, obtaining DDU geothermal hot water involves extracting hot water from a deep geothermal reservoir at a temperature between 38 °C and 130 °C. For civil engineering purposes, and considering pipe materials and efficiency, the suggested temperature for snow-melting or de-icing is between 20 °C and 55 °C according to the United States Department of Energy. Hence, based on the locations of accessible geothermal hot water, urban populations, and existing co-production wells used in the petroleum industry, two locations in western North Dakota, i.e., Minot and Bismarck, were studied for this paper. The available geothermal resources are well-documented for these studied locations. Near Minot, the temperature is 40 °C at 700 m and 60 °C at 1,300 m deep. Near Bismarck, the temperature is about 40 °C at 800 m and 60 °C at 1,300 m deep. The hot water can be accessed either by drilling horizontal wells in these aquifers or via co-production wells used in the oil industry or abandoned wells [2]. According to previous research conducted by Lund [6], if hot water is easily accessible, using DDU geothermal hot water may be the most economical snow-removal method because installation and maintenance costs will be lower than for other methods.

2 Available Temperatures of Geothermal Hot Water

In western North Dakota, geothermal resources are abundant and hot water can be accessed easily from shallow aquifers that are present in geological formations. Moreover, some co-production wells for petroleum purposes also exist. Table 1 presents details about these aquifers that contain available geothermal hot water and the depths of each aquifer. Note that, because the aquifer near Dickinson, ND contains geothermal hot water at higher temperatures (81 °C to 137 °C) than the Minot and Bismarck aquifers, and because this study aims to investigate the feasibility of using water temperatures at or below 60 °C, this paper focuses only on Minot and Bismarck. In other words, based on the potential demands for snow-melting purposes and easily accessible geothermal hot water, two sites, Minot and Bismarck, are the focus of this paper.

Table 1 includes the water temperatures in the Madison aquifer in western North Dakota. This geothermal reservoir is located 700 m to 3,500 m below the ground

Table 1. Temperatures (°C) at the tops and bottoms of the six aquifers beneath the studied sites (Gosnold et al. 2017)

Hydrostratigraphy	Minot		Bismarck	
	Temperature, °C	Depth, m	Temperature, °C	Depth, m
Cretaceou aquitard				
Dakota aquifer	34	638	36	671
	44	850	42	806
Jr. Tr. Perm. AQUITARD				
Pennsylvanian aquifer	58	1224	56	1212
	61	1350	59	1290
Mississippian aquitard				
Madison aquifer	61	1384	60	1347
	63	1458	62	1429
Bakken aquitard				
Devonian aquifer	67	1608	67	1556
	71	1730	71	1690
Prairie aquiclude				
Winnipegosis aquifer	77	2189	73	1974
Basal aquitard				
Basal Aquifer	82	2483	74	2097
	85	2710	77	2237

surface [2]. Two major cities, Bismarck and Minot, are located over this geothermal aquifer that has been proven to supply sufficient hot water at suitable temperatures for cascading use. The temperatures of this aquifer near Bismarck are about 60 °C at a depth of about 1,300 m and 40 °C at a depth of 700 m near Minot.

The current technique used to assess geothermal hot water is either to utilize open-hole lateral wells with horizontal drilling to access hot water or to use co-production wells with oil and gas. A successful project conducted by University of North Dakota researchers indicates that the former method can produce water at a combined volumetric flow rate of 51 l/s (800 gpm) and the latter can co-produce water at 0.2 l/s (3 gpm) on average or 3.3 l/s (50 gpm) for the average of the top 100 co-production wells.

3 Heat Requirements for Melting Snow

Most similar snow-melting designs that have been proposed are based on steady-state conditions. Variables such as the snowfall rate, ambient air temperature, and wind speed often are considered based on weather data compiled over several years. These data used for heat requirement calculations typically are based either on the average or upper intensity during a snowfall event. The Chapman and Katunich equation [1] assumes steady-state conditions and is used to compute the amount of heat that is needed per square foot to melt snow. The Chapman and Katunich equation can be expressed mathematically as shown in Eq. (1).

$$q_o = q_s + q_m + A_r * (q_h + q_e) \tag{1}$$

where q_o is the total heat flux per unit area of the surface, Btu/h·ft^2 (W/m^2); q_s is the total sensible heat flux to bring snow to a melting point, Btu/h·ft^2 (W/m^2); q_m is the melting load, Btu/h·ft^2 (W/m^2); A_r is the ratio of the snow-free area to the total area (snow-free ratio); q_h is the sum of the convection and radiation losses, Btu/h·ft^2 (W/m^2); and q_e is the evaporative loss, Btu/h·ft^2 (W/m^2).

In order to use the Chapman and Katunich equation to estimate the heat that is needed to melt snow, a few assumptions must be made concerning precipitation, area ratio, etc. Nonetheless, even though the equation may have some limitations, it still provides a good estimation of the heat required for a snow-melting system. Details regarding the use of the equation to estimate heat requirements can be found in papers by Lund [6] and Liu [7]. In this study, estimations of the required heat are based on the transient weather conditions and climatic conditions in two cities in western North Dakota, Minot and Bismarck. Details are provided as follows.

The heat transfer that is involved in the reasonable heating of snow can be expressed as Eq. (2) through (6).

$$q_s = \rho s c (t_f - t_a) \tag{2}$$

where ρ is the density of the liquid water equivalent of snow; s is the snowfall rate in cm or inches of water equivalent; c is the specific heat of ice, 2010 J/kg-°C; and t_f and t_a are the water film temperature and ambient temperature, respectively, °C.

$$q_m = \rho\ s\ h_f \tag{3}$$

where ρ is the density of liquid water; s is the snowfall rate in cm or inches of water equivalent; and h_f is the heat of fusion, 33 J/g.

$$q_e = (c_1 V + c_2)(P_{wv} - P_{av}) h_{fg} \tag{4}$$

where c_1 is a constant (0.0187 s^2/m^2); c_2 is a constant of 0.0229 s/m; V is wind speed, m/s; P_{wv} is the partial pressure of water vapor in the saturated air film on the surface, kPa; P_{av} is the partial pressure of water vapor in ambient air, kPa; and h_{fg} is the heat of the vaporization of water (2,257 J/g).

$$q_h = c_3 (c_1 V + c_2)(t_f - t_a) \tag{5}$$

where c_3 is a constant, 155.5 W/(m·s·K).

Of particular interest is the ratio of the snow-free area to the heated area. Snow has an insulating effect on the pavement, thereby reducing the amount of heat that is lost to convection. Although the proposed model assumes that the snow cover remains constant, the snow cover would, of course, be changing throughout the snow-melting process.

Heat is required to melt snow and for evaporation [8]. If the pavement surface is free of snow, the heat will be transferred from the surface to the atmosphere by convection and radiation. When the snow on the surface is being warmed, but before it has melted, the snow acts as insulation. Thus, the ratio of the snow-free area, A_f, to the total area, A_t, is defined as the free area ratio, A_r. The free area ratio equation can be written as Eq. (6) [1, 5].

$$A_r = \frac{A_f}{A_t} \tag{6}$$

For $A_r = 1$, the snow must be totally melted on the pavement surface, whereas $A_r = 0$ represents the surface completely covered with snow. In practical terms, the free area ratio can reasonably be assumed to be a value between 0.5 and 1.0.

Q_o is the total heat requirement, adjusted for 25% conductive heat loss on the bottom and edges of the boundaries of concrete pavement panels. The heat required for melting snow for design purposes, Q_O, is mathematically expressed as Eq. (7).

$$Q_0 = q_o \cdot (1.25) \tag{7}$$

4 Numerical Analysis

COMSOL software was employed to conduct the finite element analysis in this study. Given the available geothermal hot water in western North Dakota and the requirements for the heat to melt snow on pavement surfaces, the feasibility of using geothermal hot water is based on the temperature difference of the geothermal water between the inlet and outlet, reasonable water pressure control in the pipes, and the pavement surface to be heated. In this study, the temperatures of the geothermal hot water used for analysis were 30 °C, 40 °C, 50 °C, and 60 °C.

4.1 Available Geothermal Hot Water

The available geothermal hot water in the reservoirs adjacent to Minot and Bismarck in western North Dakota has been studied previously, and the water quality and quantity have been proven to be good [2]. In Minot, the available temperatures are between 34 °C and 44 °C at depths between 638 m and 850 m. At the other studied site, Bismarck, the available temperatures of the water range from 36 °C to 42 °C at depths between 671 m and 806 m. If a higher temperature is desired, the water temperature as high as 77 °C is also available at the depth about 2,200 m. (Details regarding the available hot water and depths of these aquifers can be found in Table 1.) Thus, for the finite element analysis, assuming water temperatures from 30 °C to 60 °C is reasonable, and will cost the pumping this water for use in a snow-melting system is a cost-effective solution.

4.2 Heat Requirements for Melting Snow

The heat requirements for the proposed snow-melting pavement panels take into account 15-year statistical weather data from NOAA for the studied sites. The heat requirements also depend on how the system will respond to snow events. As stated, the heat requirement calculations are based on the Chapman and Katunich equation [1]. The weather conditions considered for the proposed snow-melting system include air temperature, precipitation, dew point, wind speed, etc. Besides the weather conditions, the heat needed also includes the free area ratio, A_r. The parameter, A_r, depends on how 'clean' the pavement will be during a snow event. In this paper, a free area ratio, $A_r = 0.5$, is considered for the two studied cities. The heat requirements calculated are based on the four coldest months in a year. Table 2(a) and (b) summarize the heat requirements for Minot and Bismarck, respectively.

Table 2. Heat requirements for melting snow on pavement panels for area ratio, Ar = 0.5

(a) Minot					
	January	February	March	November	December
Temperature, °C	−10.1	−9.6	−2.3	−1.1	−8.6
Wind speed, m/s	5.5	5.0	5.0	5.1	5.0
Snowfall, mm/s	0.00072	0.00072	0.00197	0.00150	0.00086
Snow ratio, A_r	0.5	0.5	0.5	0.5	0.5
Heat required, W/m^2	404.9	386.9	700.8	522.3	422.0
q_0, W/m^2	506.1	483.7	876.0	652.9	527.5
(b) Bismarck					
	January	February	March	November	December
Temperature, °C	−9.9	−8.7	−0.8	−0.5	−7.9
Wind speed, m/s	4.0	3.9	4.4	3.8	3.9
Snowfall, mm/s	0.00089	0.00089	0.00264	0.00142	0.00158
Snow ratio, A_r	0.5	0.5	0.5	0.5	0.5
Heat required, W/m^2	428.68	412.19	894.5	482.24	642.55
q_0, W/m^2	535.9	515.2	1118.1	602.8	803.2

4.3 Finite Element Analysis

Finite element analysis was conducted in this study using COMSOL. Figure 1 shows a three-dimensional view of the embedded pipes and concrete panel used in the proposed snow-melting pavement system. The analysis mainly considers the feasibility of using geothermal hot water based on the heat transfer from the flowing water to the pipes and from the pipes to the concrete surface. The size of the concrete panel is assumed to be 24 ft by 24 ft (7.32 m by 7.32 m) with the thickness of 1 ft (0.3 m). The thermal conductivity of the concrete used in the analysis is 1.8 W/m K, and its specific heat at a constant pressure is 4,186 J/kg K. The temperatures of the water studied are 30 °C and 40 °C, which assumes that the water is pumped from the shallowest aquifer, the Dakota Aquifer. The embedded pipes in the concrete are assumed to be made of polyurethane, which is known specifically as crosslinked polyethylene (PEX). In the numerical modeling, the pipes have an outer diameter, $D_o = 28.6$ mm, and inner diameter, $D_i = 22.2$ mm, well thickness, $t_w = 3.3$ mm, thermal conductivity, k = 0.46 W/m K, and volumetric flow rate, $q_v = 0.00038\ m^3/s$. Since the geomechanical properties are not included in the analysis, the embedment depths of the pipes in the concrete is assumed to be 3 in. which was found to be most efficient in heat transfer in the analysis. The initial conditions are assumed to be that the temperature of the concrete is equal to the air temperature and the initial temperature of the water is 0 °C. As for the boundary conditions, the heat flux is allowed only on the top surface to melt the snow, and the inlet temperature of the water is defined to be the temperature of the available geothermal hot water.

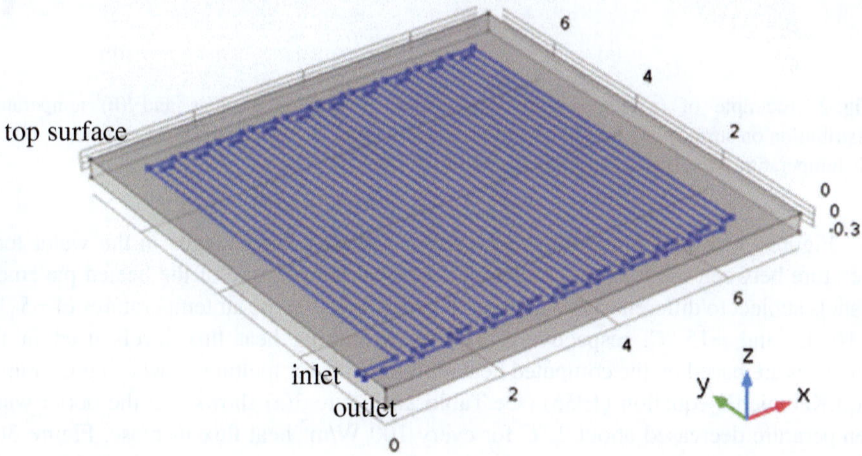

Fig. 1. Three-dimensional view of a snow-melting concrete pavement panel with embedded pipes in the finite element model

5 Analysis Results

The COMSOL analysis involves two physics concepts for heat transfer within a snow-melting pavement system: non-isothermal pipe flow and heat transfer in solids. The computed results consider only the difference between the inlet and outlet temperatures in the embedded pipe and the temperature of the heated pavement after the system has run for a few hours. Figure 2 presents analysis examples that show that the temperature of the hot water circulating in the embedded pipes decreased. Figure 2 also shows that, for the temperature distribution on the surface of the heated pavement when using the water temperature of 60 °C at the inlet, the air temperature is −10 °C and the heat supply is 400 W/m^2. The results indicate that the water temperature dropped from 60 °C to 40.5 °C, and the surface temperature of the heated pavement in most areas was about 20 °C.

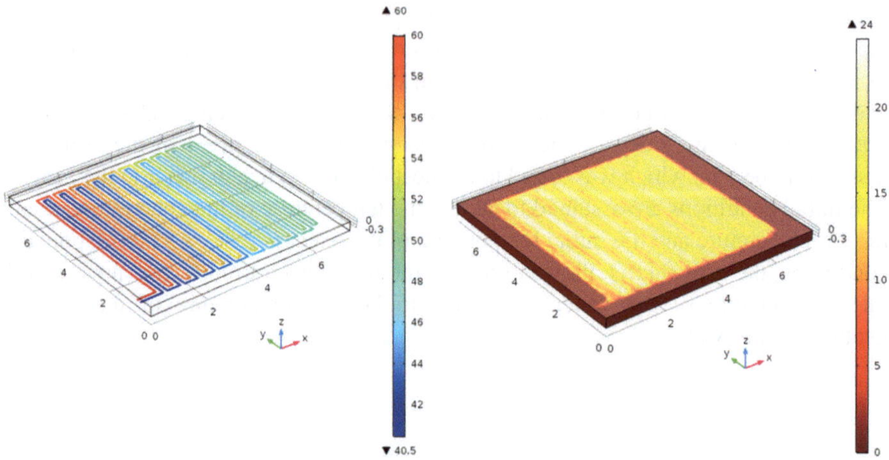

(a) (b)

Fig. 2. Example of (a) temperature distribution in embedded pipes and (b) temperature distribution on surface of a heated pavement panel in finite element analysis (inlet water = 60 °C, air temperature = −10 °C, and heat flux = 400 W/m^2).

Figures 3, 4, and 5 present the analysis results for the changes in the water temperature between the inlet and outlet and the final temperature of the heated pavement panels subject to different snow-melting heat flux levels at the air temperatures of −5 °C, −10 °C, and −15 °C, respectively. The snow-melting heat flux levels used in the analyses are based on the computed heat requirements for melting snow using Chapman and Katunich's equation (1956) (see Table 2). Figure 3(a) shows that the outlet water temperature decreased about 2 °C for every 100 W/m^2 heat flux increase. Figure 3(b) shows that, if 600 W/m^2 snow-melting heat flux is needed, the water at 40 °C lowers the surface temperature of the pavement to below 0 °C. The water temperature of 30 °C does not allow for required heat at 500 W/m^2 and 600 W/m^2. Figure 4(b) shows that the use of water at 30 °C cannot supply the heat needed for all the snow-melting heat flux levels that are assumed to melt the snow in the analysis; that is, water at 40 °C is not

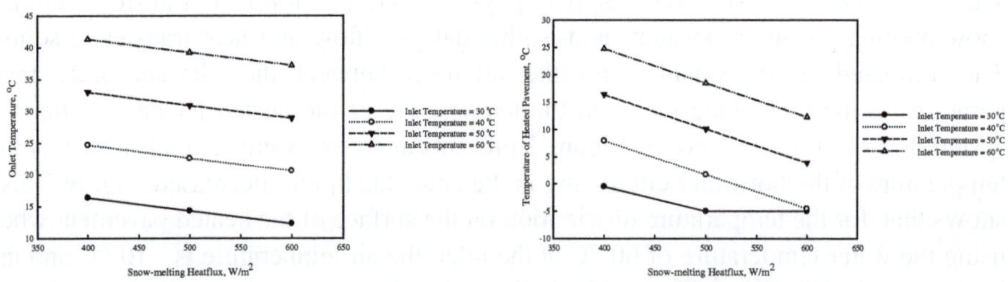

(a) (b)

Fig. 3. (a) Outlet temperature of water in pipes and (b) temperature of heated pavement panel for different snow-melting heat supply requirements at air temperature = −5 °C.

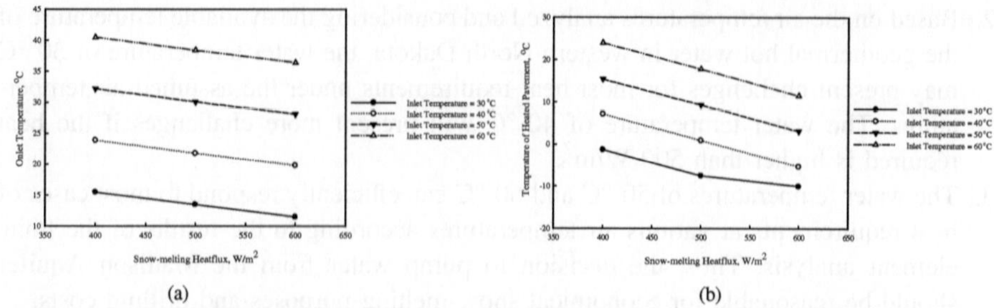

Fig. 4. (a) Outlet temperature of water in pipes and (b) temperature of heated pavement panel for different snow-melting heat supply requirements at air temperature = −10 °C.

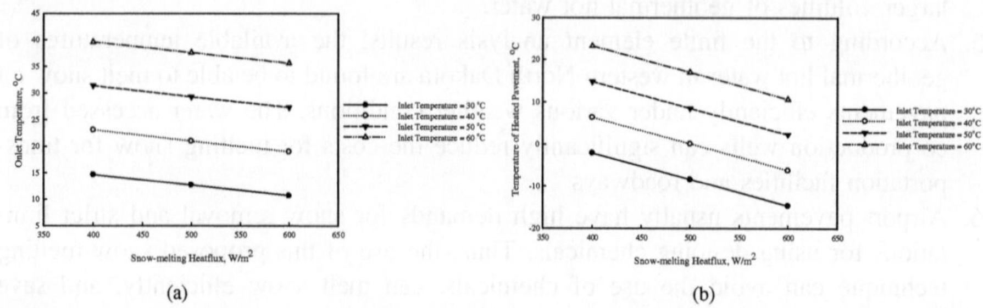

Fig. 5. (a) Outlet temperature of water in pipes and (b) temperature of heated pavement panel for different snow-melting heat supply requirements at air temperature = −15 °C.

sufficient to heat the pavement surface to a temperature above 0 °C. Although Fig. 5(a) shows that water temperatures from 30 °C to 60 °C can supply the needed heat from 400 W/m^2 to 600 W/m^2, the results shown in Fig. 5(b) indicate that the water temperature of 30 °C is unable to heat the pavement to or above 0 °C regardless of the heat flux assumed. The use of water at 40 °C can heat the pavement only up to −6.1 °C at the air temperature of −15 °C if the heat of 600 W/m^2 is required.

6 Conclusions

The finite element analysis results for the proposed snow-melting system that employs DDU geothermal hot water provide insights into the feasibility of using geothermal hot water for energy efficiency. Considering the temperature drop of hot water circulating in pipes embedded in cold concrete and the final temperatures of the heated pavement panels, several conclusions can be drawn, as follows.

1. The use of DDU geothermal hot water in western North Dakota is proven to be able to melt snow under most climatic conditions in western North Dakota. The water can be accessed economically through co-production wells or by repurposing abandoned wells.

2. Based on the air temperatures analyzed and considering the available temperature of the geothermal hot water in western North Dakota, the water temperature of 30 °C may present challenges for most heat requirements under the assumed air temperatures. The water temperature of 40 °C will present more challenges if the heat required is higher than 500 W/m^2.
3. The water temperatures of 50 °C and 60 °C can efficiently respond to most cases of heat requirements at various air temperatures according to the results of the finite element analysis. Thus, the decision to pump water from the Madison Aquifer should be reasonable for economical snow-melting purposes and drilling costs.
4. The cascading use of geothermal hot water has drawn the attention of many agencies and is suitable for traffic-related infrastructure for snow-melting purpose because it requires only a low volumetric flowrate. Other heat demands require larger volumes of geothermal hot water.
5. According to the finite element analysis results, the available temperatures of geothermal hot water in western North Dakota are found to be able to melt snow on pavements efficiently under various weather conditions. The water accessed from co-production wells can significantly reduce the costs for melting snow for transportation facilities and roadways.
6. Airport pavements usually have high demands for snow removal and strict limitations for using de-icing chemicals. Thus, the use of this proposed snow-melting technique can avoid the use of chemicals, can melt snow efficiently, and save energy.

Acknowledgements. The author would like to express appreciation for the North Dakota Established Program to Stimulate Competitive Research (ND EPSCoR) program that supported this research and would like to thank Mr. Mark Dickson for his efforts with regard to related research preparations.

References

1. Chapman, W.P., Katunich, S.: Heat requirements of snow melting systems. ASHAE Trans. **62**, 359–372 (1956)
2. Gosnold, W., Crowell, A., Nordeng, S., Mann, M.: Co-produced and low-temperature geothermal resources in the Williston Basin. GRC Trans. **39**(653–660), 2015 (2015)
3. Ho, I.H., Dickson, M.: Assessment of pavement snow-melting system using geothermal energy in cold region. In: 1st International Conference for Geo-Energy and Geo-Environmental, GeGe 2015, Hong Kong (2015)
4. Ho, I.H., Dickson, M.: Assessment of geothermal snow-melting system used in cold region area. In: Energy Geotechniques 2016, pp. 113–117. Taylor & Francis, London (2016)
5. Ho, I.H., Dickson, M.: Numerical modeling of heat production using geothermal energy for a snow-melting system. J. Geomech. Energy Environ. **10**, 42–51 (2017)
6. Lund, J.W.: Pavement snow melting. Geo-Heat Cent. Q. Bull. **21**(2), 12–19 (2000)
7. Liu, X.: Development and experimental validation of simulation of hydronic snow melting systems for bridges. Master's thesis, Tongji University, China (1998)
8. Zwarycz, K.: Snow melting and heating systems based on geothermal heat pumps at Goleniow airport. Poland Transportation Report (2002)

Procedure for Establishing a 3D Geological Model for Singapore

Xiaohua Pan[1], Wei Guo[1], Zarli Aung[1], Aung KoKo Nyo[1], Kiefer Chiam[2], Defu Wu[2], and Jian Chu[1(✉)]

[1] School of Civil and Environmental Engineering, Nanyang Technological University, Singapore 639798, Singapore
CJCHU@ntu.edu.sg

[2] Building and Construction Authority, Singapore 608550, Singapore

Abstract. A project to establish a 3D geological model in Singapore using available borehole data and other geological/geotechnical data has been initiated. This paper presents the background and procedure adopted for this project. A fence diagram and the resulting 3D geological model for a selected zone in Singapore are presented. Several issues related to the modelling are discussed. The quality and quantity of the borehole data have a direct influence on the accuracy of the 3D model created. A data-cleaning procedure is required to remove typographical errors or duplicates from the database. With a good understanding of the complete geological sequence and the likely geomorphological evolution of the area under study using the constructed fence diagrams, the 3D geological model can be built layer by layer after the geological boundaries have been constructed. The interpreted cross-sections and the constructed fence diagrams can help geologists to have a better understanding of the complex sub-surface profiles in a three dimensional way. It will become a design tool for future city planning and underground constructions.

Keywords: Singapore · 3D geological modeling · Geo-database
Gridline · Fence diagram

1 Introduction

Understanding the geological formation and the spatial distributions of soil or rock profiles is vital to geotechnical design, particularly for the underground constructions and urban planning [1, 2]. As most of the information on the subsurface is discrete on account of the nature of boreholes [3], it is necessary to establish a three-dimensional (3D) geological model using available geological and geotechnical data through some scientific procedures, in order to help the engineers and urban planners to comprehend and visualize the subsurface conditions as an integrated system. Such attempts have been made world-wide [4–13].

A simplified geological map of Singapore is shown in Fig. 1. As it can be observed, it comprises mainly igneous rock in the central part, sedimentary rock in the West, and old alluvium soils to soft marine clay in the East. A huge amount of geological borehole data has been obtained over the years from various construction projects

X. Shi et al. (Eds.): GSIC 2018, *Proceedings of GeoShanghai 2018 International Conference: Transportation Geotechnics and Pavement Engineering*, pp. 81–89, 2018.
https://doi.org/10.1007/978-981-13-0011-0_9

carried out in Singapore. These data, if processed and managed properly, can become a valuable resource for future underground development. Making full use of this resource can lead to a better understanding on the spatial organization of subsurface structures of Singapore, and therefore to save considerable amount of money and time for future underground project constructions and urban planning. Building and Construction Authority (BCA) of Singapore has already set up a GIS system to collect and compile all the borehole data into a standardized AGS (SG) format. Currently, BCA is conducting geological surveys, which comprised of seismic survey and deep borehole drilling (up to 200 m deep), to gather geological information in several study areas to build a national-wide 3D geological model. To expand the geological model to include more geological and geotechnical information for the subsurface soil above the bedrock, a research project was initiated with Nanyang Technological University (NTU) to utilize the existing shallow borehole information to build a 3D model of the sub-soil layer. BCA will combine this 3D subsoil model with information from the deep geological surveys to develop a whole 3D geological model for Singapore.

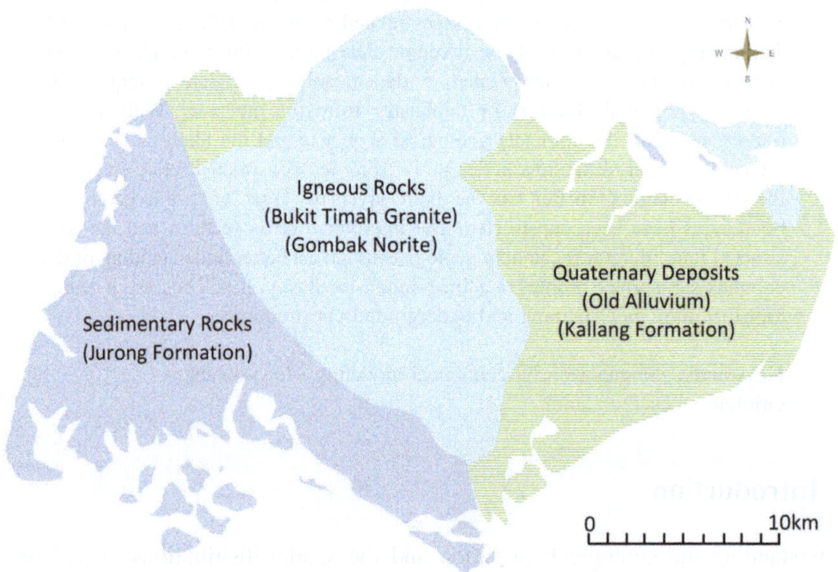

Fig. 1. Simplified geological map of Singapore [14]

This paper describe the research works related to the 3D geological modeling project. The procedures and software adopted are discussed. The resulted cross-sections and 3D geologic model for a subzone of Singapore (the East Coast area) are presented as an example. Methods for data checking and quality control are also deliberated.

2 Procedure of 3D Geological Modelling

SubsurfaceViewer was employed as the 3D geological modelling software, which was developed by INSIGHT Geologische Softwaresysteme GmbH. The software methodology of 3D geological modelling is based on a single simple philosophy—the construction of geological subsurface models has to proceed with an understanding of the complete geological sequence and the likely geomorphological evolution of the study area [9, 15]. Therefore, after preparing for the geological data which can provide useful information about both surface and underground conditions of the earth, the interpreted cross-sections were created with geological knowledge to show geological relationships and depict a representation of the geological sub-surface arrangement. The fence diagram was then constructed using the all established cross-sections which can help geologists to have a good understanding of the complex sub-surface arrangements of the study area. Finally, 3D geological models were built layer by layer based on the fence diagram and geological boundaries.

2.1 Geological Data Preparation

To build the 3D geological modelling, four main types of geological data have been collected: the borehole data, the 2D geological maps with diverse scales, the digital elevation models (DEMs), and the archival publications for geological descriptions. Borehole data, which represent the direct observations of the subsurface, are probably the most important geological data for 3D modeling. The quality and quantity of the borehole data collected originally has crucial influence on the accuracy of the resulting 3D model. Information of about 60,000 boreholes in total throughout Singapore have been collected by BCA from different departments, organizations or construction companies. A data verification procedure was adopted to process the data before they were used for the modelling, which includes removing of any errors/duplications from the database, and revising the descriptions if there is any inconsistency.

The East Coast area of Singapore has been selected as an example for explaining the 3D geological modelling work carried out so far. The distribution of borehole data of this area is shown in Fig. 2. The 2D geological maps helped to identify the extent, distribution, and borderlines of each geological formation on surface. DEMs were used to determine the top surface of the top geological formation. Archival publications for geological descriptions were used to gather geological information for the areas dealt with. To allow all the data to be integrated in one 3D geo-modelling package, the easting and northing coordinates of the collected geological data were adopted.

2.2 Data Formats Adopted

Geological data cannot be entered directly into SubsurfaceViewer_6_MX. Five necessary files in desired formats need to be prepared, which are "*.bid", "*.blg", "*.gvs", "*.gleg", and "*.asc". The "*.bid" format file is a master data file with borehole names and coordinates (Easting, Northing, and reduced level), which shows the location where the section to be drawn. The "*.blg" format file contains the borehole log information, especially the Lithostratigraphical classification code (GEO3, AGS (SG)),

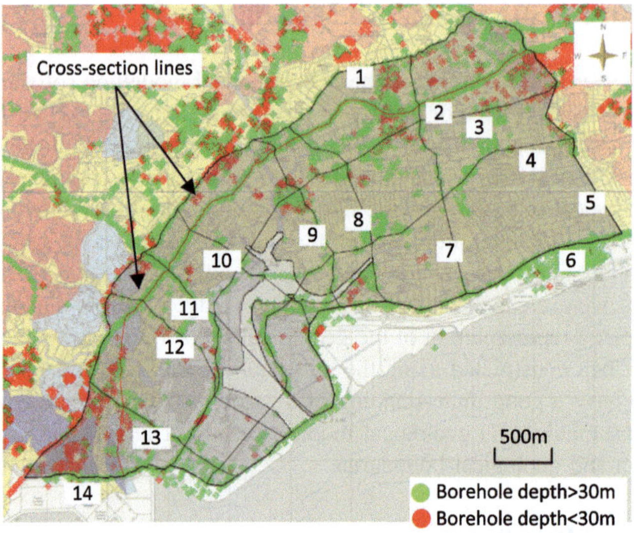

Fig. 2. The distribution of borehole data and geological cross-sections of the East Coast area (Gray color covered area) of Singapore

which will be used to interpret the Singapore formation layers. The "*.gvs" format file contains the information about the vertical sequence the layers in the cross-sections and the 3D geological models. The "*.gleg" format file is the layer legend file to set the colors for the representation of layers and borehole logs. The "*.asc" format file is an ASCII grid file; it presents the surface elevation and is used as a digital elevation model (DEM) to determine the top surface of the top geological formation.

2.3 Creation of Cross-Sections

After the five files as mentioned in the last Sect. 2.2 have been loaded into SubsurfaceViewer_6_MX, cross-sections can be created using the following procedures. Firstly, cross-section lines are established based on the quality and quantity of the boreholes data. As shown in Fig. 1, fourteen cross-section lines had been defined along the boreholes marked in green (with depths larger than 30 m). Secondly, check the integrity and validity of the boreholes data along the cross-section lines using borehole log profiles, especially the Lithostratigraphical classification code in the "*.blg" format file, which is used to interpret the Singapore formation layers. Thirdly, boreholes with high quality information along the defined cross-section lines are selected and presented in Section-Window. The cross-section lines of different formation layers are interpreted and constructed from the top to bottom layers. The top cross-section line is controlled by the trace of the DEM line. Cross-section along the No. 2 grid line is shown in Fig. 3 as an example; different color represents different geological formation. In the interpretation of the geological boundary lines, the published Singapore Geological Map (Fig. 1) is used.

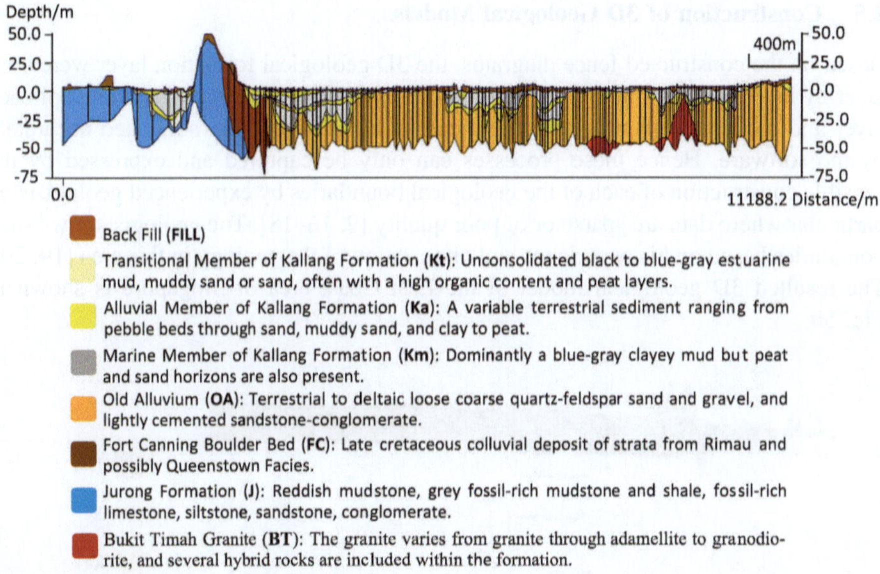

Fig. 3. Geological cross-section along the No. 2 cross-section line of the East Coast area of Singapore

2.4 Construction of Fence Diagrams

In the East Coast area of Singapore, a total of fourteen interpreted cross-sections were created. Using these cross-sections, the fence diagram of the East Coast area of Singapore were constructed as shown in Fig. 4.

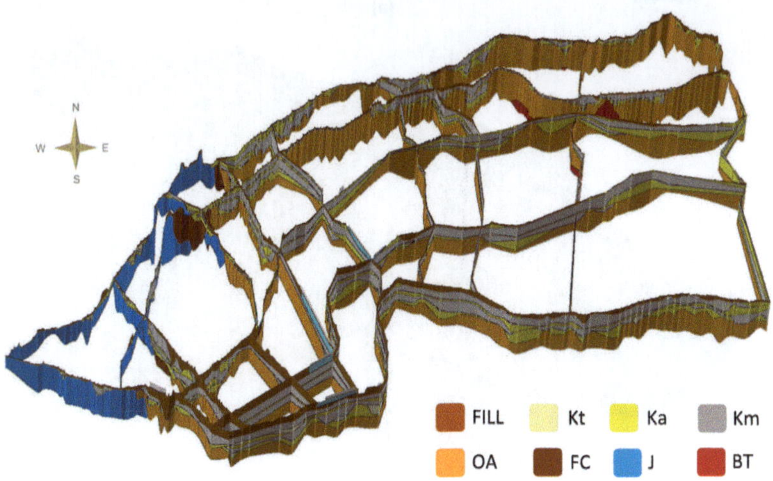

Fig. 4. The fence diagrams of the East Coast area of Singapore

2.5 Construction of 3D Geological Models

Based on the constructed fence diagrams, the 3D geological formation layer were built layer by layer as shown in Fig. 5a. The processes that form the geological units of each layer and their subsequent arrangement cannot be automatically simulated accurately by the software. Hence these processes can only be captured and expressed by the sensible construction of each of the geological boundaries by experienced geologists, in particular where data are sparse or of poor quality [9, 16–18]. The geologist draws such boundaries based on his experience and observation of the geology in this area [19, 20]. The resulted 3D geological model of the East Coast area of Singapore is shown in Fig. 5b.

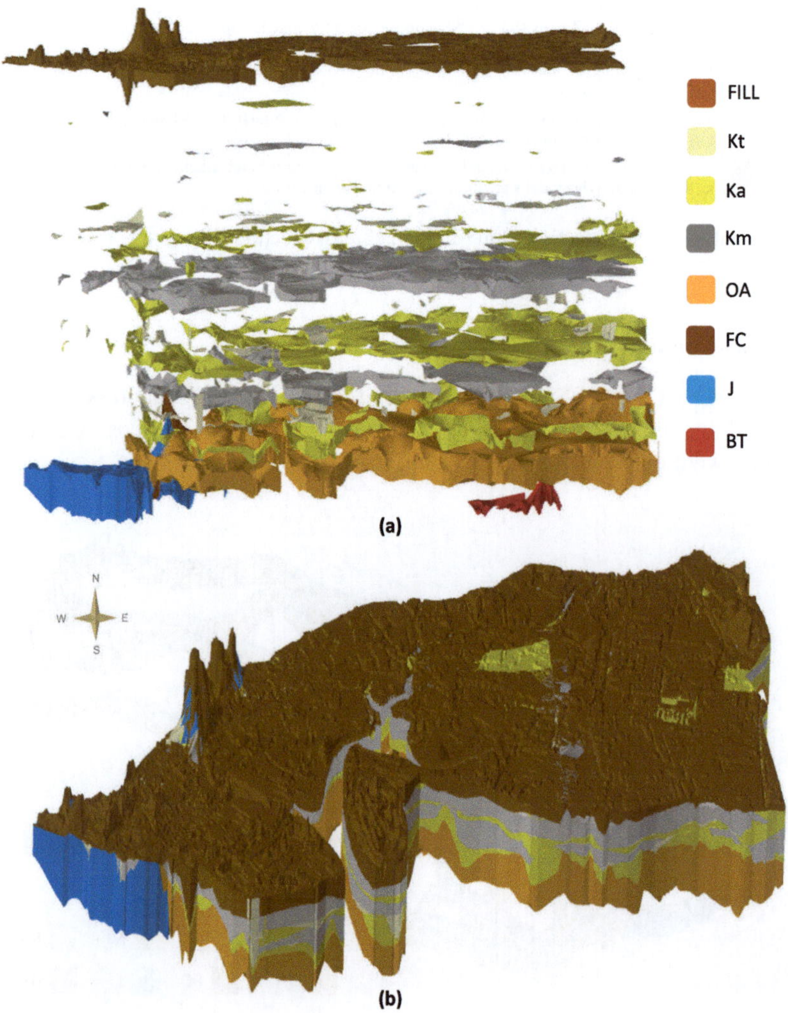

Fig. 5. (a) 3D explode model of the East Coast area of Singapore, and (b) 3D geological model of the East Coast area of Singapore

3 Discussion

The 3D geological model of the East Coast area of Singapore has been established using shallow borehole data (<60 m). There are gaps along and between the cross section lines due to the limitation of the amount of borehole data or their poor quality. Thus, some methods have been used during modelling procedures to improve the accuracy of the 3D geological model, which are briefly illustrated as follows.

As the methodology of 3D geological modelling using SubsurfaceViewer software is based on the cross-section creation, it is important to control the quality of the data and the cross-section construction during the interpreting process by experienced geologists. A geological consistency method based on the consistency of the neighboring borehole information and the consistency with the existing geological maps is adopted to check the validity of the borehole data, and to ensure the accuracy of the resulting cross-sections.

The accuracy of the 3D geological model is directly related to the alignment and frequency of the cross-sections that are put together to build the fence diagram. A method to add sub-cross-sections is used to increase the frequency of the cross-sections of the fence diagram, where more high quality borehole are existed between the cross-sections.

The 3D geological formation stratum are built layer by layer after the fence diagram has been constructed. The geological boundaries of layers are sensible constructed by experienced geologists using borehole and geological map based combining method. Two steps are used to ensure the quality control during geological boundary construction process: constructing rough geological boundaries of all formation layers, and revising the boundaries towards higher accuracy by checking the boreholes nearby or referring to the geological boundaries of geological map. The detail of the validity procedure will be discussed in a separate paper.

4 Conclusions and Recommendations

A 3D geological modeling procedure was presented in this paper. An example of the resulted 3D geological model construction for the East Coast area in Singapore was illustrated as well. The following conclusions or recommendations can be drawn:

(1) The quality and quantity of the borehole data have a direct influence on the accuracy of the 3D model created. A data-cleaning procedure is required to remove typographical errors or duplicates from the database.

(2) The constructed fence diagrams using the established cross-sections can help geologists to have a better understanding of the complex sub-surface profiles and local geology of the area of interest.

(3) With an understanding of the complete geological sequence and the likely geomorphological evolution of the target area using the constructed fence diagrams, the 3D geological model can be built layer by layer after the geological boundaries being determined.

(4) The presented 3D subsurface geological model has been developed using shallow borehole data (<60 m). There are gaps between the cross section lines due to lack of data. Although the cross sections can be interpreted manually by an experienced geologist, the geological layers are generated automatically by the software. Therefore, it is important to verify the model constructed. The geological model should not be used as a replacement of site investigation work for any construction project. Instead, it should be used as a tool for more effective site investigation plan to be made. The new borehole data in turn can be used to verify or enhance the established subsurface model. In this way, the uncertainties in ground model can be largely reduced, and the reliability of future geotechnical design can therefore be improved.

Acknowledgement. We would like to acknowledge that this study is supported by the project "Web-based 3D geodata modelling and management system (GeM2S) (Grant No. L2NICCFP2-2015-1)" from the Ministry of National Development, Singapore, under the Land and Livability National Innovation Challenge (L2NIC) program.

References

1. Hack, R., Orlic, B., Ozmutlu, S., Zhu, S., Rengers, N.: Three and more dimensional modelling in geo-engineering. Bull. Eng. Geol. Environ. **65**(2), 143–153 (2006)
2. Dong, M., Neukum, C., Hu, H., Azzam, R.: Real 3D geotechnical modeling in engineering geology: a case study from the inner city of Aachen, Germany. Bull. Eng. Geol. Environ. **74**(2), 281–300 (2015)
3. Kingston, R., Carver, S., Evans, A., Turton, I.: Web-based public participation geographical information systems: an aid to local environmental decision-making. Comput. Environ. Urban Syst. **24**, 109–125 (2000)
4. Caumon, G., Collon-Drouaillet, P., De Veslud, C.L.C., Viseur, S., Sausse, J.: Surface-based 3D modeling of geological structures. Math. Geosci. **41**(8), 927–945 (2009)
5. Chang, Y.S., Park, H.D.: Development of a web-based geographic information system for the management of borehole and geological data. Comput. Geosci. **30**(8), 887–897 (2004)
6. Culshaw, M.G.: From concept towards reality: developing the attributed 3D geological model of the shallow subsurface. Q. J. Eng. Geol. Hydrogeol. **38**(3), 231–284 (2005)
7. Jones, R.R., McCaffrey, K.J.W., Clegg, P., Wilson, R.W., Holliman, N.S., Holdsworth, R.E., Imber, J., Waggott, S.: Integration of regional to outcrop digital data: 3D visualisation of multi-scale geological models. Comput. Geosci. **35**(1), 4–18 (2009)
8. Calcagno, P., Chilès, J.P., Courrioux, G., Guillen, A.: Geological modelling from field data and geological knowledge: part I. Modelling method coupling 3D potential-field interpolation and geological rules. Phys. Earth Planet. Inter. **171**(1), 147–157 (2008)
9. Kessler, H., Mathers, S., Sobisch, H.G.: The capture and dissemination of integrated 3D geospatial knowledge at the British Geological Survey using GSI3D software and methodology. Comput. Geosci. **35**(6), 1311–1321 (2009)
10. El May, M., Dlala, M., Chenini, I.: Urban geological mapping: geotechnical data analysis for rational development planning. Eng. Geol. **116**, 129–138 (2010)
11. Merritt, J., Entwisle, D., Monaghan, A.: Integrated geoscience data, maps and 3D models for the City of Glasgow, UK. IAEG paper, no. 394, pp. 1–10 (2006)

12. Self, S., Entwisle, D.C.: The structure and operation of the BGS National Geotechnical Properties Database. British Geological Survey Internal Report, IR/06/092 (2006)
13. Self, S., Entwisle, D., Northmore, K.: The structure and operation of the BGS National Geotechnical Properties Database. Version 2 (2012)
14. PWD: Public Works Department: The geology of the Republic of Singapore. Public Works Department, Singapore (1976)
15. Fookes, P.G.: The first Glossop lecture: geology for engineers: the geological model, prediction and performance. Q. J. Eng. Geol. **30**, 293–431 (1997)
16. Lemon, A.M., Jones, N.L.: Building solid models from boreholes and user-defined cross-sections. Comput. Geosci. **29**, 547–555 (2003)
17. Wu, Q., Xu, H., Zou, X.: An effective method for 3D geological modeling with multi-source data integration. Comput. Geosci. **31**, 35–43 (2005)
18. Kaufmann, O., Martin, T.: 3D geological modelling from boreholes, cross-sections and geological maps, application over former natural gas storages in coal mines. Comput. Geosci. **34**, 278–290 (2008)
19. Kessler, H., Bridge, D., Burke, H., Butcher, A., Doran, S.K., Hough, E., Lelliott, M., Mogdridge, R.T., Price, S.J., Richardson, A.E., Robins, N., Seymour, K.: EA urban Manchester hydrogeological pathways project, p. 65. Commissioned Report CR/04/044, British Geological Survey, Keyworth, UK (2004)
20. Kessler, H., Lelliott, M., Bridge, D., Ford, J., Sobisch, H.-G., Mathers, S., Price, S., Merritt, J., Royse, K.: 3D geoscience models and their delivery to customers, In: Three-Dimensional Geologic Mapping for Groundwater Applications Workshop, Extended Abstracts, pp. 39–42, Salt Lake City, UT. Geological Survey of Canada, Ottawa (2005). Open File 5048

Geological Data Extension for Metro Tunnel BIM Model with the Linked Data Approach

Yi An[1(✉)], Yan Liu[2], and Hanbin Zheng[3]

[1] Eindhoven University of Technology, 5612 AZ Eindhoven, Netherlands
y.an.student.tue@gmail.com
[2] Delft University of Technology, 2628 CN Delft, Netherlands
[3] Shanghai SiHanFu Information Technology Co., Ltd., Shanghai 200093, China

Abstract. A lot of previous researches have indicated that among the common extension mechanisms for extending geological data into metro tunnel, the Linked Data is evaluated as the most appropriate approach. This paper particularly emphasizes the technical development process for the Building Information Modelling (BIM) model extension using Linked Data. The proposed process has been demonstrated and verified using a functional case study as well.

Keywords: Linked data · BIM · Metro tunnel · Geological data

1 Introduction

The semantic web creates possibilities for the support of large scale information sharing in the architecture, engineering, construction and facility management (AEC/FM) industry [1]. These possibilities of large scale information sharing allow the semantic web to offer a solution to one of the main obstacles of BIM implementation, which is the interoperability between BIM systems [2]. Previous researches proved that the Linked Data is a desirable mechanism to extend the BIM metro tunnel model. However, IFC databases are not based on the Semantic Web data sets like Resource Description Framework (RDF) or Web Ontology Language (OWL) data, but on the EXPRESS modelling language. IFC databases can therefore not be directly connected to the Semantic Web technology without converting an IFC schema into an OWL ontology, which is considered to be straightforward except for some minor complications. Still, IFC can benefit from the advantages of the Semantic Web related technologies due to the possibility of converting IFC into OWL files.

To achieve and create Linked Data, technologies should be available for a common format (RDF), to make either conversion or on-the-fly access to existing databases (relational, XML, HTML, etc.). It is also important to be able to setup query endpoints to access that data more conveniently. W3C provides a palette of technologies (RDF, GRDDL, POWDER, RDFa, the upcoming R2RML, RIF, Simple Protocol And RDF Query Language (SPARQL)) to get access to these data. To make the application process more detailed and elaborate, this paper systematically introduces a process of

X. Shi et al. (Eds.): GSIC 2018, *Proceedings of GeoShanghai 2018 International Conference: Transportation Geotechnics and Pavement Engineering*, pp. 90–99, 2018.
https://doi.org/10.1007/978-981-13-0011-0_10

extending geological data into metro tunnel using linked data approach. In order to verify the process, a case study based on metro station is presented as well.

2 Geological Data Extension for Metro Tunnel BIM Model with the Linked Data Approach

This part specifically introduces the process of applying the Linked Data approach to extend geological data into metro tunnel BIM model. The process contains five main steps, data collection, data analysis, format conversion, data linking, and reference block development respectively. Also, there are some sub procedures within these main steps, which are discussed within the framework of corresponding linked data step.

2.1 Data Collection

As the first step of extending IFC shield tunnel with the linked data mechanism, data collection is crucial since it administer known effect on the not only results but also process. For instance, different data sources will determine the use scenarios and conversion tool, the complexity of which will be determined when the data source is certain.

2.2 Data Source Selection

Normally the main criterion for data selection is meeting end user's information needs, in this case, it is supposed to be the designers' who are going to design a new project in vicinity of an existing building project. For instance, in case of underground metro tunnel another building might be constructed above or in close proximity. Since the geological data is an open source data, it may be easily accessible for interested suitors. However, the foundation treatment or the monitoring plan which are made for the "existing tunnel" by a certain metro tunnel contractor are not "open sources". In fact, the designers of the proposed building in vicinity of underground tunnel normally would check the metro tunnel model to integrate their design. In this specific case, the monitoring data of ground settlement could meet designers' needs. Apart from that, the accuracy, continuity and stability of the data are crucial criterion beyond question. The synthesis of these criteria would help selecting suitable data sources for later integration. To well serve the end user's needs, not only requires direct data, but also auxiliary data that supports those direct data should be collected.

2.3 Data Obtaining

Data obtaining refers to finding correct channels to receive the data. Most of public domain data can be gathered without hindrances, but some private or organizational owned data may require extreme identity requirements. Also, the information can be via different sources, a file, online, database, etc. Then the key point is how to find suitable data-saving method based on different forms. Data obtaining also means collecting data with legal clearance. Authoritative data publishers possibly provide different applicable licenses for different end users, serving various purposes and supporting distinct processing method. In case the conflicts and liability, it is necessary to identify the authoritative data publishers and related data using license.

2.4 Data Analysis

After the data resources are ready, looking into datasets thoroughly as preparation for data conversion is the second step.

2.5 Data Structure Analysis

Data analysis should always start with what the structure of datasets is and how it is organized. The obtained datasets may have heterogeneous formats. Also, to get more insight about provided information form. For later use with linked data extension mechanism, some files need to be converted so that they can be processed further, such as csv to ttl or RDF/xml, rvt to ifc and to RDF/xml, etc. The preferred data formats will be determined based on data organizing structure and other linked data sources.

2.6 Obtaining Data Schema

Obtaining data schema aims at having a basic understand of the various data-sets major concepts, as well as information overlap and expression differences. Analyze the "links" among different major concepts to develop a combined relational graph—ontology. This could help perform the conversion and linking tasks, meanwhile evaluate if joint data can meet the end users' specific information needs.

2.7 Format Conversion

The data conversion is actually about converting data into linked data way, which is RDF format, for the purpose of data linking preparation.

2.8 Uniform Resource Identifier

For the Uniform Resource Identifier (URI) naming, there are several forms and guidelines. For instance, Radulovic et al. mentioned two basic forms of URI-hash URI and slash URI [3]. There are also different guidelines for helping URI design, like 10 rules for persistent URIs, and Providing and Discovering URI Documentation [4]. However, for selecting a basic form of URI, the advantages and disadvantages of each basic form and also the preferred URI forms applicable for the data transform tools that will be used need to be considered. Besides the URI forms, clear and accurate mapping of relationships and hierarchy between various resources is essential for URI design as well.

2.9 Ontology Development

Ontology is a formal specification of a shared conceptualization [5]. In computer and information science, it is used to name and define types, properties, and interrelationships of entities from a particular domain. Radulovic raised seven steps to develop a well-designed ontology [3]. The main idea of his development method is referenced in this ontology development step.

Developing ontology to describe resources and relationships from a merged domain starts from considering the overall data content and structure in order to grasp the main concepts forming the ontology and their connections. After that, a search for existing ontologies and the selection of some of them for reuse should be conducted. Reusing existing ontologies can help build a new ontology structure and evolve it into more accessible. For the information not defined in the existing ontologies, new classes or properties should be created to represent it.

2.10 Convert File Format and Merge Common Format File

The resources naming and ontology developing are all steps for preparation for the transformation step. As Radulovic et al. suggested, select a RDF serialization (RDF/XML, Turtle, N-Triples etc.) as the transformation format. Then select the suitable transformation platform (converter) to achieve the RDF conversion based on input and output data formats that the IT tools supports. Since data sources may be collected from different formats like spreadsheet, XML, and IFC, several transformation tools may be required to normalize all cross platform data into one. After selecting the tools, transform those to RDF format, and merge the transformed RDF files into one model. This can be produced through Java RDF API or other software like Google Refine. The evaluation of the converted file format and merged common format file can be executed by SPARQL queries. If the SPARQL queries go smoothly and get correct results, the file format converting and merging works well.

2.11 Linking

Linking aims at creating links between the RDF data that comes from different data sources. The data resources content analysis and schema extraction have been performed in the data conversion part.

2.12 Identify Linking Objects

The RDF data from different data sources are isolated in the merged model. Selecting linking objects depends upon the goal of performance tasks needed by end users-designers. The created links would help to navigate end users travel through different information islands to reach the required information. With reference to ontology, direct and simple links among dissimilar RDF data is basic rule for identification of linking objects.

2.13 Relations Clarification

After identifying linking objects, clarifying resource relations through suitable properties would be the following step. The property selection can take a reference of existing ontology vocabularies. If existing properties fail to achieve the satisfaction, then new properties would be created into existing class.

2.14 Semantic Web Applications Constructing Tools Selection

There are certain tools creating links among RDF data, such as Google Refine, Jena Apache API, Fuseki, TDB, etc. Different tools have different data requirements and function configurations, like RDF serialization limitation, data structure requirements and data size requirements. Selecting a suitable tool based on the data attributes can simplify the linking process.

2.15 Developing Reference Blocks

The reference design reflects end users' demands, in this specific case it is the monitoring information of surrounding metro projects. It is designed based on available data and user demands that help managers to grasp monitoring plan, drill hole types and sensitive area. Every query performs a specific topic. All of composed coding components for this reference topic are collected as a whole and called 'reference block', as a packaged block, the package is more convenient than shattered codes for designers to use. The reference blocks conduct SPARQL queries.

SPARQL queries can be regarded as a form of evaluation for the results from data conversion and data lining parts. It could help testing whether the ontology design is feasible, whether the RDF data transformation performs well enough to show the needed data structure, whether the merging and linking works well, etc. As long as this test reflects problems regarding the previous parts, then back iteration to the conversion and merging is necessary.

2.16 Topics Selection

The data sources in this research are sensor data, drill hole description and a BIM model. The end users are the new project designers. The research aims at helping new project designers referencing shared monitoring plan and new project design. It also aims at helping project manager to monitor the building's structural features during construction of surrounding project. Designer's possible information demand based on sensor data and building information can be the basement for designing reference block. After detailed demand analysis, several new-project-designer-caring topics are selected.

2.17 Developing SPARQL Blocks

After reference topics are selected, developing the SPARQL blocks would be necessary. The blocks include the SPARQL query part and the query results processing part. The SPARQL query part extracts information among direct or indirect resources relationships. The query results processing part performs further data deductions with the supports from other domain or data processing methods to help designers starting new projects. The different blocks are based on different reference query topics.

3 Case Study

This part is going to illustrate an actual case application with the aforementioned methodology, which aims at achieving extended geological survey data into IFC shield tunnel model through Linked Data approach. Due to non-disclosure agreement, the project is presumed based on actual case, names and data have been slightly changed considering not to hamper the final result.

The adopted project is called 'AA Station'. It is a metro station of metro line X in city B, located in C District, which has tremendous opportunity for building development due to rapid urbanization. The AA station surroundings holds a huge vacant area that needs to be developed later on. As matter of fact, the government did have plans to set more commercial, educational and living facilities there as part of city urbanization. The metro station is assumed to be present already at location and for that there has been huge information already collected regarding the geological survey, experiment and monitoring. Such information can be shared with will-be-developing projects in the surrounding area. In this case, the station is selected as the research object; named AA station BIM model.

For the AA station case, the new ontology and linked merged model is designed to facilitate designers of surrounding project to design their building with the metro station monitoring information. Different types of new building project with potential influence from dynamic geological data are assumed to be the user information demands for new designers. So, sensor data and geological data are supposed to be both general and specific and a BIM model is used to present the related structural components context.

Water level monitoring and ground settlement monitoring are selected as data sources since they are different from other monitoring items that are limitedly tightly linked to the project itself and hard to connect to other surrounding building, and these 2 monitoring items are very directly linked to the new surrounding projects, in other words, sensors measurement meet the designer's information needs. Therefore, there are 31 monitoring points with 31 displacement sensors and 18 monitoring wells with 18 water level sensors. Since the water level monitoring plan and ground settlement data from the AA station are already available, these collected data could help designers. The sensors of these two items are equipped with the drill holes. Moreover, to make this information easy to understand and query (referenced), drill hole description regarding the types, locations and etc. is added as supplementary data for those sensor values.

Meanwhile, a BIM model is needed to present the station space context. The model is requested from the involved consulting company. Sensor values of ground settlement and water level are obtained directly from consulting company, the sensor data is saved as a local excel. Due to the confidential issue, the access to their database was denied, instead the sensor data was sent directly by the concerned company. The drill hole description file is generated from project manager's monitoring plan, which is also an excel file. The Excel file mainly focus on location descriptive data, which is the necessary information in the domain of geological data.

The data analysis and data conversion are combined together in this part. These two procedures are coordinated for the aim of creating a well merged RDF model. Open refine is selected to convert the csv file to RDF file. OpenRefine is a powerful tool for working with messy data: cleaning it; transforming it from one format into another; and extending it with web services and external data. In this case, an ontology describing drill hole description would be published, so the chosen name is Drill Hole Description ontology (DHD to keep it short). The technical support for defining URI is given as following.

Since there are certain potentials to publish the ontology, the chosen URI should be permanent and defined in a domain I control. To prevent certain usual case, somebody is reusing the concepts defined in my ontology and I changed its URI. The person reusing my ontology will no longer know the proper definitions and semantics of the reused term. And I assume that most of people reading this are not willing to pay for a new domain each time a new ontology is published. Therefore, I define the URI of my ontology in http://purl.org. PURL stands for persistent uniform resource locator; they are widely used to give persistent URIs to resources. The assumption of this part is that I registered in the page, and I started directly defining a new domain, waiting for the approval and create the URI for my ontology. As a consequence, in my case it is http://purl.org/net/DHD. I created the name under the /net/ domain; things would go faster since it is the default domain. Otherwise they will have to approve the domain and the name of my ontology. Then hash URI is chosen due to the fact that information is easy to publish using an editor for RDF files; with slash, the server may need to be set up to do a 303 redirect from one URI to another URI for documentation purposes. This involves control of the web server which many people do not have, or not aware of how to use it; also run time is an another factor which need to be considered, the client looking up the URI just strips of the right part, and performs a single access to get the document whatever it is about. This will in many cases also give information about other related things, with URIs starting with the same document URI. Further fetches will not be necessary at run time. The naming path for the drill hole description data begins with 'http://purl.org/net/DHD#', and concatenates with described items. The naming path for the geo sensor data begins with 'http://purl.org/net/Geosensor#', then continues with the items to be described. The resources and properties related to the BIM model will be named automatically by the RDF transformation tool. All these sensors and BIM model's resource and property names are only used as unique identifiers.

Since the main idea of the collected data is about sensor data and building geological data, the developed ontology should also focus on describing the entities and relationships between building site and sensor data. The Smart Appliances REFerence (SAREF) ontology is a shared model of consensus that facilitates the matching of existing assets model of consensus that facilitates the matching of existing assets in the smart appliance domain. The SAREF ontology provides building blocks that allow separation and recombination of different parts of the ontology depending on specific needs. Also, IfcOwl ontology is used to describe IFC-based vocabularies. To conclude, including the basic OWL, there are three existing ontologies applied in the case study's ontology for the geological survey data extension.

Additionally, these three existing ontologies fail to suit the case perfectly, since the related drill hole descriptive data and sensor descriptive data are not defined. Therefore, the new ontology which contains the new classes and properties to illustrate the related sensor values and drill hole description data. For instance, Data property Drill hole: drillholeDepth is added to the ontology to introduce the different depths in the drill hole description file. Class drill hole: drillhole is added to the ontology to introduce a narrow shaft bored in the ground, either vertically or horizontally for specific engineering use, in the case, we considered it as a container for sensors. Property sensor: accommodate represents the relation between a sensor and the location (drill hole). Ontology before linking has been created to explain the isolated information island, and it is based on the data sources.

The RDF serialization used in this case is Turtle, because of its readable attribute, which is much easier for people to read. Since the original data source formats in this case are heterogeneous, two transformation tools and python script converter are selected to perform the turtle transformation process. The tool used for the data merging is Apache Jena Fuseki, which is a SPARQL server. It provides RDF writing, reading, simple query and other Java-based methods. In the merging data section, it is used to create a model which combines all the turtle data. Even the merged model is created, the data from different sources are still isolated. No path is available to navigate through different information islands. The Linked Data approach aims at connecting related data that has not been connected, to help explore connected information world. As the ontology development section shows, there are four parts of data in the merged model: the sensor value data, building model, drill hole description and sensor description. In the linking process, drill hole description is in the middle to generate links, as the drill hole description contains resources that have relationships with sensor value and sensor description. The Drill hole description has a class "DHD:DHD" expressing the accommodate attribute in string format, which can make a link with "Geosensor" class in the geosensor ontology, since the "Geosensor" is located in the drill holes. This relationship between these classes is presented by a property named "DHD:accommodate". This link is set to connect according sensors and drill holes. The "SD:sensordescription" class express descriptive data of the geosensors, and these sensors do introduce certain value in the Geosensor ontology. The logic relationship to link sensor description (SD) with Geosensor (GS) is presented by a property named "SD:describe". On the other hands, geosensor is not part of building but it remotely contribute in building. In this situation, the class "GS:geosensor" is linked to "Ifcowl: Ifcproduct" directly to reveal the logic relationship between the geosensor and building model, this relation is set as "GS:served". Though, the ontology "Geosensor" expresses the ontology outside of the building project, it is not located within the building from the perspective of spatial relationship. It still serves the structural element that is within the building project indirectly. Thus, the property "saref:hassubclass" is set as a link between saref ontology and geo sensor, the set property would connect class "Saref: Device" and "Geosensor:Geosensor". After these connections, the five information islands are integrated as one resource interrelated model. This is preparation for the data query process. Noticeably, the original links served for sensors and project are supposed to be created between water level sensor and IfcGround, as for ground settlement sensors can be attached to IfcBorehole directly like the aforementioned developed extension. However, in the BIM model offered by involved consulting company, neither IfcGround nor IfcBorehole is created within the model. Still, model validation which is on purpose of verifying the Linked Data approach and the actually developed model needs to be conducted. Thus, the assumption is that the sensors are remotely contributing to a certain structural elements. For example, the drill hole "ZK13", ZK13 accommodates the ground settlement sensor that has sensor code "W30". And in the model, there is a drilling pile with Guid "0f5PYVtSrAQgQb-S3YOD0$q" and tag "785871". The sensor data and drill hole description documented for integration are about water level monitoring and ground settlement. All the related information could help project manager for surrounding project to manage their own project process, not only early design stage but also the construction stage. For the purpose of drill hole, share related data between the existing project and new developed project, during the stage of design. The shared geological information could facilitate designer's process. Apart from offering related data to designers, but also check the validity of merged model. To achieve query, the merged model called "new model.ttl" is uploaded to fuseki server. Firstly, a new dataset is added. Then merged model is uploaded in the created dataset. In the SPARQL query, SELECT is used to signify that SELECT is required to obtain certain information and WHERE to signify your conditions, restrictions, and filters. In the situation, as a designer who is going to design the building nearby the AA Station, I would like to know what drill holes are already there serving the AA station, and where exactly they are positioned, and what sensors are in the holes. This information is given in the drill hole description, therefore, the prefix DHD is set in front as "PREFIX DHD: http://purl.org/net/DHD#". This query is for a project manager to monitor the current structural properties under the situation that there are construction operation of other buildings being conducted in the neighborhood, or to monitor the structural properties in the certain time in case the construction behavior causes the unexpected disturbance. For precautions, better monitoring will be necessary. Here the date is assumed 2016/01/07, corresponding assumption is that on this day, the excavation of surrounding office building's substructure is conducted. It is quite possible to cause disturbance at this time. Moreover, the most potentially infected building structural elements are shown as well. If the sensor data could project which things can go wrong, appropriate action can be taken in time. The code to query the specific information in this case is accordingly developed.

4 Conclusion

The process of applying the Linked Data approach to extend geological data into metro tunnel BIM model has been thoroughly introduced here. Even though the case study explained isn't conducted first hand, it still succeeded in exploring potential of integrating geological data of metro tunnel with GIS related application for better visualization. On the other hand, based on such approach, a system that could integrated indicators of geo-sensors and recommendation platforms relying on RDF processing can be a promising product. Additionally, SPARQL provides efficient ways of accessing the Linked Data by following users to write flexible queries. However, in order to use the SPARQL query language, the user has to be familiar with the designed ontology. Therefore, user friendly query interfaces are required to allow general users to use SPARQL without the need to understand the SPARQL query language or the designed ontology. Such a faceted browsing, content-specific browsing, full-text search, navigation mechanisms, statistical analysis, etc. need to be further examined so that the user could easily define subclass concepts related to certain super classes through the inheritance capacity of classes in a more user-friendly manner.

Acknowledgement. I would like to thank my Prof. Dr. J. Beetz for the continuous support of my BIM related research, for his patience, motivation, enthusiasm, and immense knowledge. His guidance helped me in all the time of research and writing of this paper. I could not have imagined having a better advisor and mentor for my research.

Besides, I would like to express my sincere gratitude to my advisor T.F. (Thomas) Krijnen. The "door" to Thomas was always open whenever I ran into a trouble spot or had a question about my research or writing. I am gratefully indebted for his very valuable comments on this paper.

My sincere thanks also go to Prof. Bai Yun from Geotechnical Department of Tongji University, Dr. Chen Hong, Xia Haibing from STEDI. These experts were involved in the helping research development. Without their passionate participation and input, the whole research could not have been successfully conducted.

References

1. Beetz, J., Van Leeuwen, J., De Vries, B.: IfcOWL: a case of transforming EXPRESS schemas into ontologies. Artif. Intell. Eng. Des. Anal. Manuf. **23**, 89–101 (2009)
2. Volk, R., Stengel, J., Schultmann, F.: Building information modeling (BIM) for existing buildings—literature review and future needs. Autom. Constr. **38**, 109–127 (2014)
3. Radulovic, F., Poveda-Villalon, M., Vila-Suero, D., Rodriguez-Doncel, V., Garcia-Castro, R., Gomez-Perez, A.: Guidelines for linked data generation and publication: an example in building energy consumption. Autom. Constr. **57**, 178–187 (2015)
4. Rees, J.: Providing and discovering URI documentation. W3C: 2 February 2012. Accessed 24 Dec 2016. https://www.w3.org/2001/tag/awwsw/issue57/latest/
5. Gruber, T.: A translation approach to portable ontology specification. Knowl. Acquis. **5**, 199–220 (1993)

Three-Dimensional Simplified Method for Predicting Settlement Along Railway Due to Excavation

Yiqun Tang[1,2](✉), Siqi Xiao[1], Jie Zhou[1], and Yangjie Zhan[1]

[1] Department of Geotechnical Engineering, Tongji University, Shanghai 200092, China
tangyiqun2@tongji.edu.cn
[2] Key Laboratory of Geotechnical Engineering, Tonji University, Shanghai 200092, China

Abstract. More excavation projects are arising near existing buildings and structures due to large-scale urban construction, in which the excavation unavoidably brings settlement influence and potential safety danger to surrounding construction and buildings. Settlement profile vertical to the excavation edge is a widely discussed topic. But for linear traffic facilities parallel to the excavation, settlement profile parallel to excavation, namely, settlement along the traffic line, should also be considered. Besides, the precise control of differential settlement along the traffic line also plays very important role. Thus, it is necessary to establish a quick prediction model, which is able to consider both vertical and parallel settlement profile, on the basis of basic information of excavation. A simplified method for predicting ground settlement along railway induced by excavation is proposed. According to characteristics of settlement profile, Rayleigh and Gauss distribution functions are adopted to calculate settlement distribution. Settlement modified parameters are introduced through three-dimensional perspective. A comparison between calculated results and monitoring data is given to verify the feasibility of proposed method. Good agreement of comparison indicates proposed method can be employed to predict settlement along railways due to adjacent excavation.

Keywords: Excavation · Settlement · Simplified method · Railway

1 Introduction

With rapid urban development, high dense buildings and advanced transport network gets increasingly interlaced. Surrounding existing buildings or linear transportation structures have to be encountered in more and more excavation engineering. Attention has been paid in several circumstances, such as railway and subway [1–3]. The existence of those linear transportation infrastructure increases the importance of study on settlement prediction since railway and subway generally have strict requirement of settlement control.

Research shows that ground settlement due to excavation is relevant to pit geometry, excavating depth, retaining structure, geological conditions, construction steps

X. Shi et al. (Eds.): GSIC 2018, *Proceedings of GeoShanghai 2018 International Conference: Transportation Geotechnics and Pavement Engineering*, pp. 100–108, 2018.
https://doi.org/10.1007/978-981-13-0011-0_11

and quality [4]. Empirical statistics and numerical simulation are two main methods used for calculating settlement due to excavation. Peck method and Rayleigh curve method are widely accepted empirical methods. Since practical projects are always sophisticated and varied, it costs a lot and consume much time to adopt numerical simulation to get good results, such as expensive computing hardware or very slow computation. Besides, rich experience to select parameters and simplify numerical model is also crucial. Therefore, it is necessary to propose a simplified method to predict settlement due to excavation.

In the aspect of empirical method, Peck [5] drew settlement profiles according to the monitoring data in Chicago. Bowles [6] proposed calculating method based on lateral deformation of retaining structure. Clough and O'Rourke [7] drew settlement profile in accordance with different geological conditions. Ou and Heish [8] developed simplified method based on the concept of primary influence zone and secondary influence zone. Yang and Lu [9] proposed an estimation method based on the four basic deformation modes of the retaining structures. However, methods mentioned above only focus on the settlement vertical to excavation pit edge, in which the excavation is reducible to plane strain problem, so that the settlement parallel to the excavation is not taken into account. But as we explained above it could not be ignored and should be precisely controlled if linear transportation infrastructure existed in the surrounding is parallel to excavation edge.

This paper develops a simplified approach to consider settlement both vertical and parallel to excavation pit edge. The feasibility of proposed method is demonstrated by comparison of calculation results and field statistics. It provides significant reference for the construction area of high dense buildings and advanced transport network.

2 Calculation Method

Settlement caused by excavation is considered through two viewports, settlement vertical and parallel to excavation. Firstly, settlement vertical to excavation can be simplified to plane strain problem. The location of the settlement analysis is set up at the center line of excavation, where best fits the hypothesis of plane strain. Secondly, settlement profile parallel to excavation is established based on the result of settlement profile vertical to excavation. The calculation method is expressed as below in detail.

2.1 Settlement Profile Vertical to Excavation

Research shows that maximum ground surface settlement δ_{vm} generally increases with the increase of excavation depth H.

$$\delta_{vm} = k_{\delta}\delta_{hm} \tag{1}$$

where H is excavation depth, k_{δ} is proportional coefficient. According to engineering experience in Shanghai, k_{δ} ranges from 0.1% to 0.8% with an average value of 0.42% [10]. For a relative stiff wall, the value of k_{δ} tends to be significantly smaller.

Combined with monitoring data, Rayleigh distribution function is suggested to estimate settlement in vertical direction [11]. Settlement can be derived from Eq. (2)

$$k_s = \frac{\delta_v(d/H)}{\delta_{vm}} = \frac{(d/H)+0.75}{0.9} \cdot \exp\left\{-\frac{[(d/H)+0.75]^2}{4.5}\right\} \tag{2}$$

where d is distance from calculating point to excavation, k_s is ratio of ground settlement δ_v at different d to max ground settlement δ_{vm}.

2.2 Settlement at Different Depth

According to settlement profile vertical to excavation, soil outside the excavation can be divided into several zones to figure out the major and secondary affected areas. For soil at different depths, modified coefficient is introduced for calculating its settlement based on three zones for soil outside the excavation [12]. The three zones include fold-line attenuation zone $(0 \le d \le H)$, linear attenuation zone $(H \le d \le 2H)$ and unaffected zone $(d \ge 2H)$. Equation (3) is used for soil located in fold-line attenuation zone while Eq. (4) is used for soil in linear attenuation zone.

$$k_1 = \frac{\delta_y}{\delta_v} = \begin{cases} 1 & \left(0 \le \frac{y}{H} \le 0.5\right) \\ -2 \cdot \frac{y}{H} + 2 & \left(0.5 \le \frac{y}{H} \le 1\right) \end{cases} \tag{3}$$

$$k_2 = \frac{\delta_y}{\delta_v} = \left\{ -\frac{y}{H} + 1 \quad \left(0 \le \frac{y}{H} \le 1\right) \right. \tag{4}$$

where k_1, k_2 are ratio of settlement at certain depth δ_y to ground settlement δ_v at the same *d/H* of broken-line reducing zone and linear reducing zone, respectively. y is certain depth from the ground surface to calculate.

Adjusted coefficient α is introduced to solve continuum problem at cut-point of influence zones as illustrated in Eq. (5).

$$\left.\begin{aligned} \alpha_1 &= -\frac{1}{2H} \cdot d + 1 \\ \alpha_2 &= \frac{1}{2H} \cdot d \end{aligned}\right\} \tag{5}$$

According to Eqs. (4)–(5), for soil at distance of d m away from excavation, reducing coefficient k_y for soils at different depths can be derived as follows.

$$k_y = \alpha_1 k_1 + \alpha_2 k_2 \tag{6}$$

Specially, for settlement on the ground surface and settlement in the unaffected zone, $k_y = 1$. In conjunction reducing coefficient k_y with ground settlement δ_v, combined with stiffness of existing structure at aimed position, soil settlement at a certain depth y can be derived out by Eq. (7),

$$\delta_y = \beta k_y \delta_v \tag{7}$$

where β is the empirical coefficient related to foundation modes and stiffness. For rigid foundation and strong ground improvement, $\beta = 1$.

2.3 Settlement Profile Parallel to Excavation

According to characteristics of settlement profile parallel to excavation, the most dangerous position is usually at the center part of excavation. Through analyzing settlement statistics from references [13–15], the settlement profile is found to be largest in the middle area. The settlement decreases with increasing distance from the center line of excavation. Thus, Gauss distribution function is chosen for depicting settlement profile parallel to excavation pit edge as shown in Eq. (8).

$$k_x(x/l) = \frac{\delta_{vl}(x/l)}{\delta_y} = \exp\left(-\frac{(x/l)^2}{0.75}\right) \tag{8}$$

where x is the distance from calculating point to center line of excavation, l is half of excavation length, δ_{vl} is the settlement at different distance from the excavation center line.

For a certain excavation pit, excavation depth and width vary with different positions. Thus, geometry modified parameter θ is introduced to consider excavation geometry effect as shown in Eq. (9) under the condition of $-1<x/l<1$. The center geometry of excavation is chosen as reference for modification. If $x/l \in (-\infty, -1) \cup (1, +\infty)$, namely, the calculated point is located outside the length of excavation, $\theta = 1$. The value means the excavation geometry is not considered for the settlement outside the length of excavation.

$$\theta=\sqrt{\frac{B(x/l)^2+H(x/l)^2}{B_m^2+H_m^2}} \quad -1<x/l<1 \tag{9}$$

where B_m is the width of excavation at central line; H_m is the excavation depth of excavation at central line.

Thus, the final modified settlement δ_{vf} can be derived from Eq. (10)

$$\delta_{vf}(d/H) = \theta \cdot \delta_{vl}(d/H) \tag{10}$$

The comparison of calculated curve parallel to the excavation according to Eqs. (8) and (9) and monitoring data is shown in Fig. 1. The coordinate y refers to the ratio of measured settlement δ_v to the product of maximum measured data δ_{vm} and geometry modified parameter θ. It is demonstrated that monitoring data are mostly enveloped in calculated curve with similar curve shape. The comparison results verify the feasibility of adopting Gauss distribution curves to calculate settlement profile parallel to the excavation.

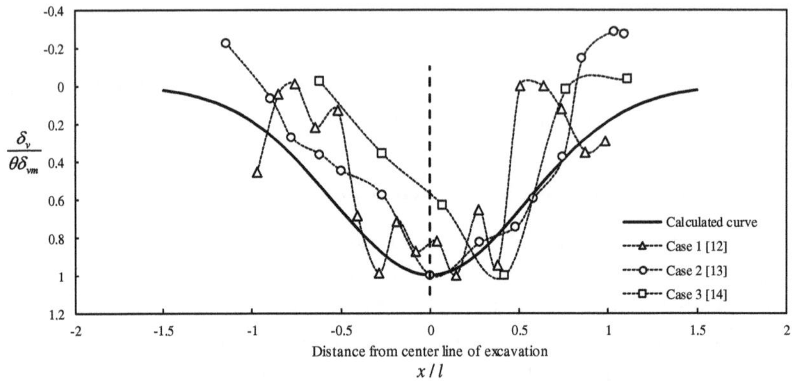

Fig. 1. Comparison of calculated curve and monitoring data

3 Procedure for Calculating Settlement

According to the preceding sections, the proposed method can be applied based on the following procedure:

(1) Estimate proportional coefficient α according to design parameters and geological conditions. Then calculate max ground settlement δ_{vm} based on Eq. (1);
(2) Obtain vertical settlement profile according to Eq. (2) to get ground settlement δ_v at different position in vertical direction.
(3) Get reducing coefficient k_y based on Eqs. (3)–(6). Then calculate soil settlement at designated depth δ_y based on Eq. (7);
(4) Obtain parallel settlement profile according to Eq. (8). Calculate geometry modified parameter θ based on Eq. (9). Then get final modified settlement δ_{vf} based on Eq. (10).

In summary, settlement due excavation in arbitrary position can be derived as Eq. (11). The formula shows that the proposed method consider settlement distribution in three dimensions.

$$\delta_{vf}(d/H, y, x/l) = \alpha H \cdot k_s(d/H) \cdot k_y(y) \cdot k_x(x/l) \cdot \beta\theta \tag{11}$$

4 Case Study

Excavation of Xinzhuang transportation hub is located at southern Shanghai. Diaphragm wall with 1 m thickness and 33 m effective wall depth is used as retaining structure. Excavation pit area is 3300 m^2. Excavation depth is 14.85 m with part depth of 15.85 m for towers as shown in Fig. 2. The excavation depth of shaded area in Fig. 2 is 15.85 m while the rest of pit is 14.85 m. The geological conditions are shown in Fig. 3. Four concrete supports are set at the depth of 0.5 m, 4.5 m, 8.2 m and

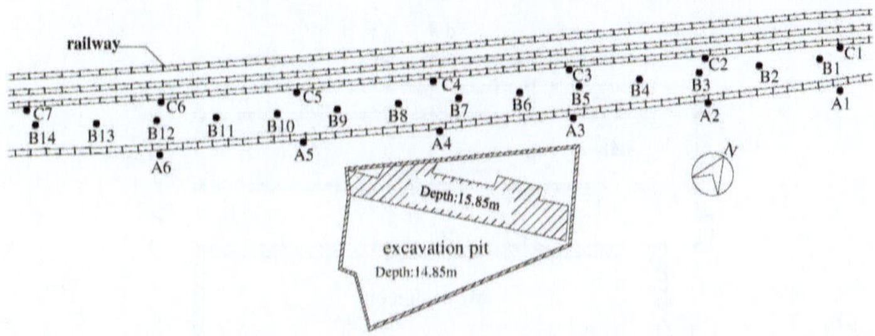

Fig. 2. Plane sketch of excavation

11.4 m. Existing railways are located at northern excavation with closest distance of 10 m. 27 monitoring point are set in field to monitor ground settlement along railway due to excavation. Monitoring points are divided into three groups. Monitoring series A, B and C are 10 m, 23 m and 28 m away from excavation respectively. The length of excavation close to the railway is 75 m. The length of eastern excavation edge is 37 m

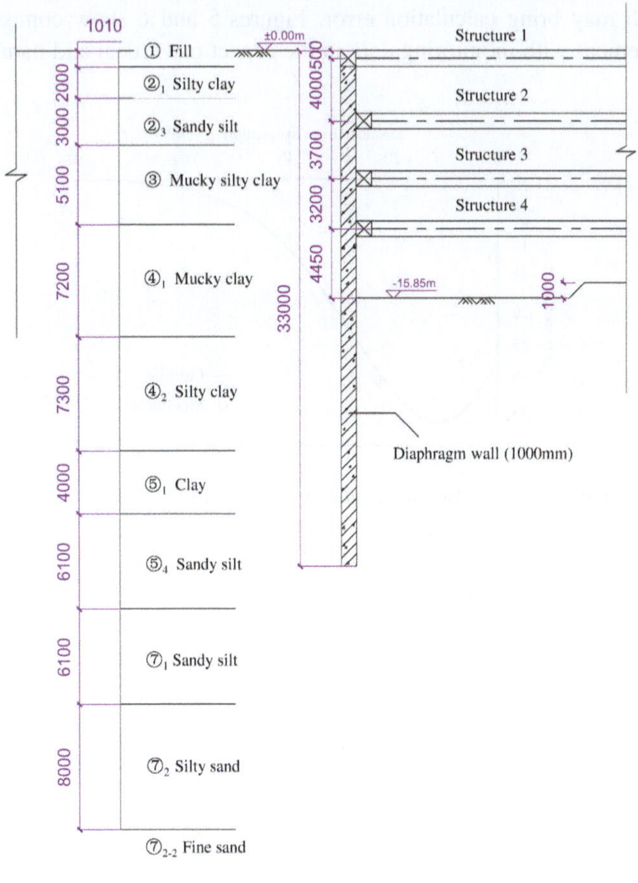

Fig. 3. Section of distribution of soil layers

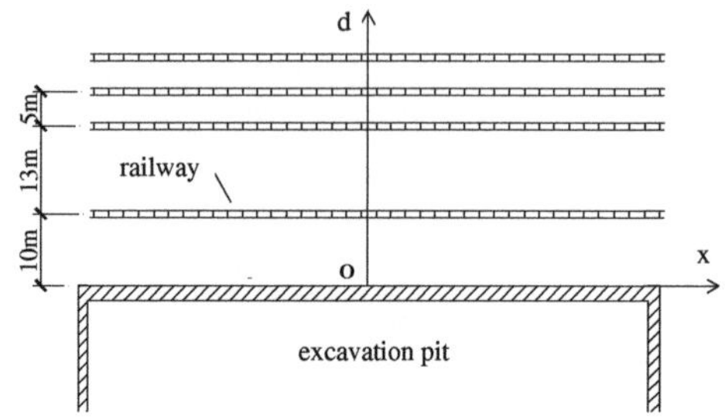

Fig. 4. Coordinate sketch of excavation

while of western excavation edge is approximately 50 m. The excavation width at center line is 44 m. The calculated coordinate is drawn in Fig. 4.

According to the method proposed above, the final settlement can be predicted due to field data. In this project, proportional coefficient α is set as 0.3%. As for railway foundation, strong foundation improvement is usually used. In this case, empirical coefficient β is set as 1. The excavation depth is 15 m as an average value for calculation, which may bring calculation error. Figures 5 and 6 show comparison of calculated settlement with monitoring data in the aspect of vertical and parallel direction,

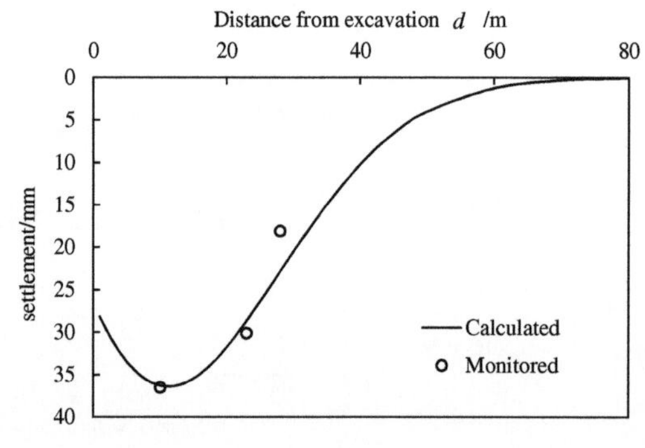

Fig. 5. Comparison of between calculated result and field data in vertical direction

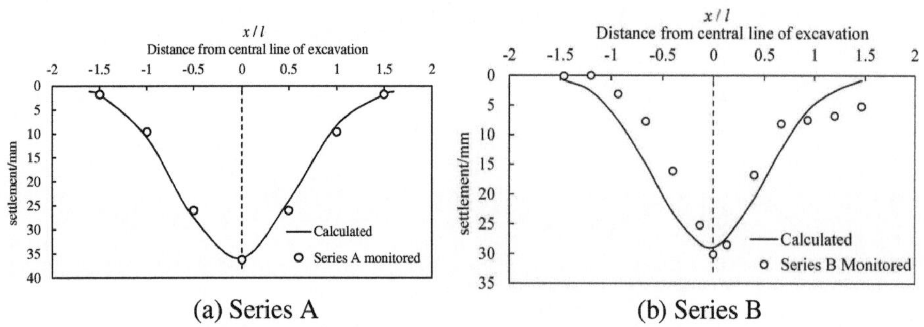

(a) Series A (b) Series B

Fig. 6. Comparison of between calculated result and field data in parallel direction respectively. Figure 6(a) and (b) show the comparison between calculated results and monitored data of series A and series B respectively. It is demonstrated that the proposed method for calculating ground settlement along railway due to excavation is feasible.

5 Conclusion

This paper proposed a simplified method to predict settlement due to excavation based on Rayleigh and Gauss distribution functions. Settlement profile vertical to excavation edge adopts Rayleigh distribution functions while settlement profile parallel to excavation edge uses Gauss distribution functions. Through settlement analysis from two direction, a three-dimensional settlement function is established with usage of excavation depth and width for modification. Comparison of calculated results and field data shows acceptable and applicable prediction of proposed method. More data analysis based on engineering projects have to be conducted to verify the value of proportional coefficient k_δ and empirical coefficient β for further study.

Acknowledgement. This work was supported by the National Natural Science Foundation of China (Grant No. 41572285), the National Natural Science Foundation of China (Grant No. Grant No. 41072204) and Shanghai Sailing Program (Grant No. 02302360185). The authors are deeply indebted to these financial supporters.

References

1. Hou, Y.M., Wang, J.H., Zhang, L.L.: Three-dimensional numerical modeling of a deep excavation adjacent to Shanghai metro tunnels. In: International Conference on Computational Science - ICCS 2007, Beijing, China, pp. 1164–1171 (2007)
2. Liu, C., Guo, X., Gao, W., Jia, J.: Impact study of deep foundation pit excavation on operating railway trestle in soft clay region. In: Geo-Shanghai 2014, Shanghai, pp. 332–342 (2014)
3. Yu, X.L., Yan, Q.S., Wang, W.F.: Analysis and study of safety influence of foundation pit excavation on adjacent existing subway structure. In: Tenth International Conference of Chinese Transportation Professionals, Beijing, pp. 2829–2838 (2010)
4. Hsieh, P.-G., Ou, C.-Y.: Shape of ground surface settlement profiles caused by excavation. Can. Geotech. J. **35**, 1004–1017 (1998)
5. Peck, R.B.: Deep excavation and tunneling in soft ground. In: Proceedings of the 7th Soil Mechanics and Foundation Engineering, State-of-the-Art-Volume, Mexico City, pp. 225–290 (1969)
6. Bowles, J.: Foundation Analysis and Design. McG-H, pp. 803–806 (1990)
7. Clough, G., O'Rourke, T.D.: Construction induced movements of insitu walls. In: Proceedings of the Design and Performance of Earth Retaining Structures, Geotechnical Special Publication, vol. 25, pp. 439–470. ASCE, New York (1990)
8. Ou, C.-Y., Hsieh, P.-G.: A simplified method for predicting ground settlement profiles induced by excavation in soft clay. Comput. Geotech. **38**, 987–997 (2011)
9. Yang, M., Lu, J.Y.: Estimation ground settlement aroused by deep excavation. Chin. J. Geotech. Eng. **32**(12), 1821–1828 (2010)

10. Wang, J.H., Xu, Z.H., Wang, W.D.: Wall and ground movements due to deep excavations in shanghai soft soils. J. Geotech. Geoenviron. Eng. **136**(7), 985–994 (2010)
11. Ding, Y.C.: Excavation-induced deformation and control in soft deposits. Master thesis, Shanghai Jiaotong University (2009)
12. Fan, F., Chen, J.J., Zhang, H.B., Wang, J.H.: Simplified method to predict settlements of adjacent buildings due to deep excavation. Chin. J. Geotech. Eng. **37**(S2), 61–64 (2015)
13. Zhang, X.D.: Study on the retaining structure of deep foundation pit and safety control of subgrade close to railway. Master thesis, Beijing University of Technology (2015)
14. Shi, Y.F.: Stability study on the retaining structure of excavation pit close to railway and subway station. Master thesis, Central South University (2010)
15. Li, M.G., Chen, J.J., Xu, A.J., Dong, F., Wang, J.H.: Interactive behavior between the deep excavation and close operating railway. Chin. J. Undergr. Space Eng. **11**(02), 435–439 (2015)

Local Stiffness Quantification of Geogrid Stabilized Aggregates in Relation to Deformation Behavior

Yong-Hoon Byun[1] and Erol Tutumluer[2(✉)]

[1] Kyungpook National University, Daegu 41566, Korea
[2] University of Illinois at Urbana-Champaign, Urbana, IL 61801, USA
tutumlue@illinois.edu

Abstract. Permanent deformation accumulation is one of the most critical damage mechanisms of unbound aggregate base layer. However, the effect of stress state on permanent deformation behavior of flexible pavement is not clearly known. This study presents the variation in the shear modulus of geogrid reinforced/unreinforced aggregates according to permanent deformation behavior. To characterize the local stiffnesses near and far away from geogrid, two pairs of bender elements are inserted in triaxial specimens. As the number of repeated loading increases, the variation in the shear waves at two different heights is monitored. The test results show that, after the specimen preparation, the shear modulus near the geogrid is greater than that far away from the geogrid. The shear modulus estimated at both levels of unreinforced specimen is similar to that estimated far away from geogrid in the reinforced specimen. This study demonstrates that the local stiffness of aggregate can be monitored by using the bender elements under permanent deformation behavior, and suggests the bender element systems can be effectively used to validate the effect of geogrid reinforcement under permanent deformation behavior.

Keywords: Aggregate · Bender element · Geogrid · Permanent deformation Shear wave

1 Introduction

Permanent deformation accumulation (or rutting) is the primary load-associated distress in unbound aggregate layer and thus, the rutting resistance of the unbound aggregate layer is regarded as a main aspect for the design of multi-layered pavement structure. Many researchers reported that the permanent deformation of unbound aggregate layers may contribute the total accumulation of permanent deformation in the pavement structure [1–3]. The proper rutting resistance of unbound aggregate layer is required more importantly for unsurfaced or thinly surfaced pavements.

The geogrid is widely used for base reinforcement in flexible pavement, and the primary function of geogrid is known as lateral confinement, which is expected to enhance the horizontal stiffness in the mechanically stabilized base layer [4]. The geogrid can be placed within base/subbase layers or at the interface between base layer and subgrade to improve the rutting resistance of pavement structure [5].

X. Shi et al. (Eds.): GSIC 2018, *Proceedings of GeoShanghai 2018 International Conference: Transportation Geotechnics and Pavement Engineering*, pp. 109–114, 2018.
https://doi.org/10.1007/978-981-13-0011-0_12

Many researchers have studied about the quantification of lateral confinement induced by geogrid. Perkins et al. [6] reported that the influenced zone is formed above and below the geogrid by measuring the radial displacement along the triaxial specimen height. Chen et al. [7] also showed the influenced zone around the geogrid in the ballast specimen using discrete element method. Recently, Byun and Tutumluer [8] observed the local stiffness enhancement by geogrid in aggregate specimen based on shear wave measurement. Compared to resilient modulus, the shear modulus obtained from using shear wave transducers was useful for quantifying the local stiffness. This paper describes the local stiffness characteristics of geogrid-reinforced and unreinforced aggregates subject to the repeated loading using embedded shear wave transducers called bender elements.

2 Experimental Study

The aggregates sampled from a quarry located in North Wilkesboro, North Carolina were used in this study. After sieving, the grain size distribution curve was obtained as shown in Fig. 1. The source gradation of the aggregates satisfies the requirement for base course material with regard to the upper and lower limits of North Carolina DOT specification. The mean diameter (D_{50}) and percent passing the No. 200 sieve are 8.4 mm and 7.6%, respectively.

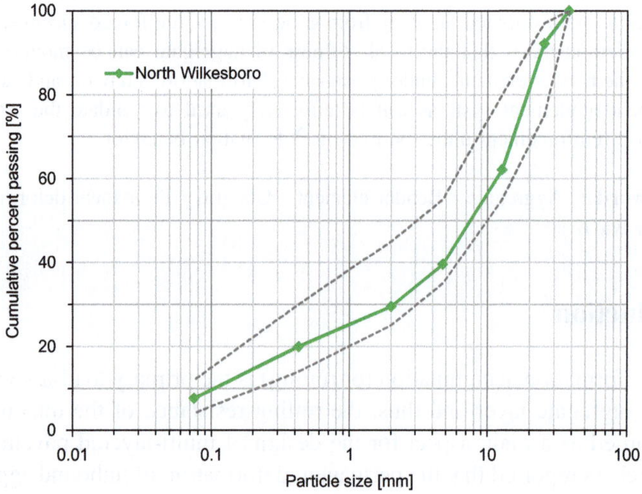

Fig. 1. Grain size distribution of aggregate used in this study.

The geogrids with rectangular aperture designed for multidirectional confinement within base layer were used in this study. The detailed information about the geogrid dimensions are summarized in Table 1. The rib pitch and mid depth of geogrid are known to be related to the shear strength of the reinforced aggregates [9].

Table 1. Dimension of rectangular aperture geogrid used in this study.

Rib pitch [mm]	Mid depth [mm]
33	2.8

Using the aggregates and geogrid, the reinforced and unreinforced specimens were prepared for permanent deformation tests. The two specimens which have 150 mm-diameter and 300 mm-height were compacted according to the Proctor test procedure [ASTM D1557] using the 56 blows per lift and the modified compactive effort of $2,700\ \mathrm{kN} \cdot \mathrm{m/m^3}$. The compaction process was completed to 6 different lifts. For the reinforced specimen, the geogrid was placed at the middle of the specimen height. Regardless of geogrid reinforcement, two sets of bender elements used as shear wave transducers were embedded around the top and middle of the specimen height, as shown in Fig. 2 (refer to Byun and Tutumluer [8] for detailed information about the bender element arrangement). Based on the bender element arrangement, the horizontal shear modulus can be estimated at two different levels of specimens.

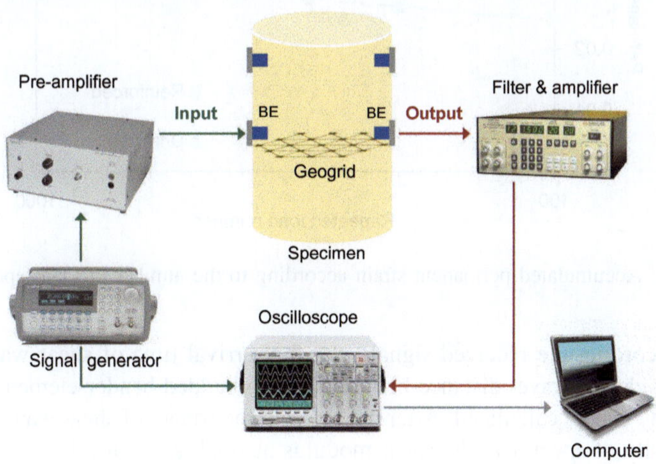

Fig. 2. Triaxial specimen including two bender elements and shear wave measurement system.

After the specimen preparation, the confining pressure of 103.4 kPa was applied to both specimens and then, the axial deviator stresses of 103.4 kPa were repeated up to the 1,000 cycles in the shape of haversine pulse. The waveform of dynamic deviator stress consists of 100 ms-loading period and 900 ms-resting period.

The measurement system for shear wave was configured as shown in Fig. 2. Signal generator and pre-amplifier were used to excite and intensify the input signals. To minimize the noise on the received signals, filter-amplifier was used. The signals recorded in oscilloscope were saved in a personal computer. Using the measurement system, the shear waves at two different levels of specimens were recorded at the resting period of load pulse after every 100 loading cycles.

3 Results and Discussion

The accumulated permanent strain of reinforced and unreinforced specimens along the number of repeated loading is plotted in Fig. 3. After 100 load repetitions, the accumulated permanent strains for reinforced and unreinforced specimens are 0.025% and 0.046%, respectively. The accumulated permanent strains for both specimens increase with an increase in the repeated load applications. The final accumulated permanent strains for two specimens reach to 0.033% and 0.055%, respectively. After all, the accumulated permanent strains for reinforced specimen are smaller than those for the unreinforced specimen.

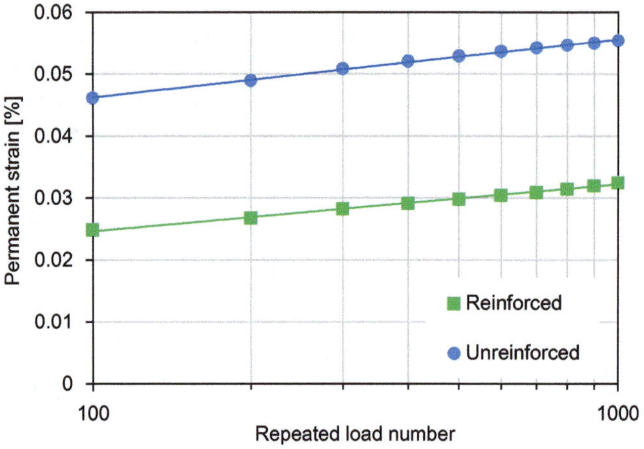

Fig. 3. Accumulated permanent strain according to the number of load repetition.

From recording the received signals, the first arrival time of shear wave is determined. Considering travel distance between two embedded bender elements, the shear wave velocity can be calculated. After multiplying the square of shear wave velocity by the dry density of specimen, the shear modulus at small strain level can be estimated. The shear modulus obtained from two triaxial specimens for different load cycles is plotted in Fig. 4. Figure 4 also shows the variation in shear modulus at two different heights in the reinforced specimen. The shear modulus at the middle of specimen is much greater than that at the top of specimen. After 1,000 load repetitions, the shear modulus at both locations slightly increases compared to that at the beginning. In contrast, the variation in shear modulus at two different levels of unreinforced specimen remains almost constant. The shear modulus after 100 load cycles is approximately 125 MPa for both levels of the specimen. The shear modulus at the top of unreinforced specimen slightly increases after 1,000 load cycles.

On the other hand, the shear modulus estimated from the unreinforced specimen was similar to that estimated at the top of reinforced one. These results support the concept of residual stresses assigned around a geogrid in the compacted aggregate layer suggested by Kwon et al. [10]. Considering that the target density for both specimens is

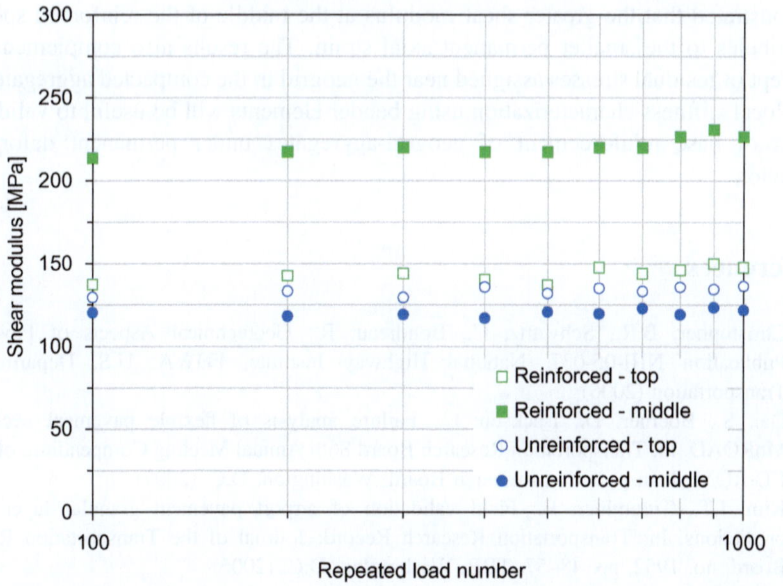

Fig. 4. Variations in shear modulus at two different specimen heights changing with the number of load repetitions.

consistent by the same water content and compaction procedure, the greater shear modulus at the middle of the reinforced specimen contributes to the smaller permanent axial strain. Note that these tests were performed as the conditioning for resilient modulus test and thus, these results are only valid for up to the 1,000 load repetitions. For future study, more test data are required to validate the effect of local stiffness enhancement by geogrid on the permanent deformation behavior.

4 Summary and Conclusions

This paper presented the use of bender elements for reinforced/unreinforced triaxial specimens to characterize the local stiffness enhancement by geogrid for the permanent deformation behavior. Two sets of bender elements were installed at the middle and top of the specimen height. The reinforced/unreinforced specimens were prepared at the same conditions of water content and compaction. The permanent deformation tests were performed under the 103.4 kPa confining pressure and 103.4 kPa dynamic axial deviator stress until the number of load repetitions reached to 1,000. The accumulated permanent strains for the reinforced specimen were smaller than those for the unreinforced one. After every 100 load repetitions, the shear waves were recorded at each specimen height. The shear modulus at the middle of reinforced specimen was greater than that at the top of reinforced specimen, while the shear moduli at both heights of the unreinforced specimen was similar to each other. Furthermore, the shear moduli at both heights of the unreinforced specimen was similar to that at the top of reinforced specimen. These results demonstrated that the greater shear modulus at the middle of the reinforced specimen contributes to the smaller permanent axial strain. The results also complemented the concept of residual stresses assigned near the geogrid in the compacted aggregate layer. The local stiffness characterization using bender elements will be useful to validate the effect of base reinforcement of geogrid-aggregates under permanent deformation behavior.

References

1. Christopher, B.R., Schwartz, C., Boudreau, R.: Geotechnical Aspects of Pavements. Publication NHI-05-037. National Highway Institute, FHWA, U.S. Department of Transportation (2006)
2. Dai, S., Boerner, D., Isackson, C.: Failure analysis of flexible pavement section on MnROAD. In: Transportation Research Board 86th Annual Meeting Compendium of Papers CD-ROM. Transportation Research Board, Washington, D.C. (2007)
3. Kim, I.T., Tutumluer, E.: Field validation of airport pavement granular layer rutting predictions. In: Transportation Research Record: Journal of the Transportation Research Board, no. 1952, pp. 48–57. TRB, Washington, D.C. (2006)
4. Giroud, J.P.: An assessment of the use of geogrids in unpaved roads and unpaved areas. In: Proceedings of the Jubilee Symposium on Polymer Geogrid Reinforcement, London, UK, 8 September 2009 (2009). http://www.tensar.co.uk/jubilee-symposium/index.html
5. Kwon, J., Tutumluer, E.: Geogrid base reinforcement with aggregate interlock and modeling of associated stiffness enhancement in mechanistic pavement analysis. In: Transportation Research Record: Journal of the Transportation Research Board, no. 2116, pp. 85–95. Transportation Research Board of the National Academies, Washington, D.C. (2009)
6. Perkins, S.W., Christopher, B.R., Cuelho, E.L., Eiksund, G.R., Hoff, I., Schwartz, C.W., Svano, G., Want, A.: Development of Design Methods for Geosynthetic Reinforced Flexible Pavements. FHWA report DTFH61-01-X-00068 (2004)
7. Chen, C., McDowell, G.R., Thom, N.H.: Investigating geogrid-reinforced ballast: experimental pull-out tests and discrete element modelling. Soils Found. **54**(1), 1–11 (2014)
8. Byun, Y.H., Tutumluer, E.: Bender elements successfully quantified stiffness enhancement provided by geogrid-aggregate interlock. Transp. Res. Rec.: J. Transp. Res. Board **2656**, 31–39 (2017)
9. Jewell, R.A., Milligan, G., Sarsby, R.W., Dubois, D.: Interaction between soil and geogrids. Polymer grid reinforcement. In: Proceedings of a Conference sponsored by the Science and Engineering Research Council and Netlon Ltd., London, 22–23 March 1984
10. Kwon, J., Tutumluer, E., Konietzky, H.: Aggregate base residual stresses affecting geogrid reinforced flexible pavement response. Int. J. Pavement Eng. **9**(4), 275–285 (2008)

Research on Micro-geometry of Sea Sands Using Scanning Microscope and Particle Analyzer

Zhiyuan Li[1], Pei Cao[2,3](✉), and Junchao Gui[3]

[1] Department of Civil and Environmental Engineering, Stanford University, Palo Alto 94305, USA
[2] Key Laboratory of Failure Mechanism and Safety Control Techniques of Earth-Rock Dam of the Ministry of Water Resources, Nanjing 210029, China
[3] Department of Geotechnical Engineering, College of Civil Engineering, Tongji University, Shanghai 200092, China
pcao@tongji.edu.cn

Abstract. The micro-geometry of soil plays a vital role in its mechanical, osmotic and deforming properties. This paper presents micro-geometrical analysis of sea sand by applying Scanning Electron Microscopy (SEM) and Laser Particle Size Analyzer (LPSA). After image processing by Matlab, it can be concluded that with the augment of particle size, elongation and convexity increases while roundness and fractal dimension dwindles. Hence, the particle shape will be flatter and the surface will be smoother as well as less complicated with the growth of size. Besides, a majority of the grains are classified as rounded, and shapes of the particles turn to be less diverse for larger groups. Moreover, results of coefficient analysis show that convexity and roundness have a strong negative linear relation, while contour factor behaves little correlation with other parameters. Last by as significant, data provided by LPSA presents smaller deviation standard and consistently smaller elongation than SEM. This paper aims to provide reference regulations for experiments and simulations.

Keywords: Micro-geometry · Sea sand · Digital image processing

1 Introduction

With the development of digital image processing (DIP), it has been known that the micro-geometry of sands has an essential influence in its geotechnical properties [1]. To quantitatively describe particle shapes, Wadell [2] firstly defined roundness and spherity in 1935. Two years later, Russell and Taylor [3] proposed classification of particle shape. Then in 1953, Powers [4] proposed a standard roundness chart.

Since then, many scholars have devoted to the topic. Mora et al. [5], Kwan [6], Zheng and Hryciw [7] proved the effectiveness of DIP. Mora et al. [8] summarized the relation between shape factors and angularity number. Geri and Mari [9] depicted fractal dimension as a function of internal frictional angle. Alshibli et al. [10] concluded that frictional and dilation angle augment with roughness. Man et al. [11] found the relation between shear wave velocity and roundness. Charpentier et al. [12] used a

X. Shi et al. (Eds.): GSIC 2018, *Proceedings of GeoShanghai 2018 International Conference: Transportation Geotechnics and Pavement Engineering*, pp. 115–123, 2018.
https://doi.org/10.1007/978-981-13-0011-0_13

modular Matlab tool and performed Flourier analysis. Yang and Luo [13] presented the correlation between critical state and granular shape.

Moreover, particle shapes have non-negligible impacts in discrete element analysis (DEM) [14]. In 1988, Ting et al. [15] proposed that circular elements present lower shear resistance than real materials in DEM simulation. Ting et al. [16] concluded that elliptical grains have a better simulating result than standard circles do. Then, Antony and Kuhn [17] employed elongated granules in DEM. Later, Kuhn [18] pointed out that local weakening or strengthening behavior has strong relation to shapes. Hence, a further understanding of micro-geometry is necessary for the development of DEM.

This paper presents micro-geometry analysis with Scanning Electron Microscopy (SEM) and Laser Particle Size Analyzer (LPSA). Moreover, shape classification, correlation analysis, and device comparison are performed.

2 Micro-geometrical Characters Measurement

2.1 Experiments Using SEM and LPSA

The soil is classified as sea sands and was divided into seven groups by manual sieving, the specific gravity (G_s = 2.65), minimal dry density (ρ_{dmin} = 1.51) and maximal dry density (ρ_{dmax} = 1.32) were gauged.

The SEM (type: Hitachi SU1510) and LPSA (type: Lixell) in Key Laboratory of Geotechnical & Underground Engineering, Tongji University, were applied to analyze microstructure of the sand. The photos obtained by them were shown in Fig. 1.

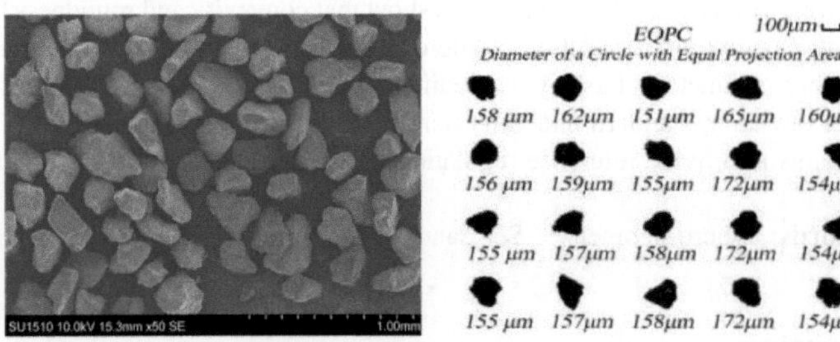

(a) SEM photo (amplitude: 50) (b) LPSA photo (size: 0.15mm-0.2mm)

Fig. 1. SEM and LPSA image

2.2 Definition of Micro-geometrical Characters

According to previous researches [5–13], micro-geometry can be described by:

(i) Overall shape, and the three parameters are defined by Eqs. (1), (2) and (3).

$$\text{elongation} = E = \frac{F_{max}}{F_{min}} \tag{1}$$

$$\text{contour factor} = CF = \frac{\pi D_1}{P_1} \tag{2}$$

$$\text{circularity} = C_s = \frac{4S_1}{\pi \times EQPC} \tag{3}$$

(ii) Surface roughness, namely convexity C_v (Eq. (4)) and roundness R_w (Eq. (5))

$$\text{convexity} = C_v = \frac{S_1}{S_2} \tag{4}$$

$$\text{roundness} = R_w = \frac{\sum r_i}{n \times R_{max}} \tag{5}$$

(iii) Surface texture is scaled by fractal dimension f (see Eq. (6)).

$$\text{fractal dimension} = f = 2 \times \lim \frac{\text{lin } D_1}{\text{lin } S_1} \tag{6}$$

All basic elements from Eqs. (1)–(6) are listed in Table 1 and depicted in Fig. 2.

Table 1. Basic elements

Element	Definition	Element	Definition
S_1	Area of the grain	S_2	Area of the polygon envelope
P_1	Perimeter of the grain	P_2	Area of the polygon envelope
D_1	Diameter of the grain	$EQPC$	Diameter of the converted circle
R_{max}	Radius of the inscribed circle	R_i	Radius of inflection points
F_{max}	Maximum distance between a pair of parallel lines tangent to the grain outline		
F_{min}	Minimum distance between a pair of parallel lines tangent to the grain outline		

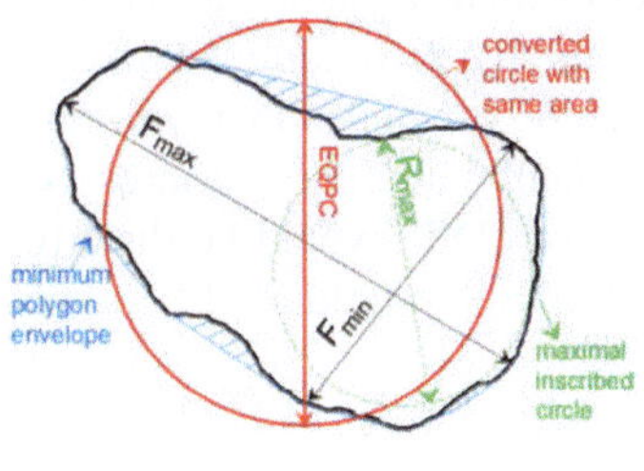

Fig. 2. Explanation for basic elements

2.3 Image Processing and Accuracy Validation

The calculation of R_w is served as an example, and as Fig. 3 illustrates, the image processing by Photoshop and Matlab has mainly three steps.

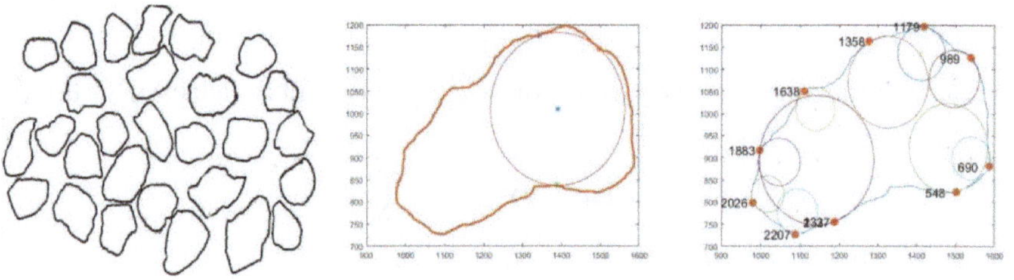

(a) Contour extraction (b) Seeking maximal inscribed circle (c) Finding angular points

Fig. 3. Image processing using Photoshop and Matlab

Image Pro Plus (IPP), one of the most widely used particle analytical software, is utilized to validate the accuracy of proposed method. The smallest and largest size groups were compared according to normalized L-2 norm shown in Eq. (7).

$$||E||_2 = \frac{\sqrt{e_1^2 + e_1^2 + \ldots + e_n^2}}{\sqrt{x_1^2 + x_1^2 + \ldots + x_n^2}} \tag{7}$$

where e is subtraction of measurements by IPP and proposed method, and x is IPP measured data. From Table 2, it can be found that the L-2 norm of error is pretty small, which validates the effectiveness of proposed method.

Table 2. L-2 norm of error by proposed method

Grain size	S_1	S_2	F_{max}	F_{min}
0.1–0.15 mm	3.6%	3.9%	2.5%	2.8%
0.4–0.5 mm	8.4%	8.9%	7.4%	5.3%

The grain boundary photo in Fig. 3(c) was transformed in to vector graph, and was changed into different pixel elementary area varying from 1500 × 1500 to 50 × 50. The Roundness R_w of the same figure in different sizes was measured and the results were shown in Table 3. It can be concluded that with pixel area decreasing, the details of boundary will be lost gradually, and the measured boundary tends to be a sphere. Moreover, the proposed method (DIP) would be distortion when pixel length is 100, which is much less than 500, the threshold of IPP software.

Table 3. Measured roundness in different pixel area

Method	1500 × 1500	1000 × 1000	500 × 500	250 × 250	100 × 100	50 × 50
R_w by IPP	0.556	0.560	0.618	0.617	0.714	0.824
R_w by DIP	0.555	0.558	0.560	0.566	0.638	0.769

3 Discussions and Results

3.1 Geometrical Parameter Analysis

The overall shape contains elongation E, contour factor CF and circularity C_s. As illustrated from Figs. 4, 5 and 6, with the ascending of grain size, elongation increases while contour factor and circularity change in parabolic shape. It can be concluded that the large-size and small-size grains are more closed to circle than medium-size particles, and the flatness of the sand slightly increases when size upraises.

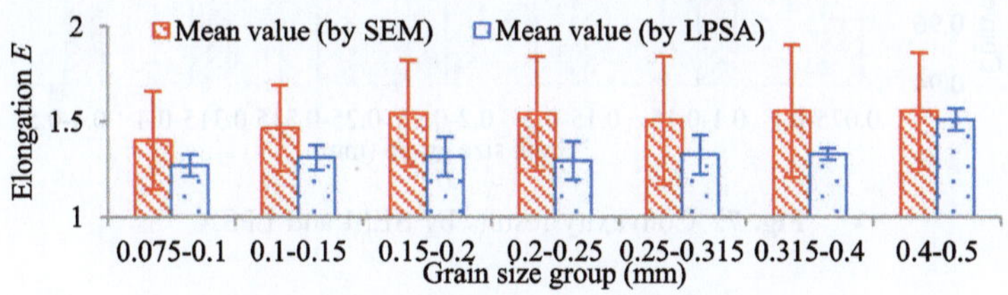

Fig. 4. Elongation results by SEM and LPSA

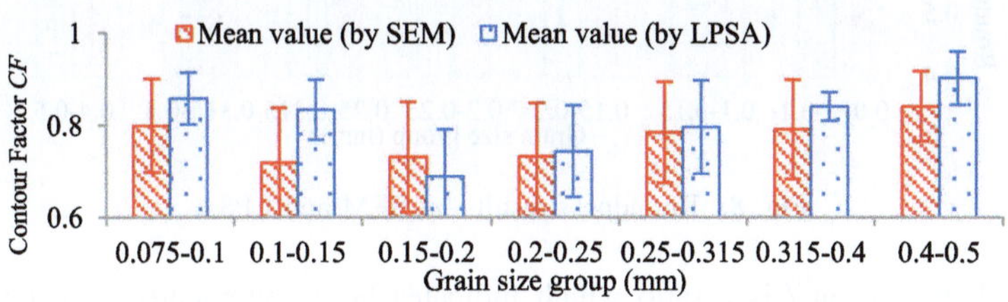

Fig. 5. Contour Factor results by SEM and LPSA

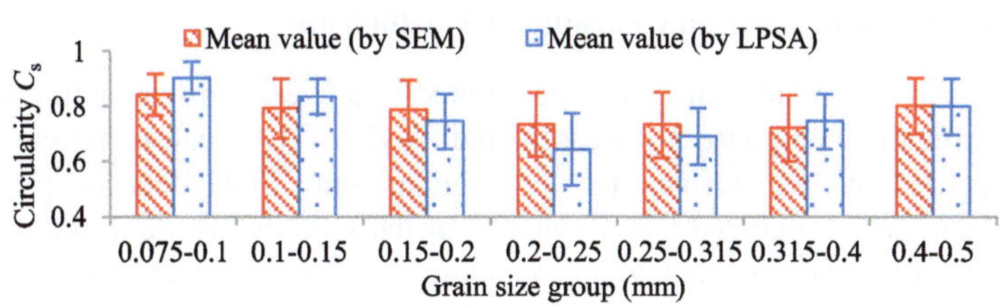

Fig. 6. Circularity results by SEM and LPSA

Moreover, it can be deduced that SEM method usually has a more turbulent result, because of a larger deviation. Besides, the elongation measured by SEM is consistently higher than that by LPSA. The difference can be explained by operating approaches. In LPSA, several grains are rotated by water screw, and their images in various perspectives are documented and averaged. Hence, LPSA shows more rounded and less elongated results. In SEM analysis, only one perspective is measured and a more extreme result may be presented by SEM.

The surface roughness includes two parameters, convexity C_v and roundness R_w. As plotted in Figs. 7 and 8, for these two cases SEM method also has a bigger deviation standard than LPSA. With the augment of grain size, convexity rises while roundness declines. If convexity is close to one, the grain boundary will be a regular polygon, thus becoming smooth. Also, smaller roundness connotes straight outline. Thus, the sandy surface became more rounded with the increase of grain sizes.

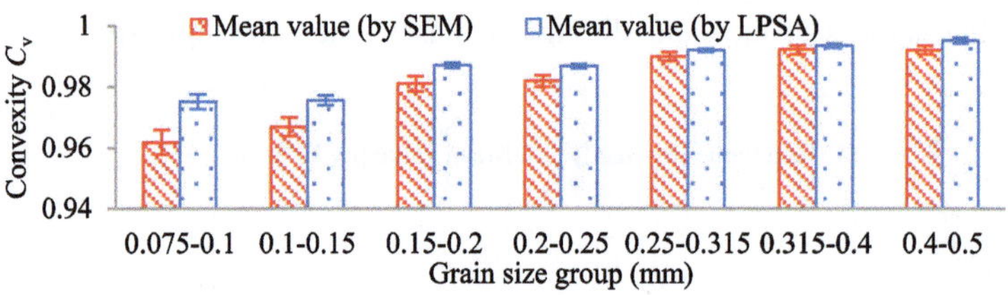

Fig. 7. Convexity results by SEM and LPSA

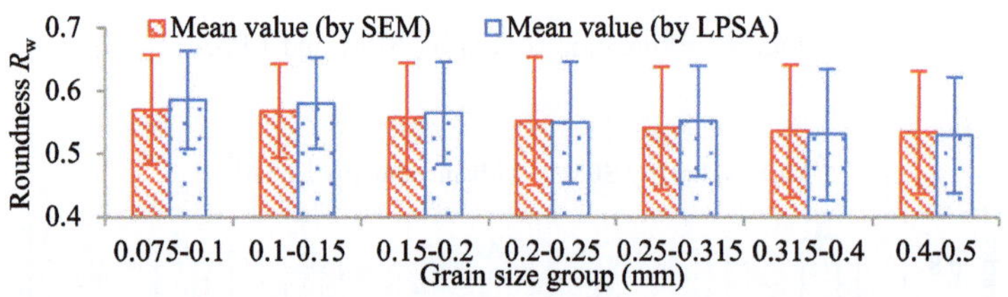

Fig. 8. Roundness results by SEM and LPSA

Fractal dimension f is a ratio which indicates how complicated a pattern is and bigger f denotes more intricate surface texture. From Fig. 9, the surface texture tends to be more simplified when grain size becomes larger.

3.2 Shape Classification and Parameter Coefficients

As Table 4 shows, granular shape can be divided into five groups according to particle shape classification proposed by Russell and Taylor [3]. Because the classification results procured by two devices are very similar, only SEM data are presented in Fig. 10. It can be summarized that a majority of the sand grains are in rounded shape, and the proportion of rounded grains increases with the augment of particle size. Moreover, shapes of the particles turn to be less diverse for larger particles.

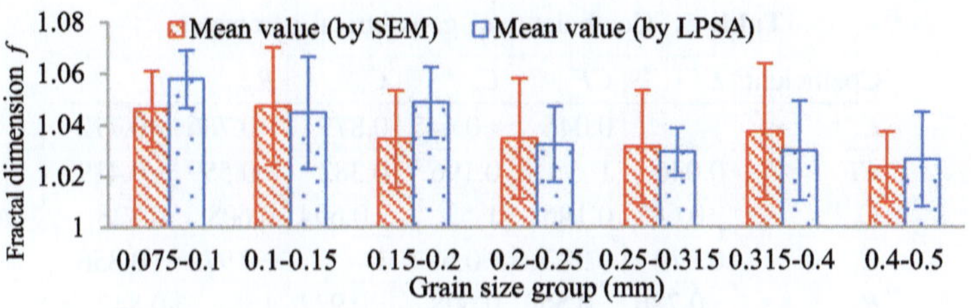

Fig. 9. Fractal Dimension results by SEM and LPSA

Table 4. Shape classification (Russell and Taylor [3])

Classification	Roundness range	Classification	Roundness range
Angular	0–0.15	Sub-angular	0.3–0.5
Sub-rounded	0.3–0.5	Rounded	0.5–0.7
Well rounded	0.7–1.0		

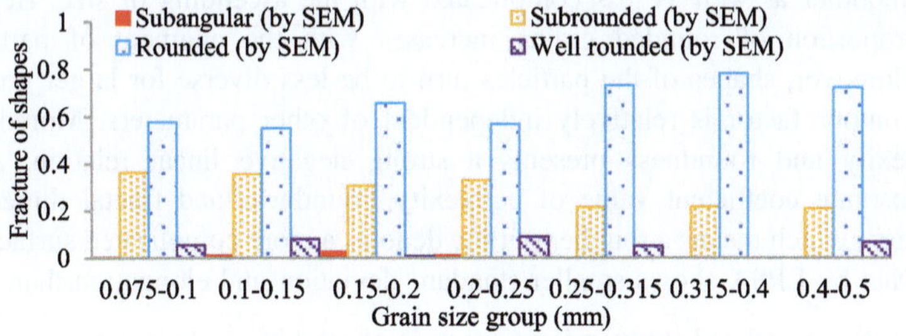

Fig. 10. Shape classification by SEM

The linear correlation between each parameters can be quantitatively scaled by coefficient, which can be calculated as Eq. (8).

$$\mathrm{r}(H_1, H_2) = \frac{\mathrm{cov}(H_1, H_2)}{\sqrt{\mathrm{Var}(H_1) \cdot \mathrm{Var}(H_2)}} \tag{8}$$

In Eq. (8), Cov(H_1, H_2) is covariance of parameter H_1 and H_2, and Var(H_1) as well as Var(H_2) are variance of H_1 and H_2, respectively.

As Table 5 shows, with the growth of size, elongation E and convexity C_v rises, while roundness R_w and fractal dimension f dwindles. Moreover, contour factor CF shows little correlation with other parameters, so it is relatively independent. Besides, convexity C_v and roundness R_w presents a strong negative linear relation. Hence, merely choosing one parameter to describe surface roughness is applicable. Last but as significant, the absolute coefficient value of C_v, R_w and f are high. Therefore, a rougher outline denotes a more complicated surface texture.

Table 5. Coefficients of geometrical parameters

Coefficient	E	CF	C_s	C_v	R_w	f
E	1	0.046	−0.647	0.879	−0.786	−0.772
CF	0.046	1	0.196	0.382	−0.559	−0.489
C_s	−0.647	0.196	1	−0.694	0.608	0.328
C_v	0.879	0.382	−0.694	1	−0.972	−0.856
R_w	−0.786	−0.559	0.608	−0.972	1	0.842
f	−0.772	−0.489	0.328	−0.856	0.842	1

4 Conclusion and Future Study

This paper presents variation regulation of sandy micro-geometry using SEM and LPSA. The conclusions are summarized as followings.

(i) With the growth of particle size, elongation and convexity rises while roundness and fractal dimension decreases. Therefore, particle outline will be flatter and smoother as well as less complicated with the ascending of size. Besides, the proportion of rounded grains increases with the augment of particle size. Moreover, shapes of the particles turn to be less diverse for larger groups.
(ii) Contour factor is relatively independent of other parameters. Moreover, convexity and roundness presents a strong negative linear relation. Also, the absolute coefficient value of convexity, roundness and fractal dimension are high, which means a rougher outline denotes a more complicated surface texture.
(iii) Data by LPSA shows smaller standard deviation and elongation than SEM.

According to related study, micro-structure of sand has a strong correlation with its mechanical behavior, such as internal frictional angle [9], dilatantion angle [10] and shear wave velocity [11]. In future study, remolded soil sample made by different grain size group will be tested in oedometer tests, permeability tests and triaxial tests, and the effects of micro-structure will be tested in DEM simulations.

Acknowledgments. The authors thank the National Natural Science Foundation of China (Grant No. 51639008) and the Open Foundation of Key Laboratory of Failure Mechanism and Safety Control Techniques of Earth-Rock Dam of the Ministry of Water Resources (Grant No. YK915004) for their financial support.

References

1. Maiti, A., Chakravarty, D., Biswas, K., Halder, A.: Development of a mass model in estimating weight-wise particle size distribution using digital image processing. Int. J. Min. Sci. Technol. **27**(3), 435–443 (2017)
2. Wadell, H.: Volume, shape, and roundness of rock particles. J. Geol. **40**(5), 443–451 (1932)
3. Russell, R.D., Taylor, R.E.: Roundness and shape of Mississippi River sands. J. Geol. **45**(3), 225–267 (1937)

4. Powers, M.C.: A new roundness scale for sedimentary particles. J. Sediment. Res. **23**(2), 117–119 (1953)
5. Mora, C.F., Kwan, A.K.H., Chan, H.C.: Particle size distribution analysis of coarse aggregate using digital image processing. Cem. Concr. Res. **28**, 921–932 (1998)
6. Kwan, A.K.H., Mora, C.F., Chan, H.C.: Particle shape analysis of coarse aggregate using digital image processing. Cem. Concr. Res. **29**(9), 1403–1410 (1999)
7. Zheng, J., Hryciw, R.D.: Roundness and sphericity of soil particles in assemblies by computational geometry. J. Comput. Civ. Eng. **30**(6) (2016)
8. Mora, C.F., Kwan, A.K.H.: Sphericity, shape factor, and convexity measurement of coarse aggregate for concrete using digital image processing. Cem. Concr. Res. **30**(3), 351–358 (2000)
9. Gori, U., Mari, M.: The correlation between the fractal dimension and internal friction angle of different granuiar materials. Soil Found. **41**(6), 17–23 (2001)
10. Alshibli, K.A., Alsaleh, M.I.: Characterizing surface roughness and shape of sands using digital microscopy. J. Comput. Civ. Eng. **18**, 36–45 (2004)
11. Man, T.B., Priest, J.A.: Discussion of ‘particle shape effects on packing density, stiffness, and strength: natural and crushed sands’ by Gye-Chun Cho, Jake Dodds, and J. Carlos Santamarina. J. Geotech. Geoenviron. Eng. **133**(11), 1473–1474 (2007)
12. Charpentier, I., Sarocchi, D., Sedano, L.A.: Particle shape analysis of volcanic clast samples with the Matlab tool MORPHEO. Comput. Geosci. **51**, 172–181 (2013)
13. Yang, J., Luo, X.D.: Exploring the relationship between critical state and particle shape for granular materials. J. Mech. Phys. Solids **84** (2015)
14. Cundall, P.A., Strack, O.D.L.: A discrete numerical model for granular assemblies. Géotechnique **30**(30), 331–336 (1979)
15. Ting, J.M., Corkum, B.T.: Strength behavior of granular materials using discrete numerical modelling, In: 5th International Conference on Numerical Method in Geomechanics, Innsbruck, Austria, pp. 305–310 (1988)
16. Ting, J.M., Meachum, L., Rowell, J.D.: Effect of particle shape on the strength and deformation mechanisms of ellipse-shaped granular assemblage. Eng. Comput. **12**(2), 99–108 (1995)
17. Antony, S.J., Kuhn, M.R.: Influence of particle shape on granular contact signatures and shear strength: new insights from simulations. Int. J. Solids Struct. **41**(21), 5863–5870 (2004)
18. Kuhn, M.R.: Are granular materials simple? an experimental study of strain gradient effects and localization. Mech. Mater. **37**(5), 607–627 (2005)

Research on the Effect of Micro Particle Shape of Marble Using Clump Parallel-Bond Models

Dan Huang(✉) and Xiao-Qing Li

Department of Civil Engineering and Mechanics, HUST, Wuhan, China
danni_452@126.com

Abstract. The deformation and failure characteristics of brittle-ductile-plastic transition, which is shown in the post-peak stage for Jinping deep marble. Using boned-particle model (BPM) to simulate the marble could not show this property. Therefore, it is simulated by clumped-particle model (CPM) to studied from the micro view of the particle flow code (PFC). In the PFC version 5.0, the different shape of particle can be created by the clump templates, it could be templates which combined with the single or multiple particles, and it could be also mixed ball and clump to generate different rock model. In this paper, different clump ratio, different clump components, different clump components position are considered. And then models were applying the uniaxial compression test to research macroscopic mechanical properties of the synthetic rock. The related numerical results have demonstrated that: (1) The effect of particle shape on the rock is very significant, especially in the case of high confining stress. (2) clump template is very important for changing the mechanical behavior of the model. The ratio of the clump is higher, the larger the number of clump particles in the model is, the greater the strength of the model will be. For different clump template, the more closely linked of the pebble is, the greater strange of the model will be. For the same number of the pebble, the larger intersection area is, the greater the "lock-in" effect, the stronger the whole model will be.

Keywords: Discrete element method · Brittle-ductile-plastic transition
Effect of particle shape · Bonded-particle model (BPM)
Clumped-particle model (CPM)

1 Introduction

When the rock material is subjected, the meso-structure will produce meso-damage or micro cracks. Initiation, growth, expansion and convergence of micro cracks will lead to the destruction of particles or bonding, and give rise to the stress redistribution in the local range, and possibly cause more meso-structure failure. The ongoing process will eventually result in macroscopic damage to the material [1]. Undoubtedly, whether analytical method or experimental method is very difficult for the research of this process [2]. The particle flow code (PFC) [3] is based on discrete element method describes the basic mechanical properties of rock, starting with the contact state between particles and its evolution characteristics. PFC reflects the change of macro mechanical properties of materials through the formation, development and confluence

X. Shi et al. (Eds.): GSIC 2018, *Proceedings of GeoShanghai 2018 International Conference: Transportation Geotechnics and Pavement Engineering*, pp. 124–135, 2018.
https://doi.org/10.1007/978-981-13-0011-0_14

of micro cracks. In this process, complex constitutive relation and flow rule are not needed, and the difficulty of description due to the microscopic characteristics of rock itself can be overcome, so that the complex mechanical characteristics of rock can be simulated and reproduced.

In the calculation of discrete element method, the discrete body allows finite displacements and rotations, and can automatically disengage and identify new contacts according to Newton's second law [4]. It is the development of these theories that make the discrete element method more software, and thus get more application and development. Potyondy et al. [5] studied the particle bonding model of rock, and analyzed the micro mechanical parameters which affect the BPM (Bonded Particle Model). BPM media uses circular particles. The reason why the internal friction angle of BPM medium is low is that the particles considered are spherical, and the contact surface between particles is smaller, so that the friction between particles is smaller. Clump can reflect the change of macro mechanical behavior of medium caused by different expansion modes of micro fracture, and has unique advantages in complex rock and stone study. Cho et al. [6] studied the aggregate model of rocks and explained the Clump theory, model establishment and its role in increasing particle friction.

When using discrete element method to simulate rock, besides the micro mechanical parameters of particles, the shape of particles also has great influence on the mechanical properties of rock. It is revealed in the round/spherical particles that it cannot exhibit the characteristics of brittle-ductile-plastic transition at the post-peak. To change the shape of the particles, the first step is to generate a clump template, which is generated by fixing the size and position of several pebble particles. Then, the original particles are replaced with clump particles, resulting in irregularly shaped particles.

In this paper, three different cases were used to research the substitution of clump for round particles. First, clump uses 2pebble templates while the mix proportion of clump is different. Mixture ratios are as fellows, 0.2; 0.4; 0.6; 0.8. Second, clump uses 2pebble, 3pebble and 4pebble templates respectively while all clump uses a mix proportion of 0.5. Third, clump uses 2pebble templates while there is a gap between the 3 templates, in other words, the locking degree between two particles is different.

2 Mechanical Properties of Marble

2.1 Mechanical Properties of Marble in Jinping

In order to deepen the understanding of the mechanical properties of marble in Jinping Baishan Group, Chu [7] carried out borehole sampling and laboratory rock tests at the eastern end of the auxiliary tunnel A and distance the horizontal tunnel 11# 45 m. The buried depth of sampling location is over 2000 m, which is located in marble of Baishan formation. 30 pieces of rock sample were processed by drilling core. 12 of them were subjected to uniaxial compression tests and the number of acoustic emission during compression was monitored. The other 18 rock blocks were subjected to triaxial compression tests. The confining stress is set between 2–50 MPa. The tests were carried out on a MTS of Wuhan geotechnical institute and loaded at typical strain loading rates. Stress and strain data were collected every 0.5 s. The acoustic emission probe is placed directly on the rock sample.

The damage threshold and strength results of each sample are summarized in Table 1.

Table 1. The results of each sample of uniaxial compression test and triaxial compression test

Sample	Peak strength σ_c(MPa)	Elastic modulus E(GPa)	σ_{ci}/σ_f	σ_{cd}/σ_f
A38	105	48	0.45	/
A45	122	50	0.4–0.5	0.74
A47	124	55	0.4	0.8
A48	105	35	0.5	0.95
A49-1	105	40	0.5	0.78
A49-2	101	28	0.6	0.82
B11	123	45	0.45	0.81
B34	99	32	0.56	0.82
B35	98	38	0.45	0.84
C29	144	60	0.55	0.78
C35	121	50	0.3–0.5	0.9

Notes: σ_c represents uniaxial compressive strength, σ_{ci} represents crack initiation strength, σ_{cd} represents damage strength.

The result of uniaxial compression test of marble in Baishan group are as follows. The dispersion of uniaxial peak strength is larger, the lowest 98 MPa, the highest 144 MPa and the average 113 MPa. The strength of marble measured in the test is slightly lower than that on site. The strength loss during the transport of rock samples is probably an important factor. Considering comprehensively, the uniaxial strength of marble in Baishan formation should be above 120 MPa. The threshold of crack growth is generally between 0.4–0.55 times the peak intensity. The damage threshold of deep marble in Jinping is slightly higher than granite. Generally, the unstable expansion threshold of crack energy is slightly higher than 0.8 times the peak intensity, with an average value of 0.824.

The result of triaxial compression test of marble in Baishan formation are as follows. The peak strength of rock block is higher than 125 MPa after applying confining pressure of 2 MPa, the highest is 143 MPa, and the average is 137 MPa. After confining pressure is applied, obvious residual strength exists in the rock block. When the confining pressure exceeds 6 MPa, the confining pressure of 5–6 MPa may be the threshold of the transition from brittle failure to ductile failure of marble. When the confining pressure exceeds 40 MPa, the yield failure of marble shows an ideal plastic feature. These laws have important reference value to judge the brittle failure of intact surrounding rock.

2.2 Micro-Parameter Calibration of BPM Model

In the current study, the BPM was constructed and calibrated against the laboratory test results of marble samples extracted from the site of Jinping II hydropower station located at Yalong River of Sichuan, China [8]. Involving the construction of four 16 Km-long headrace tunnels with the maximum depth of 2,525 m, the Jinping II hydropower station is the largest of its kind in the world.

The establishment of BPM was done by following the material-genesis procedure proposed by Potyondy and Cundall, which allowed the generation of a homogenous sample with a relatively uniform locked-in stress field. The model has a dimension of 50 × 100 mm^2, which is the same size as that of the real marble sample. The macro-properties of Jinping marble used for model calibration includes Young's modulus E, unconfined compressive strength and Poisson's ratio. This parametric study yielded a set of best-fit parameters as listed in Table 2.

Table 2. The micro-parameters of BPM numerical model

Particle parameters		Parallel bond parameters	
E_c	45 GPa	$\bar{E}_c$	45 GPa
k_s/k_n	2.5	$\bar{k}_n/\bar{k}_s$	2.5
μ	2.5	$\bar{\sigma}_c$	120e6 ± 24e6
R_{max}/R_{min}	1.66	$\bar{\tau}_c$	120e6 ± 24e6
R_{min}	1 mm	$\bar{\lambda}$	1.0

The corresponding stress-strain relationship of uniaxial compression test and triaxial test are shown in Fig. 1. The numbers of confining stress are set as 2,4,6,8, 10,30,40, consistent with the triaxial test.

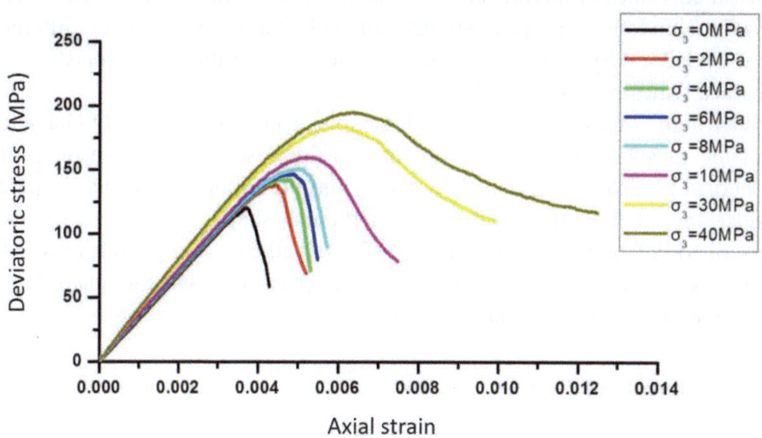

Fig. 1. Uniaxial compression test and triaxial test of BPM

It is seen in Fig. 1 that the elastic modulus, peak intensity and brittle fracture characteristics of uniaxial compression test are consistent with laboratory experiments. And in the triaxial compression test, with the increase of confining pressure, the peak strength of triaxial compression test increases. However, the post-peak characteristics do not exhibit mechanical characteristics of brittle-ductile-plastic transition with the increase of confining pressure, but only exhibit ductility under high confining pressure.

In order to further analyze the failure characteristics of marble, the rock peak strength characteristics based on Hoek-Brown criterion are shown in Fig. 2. Figure 2a is the peak strength characteristic of group A marble, Fig. 2b is the peak strength characteristic of group B marble, and Fig. 2c is the peak intensity characteristic of triaxial compression test using BPM model.

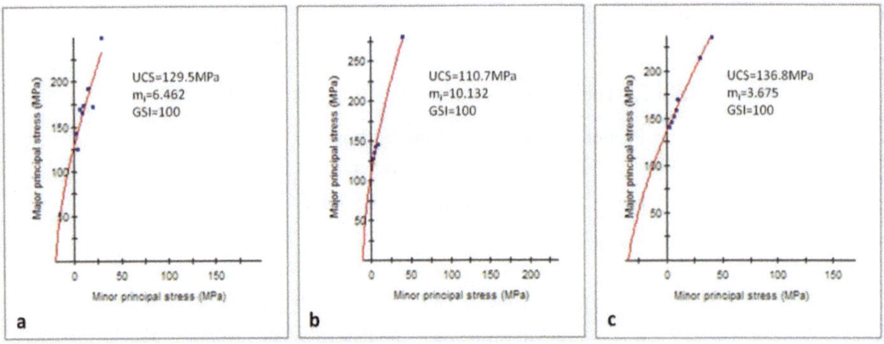

Fig. 2. Samples sketch and stress-strain curve at SRM with different inclination angle under uniaxial compression test.

It can be seen from Fig. 2 that:

(1) The m_i value of the HB criterion fitted by the BPM model is less than the measured values of marble, and UCS is larger than marble measured value. In the triaxial simulation, the peak strength of rock sample is not much different from that of laboratory test, but it cannot reflect the brittle-ductile-plastic transition characteristics of marble well;
(2) Under high confining pressure, rock samples can exhibit ductility characteristics, but the plastic properties can not be well represented. The reason is that the spherical particles are relatively smooth, and the friction resistance is not enough under the high confining pressure. Different shapes may be considered, As described in Cho [6]. The problem can be solved by using particle cluster Clump, which can increase the friction between particles. Specific analysis will be described in detail in the next section.

2.3 The Principle of Clump

A clump is a rigid collection of n rigid spherical pebbles. The surface is defined by the pebble positions and radii $\left(\{x^{(i)}, R^{(i)}\}, i = 1, 2...n\right)$. The surface properties of a clump can be specified independently for each pebble. Clumps can translate and rotate (defined by $\{v, w\}$, the generalized velocity and angular velocity, or spin of the clump centroid). Clump motion obeys the equations of motion. This requires that mass properties, loading conditions, and velocity conditions exist. Mass properties are mass, centroid position, and inertia tensor ($\{m, x, I\}$). Mass properties are defined below. Loading conditions are defined by: the force and moment resulting from interactions with other pieces; gravity (defined by g, the gravitational acceleration vector); and an externally applied force and moment acting on the clump (defined by F_A and M_A). Velocity conditions are defined by velocity-fixities V_f with {three values in 2D; six values in 3D} of the clump.

Mass properties are defined by: (1) the reference configuration and clump density (ρ); (2) the reference configuration and ($\{m, x, I\}$) for the reference configuration; or (3) a closed surface and associated mass properties given by the clump density. In all three cases, the reference configuration defines the clump surface used for contact detection/resolution and must be provided. The mass properties are determined for the first case by the positions and sizes of the pebbles in the reference configuration accounting for sphere overlap. The mass properties are determined for the second case by direct specification; this separates the mass and pebble distributions so that pebbles are used only for surface representation. The mass properties are determined for the third case by automatic computation assuming that the reference configuration is consistent with the closed surface.

The preferred clump work flow is to first define a set of clump templates that represent the desired particles. A surface description, in the form of a closed geometry of {line segments in 2D; triangular facets in 3D} that is manifold and orientable may be used to (1) create the pebble distribution and (2) calculate the mass properties. The clump template create bubblepack command is used to automatically create the pebbles to represent the surface to a desired fidelity. The surface descriptions may be either convex or concave. Once the set of clump templates have been defined, one can insert clumps into the domain in three ways: by replicating clumps one at a time; by generating non-overlapping sets of clumps; or by distributing overlapping clumps to match a specified size distribution.

3 Simulation Analysis of the Particle Shape Effect

3.1 Different Clump Ratio

Using the 2pebble template to construct clump and replace the ball particles, the clump particles of different proportions in the model were studied. The proportion coefficient R (the number of clump particles/the number of original ball particles) are 0.2, 0.4, 0.6 and 0.8 respectively. The BPM model which mixed with ball and clump are list in the Fig. 3, a, b, c and d represent the scale coefficients 0.2, 0.4, 0.6 and 0.8 respectively. The ball particles are expressed in blue, and the clump particles are expressed in red.

The four models were subjected to uniaxial compression test. The stress-strain curves are shown in Fig. 4. It can be seen, when the ratio is 0.8, the strength is the largest. When the ratio is 0.2, the strength is the lowest. The larger the scale factor of a

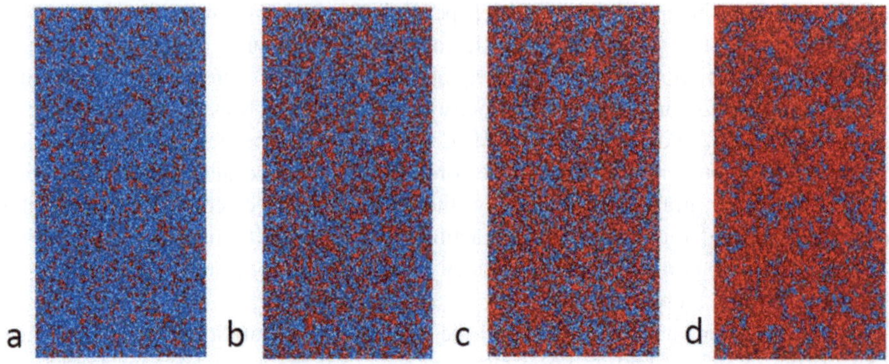

Fig. 3. BPM models mixed ball particle with clump particle

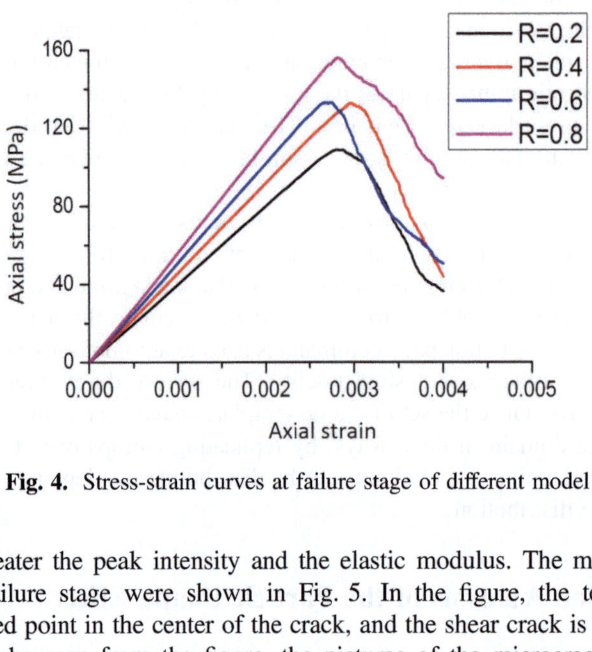

Fig. 4. Stress-strain curves at failure stage of different model

model, the greater the peak intensity and the elastic modulus. The microcracks distribution at failure stage were shown in Fig. 5. In the figure, the tensile crack is indicated by red point in the center of the crack, and the shear crack is represented by black. As can be seen from the figure, the pictures of the microcracks of the four models show that the number of tensile cracks is greater than the number of shear cracks. When R is 0.2, the model is splitting failure. When R is 0.4, the model is conjugate damage. When R is 0.6, the model is splitting damage. And when R is 0.8, the model is stair-step damage.

It is concluded that the 2pebble clump particles in the model are stronger than the ball particles in the case of the same microcosmic parameters. The larger the number of clump particles in the model, the greater the strength of the model.

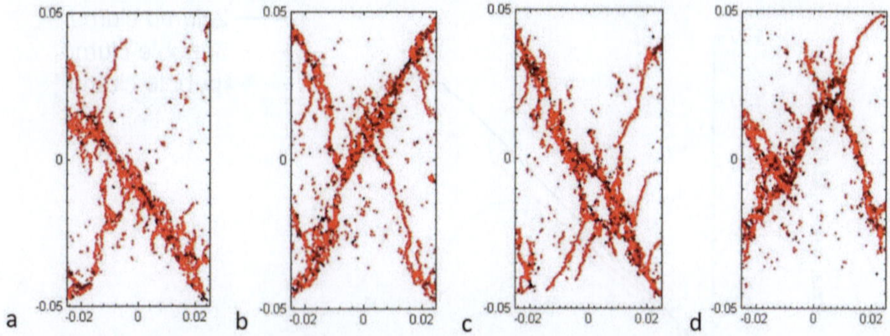

Fig. 5. Microcrack distribution at failure stage of different model

3.2 Different Clump Components

Using different clump template to construct the clump particle, 2pebble clump, 3pebble clump and 4pebble clump are considered in this paper. The shapes of clump are respectively 2pebble clump, 3pebble clump and 4pebble clump, and the ratio of the number of clump and ball ratio is 1:1. The specimens of the different models are shown in Fig. 6. In the figure a, b, c, respectively on behalf of 2pebble, 3pebble, 4pebble clump, enlarge figure of the clump is list on the right. The ball particles are expressed in blue, and the clump particles are expressed in red.

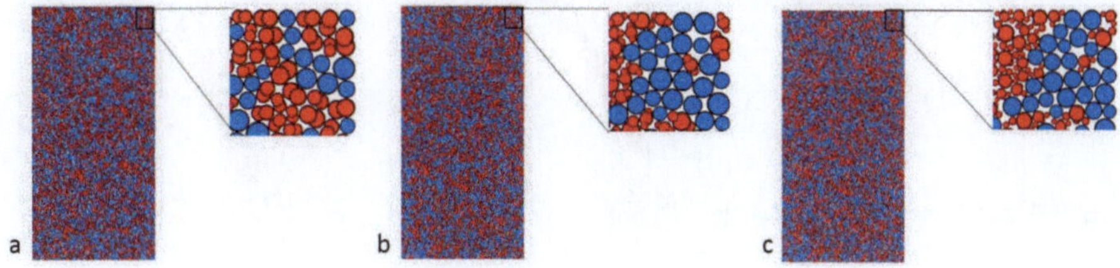

Fig. 6. BPM model with different clump template

The three models were subjected to uniaxial compression test. The stress-strain curves were shown in Fig. 7. It can be seen, the 2pebble clump model has the largest peak strength and shows obvious brittle characteristics, the 4pebble clump model has the least peak strength and shows ductility characteristics. The microcrack distribution at break were shown in the Fig. 8. In the figure, the tensile crack is indicated by red, and the shear crack is represented by black. As can be seen from the figure, the microcrack distribution figure, the number of tensile cracks is greater than the number of shear cracks. 2pebble clump model microcracks tilt distribute on both sides, 3pebble clump model microcracks are conjugate distribution, the distribution range of 4pebble clump model microcrack is the largest.

It is concluded as the 2pebble clump model has the largest strength, the second is the all ball model, the third is 3pebble clump model, the 4pebble clump model has the

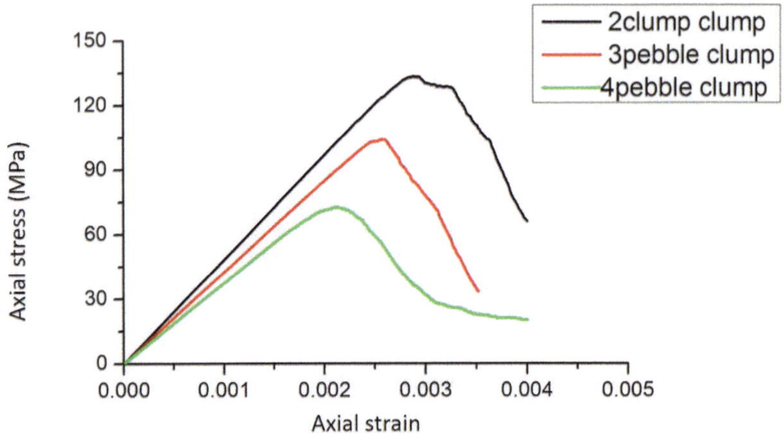

Fig. 7. Stress-strain curves at failure stage of different model

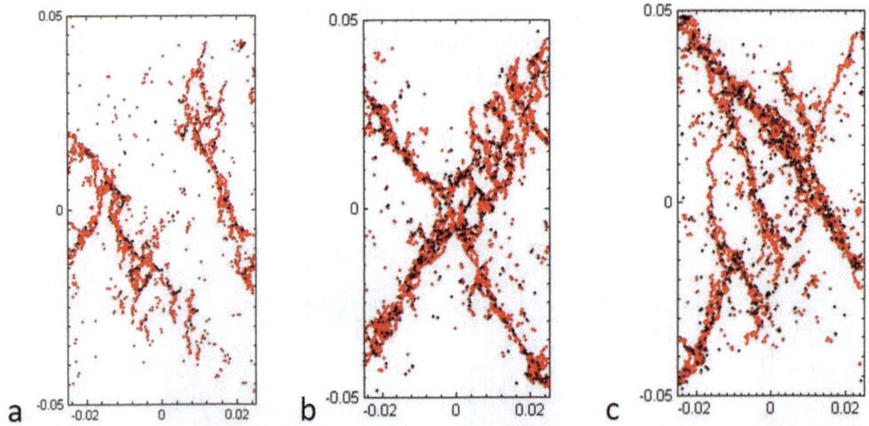

Fig. 8. Microcrack distribution at failure stage of different model

least strength. This is because, the degree of handover of the pebble which compose 3pebble and 4peeble clump particles is not enough. Although the particle shape has changed, but the composition of the particles is disperse, the strength of the particles reduce instead.

3.3 Different Clump Components Position

For the above situation, we took further verification. We proposed parameters K (the intersection area of the pebble/the sum of the area of the pebble) to quantifying the impact. When the interaction value of pebble is different, the degree of "lock-in" among pebble is different, we analyzed the mechanical behavior of the model. The three clump models we used are shown in the Fig. 9. In the figure, a, b, and c respectively represent the change in the degree of "lock-in" among pebble.

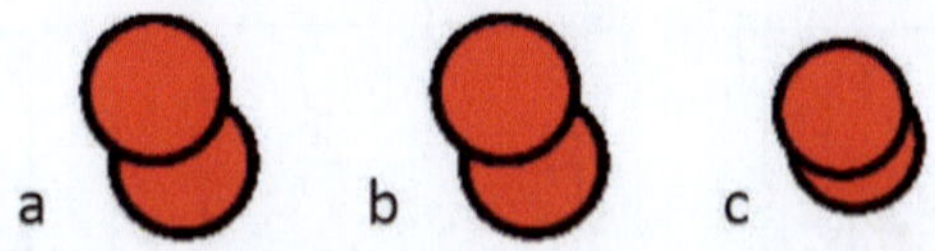

Fig. 9. Clump template of 2pebble clump with different intersect area

The three models were subjected to uniaxial compression test. The stress-strain curves were shown in Fig. 10. It can be seen, the greater the value of K, the greater the strength of the model. Note that the greater the value of K, the greater the strength of clump particle when it seen as a single particle. The microcracks distribution at failure stage were shown in Fig. 11. In the figure, the tensile crack is indicated by red, and the shear crack is represented by black. It can be seen from the microcrack distribution figure, the number of tensile cracks is greater than the number of shear cracks. When K is small, the model is shear failure. When K is medium, the microcracks tilt distribute on both sides, and when K is large, the distribution range of microcracks is larger.

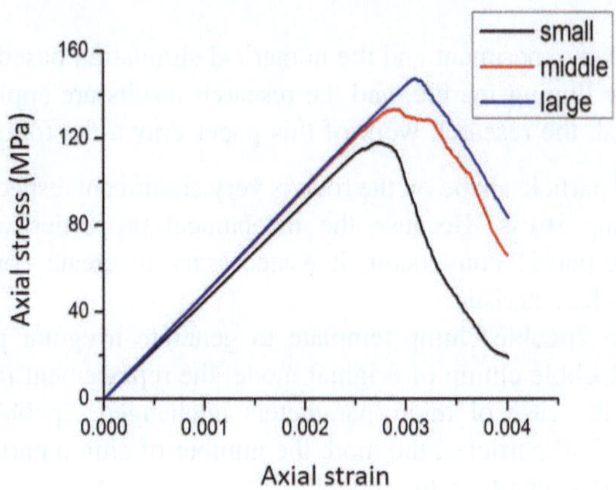

Fig. 10. Stress-strain curves at failure stage of different model

It is concluded as the "lock-in" effect influence the mechanical behavior of the whole model. The number of pebble and the relative position of the pebble are all the important aspect of the clump particle. The more proximity of the pebble, the stronger of the "lock-in" effect of the pebble, the greater strange of the whole model. The bonding force are exist between the ball in the BPM, there are two directions of the parallel bond. When the particle is instead by the clump, the bonding force are exist between the clump. In a clump particle, the pebble is solid, the relative position of each pebble is fixed, so the bonding force consistent regardless of the overlap areas.

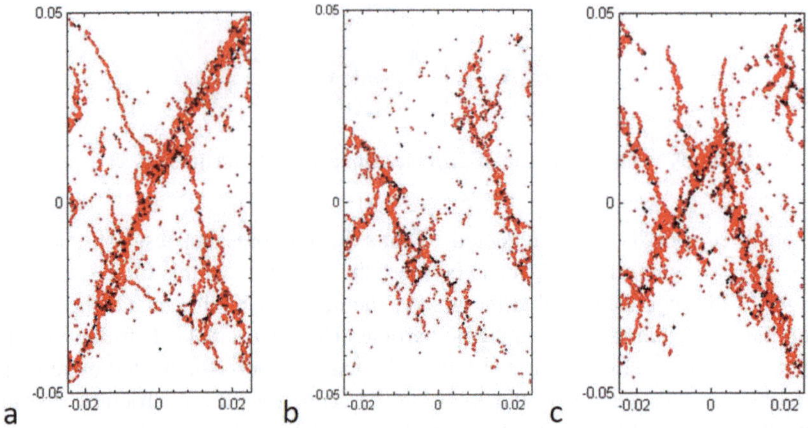

Fig. 11. Microcrack distribution at failure stage of different model

4 Conclusion

In this paper, the lab experiment and the numerical simulation based on laboratory test are carried out for Jinping marble, and the research results are applied to engineering practice. In general, the research work of this paper drawn the following conclusions:

(1) The effect of particle shape on the rock is very significant, especially in the case of high confining stress. Because the mechanical properties of the marble has brittle-ductile-plastic conversion, it is necessary to create complex particles to express this characteristic.

(2) According to 2pebble clump template to generate irregular particles, and then replace the 2pebble clump of original mode, the replacement ratio are 0.2,0.4,0.6 and 0.8. In the case of micro-parameters unchanged, 2pebble clump particles stronger than ball particles, the more the number of clump particles in the model, the greater the strength of the model.

(3) According to the template to generate clump particles, different particle shape effect the rock's overall mechanical behavior. It is concluded as the 2pebble clump model has the largest strength, the second is the all ball model, the third is 3pebble clump model, the 4pebble clump model has the least strength. This is because, the degree of handover of the pebble which compose 3pebble and 4peeble clump particles is not enough. Although the particle shape has changed, but the composition of the particles are disperse, the strength of the particles reduce instead.

(4) Through analysis of the effects of different clump templates, it can be found that the position of pebble is also equally important. We used model composition of 2pebble clump particle, changing the position among pebble, thus changing the degree of "lock-in" among pebble. The greater the degree of "lock-in" among pebble, the greater the strength of the model.

(5) Using clump to generate complex particle shape, and it can make a good foundation for the future research, which means that we can use mathematical methods to generate irregular particle shape to simulate the real sand and rock particles and then do research.

References

1. Hazzard, J.F., Young, R.P.: Simulating acoustic emissions in bonded-particle models of rock. Int. J. Rock Mech. Min. Sci. **37**(5), 867–872 (2000)
2. Feng, X.T., Chen, S.L., Zhou, H.: Real-time computerized tomography (CT) experiments on sandstone damage evolution during triaxial compression with chemical corrosion. Int. J. Rock Mech. Min. Sci. **41**(2), 181–192 (2004)
3. Itasca Consulting Group, Inc. (ICG): PFC (Particle Flow Code in Three Dimensions), Version 5.0-Pre-release. ICG, Minneapolis (2015). Cundall, P.A., Hart, R.: Numerical modeling of discontinua. J. Eng. Comp. **9**, 101–113 (1992)
4. Li, Y.P., Chen, L.Z., Wang, Y.H.: Experimental research on precracked marble under compression. Int. J. Solids Struct. **42**(9–10), 2505–2516 (2005)
5. Potyondy, D.O., Cundall, P.A.: A bonded-particle model for rock. Int. J. Rock Mech. Min. Sci. **41**, 1329–1364 (2004)
6. Cho, D., Martin, C.D., Sego, D.C.: A clumped particle model for rock. Int. J. Rock Mech. Min. Sci. **44**, 997–1010 (2007)
7. Chu, W.: The stability and structural safety assessment of tunnel surrounding rock under buried deep conditions. Postdoctoral report (2009)
8. The Itasca (Wuhan) Consulting Co., Ltd.: Jinping hydropower tunnel surrounding rock stability, dynamic supporting design of case studies. 2008–2009 mid-phase summary report (2009)

Coupled ICT and Dynamic Optimization Tools Toward an Integrated Earthwork Management System

António Gomes Correia[1(✉)] and Manuel Parente[2]

[1] ISISE – Institute for Sustainability and Innovation in Structural Engineering, University of Minho, Campus Azurém, 4800-058 Guimarães, Portugal
agc@civil.uminho.pt

[2] INESC TEC – Institute for Systems and Computer Engineering, Technology and Science, Campus da FEUP Rua Dr Roberto Frias, 4200-465 Porto, Portugal

Abstract. The main challenges of earthwork management are to optimize, in terms of time, costs and sustainability, the integration of several different tasks involved in the process. These tasks have to deal mainly with the geomaterials management (influencing excavation and compaction equipment as well as its use as a construction material) and equipment productivity (excavation, transport and compaction), under real environment conditions. To solve this complex management problem, a theoretical integrated system has been developed and evaluated for a database available of a real construction site.

This paper describes how the information and communication technologies available in equipment (GPS-sensors supported by tachymeter and laser scanning, derived volume of materials, intelligent compactors, maintenance time) can improve the dynamic optimization process. Indeed, it will be shown in cases where the monitored productivity for the available equipment falls short of or exceeds the estimates performed in design phase, how the necessary actions can be taken towards an optimal redistribution of resources. In summary, this paper presents a framework that combines information and communication technologies and dynamic optimization tools, in order to monitor and optimize the allocation of active equipment throughout a construction site in real-time, allowing savings and benefits for the environment (e.g., carbon emissions).

Keywords: Earthworks · Resource allocation · Dynamic optimization ICT · Management

1 Introduction

Earthworks are complex processes based on different types of resources (i.e., mechanical equipment, such as excavators, dumper trucks, bulldozers and compactors) and dependency relations between tasks (i.e. excavation, transportation, spreading and compaction). From an Operations Research point of view, this process is translated into several production lines that work simultaneously in a dynamic environment.

While some research has been aimed at finding optimal solutions for the earthworks process during design phase [1, 2], it is very challenging to keep the optimal status of

X. Shi et al. (Eds.): GSIC 2018, *Proceedings of GeoShanghai 2018 International Conference: Transportation Geotechnics and Pavement Engineering*, pp. 136–143, 2018.
https://doi.org/10.1007/978-981-13-0011-0_15

resource usage through the whole execution of the project, due to the inherently unpredictable and dynamic nature of earthworks. In fact, one of the main challenges of earthworks design is to accurately determine the on-the-job productivity of equipment during execution phase. A solution for this problem can be related to a better management of the available resources during construction phase itself. In this context, recent developments on the area of remote information and communication technologies (ICT) have originated innovative tools and instrumentation targeted at monitoring active construction equipment. These real-time technologies range from excavator bucket instrumentation, for determination of bucket payload and productivity [3], to continuous compaction control instrumentation [4, 5] and intelligent compaction technologies [6, 7], which allow for a real-time control of the compaction process, and have been target of increased research and development in recent years.

The objective of this work is to explore the framework for a real-time monitoring and optimization system. The system is based on the integration of data-driven and optimization techniques with remote ICT for real-time equipment monitoring. While the latter are responsible for monitoring and assessing real-time productivity of mechanical equipment during activity, the former use this gathered data to constantly update and re-optimize the construction process. Given that the effectiveness of optimal design has already been validated in previous work [1, 2], this enhances the system to keep up with the highly dynamic and volatile environment that characterizes earthwork construction.

The structure of this paper includes a description of the technologies used to support both the optimization and the monitoring processes in Sect. 2. Section 3 depicts the architecture of the optimization system, climaxing on its integration with the described field technologies. The subsequent conclusions are drawn in Sect. 4.

2 Technologies to Support Real-Time Earthwork Optimization

2.1 Optimization Technologies

Conventional and Operations Research optimization methodologies (e.g. linear programming) are, in most cases, unable to deal with the highly complex earthworks allocation problem. Metaheuristics comprise a viable alternative, as shown by a high number of successful applications [1, 8–10]. These modern optimization methods can search large search space regions under a reasonable use of computational resources. Genetic algorithms (GA) are one of such techniques, consisting of stochastic algorithms whose search methods model natural phenomena, such as genetic evolution and the concept of Darwinian natural selection [11]. Based on artificial intelligence and natural selection processes, GA start by generating random solutions for a problem, which are iteratively improved, tending towards optimal or near-optimal solutions.

However, optimization methods like GA alone are not enough to allow an effective optimization of earthworks in initial design phases. In fact, during this phase, it is very challenging to estimate the actual productivity of equipment, since it varies with a high number of variables (e.g., material types, available equipment, work conditions, meteorological conditions). Other than the experience of the designer, which may not be enough in many cases, data-driven prediction models are some of the few methods that have stood out as capable of providing realistic estimates for this problem [12–15]. These are mostly based on data mining (DM) methods, which consist of the overall process of discovering useful knowledge from data. This process involves several steps, such as selection, pre-processing and processing of data, application of DM algorithms, interpretation and processing of knowledge [16]. In the context of earthworks, DM algorithms are used to analyze large databases for patterns and trends in data to infer trends and rules among the variables, being able to handle large volumes of data in a short time and transforming them into knowledge. The extracted knowledge can be used to estimate the productivity of the available equipment under similar conditions.

2.2 Field Technologies

The management and allocation of resources in highly dynamic environments, such as earthworks or mining operations, are traditionally based on a manager's experience and guesswork. However, the development of remote ICT in recent years has generated ways to tackle this problem by providing the managers a number of tools to obtain information on resource activity in real-time, in an attempt to reduce the guesswork involved with this type of activity. In order to achieve this purpose, different real-time monitoring instrumentation has been integrated in the heavy mechanical equipment that comprises the available resources, depending on the work that each piece of equipment performs (Table 1).

In this context, most developments in productivity monitoring are supported by global positioning system (GPS) technology, while some complement it with other types of instrumentation. Excavator and dumper truck instrumentation has had some

Table 1. Heavy machinery instrumentation and monitored features

Equipment	Instrumentation	Monitored features
Excavation	GPS and/or bucket sensors	Bucket payload per cycle and number of cycles in a timeframe (productivity); material loaded per truck; idle time
Transportation	GPS or radio signal	Number of trips in a timeframe (productivity); machine speed; idle time
Spreading/grading	GPS, laser and/or blade sensors	Blade slope and elevation; machine speed; idle time
Compaction	GPS, drum sensors (continuous compaction control), and/or machine controls (intelligent compaction)	Number of passes in a timeframe; machine speed; idle time; continuous stiffness measurement of material being compacted and generation of compaction map (continuous compaction control); automatic adjustment of compaction effort (intelligent compaction)

development not only from the earthworks construction industry, but also on account of their usage in the mining industry [17]. These developments include capabilities such as measuring excavator bucket payload in real-time [3], tracking dumper trucks during their activity to determine the number of haul trips in a given timeframe [18], or monitoring and automatically adjusting blade slope and elevation in spreading and grading activities [19], ultimately providing advanced insight regarding the on-the-job productivity of these types of equipment.

Regarding the compaction process, which is significantly more complex that those associated with the preceding tasks, industry and researchers have developed ways to guarantee the quality of the process via continuous monitoring of both the machinery and the compacted material. Noteworthy advances include continuous compaction control [4, 5] and intelligent compaction [6, 7] technologies. While they are aimed at real-time control and guaranteeing an improved final result of the compaction process, the data gathered by these instruments is directly related to the productivity of the associated machinery.

3 Framework for Real-Time Earthwork Optimization

Based on previous work [1, 2], the implemented system architecture integrates 3 main modules (equipment, spatial and optimization modules) with capabilities to acquire and manipulate data from each phase of an earthwork project. Figure 1 provides a description of the optimization system, depicting the integration of the separate modules and information flow.

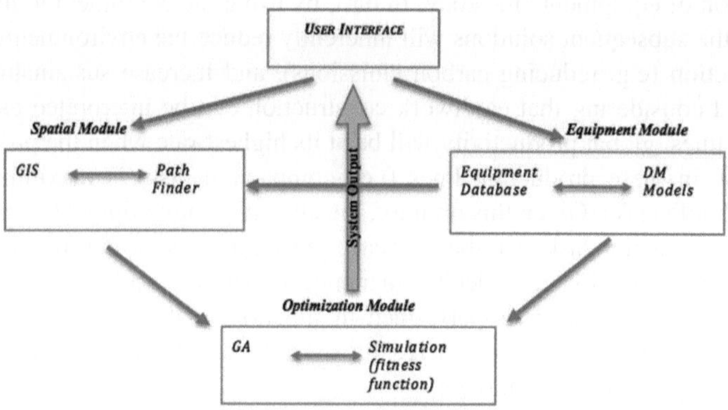

Fig. 1. System architecture

The equipment module is responsible for receiving the user input for available equipment plant. Simultaneously, this module includes the DM models used for the determination of the productivity rates for the available equipment. Together with the spatial module, they provide the optimization module with the data it requires to perform the optimal allocation of resources throughout the earthworks construction site. In the design phase, the data provided by the spatial module is mostly related to the optimization of transportation routes from excavation to embankment fronts, which may be originated from geographic information system software. The optimization module is based on an evolutionary optimization algorithm (namely a GA), which attempts to find a near optimal solution for the problem of determining the best possible equipment fleet and its optimal distribution throughout the work area. The multi-objective optimization attempts to minimize construction time and costs, subsequently reducing carbon emissions and using the available resources to their maximum potential. This translates into allocation the available resources into the earthworks production lines in such a way that productivity bottlenecks are prevented and the work rate of equipment maximized.

One of the main challenges to achieve optimal resource usage throughout the earthworks project stems from the dynamic nature of earthworks constructions. Indeed, considering that the optimum equipment locations are not static over time, equipment from one work front should be reassigned to others as their initial tasks are completed. In order to achieve this, the allocation is initially focused on the compaction equipment only (i.e. associated with the last task of the production line). By adopting this approach, it becomes possible to allocate the equipment associated with the remaining tasks in function of the productivity that has been allocated in the embankment fronts (i.e. compaction teams), guaranteeing that the productivity in each task is equals or above that value. The steps followed to determine the objective functions (total construction time and cost) are summarized in Table 2.

As the allocation of equipment dictates the resulting construction time and cost, the usage of equipment to its full potential is paramount. In other words, this allocation not only takes into account the minimization of construction time and cost, but also the maximization of equipment efficiency. In turn, by using the equipment to its maximum efficiency, the subsequent solutions will inherently reduce the environmental impact of the construction (e.g. reducing carbon emissions), and increase sustainability. In this context, and considering that earthwork construction can be interpreted as a series of production lines, global productivity will be at its highest rate when the productivity of the last task in these production lines (i.e., compaction task) is maximized and not limited by bottlenecks. Given this premise, the allocation of equipment is firstly carried out for the compaction task, and then for each preceding task, as described in steps 1–6 of Table 2. Additionally, in order to guarantee maximum equipment efficiency, the allocation of the other tasks is performed in function of the productivity verified in embankment fronts, which is estimated by the DM algorithms in function of work conditions, as previously mentioned.

Having assembled the resulting production lines, the compaction time of each embankment front for present conditions can be determined. However, as previously referred, the earthworks optimization problem include a dynamic aspect, translating into a need to perform a new reallocation at the point in time when one of these production lines completes its work. In other words, when a team finishes its excavation or embankment work, the associated equipment becomes idle/available once again, thus changing optimization conditions. As such, a new construction phase ensues, beginning with the update of the remaining material volumes in the work fronts that have yet to be completed. These correspond to steps 7–9 in Table 2. Moreover, having the knowledge

Table 2. Steps for determination of total cost and duration of equipment allocation solutions

Step	Description	Associated variables/factors
1	Allocation of compactors to embankment fronts	Available compactors (type and quantity); available embankment fronts
2	Determine individual productivity of the allocated compactors for each case	Compactor type; material type; compaction conditions (e.g., layer thickness, meteorological conditions)
3	Calculate total productivity in each active compaction front	Number of compactors of each front and individual productivity of each compactor
4	Allocate spreading equipment (LP model)	Total productivity in compaction task for associated embankment front; available spreaders/bulldozers (type and quantity); material type; work conditions
5	Allocate transportation equipment (LP model)	Minimum productivity in spreading and compaction tasks for associated embankment front; available trucks/dumpers (type and quantity); transportation distance; work conditions
6	Allocate excavation equipment (LP model)	Minimum productivity in transportation, spreading and compaction tasks for associated embankment front; available excavators (type and quantity), material type; work conditions
7	Calculate compaction duration in each embankment front	Productivity of production line (minimum productivity amongst all tasks in a production line); required material volumes for completing each embankment front
8	Verify fastest production line to complete its work (corresponds to the duration of the current construction phase)	Compaction duration of each production line; total volume of material required to complete each active embankment front
9	Calculate volumes of materials which have been excavated and compacted in each front during current phase	Duration of current construction phase; productivity of each production line; volume of material available/required in each active front
10	Calculate cost according to the used equipment and the duration of current construction phase	Direct and indirect costs of active equipment; duration of current construction phase
11	Verify if all embankment fronts have been completed. If not, initiate new construction phase (step 1), taking into account updated material volumes (calculated in step 9). Otherwise, output accumulated cost and duration	Available embankment fronts (if initiating new construction phase); individual cost and duration for each construction phase (if outputting results)

of the amount of hours each equipment has been active during the previous construction phase, it is possible to determine the time-dependent cost (e.g. fuel costs) for each active piece of equipment. By adding the result to the direct costs of active equipment (e.g. rental and manpower costs), the total cost for each piece of equipment can be calculated. The total cost for the current construction phase will correspond to the sum of the costs associated with of the active equipment (step 10 in Table 2).

Although a construction phase is considered to end as soon as a compaction front is completed, each solution evaluated by the optimization algorithm is only complete when all fronts have been compacted. As such, this process is repeated for each construction phase, calculating the associated time and cost until all fronts have been compacted (step 11 in Table 2). Ultimately, the total time and costs are determined in the end of the process by adding the values for each construction phase.

The offline application of this methodology in design phases has been shown to potentially reduce execution cost and duration up to 50% when compared with the conventional design methodologies [1, 2]. However, during construction phase, the volatility of earthwork constructions, in which unpredictable occurrences such as equipment malfunction, changes in atmospheric conditions, or low productivity rates are common, a purely predictive and offline approach is not effective enough. Instead, a more reactive, online approach is necessary, in order to be able to tackle that volatility. As such, it is possible to leverage on the capabilities of the aforementioned field technologies (see Sect. 2.2) to constantly monitor and assess the productivity of the active equipment during execution. This allows the spatial and optimization modules to reassess the productivity that has been estimated during design phase by the DM models. Should the actual productivity of production lines differ from what was originally estimated for whatever reason, the first priority should be to identify the bottlenecks in the associated earthwork tasks, after which the actual productivity of the equipment can be updated, allowing the optimization system to readjust the allocation setup in real-time to guarantee its optimal status.

4 Conclusions

This paper presents a framework that combines remote ICT and dynamic optimization tools, in order to monitor and optimize the allocation of active equipment throughout a construction site in real-time. By leveraging on these concepts, it is demonstrated how the management of earthwork resources during construction can be greatly enhanced.

The optimization capabilities of the proposed framework stem from the integration of several technologies, namely metaheuristics, data mining and geographic information systems. These technologies comprise a system that has been previously validated for design purposes. Simultaneously, the remote ICTs and tools are used in the form of instrumentation (e.g. GPS) integrated with the mechanical equipment (i.e. excavators, dumper trucks, spreaders, compactors) that performs earthworks processes. These are used to determine the productivity of equipment during construction, feeding that data in real-time to the optimization system. This data is then used to identify potential bottlenecks in the allocated production lines, either resulting from inaccurate productivity estimates in design phase or as a consequence of unpredictable events (e.g. machinery malfunction). Ultimately, this is expected to allow the system to constantly monitor any unexpected productivity variations during construction, and automatically re-adjust the usage of resources in order to keep the optimal status of the ongoing processes.

References

1. Parente, M., Cortez, P., Gomes Correia, A.: An evolutionary multi-objective optimization system for earthworks. Expert Syst. Appl. **42**(11), 6674–6685 (2015)
2. Parente, M., Gomes Correia, A., Cortez, P.: Modern optimization in earthwork construction. In: Proceedings of the XVI ECSMGE, pp. 343–348 (2015)
3. LC Products: Titan 3330 Production Monitor Brochure (2015)
4. Gomes Correia, A., Quibel, A.: Compaction of Soils and Granular Materials. Presse Nationale des Ponts et des Chaussées (2000)
5. Thurner, H., Sandström, A.: Continuous Compaction Control, CCC. In: Gomes Correia, A., Quibel, A. (eds.) Compaction of Soils and Granular Materials. Presse Nationale des Ponts et des Chaussées, Paris, France (2000)
6. Gomes Correia, A., Parente, M.: Intelligent compaction technology for geomaterials. A demonstration project. In: TRA2014 Transport Research Arena, pp. 1–10 (2014)
7. Camargo, F., Larsen, B., Chabourn, B., Roberson, R., Siekmeier, J.: Intelligent compaction: a Minnesota case history. In: Proceedings of the 54th Annual University of Minnesota Geotechnical Conference (2006)
8. Moselhi, O., Alshibani, A.: Crew optimization in planning and control of earthmoving operations using spatial technologies. J. Inf. Technol. Constr. **12**, 1–17 (2007)
9. Xu, Y., Wang, L., Xia, G.: Research on the optimization algorithm for machinery allocation of materials transportation based on evolutionary strategy. Procedia Eng. **15**, 4205–4210 (2011)
10. Zhang, H.: Multi-objective simulation-optimization for earthmoving operations. Autom. Constr. **18**(1), 79–86 (2008)
11. Holland, J.H.: Adaptation in Natural and Artificial Systems. University of Michigan Press, Ann Arbor (1975)
12. Parente, M., Correia, A., Cortez, P.: Use of DM techniques in earthworks management: a case study. In: GeoHubei 2014 International Conference on ASCE Geotechnical Special Publication (GSP), pp. 1–8 (2014)
13. Hola, B., Schabowicz, K.: Estimation of earthworks execution time cost by means of artificial neural networks. Autom. Constr. **19**(5), 570–579 (2010)
14. Tam, C.M., Tong, T., Tse, S.: Artificial neural networks model for predicting excavator productivity. J. Eng. Constr. Archit. Manag. **9**(5–6), 446–452 (2002)
15. Cortez, P., Marques, R., Gomes Correia, A.: Artificial intelligence applied to compaction rules and management. In: International Seminar on Interaction Soil-Rail Track for High Speed Railways (2008)
16. Fayyad, U., Piatetsky-Shapiro, G., Smyth, P.: From data mining to knowledge discovery in databases. Am. Assoc. Artif. Intell. **17**(3), 1–18 (1996)
17. Caterpillar Inc.: Cat Minestar System - Robison Mine case study (2012)
18. Montaser, A., Moselhi, O.: Truck + for earthmoving operations. J. Inf. Technol. Constr. **19**, 412–433 (2014)
19. Caterpillar Inc.: System for Track-Type Tractors AccuGrade: Grade Control System for Track-Type Tractors (2006)

Mechanical Performance of Anchored-Plating Cantilever Retaining Wall

Hongbo Zhang, Liang Lu, Hongya Yue, and Xiuguang Song(✉)

School of Civil Engineering, Shandong University, Jinan 250061, China
songxiuguang@sdu.edu.cn

Abstract. Because of complex construction and larger deformation of wall, traditional anchored-plating wall had been used limitedly. A new type of retaining wall, anchored-plating cantilever retaining wall was put forward to improve the self-stability and simplify the construction. In order to analyze the mechanical performance of the new type of retaining wall, a series of model tests were performed by considering of the variation of upload and prestressing force. And some cells were installed at different position of retaining wall to monitor the variation of lateral deformation of wall facing, lateral earth pressure behind wall, foundation pressure and cable force. Test results showed that there would be obvious redistribution of lateral earth pressure behind wall after applying pre-stressed force. Stress focus would happen around anchorage point. And also, the variation of lateral earth pressure and deformation was influenced greatly by the value of prestressing force. The higher for prestressing force, the higher for earth pressure was, and vice versa. Also, with the increment of prestressing force, the earth pressure would be close to even over the line on rest. At the same time, the variation of lateral earth pressure behind wall accords with the classical earth pressure theory. By analyzing the variation of deformation of wall face and foundation pressure, the stability of this type of retaining wall could be controlled well. But the prestressing force should designed in a rational scope.

Keywords: Anchored-plating cantilever retaining wall · Lateral earth pressure Displacement · Surcharge · Pre-stressed force

1 Introduction

During the past decades, application of anchored retaining wall for slope stabilization has been proved to be the reliable and effective technique to prevent excessive slope movement (Silva et al. 2017). Anchored plate cantilever retaining wall, which takes the advantage of simplifying construction process, reducing the quantities of materials, saving farmland (Bowles 1988) is widely used as a kind of new retaining structure. However the theory study including the reinforcement mechanism, the behavior of retaining wall during construction and the effect of pre-stressed force remained lagging behind (Chaudhary et al. 2017).

About the cantilever retaining wall, Bentler (2006) carried out a series of field tests of the reinforced concrete cantilever retaining wall in Minnesota, the measured data

X. Shi et al. (Eds.): GSIC 2018, *Proceedings of GeoShanghai 2018 International Conference: Transportation Geotechnics and Pavement Engineering*, pp. 144–152, 2018.
https://doi.org/10.1007/978-981-13-0011-0_16

were compared with numerical model results (Chughashok and Labuzjoseph 2011), the comparison results indicated the inadequacy of bearing capacity of conventional cantilever retaining wall under large surcharge which reflected the defects of current design method and the limitation of structure itself. Anchored plate cantilever retaining walls is a kind of composite structure, which makes full use of the advantages of anchor as reinforcement. When the anchor block is installed across the active and stable soil, destabilizing forces in active zone would be transferred to resistant zone through reinforcement and improve the stability of retaining structure (Alam and Siddiquee 2014). However, the existence of anchor would make the earth pressure state more complicated, and the effect of the anchor is therefore a major consideration in the current study. As pointed out by Hatami et al. (2001) the use of reinforcement may give rise to pressure concentrations which would lead to redistribution of lateral earth pressure exerted on wall facing. We consider that lateral earth pressure distribution curve is more complicated than conventional prediction due to the stress concentration. Note that the application of anchor has been regarded as an efficient method to limit lateral displacement and improve the global stability of retaining wall (Anderson et al. 1983). The stress concentration induced by anchor (prestressed or non-prestressed) was not considered properly in the conventional design (Silva et al. 2017), which resulted in wide discrepancies between predicted and measured bearing capacities of these reinforce systems reported in the literature (Hanna et al. 2011).

Therefore, in order to make up for the inadequacy of existing theories, perfect the design theory, model test method was applied in the research. The lateral earth pressure redistribution induced by pressure concentration and behavior of wall under different surcharge and pre-stressed force conditions was put in-depth study in this paper.

2 Model Test Design

A series of reduced-scale model tests were conducted to investigate the behavior of the anchored plate retaining wall subjected to static loading and different pre-stressed force. The main components of the experimental apparatus included a loading frame with a platform, a test box, loading plates, and dial gauges. The box with dimensions of 3.0 m (length) × 2.5 m (width) × 2.0 m (height), was made of steel plate with 25 mm thick in three sides, as shown in Fig. 1. The loading system consisted of a pair of vertical jacks and a hollow lateral jack, which were controlled by hydraulic control handle and pressure panels to apply different loads.

The anchor bar-block systems consist of Φ32 mm rebar, 30 × 30 × 3 cm steel plate and bolts, which located at the 1.0 m height of retaining wall. And a steel hollow pipe was used to model the external mortar coverage to protect the rebar. The reinforcement tension meter was welded with the bolt to measure the tension force of anchor bolt.

Individual wall facing unit with a dimension of 2.0 m (height) 2.5 m (width) 25 mm (thick), were prepared by welded steel plate with wall heel, 2.5 m (width) × 0.6 m (length) × 25 mm (thick), and wall toe, 2.5 m (width) × 0.4 m (length) × 25 mm (thick).

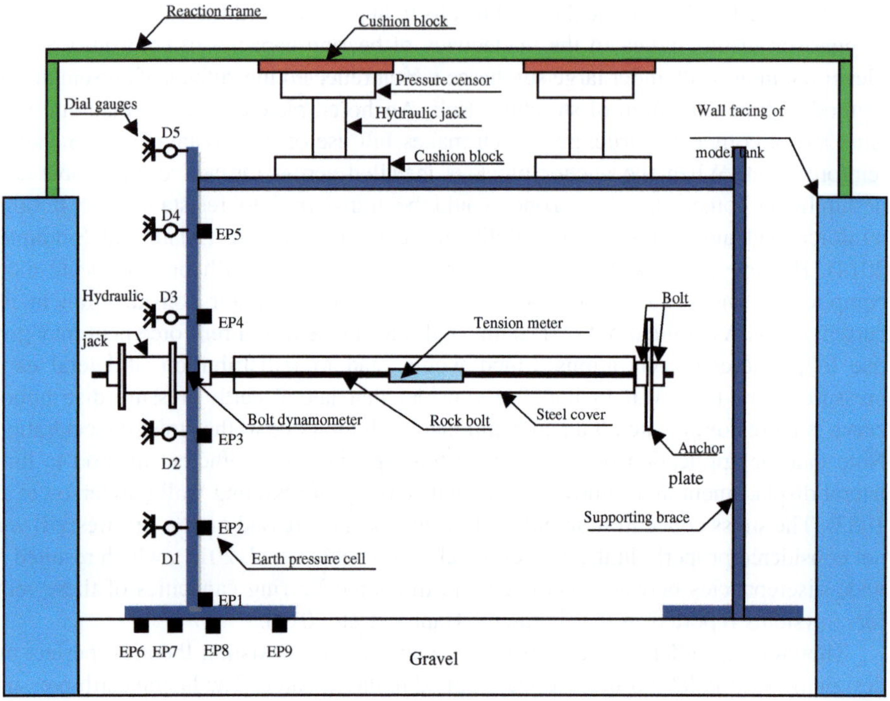

Fig. 1. Schematic diagram of the test apparatus

Five digital dial gauges placed at the height of 0.4 m, 0.8 m, 1.2 m, 1.6 m and 2.0 m in front of the wall surface to measure the lateral deformations of the wall facing. Five earth pressure cells were placed at the height of 0 m, 0.4 m, 0.8 m, 1.2 m and 1.6 m in the internal face of the wall to measure the lateral earth pressure of the wall facing. Four earth pressure cells were placed under the wall with the distance from edge of wall toe being equal to 0.1 m, 0.3 m, 0.5 m, 0.9 m to measure the foundation pressure of wall. Also, one bolt dynamometer is welded with bolt to measure the total tension force.

All the filling height in model walls had the height of 180 cm. A poorly-graded dry river sand was used in the model tests as backfill soil. The sieve analysis tests were conducted following Test Methods of Soils for Highway Engineering (JTG E40-2007). The mean grain size (D50) was 0.55 mm, the Cu and Cc of the sand were 2.53 and 0.93, respectively. The maximum and minimum dry unit weights of the sand were $\gamma_{max} = 18.75$ kN/m^3 and $\gamma_{min} = 16.29$ kN/m^3, respectively. The density of the backfill was achieved by compaction with fixed mass of sand into a pre-calculated volume of each lift. The target relative density of the backfill was 70%. The peak friction angle of the sand at 70% relative density was 40° based on the triaxial tests. Table 1 tabulates the test plan.

Table 1. Model test plans

Test amount	Pre-stressing force (kN)	Surcharge (kPa)	Note
1	0	10, 20, 30, 40, 50	
1	32.8		
1	55		
1	63		
1	80 through 220, each step was 10		Failure test of anchored plate

3 Tests Results and Discussion

3.1 Effect of Surcharge on Lateral Earth Pressure of Retaining Wall

According to the Fig. 2, the distribution of lateral earth pressure from top to bottom varied greatly with the increment of surcharge. When the surcharge was less than 30 kPa, the peak value point of lateral earth pressure located around the anchor level, about 2/5 wall height. But when the surcharge was over 30 kPa, the peak value point became lower, about 1/5 wall height. The test results were compared with classic earth pressure theory, as shown in Fig. 3. Firstly, lateral earth pressure in the bottom section was more closely to the active earth pressure line K_a, Secondly, lateral earth pressure in the top section was more closely to the static earth pressure line K_0. Thirdly, the peak value was always over than the K_0 value.

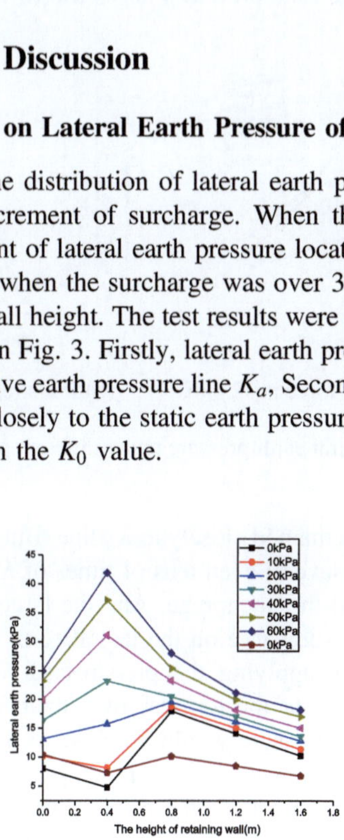

Fig. 2. Distribution of lateral earth pressure along the height of retaining wall under different surcharge (no pre-stressed force)

3.2 Effect of Pre-stressing Force on Lateral Earth Pressure of Retaining Wall

As shown in Figs. 4 and 5, two pieces of phenomenon can be concluded. Firstly, by applying the lateral pre-stressed force, the value of earth pressure increased greatly except lower part of retaining wall, about 0–20% wall height. For this lower part of

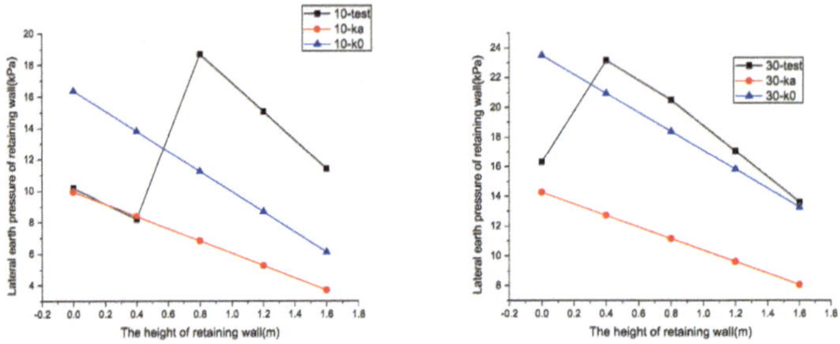

Fig. 3. Comparison of lateral earth pressure with K_0 and K_a line under different surcharge

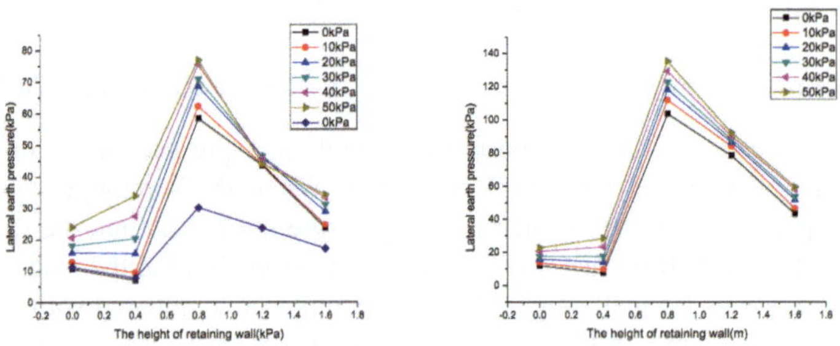

(a) Pre-stressed force is 32.8kN (b) Pre-stressed force is 63kN

Fig. 4. The distribution of lateral earth pressure under different pre-stressed force and surcharge

wall facing, earth pressure remained closely to K_a line. But earth pressure on the higher part of wall facing became several even tens of times of K_0 value. Secondly, the peak value mainly located around the anchorage. And the larger the pre-stressed force, the more seriously influence of surcharge on the lateral earth pressure was.

As shown in Fig. 5, by applying the pre-stressed force on cable, lateral earth pressure from 1/5 wall height to the top was over than K_0 line. With the increasing pre-stressed force, the earth pressure would be close to the passive earth pressure K_p line gradually. Test results showed that after the pre-stressed force was over than 200 kN, the anchor plate was nearly in the limit failure state, and earth pressure was closely to the K_p line. Therefore, by applying lateral pre-stressed force, lateral earth pressure always varies between K_0 line and K_p line. Meanwhile, there is an obvious stress concentration around the anchorage, especially within the distance ±10–20% of total wall height from the anchorage center.

3.3 Effect of Pre-stressing Force on Foundation Pressure of Retaining Wall

As shown in Fig. 6, foundation pressure increased with the increment of surcharge load. And foundation pressure at outer edge was larger than inner edge because retaining wall will rotate outside around the center of foundation slab under surcharge load.

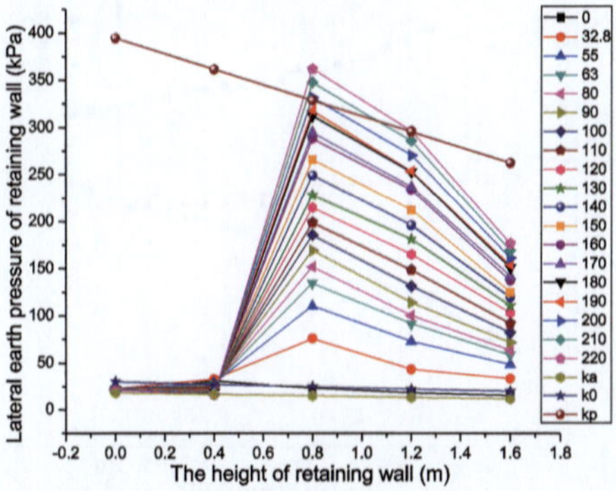

Fig. 5. Comparison of pre-stressed force on lateral earth pressure of retaining wall (surcharge load was 50 kPa)

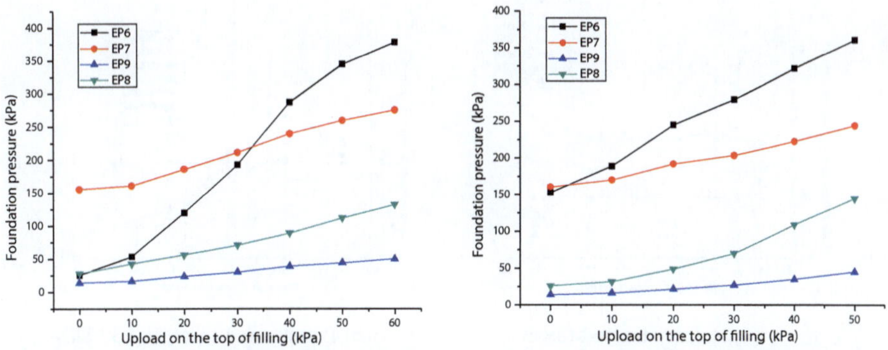

(a) No pre-stressed force (b) Pre-stressed force=32.8kN

Fig. 6. Foundation pressure variation under different pre-stressed force and surcharge load

The variation of foundation pressure subjected to different pre-stressed force was plotted in Fig. 7. Because of measurement errors, data obtained from earth pressure cell EP9 was not considered in current analysis. With the increment of pre-stressed force, foundation pressure under wall toe and wall heel showed decrease and increment separately. It showed that retaining wall rotated toward inner-side after applied higher pre-stressed force.

3.4 Lateral Displacement of Retaining Wall

As shown in Fig. 8, for the anchor plate retaining wall without pre-stressed force, the shape of lateral displacement was like an arrow. Thereinto, the maximum lateral displacement occurred at the 1/4 height. And the retaining wall has an outward movement under lateral earth pressure. However, after applied pre-stressed force, the outward movement could be decreased greatly. It showed that lateral displacement can be

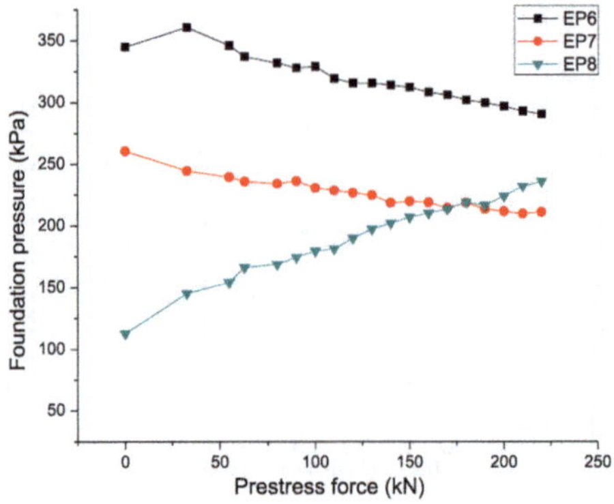

Fig. 7. Relationship of foundation pressure with the increment of pre-stressed force

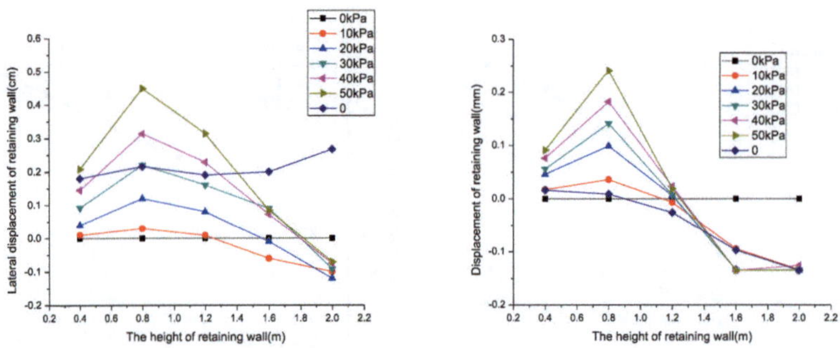

(a) No pre-stressed force (b) Pre-stressed force was 32.8kN

Fig. 8. Displacement variation under different surcharge

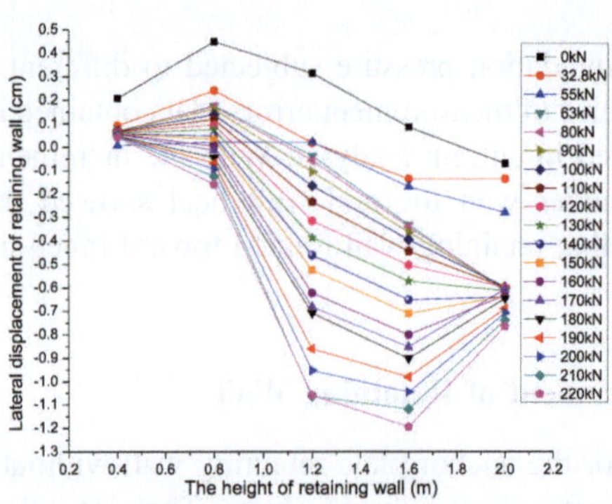

Fig. 9. Displacement of retaining wall under different pre-stressed force (surcharge was 50 kPa)

controlled well by applying pre-stressed force. Larger of pre-stressed force, greater inward displacement of wall facing, as shown in Fig. 9. The maximum lateral displacement occurred above the anchorage. And the foundation also had an inward movement under pre-stressed force. Because of the inward movement of wall facing, the lateral earth pressure on wall facing increased greatly, and even reached the passive earth pressure state.

4 Conclusion

In this paper, a series of reduced-scale model tests of anchored plate cantilever retaining wall were conducted. Base on the results of the present investigation, the following can be concluded: By applying pre-stressed force, the lateral earth pressure exerted on the wall facing increased obviously, with the increment of pre-stressing force, lateral earth pressure of top section (above 1/5 of the wall height) always varies between K_0 line and K_p line, and that of the bottom section (below 1/5 of the wall height) more closely to K_0 line. Stress concentration occurred especially within the distance ±10–20% of total wall height from the anchorage center. The retaining wall rotated toward inner-side after applied higher pre-stressed force, therefore lateral displacement can be controlled efficiently, and the maximum lateral wall displacement was less than that of a typical anchored-plate retaining wall without pre-stressed force.

Acknowledgement. This research work was conducted when the first author visited the University of Kansas as a visiting scholar, which was sponsored by the China Scholarship 610 Council. This research work was also supported by Grant No. 51208284 and 2013GGE27065 from the Natural Science Foundation of China and Shandong Province Technology Development Project. Their financial support is gratefully acknowledged.

References

Alam, J., Siddiquee, S.A.: A parametric study of anchored earth wall by finite element method. J. Civil Eng. **18**(7), 2034–2042 (2014)

Anderson, W.F., Hanna, T.H., Abdel-Malek, M.N.: Overall stability of anchored retaining walls. J. Geotech. Eng. **109**(11), 1416–1433 (1983)

Bentler, J.G.: Performance of a cantilever retaining wall. J. Geotech. Geoenviron. Eng. **132**(8), 1062–1070 (2006)

Bowles, J.E.: Foundation Analysis and Design. McGraw-Hill, New York (1988)

Chaudhary, B., Hazarika, H., Krishan, A.M.: Effect of backfill reinforcement on retaining wall under dynamic loading. In: Geotechnical Hazards from Large Earthquakes and Heavy Rainfalls. Springer, Tokyo (2017)

Chughashok, K., Labuzjoseph, F.: Numerical simulation of an instrumented cantilever retaining wall. Can. Geotech. J. **48**(9), 1303–1313 (2011)

Hatami, K., Bathrust, R.J., Di Pietro, P.: Static response of reinforced soils retaining walls with non-uniform reinforcements. Int. J. Geomech. **1**(4), 477–506 (2001)

Hanna, A., Rahman, F., Ayadat, T.: Passive earth pressure on embedded vertical plate anchors in sand. Acta Geotech. **6**(1), 21–29 (2011)

Silva, A.P.D., Festugato, L., Masuero, J.R.: A new methodology to assess the structural safety of anchored retaining walls. Thin-Walled Struct. **117**, 343–355 (2017)

Required Reinforcement Length for External Stability of MSE Walls Using Pseudo-Dynamic Method

Xiaobo Ruan[1(✉)], Ke Lin[1], Xiaofeng Han[2], and Dayong Zhu[3]

[1] School of Automotive and Transportation Engineering, Hefei University of Technology, Hefei 230009, China
xiaoboruan@hfut.edu.cn

[2] Department of Geotechnical Engineering, Nanjing Hydraulic Research Institute, Nanjing 210029, China

[3] School of Civil Engineering, Hefei University of Technology, Hefei 230009, China

Abstract. In the earthquake-prone zone, the pseudo-static limit equilibrium (LE) method is frequently applied to analyze the external stability of MSE walls. When using the pseudo-static LE method, the required reinforcement length in MSE walls may be overestimated resulting in redundancy. Thus, this paper firstly formulate the required minimum reinforcement length for the external stability of MSE walls using pseudo-dynamic LE method, considering sliding, eccentricity, and bearing capacity failure modes. And the formulation includes some parameters, such as seismic coefficient, soil amplification factor, wall height, and the properties of reinforced soil, backfill and foundation soil. Finally, the relationships between the pseudo-dynamic and the pseudo-static minimum reinforcement lengths under different failure modes are presented.

Keywords: Required reinforcement length · MSE walls
Pseudo-dynamic method · External stability · Limit equilibrium method

1 Introduction

As a retaining structure or soil resistance system, the MSE wall has a wide range of applications in the engineering project [1]. Generally, specifications [2–5] check internal and external stability to design this structure. In particular, when checking its external stability, the wall and the reinforced soil zone are usually considered as a whole, and the reinforcement length is an important parameter. That is because this parameter determines the size of the reinforced soil zone.

In the earthquake-prone zone, seismic design of the MSE wall is essential, and often uses the pseudo-static method [6, 7]. However, this method ignores the dynamic characteristics of seismic inertial forces. In order to compensate for this deficiency, Steedman and Zeng [8] proposed a pseudo-dynamic method for seismic design of retaining walls; and this method was then validated by the centrifuge model test [9].

X. Shi et al. (Eds.): GSIC 2018, *Proceedings of GeoShanghai 2018 International Conference: Transportation Geotechnics and Pavement Engineering*, pp. 153–160, 2018.
https://doi.org/10.1007/978-981-13-0011-0_17

Currently, the pseudo-dynamic method has been used for calculation of active earth pressure [10, 11], design of reinforced soil structures [12, 13] and slope stability analysis [14–16].

The computation formula of the reinforcement length required for the MSE wall can be deduced by external stability analysis. Under static conditions, Bilgin and Mansour [17, 18] got the calculation formula of the minimum reinforcement length required for MSE walls under four failure modes (i.e., sliding, overturning, eccentricity and bearing capacity failure modes). Since the failure effect of eccentricity is equivalent to overturning under seismic load [4], the overturning failure is then ignored in this paper. Ruan and Sun [6] used the pseudo-static method to introduce seismic loads, and extended the calculation formula of required reinforcement length under three failure modes (i.e., sliding, eccentricity and bearing capacity failure modes). The purpose of this paper is to further extend the calculation formula of required reinforcement length under pseudo-dynamic seismic loads.

2 Required Minimum Reinforcement Length for Failure Modes

2.1 Pseudo-Dynamic Load

An MSE wall, along with the forces acting on the wall used for external stability analysis, is shown in Fig. 1. In addition, the performance criteria is used in this paper and the required values given by reference [4] for sliding, eccentricity and bearing capacity failure modes are given in Table 1.

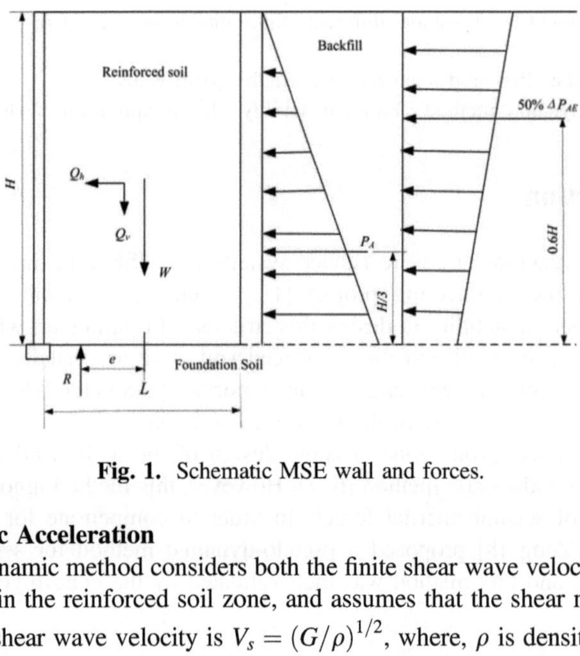

Fig. 1. Schematic MSE wall and forces.

2.1.1 Seismic Acceleration

The pseudo-dynamic method considers both the finite shear wave velocity and primary wave velocity in the reinforced soil zone, and assumes that the shear modulus, G is a constant. The shear wave velocity is $V_s = (G/\rho)^{1/2}$, where, ρ is density of reinforced

Table 1. Performance criteria used in the design of MSE walls under seismic conditions.

Failure modes	Performance criteria
Sliding	$FS_S \geq 1.125$
Eccentricity	$e \leq L/3$ for both soil and rock
Bearing capacity	$FS_{BC} \geq 1.875$

soil or backfill, and the primary wave velocity is $V_p = [2G(1-\nu)/\rho(1-2\nu)]^{1/2}$, where, ν is Poisson ratio. Taking into account the expansion of vibration, it is assumed that both horizontal and vertical seismic accelerations from the base to the surface to change linearly along the depth. Then, both horizontal and vertical seismic acceleration in reinforced soil zone of arbitrary depth, z and any time, t are as follows

$$a_h(z,t) = \left[1 + \frac{H-z}{H}(f_s - 1)\right] k_h g \sin\left[\omega\left(t - \frac{H-z}{V_s}\right)\right] \tag{1}$$

$$a_v(z,t) = \left[1 + \frac{H-z}{H}(f_s - 1)\right] k_v g \sin\left[\omega\left(t - \frac{H-z}{V_p}\right)\right] \tag{2}$$

where k_h and k_v are horizontal and vertical seismic coefficients, respectively; g is gravity acceleration; f_s is soil amplification factor; H is MSE wall height; ω is angular velocity, $\omega = 2\pi/T$, T is period of the lateral vibration.

2.1.2 Pseudo-Dynamic Load in Reinforced Soil Zone

The mass of reinforced soil-wedge unit at the arbitrary depth, z is as follows

$$m(z) = \frac{\gamma_r L_{\min}}{g} dz \tag{3}$$

where $L_{\min}$ is minimum reinforcement length; γ_r is unit weight of reinforced soil.

Then, the horizontal and vertical seismic forces in the reinforced soil zone are as follows

$$\begin{aligned} Q_h = \int_0^H a_h m(z) = \frac{\lambda \gamma_r L_{\min} k_h}{2\pi} \Bigg\{ & f_s \cos 2\pi\left(\frac{t}{T} - \frac{H}{\lambda}\right) - \cos 2\pi\left(\frac{t}{T}\right) \\ & - \frac{\lambda(f_s - 1)}{2\pi H}\left[\sin 2\pi\left(\frac{t}{T}\right) - \sin 2\pi\left(\frac{t}{T} - \frac{H}{\lambda}\right)\right]\Bigg\} \end{aligned} \tag{4}$$

$$\begin{aligned} Q_v = \int_0^H a_v m(z) = \frac{\eta \gamma_r L_{\min} k_v}{2\pi} \Bigg\{ & f_s \cos 2\pi\left(\frac{t}{T} - \frac{H}{\eta}\right) - \cos 2\pi\left(\frac{t}{T}\right) \\ & - \frac{\eta(f_s - 1)}{2\pi H}\left[\sin 2\pi\left(\frac{t}{T}\right) - \sin 2\pi\left(\frac{t}{T} - \frac{H}{\eta}\right)\right]\Bigg\} \end{aligned} \tag{5}$$

where $\lambda = TV_s$ is wavelength of shear wave; $\eta = TV_p$ is wavelength of primary wave.

2.2 Required Minimum Reinforcement Length for Sliding Failure Mode

According to the force limit equilibrium (LE) equation, the required minimum reinforcement length satisfying sliding failure mode, $L_{\min,\ S}$, can be calculated as

$$L_{\min,S} = \frac{1.125(P_A + 0.5\Delta P_{AE})}{(\gamma_r H + Q_v^{'}) \tan\phi_l - 1.125 Q_h^{'}} \tag{6}$$

where $\Delta P_{AE} = P_{\mathrm{AE}} - P_A$, P_{AE} is pseudo-dynamic active earth pressure given by reference [10], and $P_A = 0.5\gamma_b H^2(1 - \sin\phi_b)/(1 + \sin\phi_b)$, γ_b and ϕ_b are unit weight and friction angle of backfill; $Q_h^{'} = Q_h/L_{min}$, and $Q_v^{'} = Q_v/L_{min}$; ϕ_l = friction angle of reinforced soil or foundation soil, whichever is lower.

2.3 Required Minimum Reinforcement Length for Eccentricity Failure Mode

According to the moment LE equation on centerline of foundation, the required minimum reinforcement length satisfying eccentricity failure mode, $L_{\min,\ e}$, can be calculated as

$$\left(\gamma_r H + Q_v^{'}\right) L_{\min,e}^2 - 1.5\left(Q_h^{'} H + 1\right) L_{\min,e} + (0.75H - P_A H - 0.9\Delta P_{AE} H) = 0 \tag{7}$$

2.4 Required Minimum Reinforcement Length for Bearing Capacity Failure Mode

According to Meyerhof stress distribution, the required minimum reinforcement length satisfying bearing capacity failure mode, $L_{\min,\ BC}$, can be calculated as

$$b_4 L_{\min,BC}^4 + b_3 L_{\min,BC}^3 + b_2 L_{\min,BC}^2 + b_1 L_{\min,BC} + b_0 = 0 \tag{8}$$

where $b_4 = 0.5\gamma_f N_r[\gamma_r H(\gamma_r H + Q_v^{'}) + Q_v^{'}]$, and $N_r = 2(N_q + 1)\tan\phi_f, N_q = e^{\pi \tan\phi_f}\tan^2(\pi/4 + \phi_f/2), \gamma_f$ and ϕ_f are unit weight and friction angle of foundation soil; $b_3 = c_f N_c \gamma_r H(\gamma_r H + Q_v^{'}) - 0.5\gamma_f N_r \gamma_r H(2Q_h^{'} H - Q_v^{'} H) - 1.875(\gamma_r H + Q_v^{'})^3$, and $N_c = (N_q - 1)\cot\phi_f, c_f$ is cohesion of foundation soil; $b_2 = -0.5 c_f N_c(2Q_h^{'} H - Q_v^{'} H)(\gamma_r H + Q_v^{'}) + \gamma_f N_r\ [0.125(2Q_h^{'} H - Q_v^{'} H)^2 + (P_A H/3 + 0.3\Delta P_{AE} H)(2Q_h^{'} H - Q_v^{'} H - \gamma_r H - Q_v^{'})]$, $b_1 = -2c_f N_c(P_A H/3 + 0.3\Delta P_{AE} H)(\gamma_r H + Q_v^{'}) + \gamma_f N_r(P_A H/3 + 0.3\Delta P_{AE} H)(2Q_h^{'} H - Q_v^{'} H)$, $b_0 = 2\gamma_f N_r(P_A H/3 + 0.3\Delta P_{AE} H)^2$.

3 Results and Discussion

In order to study required minimum reinforcement length against external failure modes of MSE walls, properties used for baseline case are as follow: $\gamma_b = 18\,\mathrm{kN/m^3}$, $\gamma_r = 18\,\mathrm{kN/m^3}$, $\gamma_f = 18\,\mathrm{kN/m^3}$, $\phi_b = 35°$, $\phi_r = 35°$, $\phi_f = 35°$, $c_f = 0$, $k_h = 0.2$, $k_v = 0.5k_h$. In addition, it is assumed that $H/\lambda = 0.3$, $H/\eta = 0.16$ referring to reference [10].

3.1 Required Minimum Reinforcement Length for Sliding Failure Mode

Figure 2a through d show the effect of k_h, ϕ_b, γ_b or γ_r on the $L_{\min,\ \mathrm{s}}$ under pseudo-static and pseudo-dynamic loads when considering the sliding failure mode. In the Fig. 2a through b, it can be seen that $L_{\min,\ \mathrm{s}}$ under pseudo-dynamic load is less than that under pseudo-static load when f_s is not more than 1.4. However, it is indicated in the Fig. 2c through d that $L_{\min,\ \mathrm{s}}$ under pseudo-dynamic load is less than that under pseudo-static load when f_s is not more than 1.6.

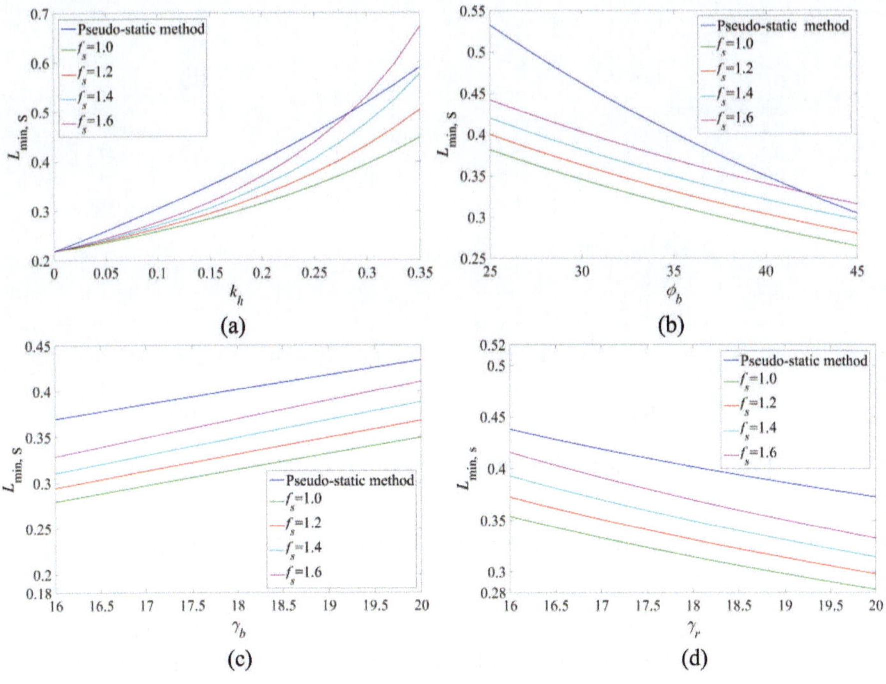

Fig. 2. $L_{\min,\ \mathrm{s}}$ under different parameters: (a) k_h, (b) ϕ_b, (c) γ_b, and γ_r.

3.2 Required Minimum Reinforcement Length for Eccentricity Failure Mode

Figure 3a through d show the effect of k_h, ϕ_b, γ_b or γ_r on the $L_{\text{min, e}}$ under pseudo-static and pseudo-dynamic loads when considering the eccentricity failure mode. In the Fig. 3a and 3c through d, it can be seen that $L_{\text{min, e}}$ under pseudo-dynamic load is less than that under pseudo-static load when f_s is not more than 1.4. Besides, it is investigated in the Fig. 3b that $L_{\text{min, e}}$ under pseudo-dynamic load is less than that under pseudo-static load when f_s is not more than 1.6.

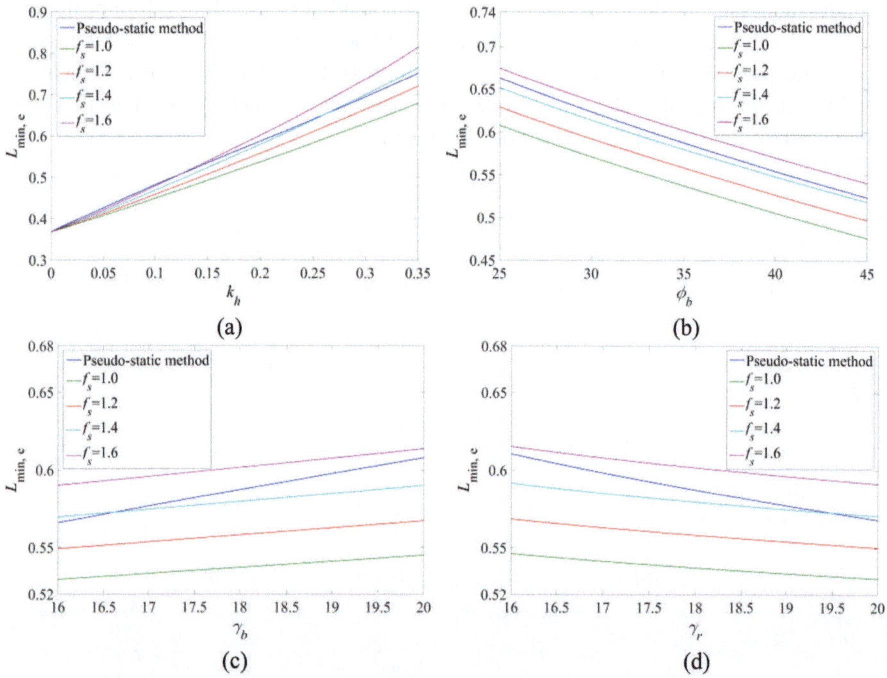

Fig. 3. $L_{\text{min, e}}$ under different parameters: (a) k_h, (b) ϕ_b, (c) γ_b, and γ_r.

3.3 Required Minimum Reinforcement Length for Bearing Capacity Failure Mode

Figure 4a through d show the effect of k_h, ϕ_b, ϕ_f or γ_b on the $L_{\text{min, BC}}$ under pseudo-static and pseudo-dynamic loads when considering the bearing capacity failure mode. In the Fig. 4a through b, it can be seen that $L_{\text{min, BC}}$ under pseudo-dynamic load is less than that under pseudo-static load when f_s is not more than 1.0. But it can be seen that $L_{\text{min, BC}}$ under pseudo-dynamic load is less than that under pseudo-static load when f_s is not more than 1.0 and ϕ_f is more than 30° in the Fig. 4c or γ_b is not more than 19.0 kN/m^3 in the Fig. 4d.

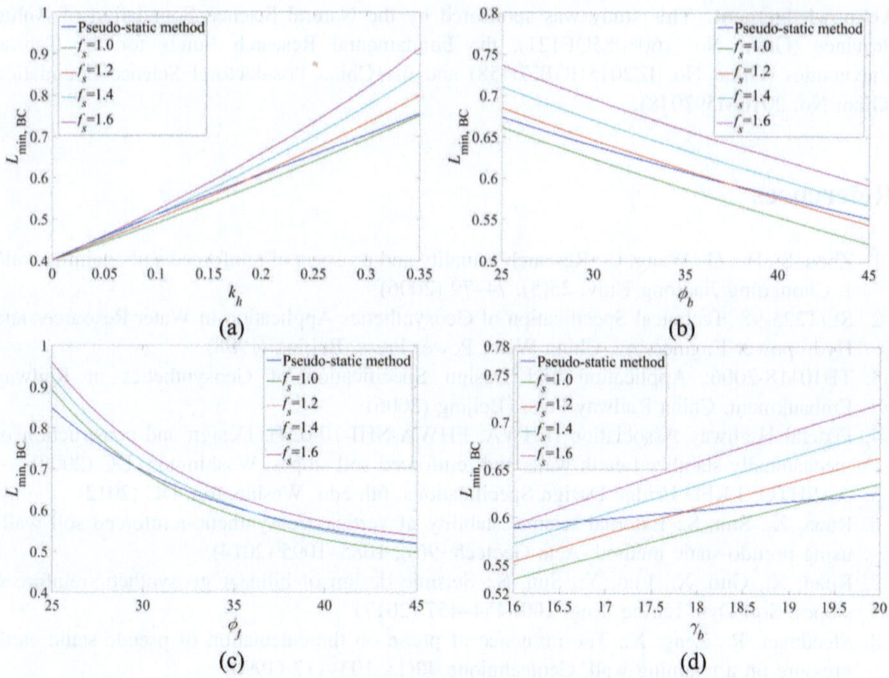

Fig. 4. $L_{min,\ BC}$ under different parameters: (a) k_h, (b) ϕ_b, (c) ϕ_f, and γ_b.

4 Conclusions

This paper develops the calculation formula of the required minimum reinforcement length satisfying different failure modes under the pseudo-dynamic seismic load. The calculation formula includes seismic coefficient, soil amplification factor, wall height, and the properties of reinforced soil, backfill, and foundation soil. The comparison of the minimum reinforcement length against different failure modes under pseudo-dynamic loads and that under pseudo-static loads is investigated.

When f_s is not more than 1.6 with the action of γ_b or γ_r in the sliding failure mode, the required minimum reinforcement length under pseudo-dynamic loads is less than that under pseudo-static loads. When f_s is not more than 1.4 with the action of k_h in the sliding failure mode or ϕ_b in the sliding or eccentricity failure mode, the required minimum reinforcement length under pseudo-dynamic loads is less than that under pseudo-static loads. When f_s is not more than 1.2 with the action of k_h, γ_b or γ_r in the eccentricity failure mode, the required minimum reinforcement length under pseudo-dynamic loads is less than that under pseudo-static loads. When f_s is not more than 1.0 with the action of k_h or ϕ_b in the bearing capacity failure mode, the required minimum reinforcement length under pseudo-dynamic loads is less than that under pseudo-static loads. Through the analysis in this paper, it can be found that the pseudo-dynamic method not only can simulate the actual situation better, but also can calculate the more economical reinforcement length comparing with the pseudo-static method.

Acknowledgement. This study was sponsored by the Natural Science Foundation of Anhui Province (Grant No. 1608085QE121), the Fundamental Research Funds for the Central Universities (Grant No. JZ2015HGBZ0458) and the China Postdoctoral Science Foundation (Grant No. 2016M592048).

References

1. Zhou, S., He, G., Wang, C.: Research actuality and prospect of reinforced soil retaining wall. J. Chongqing Jiaotong Univ. **25**(5), 74–79 (2006)
2. SL/T225-98: Technical Specification of Geosynthetics Application in Water Resources and Hydropower Engineering. China Water Power Press, Beijing (1998)
3. TB10118-2006: Application and Design Specification of Geosynthetics in Railway Embankment. China Railway Press, Beijing (2006)
4. Federal Highway Association (FHWA, FHWA-NHI-10-024): Design and construction of mechanically stabilized earth walls and reinforced soil slopes, Washington, DC (2009)
5. AASHTO: LRFD Bridge Design Specifications, 6th edn. Washington, DC (2012)
6. Ruan, X., Sun, S.: External seismic stability of vertical geosynthetic-reinforced soil walls using pseudo-static method. Acta Geotech. **9**(6), 1085–1095 (2014)
7. Ruan, X., Guo, X., Luo, Y., Sun, S.: Seismic design of bilinear geosynthetic-reinforced slopes. Soil Dyn. Earthq. Eng. **100**, 454–457 (2017)
8. Steedman, R., Zeng, X.: The influence of phase on the calculation of pseudo-static earth pressure on a retaining wall. Geotechnique **40**(1), 103–112 (1990)
9. Zeng, X., Steedman, R.: On the behaviour of quay walls in earthquakes. Geotechnique **43**(3), 417–431 (1993)
10. Ruan, X., Yu, R., Sun, S.: Analysis of seismic active earth pressure on retaining wall based on pseudo-dynamic method. J. Highw. Transp. Res. Dev. **29**(8), 25–30 (2012)
11. Ruan, X., Cheng, Q.Q., Sun, S.: Active seismic pressure against retaining wall backfilled with cohesive soils. Soil Mech. Found. Eng. **50**(3), 116–122 (2013)
12. Ruan, X., Sun, S.: Seismic stability of reinforced soil wall under bearing capacity failure by pseudo-dynamic method. J. Central S. Univ. **20**(9), 2593–2598 (2013)
13. Cheng, Y., Sun, S., Ruan, X., et al.: Pseudo-dynamic analysis of seismic stability of reinforced soil walls. Rock Soil Mech. **34**(12), 3587–3593 (2013)
14. Ruan, X., Sun, S., Liu, W.: Seismic stability of anchored rock slope using pseudo-dynamic method. Rock Soil Mech. **34**(Supp. 1), 293–300 (2013)
15. Ruan, X., Sun, S., Liu, W.L.: Effect of amplification factor on seismic stability of expanded MSW landfills using pseudo-dynamic method. J. Zhejiang Univ.-Sci. A **14**(10), 731–738 (2013)
16. Ruan, X., Lin, H.: Relationship between shear wave wavelength and pseudo-dynamic seismic safety factor in expanded landfill. Arab. J. Sci. Eng. **40**(8), 2271–2288 (2015)
17. Bilgin, Ö.: Failure mechanisms governing reinforcement length of geogrid reinforced soil retaining walls. Eng. Struct. **31**(9), 1967–1975 (2009)
18. Bilgin, Ö., Mansour, E.: Effect of reinforcement type on the design reinforcement length of mechanically stabilized earth walls. Eng. Struct. **59**(2), 663–673 (2014)

Observed Performance of an Ultra Large Deep Excavation in Shanghai Soft Clay

Ping He[1(✉)], Zhonghua Xu[2], Weidong Wang[1,2], and Zili Li[3]

[1] Department of Geotechnical Engineering, Tongji University, Shanghai, China
10heping@tongji.edu.cn

[2] Shanghai Underground Space Engineering Design and Research Institute, East China Architecture Design & Research Institute Co., Ltd., Shanghai, China

[3] Civil and Environmental Engineering, University College Cork, Cork, Republic of Ireland

Abstract. This paper conducted a case study of an ultra large deep excavation of approximately 70,500 m^2 (about 340 m in length and 200 m in width) in soft clay. The excavation depth is mostly located in Shanghai soft clay about 10.3–15.9 m below the ground surface for basements of two or three levels. In the interest of safety and convenience, the construction site was divided into three zones (I, II, III) and excavated individually by the zoned excavation method. The pits at all the three zones were excavated using the bottom-up method with embedded continuous bored pile walls as earth retaining structures. Zone I and III construction started first until the completion of their top slabs, followed by Zone II excavation. In this study, an extensive instrumentation program was carried out to monitor the performance of the deep excavation, including (1) lateral wall deflections, (2) ground surface settlements, (3) axial forces in the struts. The field measurements indicate that at the end of the construction phase, the maximum excavation-induced deflection and ground surface settlement was equal to 0.83% to 0.9% of the excavation depth (H_e), respectively. These measurements are greater than empirical values for typical size excavations in Shanghai clay. The findings obtained in this case study provide some guidance for the design and investigation of similar excavation projects.

Keywords: Ultra large deep excavation · Zoned excavation method
Soft clay

1 Introduction

Rapid urban development in congested city environments necessitates a growing number of underground infrastructures such as deep basements, pipelines and metro tunnel systems. In soft clay regions, deep underground excavation is likely to induce considerable ground deformation and lead to a significant impact on adjacent structures and facilities. It is therefore important to develop a comprehensive understanding of the performance of deep excavations in soft clay for construction safety and compliance.

Field monitoring of excavation, the practical approach to ensure construction safety, has been extensively investigated in recent decades (e.g., Peck [1], Mana and Clough [2], Clough and O'Rourke [3]). Past research has proposed a series of empirical

X. Shi et al. (Eds.): GSIC 2018, *Proceedings of GeoShanghai 2018 International Conference: Transportation Geotechnics and Pavement Engineering*, pp. 161–170, 2018.
https://doi.org/10.1007/978-981-13-0011-0_18

or semi-empirical graphics to estimate wall deflection and ground settlement due to excavation in different soil strata. Recent studies have been dedicated to summarizing the behavior of retaining wall and ground movements based upon numerous worldwide database case histories (e.g., Long [4], Moorman [5], Wang et al. [6]), providing valuable information and details for further theoretical or numerical analysis.

In practice, there are some different excavation methods and associated technical studies to cater for different project time schedules and construction environments, such as top-down methods (Ou et al. [7, 8], Tan and Li [9]), bottom up methods (Tan and Wang [10]), or combination of these two methods (Finno et al. [11]). However, the majority of the past studies were limited to typical scale excavation within hundreds and thousands of square meters, whereas a pit up to 20,000 m^2 was seldom reported in literature. Hence, the ultra large size deep excavation in soft clay remains to be investigated, and it is widely acknowledged that the excavation size plays a crucial factor on the behavior of a deep excavation. Such academic perception is confirmed by site engineers' experience: ultra large deep excavations need to be divided into several smaller zones and excavated relatively independently using the zoned excavation method so as to minimize ground displacements. It is therefore essential to evaluate and assess the effectiveness of the zoned excavation method during ultra large deep excavations.

In this paper, the performance of an ultra large deep excavation (approximately 70,500 m^2) in Shanghai soft clay using zoned excavation method was investigated through an extensive field monitoring programme. For brevity, this paper mainly focuses on the field measurements of the lateral wall deflections, ground surface settlements and axial forces in the struts. The well documented construction records and comprehensive field data will provide valuable reference for researchers and engineers involved in a geotechnical project of similar construction conditions.

2 Project Description

In this paper, the excavation site-Meilong South Business Project (MSBP) was located in the Minhang district of Shanghai. The main structures of MSBP consist of three blocks of 30-storey office building (136.5 m in height), one block of 19-storey hotel (80 m in height) and 3–5 storey street commercial buildings. Figure 1 presents the site plan of the project along with the lateral wall deflection and ground surface settlement instrumentation layouts for field monitoring. The site was bounded by Wanyuan Road on the east side, West Gumei Road on the north side, Hechuang Road (under planning) on the west side and Minhong Road (under planning) on the south side. The shape of the excavation resembles a foot with slightly narrow at the ankle. The area of the excavation is approximately 70,500 m^2 with an excavation depth of 10.5 m and 14.5 m for two and three level basement, respectively.

The surrounding environment of the MSBP was complex: (1) High-rise residential buildings were under construction on the north side of the three-story basement zone. The buildings were sustained by pile raft foundations with a deep pile length up to 33 m, whilst the minimum distance to the excavation was only 22 m. (2) A 6-story brick structures supported by shallow foundations locates on the north side of the

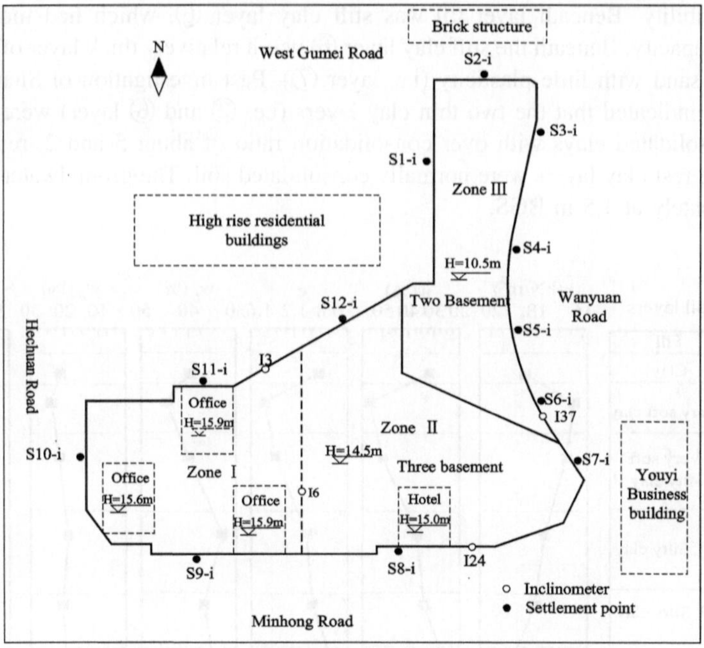

Fig. 1. Layout and instrumentation of the excavation

two-story basement zone. (3) On the east side of the excavation was Youyi business building only 48 m away from the site. (4) Moreover, there was also an extensive network of utility pipelines around the excavation site, for example water and power cable pipeline. The minimum distance between the pipelines and the excavation was within 19 m.

3 Subsurface Conditions

Prior to basement excavation, soil properties at MSBP had been comprehensively investigated by a series of in situ tests and laboratory tests (e.g., standard penetration test, field vane shear test, oedometer tests and triaxial tests). Figure 2 shows a typical soil profile with associated geotechnical properties along the depth. The subsurface soils consist of fill (layer ①) in the upper 1.3 m BGS (below ground surface), which was composed of construction debris, gravel and plant deposits. The following layer to a depth of 3 m (layer ②) was brownish yellow clay with medium-soft plasticity. Below layer ② were very soft silty clay (layer ③) and very soft clay (layer ④) to a depth of 13 m BGS. These two layers were generally with high water content (about 45%), large void ratio (about 1.2) and high compressibility but with low shear strength and permeability. There were the main layers affecting the excavation behavior. Underlying was the fifth layer, which was divided into two sub-layers, namely, layer ⑤$_1$, layer ⑤$_3$. The fifth layer was mainly grayish silty clay with medium plastic and medium to high compressibility. Beneath layer ⑤ was stiff clay layer ⑥, which had much higher bearing capacity. Beneath the stiff clay layer ⑥ was a relatively thick layer of sandy silt and silty sand with little plasticity (i.e. layer ⑦). Past investigation of Shanghai soft clay [12] indicated that the two thin clay layers (i.e. ② and ⑥ layer) were generally over-consolidated clays with over consolidation ratio of about 5 and 2, respectively, whilst the rest clay layers were normally consolidated soil. The groundwater table was approximately at 1.5 m BGS.

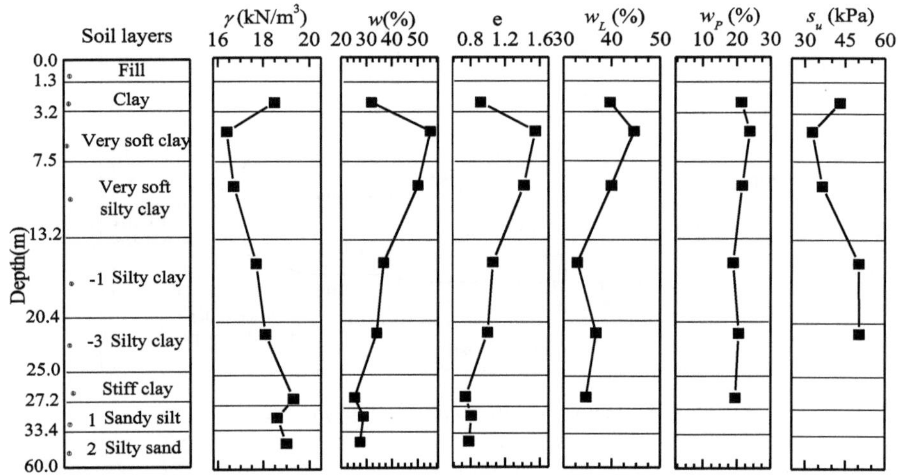

Fig. 2. Geotechnical profile and soil properties (γ = unit weight; w = water content; e = void ratio; w_L = liquid limit; w_P = plastic limit; S_u = field cane shear strength)

4 Support System and Construction Sequences

Abundant local site engineering practices suggest using the zone-excavation method for such a large scale excavation so as to minimize the adverse effects on adjacent facilities. As shown in Fig. 1, the pit was divided into three zones for separate excavation: zone I (office building zone), zone II (hotel building zone) and zone III (street commercial zone). A three-level car park basement was built for the zone I and zone II, while the street commerce zone III included a two-level basement underneath.

Figure 3 shows a typical cross section of the excavation support system for the three level basement zone III. The three levels basement zone was divided into general area (excavation depth was 14.5 m), office area (excavation depth was 15.9 m) and hotel area (excavation depth was 15 m) as shown in Fig. 1. Prior to the commencement of excavation, the soils adjacent to the retaining piles were mixed with three-shaft cement soil as waterproof curtains. The general area was supported by 1,000 mm-thick and 27.6 m-high contiguous bored piles, while the office and hotel area was supported by the same type of piles of 1,100 mm in thickness and 28.9 m in length. A group of temporary partition piles were constructed to separate Zone I and Zone II with a thickness of 950 mm and 27.6 m in length. Between zone II and zone III were

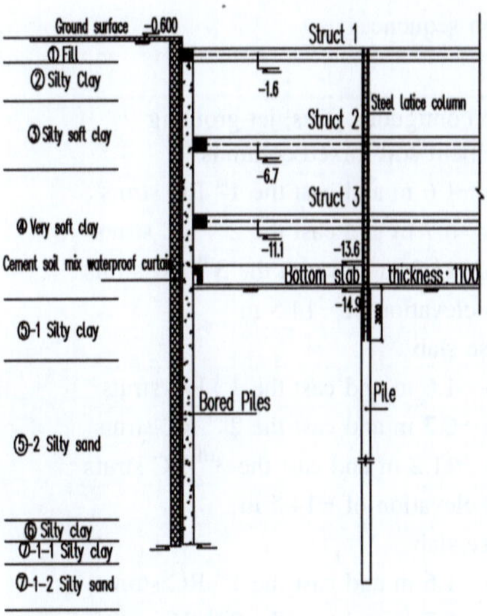

Fig. 3. Section of support system at Zone III

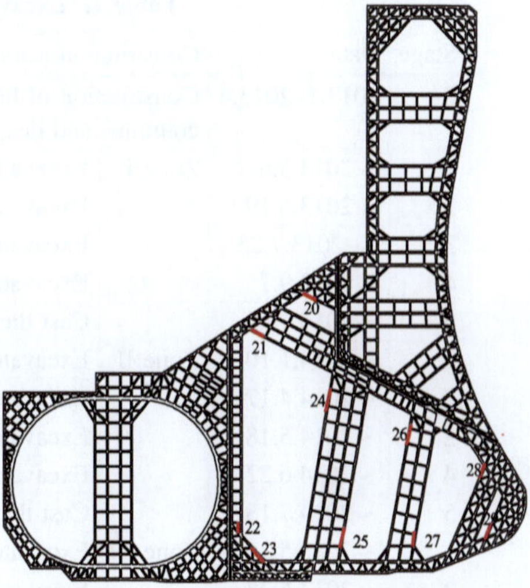

Fig. 4. Configuration of struts system

separated by the partition walls consisting of 1,100 mm diameter and 27.6 m in length. The retaining piles were supported by three levels of struts located at depths of 1.6, 6.7 m and 11.1 m BGS, respectively. The configuration of the struts system and the location of strut for monitoring axial forces was shown in Fig. 4. They were supported by 640 × 640 mm steel columns, which were socketed 1 m into the underlying bored piles.

The excavation of zone III (two-level basement) was supported by 850-mm thick and 19.5-m high contiguous bored piles. The retaining piles were braced by two levels of steel reinforced concrete struts cast at the depths of 1.6, and 6.7 m BGS, respectively. The cross section size of the first and second level struts were 1000 mm × 700 mm, 1000 mm × 800 mm, respectively.

Immediately after excavation to the design depth, the 2-m thick base slabs were poured at the pit base followed by demolishing of the temporary concrete struts and then construction of underground basement structures.

The excavation started from zone I and zone III using bottom-up method. After construction of the top slab for the basement at the two zones, the rest zone II was excavated also using bottom-up method. The detailed construction schedules at zones I, II and III are described in Table 1.

The monitoring points location in the MSBP was shown in Figs. 1 and 4. Forty inclinometer tubes (designated as I1 to I40) were placed inside steel reinforcement cages to measure the lateral displacements of the retaining walls. For simplicity, only the points used for analysis were noted in the figure. Twelve groups of ground surface settlement markers (designated as s1 to s12) were installed at the distances of 2, 7, 12, 22, 32 m behind the retaining walls away from the excavation. Ground settlements were surveyed by level instruments with an accuracy of 0.01 mm. The axial force in

Table 1. Excavation sequences

Stage	Date	Construction activities	
0	2013.1–2013.4	Construction of bored contiguous piles, jet-grouting columns, and deep cement soil mixed columns	
1	~2013.5.6	Zone I	Excavate to −1.6 m and cast the 1st RC struts
2	~2013.6.19		Excavate to −6.7 m and cast the 2nd RC struts
3	~ 2013.7.25		Excavate to −11.2 m and cast the 3rd RC struts
4	~2013.9.7		Excavate to elevation of −14.5 m
5	~2013.10.2		Cast the base slab
1	~2014.1.10	Zone II	Excavate to −1.6 m and cast the 1st RC struts
2	~2014.4.17		Excavate to −6.7 m and cast the 2nd RC struts
3	~2014.5.18		Excavate to −11.2 m and cast the 3rd RC struts
4	~2014.6.22		Excavate to elevation of −14.5 m
5	~2014.7.13		Cast the base slab
1	~2013.5.16	Zone III	Excavate to −1.6 m and cast the 1st RC struts
2	~2013.6.23		Excavate to −6.7 m and cast the 2nd RC struts
3	~2013.8.5		Excavate to the elevation of −10.5 m
4	~2013.9.1		Cast the base slab

10 of the concrete struts at zone II at each level (designated as Z1-20–Z1-29 at level 1, Z2-20–Z2-29 at level 2, and Z3-20–Z3-29 at level 3) were also monitored.

5 Monitoring Results and Discussions

5.1 Lateral Wall Displacements

Figure 5 shows the lateral wall deflection profiles along the depth at different construction stages. The retaining walls experienced deep-seated (bulging) deflection profiles in line with typical multi-propped excavations, while all the wall deflection modes are similar at different zones. In comparison to typical excavation tolerances, the maximum displacements at the last stage were more than 100 mm in three-level basement (I3, I6, I24), which was probably attributed to the oversizing of the excavation and soft ground conditions. At each excavation stage, the maximum displacement appeared approximately at the excavation surface. Due to the smaller excavation depth in zone III, the displacement of I37 is significantly smaller than those in zone I and II.

At the same excavation stage, I3 (near the corner) was noted to have a smaller displacement than I6 (near the middle of the wall) due to the corner effect. After excavation to the design depth (Stage 4), the retaining walls continued to develop noticeable deflections at Stage 5 (base slab construction). This time-dependent post-excavation deflection might be caused by the movement of the supporting structures, ground consolidation and etc.

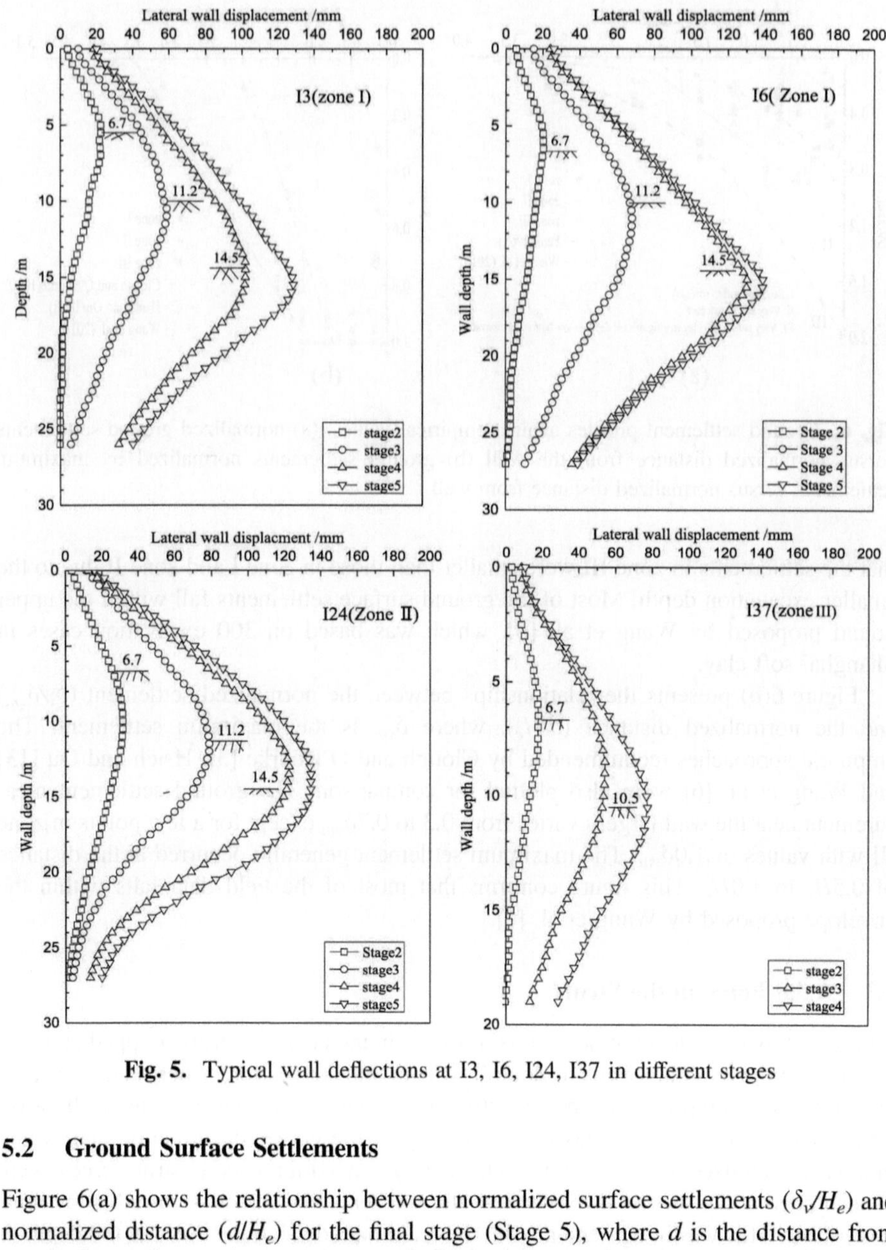

Fig. 5. Typical wall deflections at I3, I6, I24, I37 in different stages

5.2 Ground Surface Settlements

Figure 6(a) shows the relationship between normalized surface settlements (δ_v/H_e) and normalized distance (d/H_e) for the final stage (Stage 5), where d is the distance from monitoring point to the retaining wall, δ_v is the ground settlements and H_e is the excavation depth. The distribution of ground surface settlement proposed by Peck [1] and upper bound ground surface settlement proposed by Wang et al. [6] are also displayed in Fig. 6(a) for comparison. In this project, the distribution of ground surface settlements generally extends to a distance of about $3.5H_e$ away from the excavation, while most of the settlement points fell within the regions I and II defined by Peck [1]. The maximum ground surface settlement is generally less than $0.8\%H_e$. It is also noted

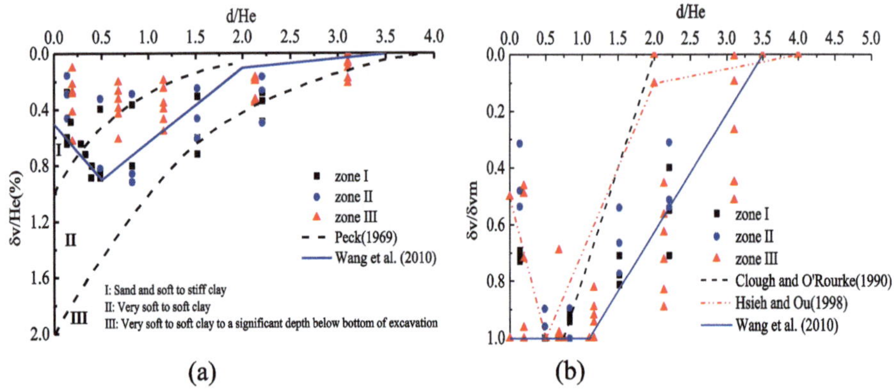

Fig. 6. Ground settlement profiles against empirical values: (a) normalized ground settlements versus normalized distance from the wall (b) ground settlements normalized by maximum settlements versus normalized distance from wall

that the settlements in zone III were smaller than those in zone I and zone II due to the smaller excavation depth. Most of the ground surface settlements fall within the upper bound proposed by Wang et al. [6], which was based on 300 excavation cases in Shanghai soft clay.

Figure 6(b) presents the relationships between the normalized settlement (δ_v/δ_{vm}) and the normalized distance (d/H_e), where δ_{vm} is the maximum settlement. The empirical approaches recommended by Clough and O'Rourke [3], Hsieh and Ou [13] and Wang et al. [6] were also plotted for comparison. The ground settlement measurement near the wall largely varies from 0.3 to $0.7\delta_{vm}$ except for a few points in zone III with values of $1.0\delta_{vm}$. The maximum settlement generally occurred at the distance of $0.5H_e$ to $1.0H_e$. This figure confirms that most of the field data falls within the envelope proposed by Wang et al. [6].

5.3 Axial Force in the Struts

Figure 7 shows the development of axial forces in the concrete struts propped at levels 1–3 against the retaining walls in Zone II. Due to unbalanced forces between excavated and soil sides, compression force (positive in the figure) was observed in all the level props. The axial forces increased dramatically at the earlier construction stages and gradually stabilized over the time. On average, maximum axial strut forces were concentrated at the second level whereas the smallest axial forces occurred at the third level struts. This was in agreement with the maximum deflections noted at the depth of second level as described earlier. The maximum axial force in the first level strut was about 580 kN at ZL1-23, the one in the second level strut was 700 kN at ZL2-20, and the one in the third level strut was 400 kN at ZL3-25.

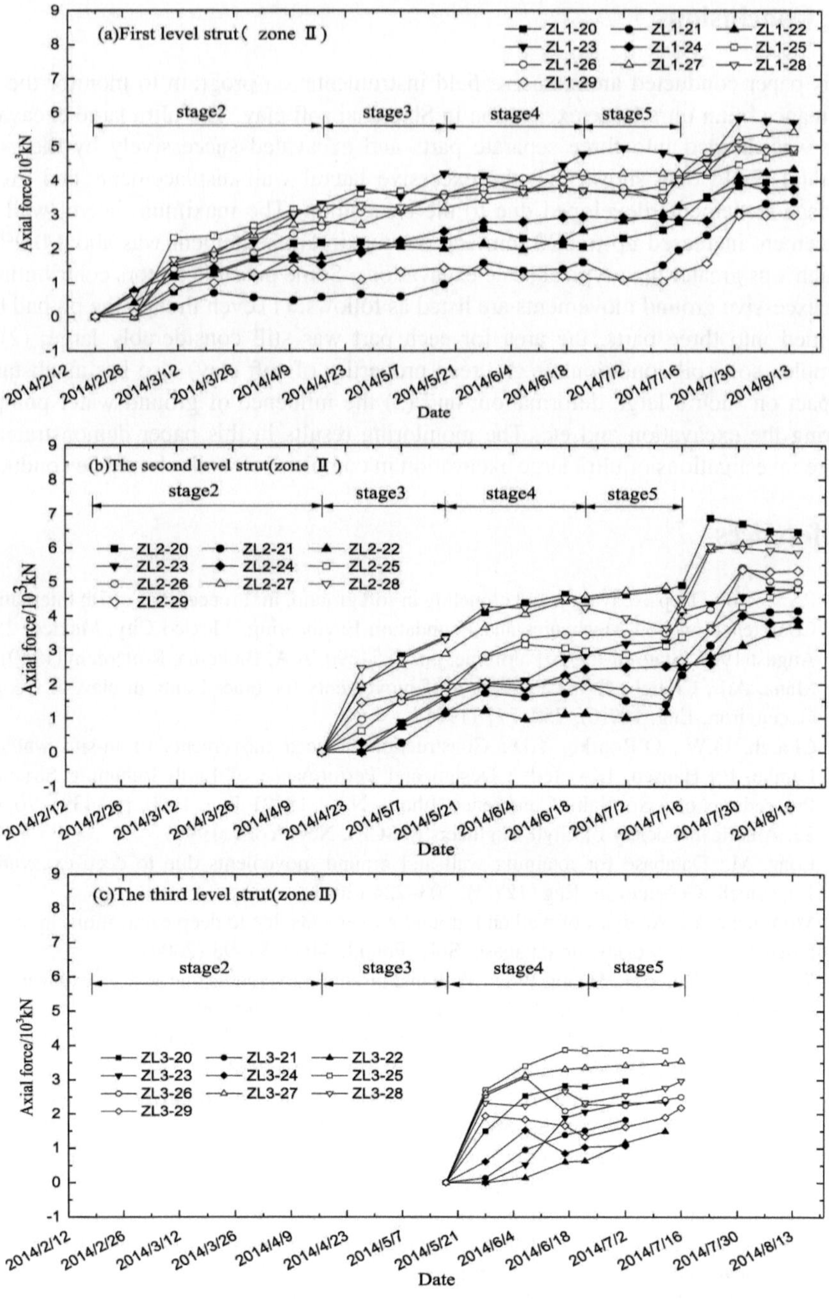

Fig. 7. Axial forces in the different levels of struts for zone II

6 Conclusions

This paper conducted an extensive field instrumentation program to monitor the performance of an ultra large excavation in Shanghai soft clay. The ultra large excavation site was divided into three separate parts and excavated successively by the zoned method. Field data shows that the excessive lateral wall displacements and ground surface settlements developed due to the excavation. The maximum lateral wall displacement increased up to 120 mm and the maximum settlement was about $0.9\%He$, which was greater than typical scale excavations. Some possible factors contributing to the excessive ground movements are listed as follows: (1) even though the pit had been divided into three parts, the area for each part was still considerably large; (2) the complex soft soil conditions (e.g., creep properties of soft clay) also has a substantial impact on such a large deformation; and (3) the influence of ground water pumping during the excavation and etc. The monitoring results in this paper demonstrate that more investigations of ultra large excavation in complex soft soils should be conducted.

References

1. Peck, R.B.: Deep excavation and tunneling in soft ground. In: Proceedings of 7th International Conference on Soil Mechanics and Foundation Engineering, Mexico City, Mexico, 25–29 August 1969. State-of-the-Art Volume, pp. 225–290. A.A. Balkema, Rotterdam (1969)
2. Mana, A.I., Clough, G.W.: Prediction of movements for braced cuts in clay. J. Geotech. Geoenviron. Eng. **107**(6), 759–777 (1981)
3. Clough, G.W., O'Rourke, T.D.: Construction induced movements of in-situ walls. In: Lambe, P., Hansen, L.A. (eds.) Design and Performance of Earth Retaining Structures, Proceedings of a Specialty Conference, Ithaca, N.Y., 18–21 June 1990, pp. 439–470, GSP 25. American Society of Civil Engineers (ASCE), New York (1990)
4. Long, M.: Database for retaining wall and ground movements due to deep excavations. J. Geotech. Geoenviron. Eng. **127**(3), 203–224 (2001)
5. Moormann, C.: Analysis of wall and ground movements due to deep excavations in soft soil based on a new worldwide database. Soils Found. **44**(1), 87–98 (2004)
6. Wang, J.H., Xu, Z.H., Wang, W.D.: Wall and ground movements due to deep excavations in shanghai soft soils. J. Geotech. Geoenviron. Eng. **136**(7), 985–994 (2010)
7. Ou, C.Y., Liao, J.T., Lin, H.D.: Performance of diaphragm wall constructed using top-down method. J. Geotech. Geoenviron. Eng. **124**(9), 798–808 (1998)
8. Ou, C.Y., Liao, J.T., Cheng, W.L.: Building response and ground movements induced by a deep excavation. Géotechnique **50**(3), 209–220 (2000)
9. Tan, Y., Li, M.: Measured performance of a 26 m deep top-down excavation in downtown Shanghai. Can. Geotech. J. **48**(5), 704–719 (2011)
10. Tan, Y., Wang, D.: Characteristics of a large-scale deep foundation pit excavated by the central-island technique in Shanghai soft clay. I: Bottom-up construction of the central cylindrical shaft. J. Geotech. Geoenviron. Eng. **139**(11), 1875–1893 (2013)
11. Finno, R.J., Arboleda-Monsalve, L.G., Sarabia, F.: Observed performance of the one museum park west excavation. J. Geotech. Geoenviron. Eng. **141**(1), 04014078 (2014)
12. Wei, D.D., Hu, Z.X.: Experimental study of preconsolidation pressure and compressibility parameters of Shanghai subsoil. Chin. J. Geotech. Eng. **2**(4), 13–22 (1980)
13. Hsieh, P.G., Ou, C.Y.: Shape of ground surface settlement profiles caused by excavation. Can. Geotech. J. **35**(6), 1004–1017 (1998)

Real Time Distress Monitoring of Soft Soil Highway Embankment

Muhammad Amjad[1(✉)], Sarfraz Ali[1], Mazhar Iqbal[1], Abdul Qudoos[2], and Ali Sarosh[3]

[1] College of Civil Engineering, NUST, Risalpur Campus, Risalpur, Pakistan
amjad8eb@gmail.com
[2] Civil Works Organization, Rawalpindi, Pakistan
[3] College of Aeronautical Engineering, NUST, Risalpur Campus, Risalpur, Pakistan

Abstract. Highway embankments constructed over soft soil foundation exhibit pronounced settlement/geotechnical distresses resulting in reduced service life of communication infrastructure; necessitates real time monitoring system. Upgradation/maintenances of road infrastructure is a continuous practice, worldwide. Construction of Swat Motorway in Pakistan is one such mega project at hand, where high water table and soft foundation strata is encountered. This paper attempts to explore and evaluate the modern procedural developments in highway embankments; acquisition, interpretation and processing of real time data associated with soft soil monitoring systems using Micro Electro Mechanical Sensors (MEMS). The authors are of the view that findings of the research would help proposing innovative methodology in reducing foundation hazards related to highways, minimizing maintenance cost and help improving roadways serviceability.

Keywords: Soft soils · MEMS · Settlement · Embankment · Highways

1 Introduction

Behavior of soft soils is unpredictable, especially in presence of subsurface water. Strength of soils and the way they behave, affects the performance of highway embankments in terms of stability. Effects of subgrade instability are reflected to the pavements surface in form of distresses and sometimes lead to permanent deformations. To ensure desired serviceability, stable pavements are necessary which comes through stable subgrade foundation. Highway engineers must be aware of soft soil foundation hazards such as high compressibility, low bearing capacity and less permeability. This leads to embankment failure, pavement distresses, settlement of approach embankments just at the beginning and end of the bridges. It adds to the maintenance cost, reduces service life, increases risk of road crashes and adds misery to the life of road user (Lei et al. 2014). To address this a comprehensive embankment management system is required. It includes continuous Realtime monitoring of the foundation soil behavior as well as the embankment. It helps in accurate planning and

X. Shi et al. (Eds.): GSIC 2018, *Proceedings of GeoShanghai 2018 International Conference: Transportation Geotechnics and Pavement Engineering*, pp. 171–178, 2018.
https://doi.org/10.1007/978-981-13-0011-0_19

forms basis of the guidelines to be followed throughout the country, whenever soft soils are encountered while constructing highways and motorways.

To flourish trade by land to the international community, huge road infrastructure is the need of time. With the inception of China Pakistan Economic Corridor (CPEC), numerous highways and motorways are being constructed in Pakistan. One of the series of motorways in Pakistan is Swat Motorway. It lies in north west frontier province Khyber Pakhtunkhwa of Pakistan. It will connect Swat, a tourism spot, with existing motorway M-1 at Karnal Sher Khan interchange (Fig. 1).

Fig. 1. General alignment - Swat Motorway.

Swat Motorway consists of numerous structures including 7 x interchanges along with toll plazas, 29 x bridges, 2 x unidirectional, 1–2 km long tube tunnels, 45 x culverts, 15 x subways for vehicles and pedestrians, 4 x cattle creeps, and 2 x service areas. Project was divided in different zones each 10 km long for purpose of geotechnical investigation. Considerable portion of the motorway passes through irrigation land which is waterlogged and inherits soft clay soil with high water table. The trial section site lies in zone 2, RD 10+800 (Fig. 2). The soil is fragile with very low bearing capacity. It necessitates application of an effective soft soil embankment management/monitoring system to help monitor behavior of foundation soil and embankment. The system will help in studying the effect of induced stresses in terms of distresses generated. A wide range of techniques is practiced internationally for this purpose. It is need of time to ascertain the techniques being practiced in Pakistan and the typical issues that exist during practical implementation of the soft soil highway embankment management/monitoring system.

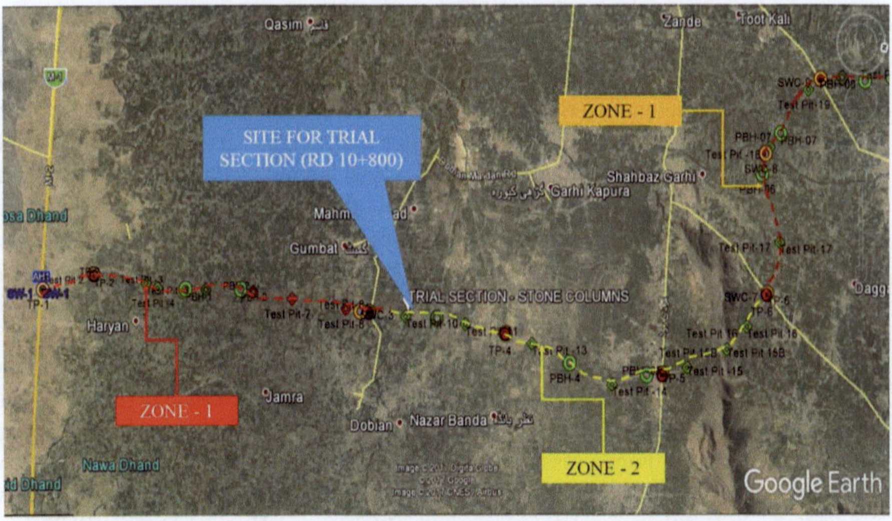

Fig. 2. Zoning of Swat Motorway – RD 0+00 to RD 30+00

2 Objectives

- To study contemporary case histories pertaining to soft soil highway embankments and draw comparison with ongoing practices in Pakistan.
- To explore soft soil highway embankment management/monitoring procedures being practiced in Pakistan.

3 Significance

Following research features signify its need and importance for highway construction industry in Pakistan.

- Determine behavior of soft soil foundation under embankment and identify its effects on pavement structure.
- Determine efficacy of stone columns.
- Recommend suitable ground improvement techniques in areas with soft soils and high water table for highway construction Soft Soil Embankment Monitoring Case Histories.

4 Case Histories

Huat (1994), while describing behaviour of five embankments on highly overconsolidated Malaysian clay found out that clays behave as undarined medium only once normally consolidated during early construction stages. Out of five two embankments were constructed as trial sections by Malaysian Highway Authorities in 1998. The soil beneath these embankments was soft to very soft marine clay up to 20 m depth. under clay existed the loose to dense and medium to coarse sand. The other three embankments were part of North-South Expressway, situated in northern state of Penang, Peninsular Malaysia. The expressway was constructed in 1992 to 1993. All embankments were instrumented with inclinometers, pneumatic piezometers and settlement plates.

Hussein and McGown (1998), used inclinometers, piezometers, settlement plate and heave markers to study the behavior of two trial embankments. One of the trial embankments was constructed within 36 days and other in one year. Slowly constructed embankment displayed higher rates of settlement, excess pore water pressure distribution and lateral displacement. Vertical and lateral displacement for both embankment was observed to be almost same. Excess pore water pressure distribution increased during construction and dissipated before becoming constant for both embankments.

Cavalcante et al. (2004), have analyzed access embankments of Jitituba river bridge at AL-413 highway, São Luiz do Quintude, Brazil and found it stable. The site consisted of 12 m thick layer of soft soils. Access embankments behavior was studied keeping in view the pore pressure, Stability, rupture, vertical and horizontal displacements by installing geotechnical instruments (MEMS) i.e. inclinometers, piezometers, and settlement plates along with Prfabricated Vertical Drains (PVDs).

Balasubramaniam et al. (2010), have highlighted the difficulties in predicting highway embankments settlements on soft soils and reclamation of marine, deltaic and estuarine lands. Ground improvement with PVDs hs been used to explain fundamentals of preloading also explains the same without PVDs. Embankments supported by Geogrid Reinforced Piles(GRPS) have been numerically analysed

Dai et al. (2017), have predicted the settlement, horizontal displacement, surcharge preloading time and variation in pore water pressure, of a freeway embankment situated in Meizhou Island with Putian City in Fujian, China, using three dimensional finite element analysis and found it to agree with that of measured actually using MEMS. The embankment passes through a shallow sea area. Sand columns were used to improve the ground. Non linear three dimensional finite element analysis was found to be reliable for complex problems of improving soft soils.

4.1 Relevance of Case Histories Versus Current Practices in Pakistan

- Realtime settlements, stability, rupture, horizontal and vertical displacement, surcharge preloading time and variation in pore water pressure are measured using MEMS.
- Soft soil hazard complexities are predicted using realtime data through numerical simulations.
- Predicted distresses can be mitigated using existing ground improvement techniques.

5 Soft Soil Highway Embankment Monitoring Practices in Pakistan

Road infrastructure in Pakistan forms major portion of the government budget and needs continuous management, specially embankment roads for better maintenance in terms of monitoring. Although ground improvement in areas with weak soil foundations is a necessary tool to ensure desired service life of a highway structure by avoiding anomalies inherited by soft soils. A series of motorways and highways in Pakistan have been constructed. All are void of a comprehensive monitoring system. Deformations at various places on Grand Trunk (GT) road, M-1, M-2, and Mardan ring road. Since design does not incorporate any provision of a comprehensive real-time monitoring system, it is impossible to evaluate distress patterns and their causes. Ground improvement techniques are used at limited scale with less or no research opportunities. Use of Geogrids and geosynthetics is also not a common engineering practice in field of construction in Pakistan. Even if, Geogrids are being used in one form or the other, no technical data on performance of the Geogrids after construction is available. Reinforced piles are used to support bridge piers foundation. Micro piling at few places has also been used. No permanent monitoring system is available to aid appropriate management of the structures, specially highway embankments constructed over soft foundation soils to gauge their performance.

Pakistan has no such system that binds the stake holders to observe the distress patterns with increase in loads on foundation soils in terms of embankment fill and traffic. Neither any record for highway embankment monitoring has been found so far. Pakistan is yet to adopt existing techniques for soft soil highway embankments management/monitoring. Stake holders need to understand the importance of the subject for long term benefits to the state. Since Swat Motorway is an expensive project, it is necessary to monitor the behavior of foundation soil and embankments using MEMS.

6 Realtime Monitoring Limitations in Pakistan

Focusing on the geotechnics of soft soils, existing monitoring system being practiced almost all over the world and Pakistan for construction of highways over soft soil foundation, following limitations were identified:

- Stereotype, old fashioned rule of thumb is in vogue once it comes to engineering practices in highway construction industry. Adoption of the modern and existing real-time monitoring techniques is silenced under umbrella of heavy expenses. People are uncertain and there is a factor of hesitation amongst authorities once it comes to application in the field of highway construction. Even an organization like FWO was hesitant in adopting measures to improve soil conditions on site using any of existing techniques and install monitoring system. They preferred using remedy, routinely being practiced i.e. Granular Material Platform GMP, for construction of road over soft soils.
- Lack of research in the field of ground improvement at industrial level renders authorities unaware of the soft soil behavior under highway embankments, and response of the structures constructed over unimproved soil foundation, during and sometimes at very early stage of their service life. Settlement of M-2 motorway near Kallar Kahar is one fine example.
- Consultancy organizations have shifted to use of modern numerical analysis and validation tools, but construction industry seems to be hesitant in adopting

recommended remedies for improving ground conditions in road construction as seen in construction of various projects.

- Tendency of using engineered methods, being practiced over time has led to lack of appropriate equipment in the local market, vision to innovate new techniques and innovative use of existing ones.

7 Formulation of Real Time Soft Soil Monitoring Framework in Pakistan - Swat Motorway a Case Study

Real time monitoring is essential once highway embankments are being constructed over soft soil foundations. No such system is being practiced in Pakistan. Generally, roads are constructed following stereotype design without regards to the behavior of foundation soils and embankments. Soft foundation soil hazards are neglected due to lack of knowledge. An effort has been made to demonstrate the importance of soft soil highway embankment management/monitoring by improving soft soil with existing ground improvement technique and monitoring the behavior of improved section and comparing it with that of unimproved section on Swat Motorway. It was planned to install MEMS including inclinometers, piezometers, and settlement gauges in order to monitor the behavior of foundation soil under different loading and climatic conditions (Figs. 3a and 3b). However only Settlement plates could be installed, and real-time monitoring results could not be included in this paper as the construction work over this specific site was suspended due some technical and administrative issues. The installed settlement plates are illustrated in Fig. 3c. The stone columns designed and analyzed using Preibie method which showed significant improvement in foundation soil strength. Due to space limitations of paper complete calculations cannot be shown, however results will be discussed in results and discussion section.

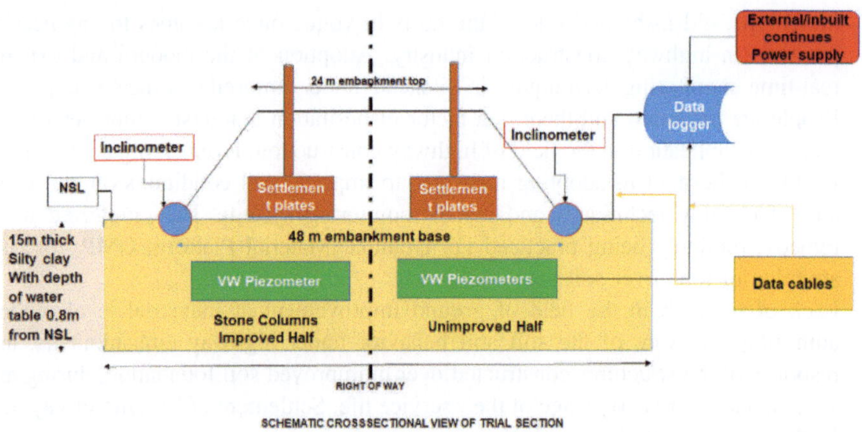

Fig. 3a. Instruments layout plan of trial section-SMP

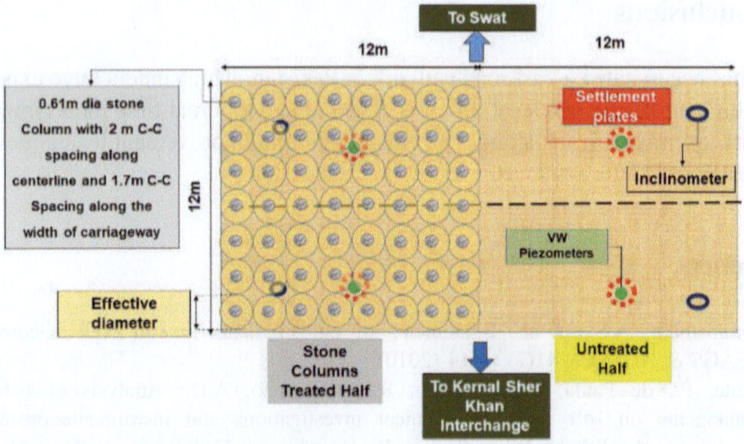

Fig. 3b. Instruments layout plan of trial section Swat Motorway-top view

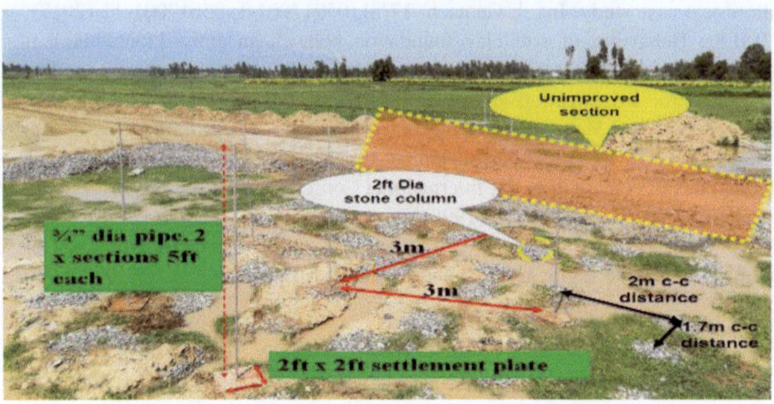

Fig. 3c. Settlement plates at trial section SMP

8 Results and Discussion

Since real-time data for 58x stone columns installed at site could not be achieved as the work was suspended for unknown period, the stone columns were analytically evaluated using Preibie method. Which Show significant improvement in bearing capacity of soil hence reduced settlement. The settlement was reduced from 66.06 cm (predicted without improvement) to 33.85 cm after installation of stone columns. Which dictates improvement of soft soils with existing ground improvement techniques is an effective measure to arrest post construction distresses.

9 Conclusions

CPEC has necessitated a vast road network in Pakistan, which incurs huge expenditure. To obtain long lasting serviceability it is essential to adopt real-time monitoring system for highway embankments along with existing ground improvement techniques such as stone columns.

References

Balasubramaniam, A.S., et al.: Settlements of embankments in soft soils. Geotech. Eng. J. SEAGS & AGSSEA **41**(2), 1–14 (2010)

Cavalcante, P., de Paula, S., Coutinho, R.Q., Gusmão, A.D.: Analysis of behavior of embankments on soft soils geotechnical investigations and instrumentation access of embankments the Jitituba River Bridge. In: International Conference on Case Histories in Geotechnical Engineering, New York (2004)

Dai, Z.-H., Chen, B.-L., Qin, Z.-Z.: Three-dimensional nonlinear numerical analysis of consolidation of soft ground improved by sand columns under a freeway embankment in shallow sea: case study. Int. J. Geotech. **17**(8), 05017001-1–05017001-12 (2017)

Huat, B.B.K.: Behaviour of soft clay foundation beneath an embankment. Pertanika J. Sci. Technol. **2**(2), 215–235 (1994)

Huat, B.B.K.: An investigation of an embankment failure in soft clay. In: International Conference on Case Histories in Geotechnical Engineering, St. Louis, Missouri (1994)

Hussein, A.N., McGown, A.: The behaviour of two trial embankments at Perlis, Malaysia with different rates of construction. In: International Conference on Case Histories in Geotechnical Engineering, St. Louis, Missouri (1998)

Lei, G., Qiu-yue, Z., Xiang-juan, Y., Zhi-hui, C.: Analysis and model prediction of subgrade settlement for Linhai highway in China. Electron. J. Geotech. Eng. **19**, 11–21 (2014)

Assessing In-situ Nonlinear Model Parameters for Material of Changheba High Earth Core Rockfill Dam Through Finite Element Analysis

Xingxing Zhang[1(✉)], Gang Deng[1], Dan Zhang[2], Shu Yu[1], Shunbin He[2], and Yanfeng Wen[2]

[1] State Key Laboratory of Simulation and Regulation of Water Cycle in River Basin, China Institute of Water Resources and Hydropower Research, Beijing 10084, China
zhangxx@iwhr.com
[2] POWERCHINA Chengdu Engineering Corporation Limited, Chengdu 610072, China

Abstract. The in-situ material engineering parameters of Changheba dam, the highest ECRD built on thick overburden in China, is studied through FE analysis. The nonlinear model proposed by Duncan is selected to simulate the stress-strain relationship of soil and rockfill materials. The dependence of the permeability coefficient of clayey material on the void ratio is considered. FE analysis based on laboratory tested parameters is performed. Meanwhile, parameters back analysis is conducted based on monitored deformation and pore pressure at the end of construction. The analyzed results show that the in-situ parameters of some dam material are quite different from in-lab tested parameters. The in-situ modulus and friction angle of blast rockfill material is smaller than in-lab tested values. The difference might be caused by scaling effect of particle size and depends on scaling ratio between in-lab samples and in-situ material. The overburden might have a higher in-situ modulus than laboratory tested samples, which might due to the in-situ long-term stress history and scaling effect. Both high pore pressure and small settlement is observed in the core wall. It is shown by back analysis results that the gravelly clay core may have lower in-situ permeability and relatively higher stiffness. The stress path in in-situ gravelly clay core is different from which in in-lab triaxial test might be an important cause.

Keywords: Earth core rockfill dam (ECRD) · Finite element analysis
Scaling effect · Nonlinear model

1 Introduction

The Changheba dam is located on the Dadu river, in Sichuan province, China, which is a 240 m high earth core rockfill dam (ECRD) built on about 60–70 m thick overburden. It is also the highest ECRD built on thick overburden in China. The design of Changheba dam is introduced by Zhang et al. [1]. The cross section of Changheba dam is shown in Fig. 1. The imperious core is built with gravelly clay, and the dam shell is filled with blast rockfill material.

X. Shi et al. (Eds.): GSIC 2018, *Proceedings of GeoShanghai 2018 International Conference: Transportation Geotechnics and Pavement Engineering*, pp. 179–188, 2018.
https://doi.org/10.1007/978-981-13-0011-0_20

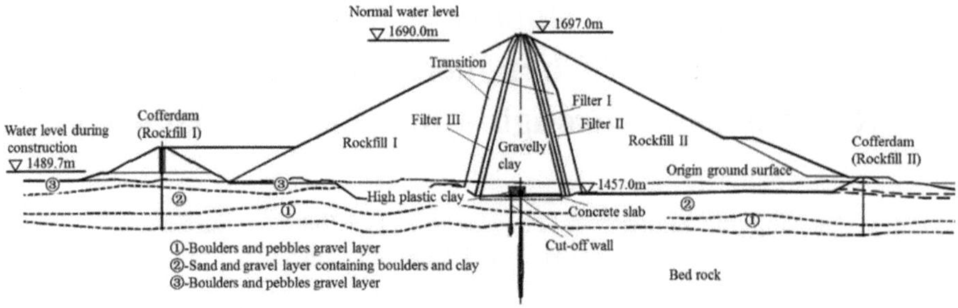

Fig. 1. Cross section of the Changheba dam.

A typical ECRD usually consists of several different kinds of materials with diverse stiffness, which often leads to inhomogeneous deformation and might cause cracks and related seepage failure of dams. Therefore, deformation prediction is very important in safety assessment of ECRDs. Finite element (FE) simulation is an effective method to predict the mechanical behavior of dams. The accuracy of numerical calculated results relies much on the constitutive model parameters. However, the in-situ mechanical properties of dam material may differ from those obtained from laboratory tests for some inevitable reasons, e.g. the scaling effect of rockfill material [2–4], and the complex stress history of dam body and overburden which cannot be fully simulated in laboratory. Therefore, case study and back analysis become effective methods to evaluate the in-situ behavior of materials in rockfill dams [5–7].

In this paper, FE simulation based on laboratory tested parameters and parameters back analysis are performed to evaluate the mechanical behavior of the Changheba dam at the end of construction. The Duncan's E-B model is selected to simulate the non-linear stress-strain relationship of soil and rockfill materials, and the nonlinear relationship between the permeability coefficient and strain state of impervious clay material is considered during simulation. The analyzed results show that the in-situ parameters of some dam material are quite different from in-lab tested parameters. The causes of difference between the in-situ and laboratory tested parameters of coarse rockfill, over burden and gravelly clay materials are discussed.

2 Description of Finite Element Analysis

2.1 Computation Methods and Constitutive Models

Originally coded by Shen and Zhang [8], and developed by the authors and other researchers, a program package applied in many earth core rockfill dams in China is implemented in this paper. A fluid-solid coupled algorithm based on the Biot's theory is adopted in the program. An E-B mode nonlinear elastic model proposed by Duncan et al. [9] is introduced to describe the stress-strain relationship of soil and rockfill materials. For clay and gravelly clay, the friction angle φ is constant. For rockfill materials, the friction angle varies with the confining pressure, which is represented by following equation suggested by Leps [10]:

$$\varphi = \varphi_0 - \Delta\varphi \log\left(\frac{\sigma_3'}{p_a}\right) \tag{1}$$

where φ and φ_0 are friction angle of rockfill when confining pressure equal to σ'_3 and p_a respectively. φ_0 and $\Delta\varphi$ is coefficient.

In this dam, rockfill, transition, filter material and overburden have much higher permeability than the impervious core, so head loss mainly occurred in the gravelly clay core. The process of consolidation of the core depends greatly on the permeability of the gravelly clay. According to previous researches [11, 12], the permeability coefficient of soil varies with the void ratio. In this paper, the variation of permeability coefficient of gravelly clay is considered, and is described with an equation suggested by Taylor [11].

$$k = k_0 \exp[\lambda(e - e_0)] \tag{2}$$

where e_0 is the original void ratio, e is current void ratio, k_0 and k are permeability coefficient when void ratio is e_0 and e respectively, and λ is an exponential. When $\lambda = 0$, the permeability coefficient k is constant, and equals to k_0.

2.2 Finite Element Model

A 3D FE mesh with 72353 elements is built to simulate the geometry of the Changheba dam, which is shown in Fig. 2. The numerical modeling of construction is divided into 30 steps. Firstly, the overburden is fully consolidated under self-weight. Then, the cofferdam is filled. The construction of cut walls is simulated in Step 3. Then the dam is filled by 27 layers form Step 4 to Step 30. The impounding of reservoir had not begun during construction of the dam, so the reservoir water level is kept near elevation 1489.7 m (lower than the crest of the cofferdam) during construction.

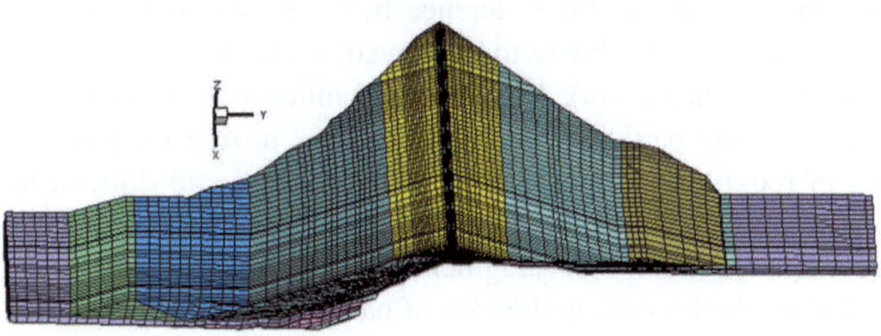

Fig. 2. FE mesh of the Changheba dam.

2.3 Model Parameters

The laboratory tested parameters of soil and rockfill materials were supplied by the designer (Chengdu Engineering Cooperation Limited) and are shown in Table 1. The nonlinear relationship between permeability coefficient and void ratio of gravelly clay and high plastic clay are considered, and the permeability coefficients of other materials are considered to be constant.

Table 1. Laboratory tested parameters of soil and rockfill materials

Material	Dry density	Permeability		Shear strength			E-B model				
	ρ_d g/cm^3	k_0 cm/s	λ	C kPa	φ_0 (°)	$\Delta\varphi$ (°)	R_f	K	n	K_b	m
Gravelly clay	2.15	2.5×10^{-6}	40	97.66	29.71	/	0.65	442	0.4	247	0.39
Rockfill I	2.24	/	/	/	51.28	7.64	0.78	1265	0.34	440	0.29
Rockfill II	2.31	/	/	/	49.98	5.60	0.76	1238	0.36	424	0.32
Filter I	2.17	5.0×10^{-3}	/	/	48.61	6.99	0.76	995	0.277	350	0.247
Filter II	2.25	5.0×10^{-2}	/	/	49.8	7.61	0.76	1166	0.327	412	0.266
Filter III	2.29	3.0×10^{-2}	/	/	50.14	7.44	0.74	1172	0.333	426	0.276
Transition	2.25	8.0×10^{-2}	/	/	50.95	8.03	0.79	1137	0.322	397	0.237
High plastic clay	1.59	1.0×10^{-6}	40	46	24	/	0.75	185	0.446	84.5	0.355
Overburden layer ①	2.33	0.5	/	/	48	7	0.8	1100	0.34	560	0.23
Overburden layer ②	2.182	0.3	/	/	46.9	6.5	0.77	1075	0.33	550	0.25
Overburden layer ③	2.112	0.3	/	/	47.9	4.1	0.79	865	0.29	450	0.25

2.4 Back Analysis Process

Usually, parameters back analysis process can be summed as two steps: (1) defining an objective function to evaluate the difference between analyzed and observed values; (2) solving the optimization problems to find a set of parameters which make the objective function minimum (or maximum). A variety of techniques, including genetic algorithm, simulated annealing and particle swarm optimization, have been applied in parameters back analysis of rockfill dams [6, 13–15]. However, it is still difficult to find a set of mathematically optimal and physically reasonable parameters purely depending on numerical techniques, and empirical judgment is still necessary in most cases. In this paper, a semi empirical method based on the idea of particle swarm is adopted. The objective function is defined as the 2-norm difference between the calculated and observed values:

$$f(p_1, p_2, \ldots, p_n) = \sqrt{\frac{1}{N}\sum_{i=1}^{N} (v_i - v_i^*)^2} \tag{3}$$

where $p_1, p_2, \ldots, p_n$ is a set of model parameters, N is the number of observed values, v_i^* is the i-th observed value, and v_i is the corresponding calculated value. Notice that the variables $v_1, v_2, \ldots, v_N$ and $v_1^*, v_2^*, \ldots, v_N^*$ should be of the same dimension.

The procedure to find optimum solution to minimize f is summarized as follows:

(1) The in-lab tested parameters is selected as initial values, and expressed as P^0 (p_1^0, $p_2^0, \ldots, p_n^0$).
(2) In the I-th iteration, a set of parameters can be generated as follows:

$$P_j^I = P_0^I + \alpha D_j^T \cdot P^0 \tag{4}$$

where P_j^I is a set of parameters, which can be seen as a particle; P_0^I is the origin of the I-th iteration, which is set as P^0 when $I = 1$; α is the relative search interval, which is fixed at 0.05 to filter out too small changes; D^j (d_1^j, $d_2^j, \ldots, d_n^j$) is a direction vector whose components d_1^j, $d_2^j, \ldots, d_n^j$ vary among −1, 0 and 1.

The particles are brought into FE analysis, and f could be obtain through Eq. (3). The particle for the minimum f is chosen as results of the iteration, and the norm of the corresponding D^j is evaluated. If $\|D\| = 0$, the particle is the optimal parameters, and iteration ends. If $\|D\| > 0$, the set of parameters is set as P_0^{I+1}, and anther iteration is performed.

At most $3n$ particles can be generated in each iteration. However, it will lead to too much FE calculation work. To avoid this, in each iteration, several most possible particles is select to performed FE simulation instead. The selection of particles will depend on the engineering experience of the researchers, so this is a "semi empirical" approach. This "semi empirical" method significantly reduces the amount of finite element calculation for each iteration, but it is possible to increase the number of necessary iterations.

Because the sensitivity of permeability coefficient and parameters of Duncan's E-B model are quite different, a step-by-step process is adopted in the optimization process:

(1) Adjust the permeability coefficient of clayey soils to fit observed pore pressure.
(2) Adjust the Duncan's model parameters to fit the observed deformation of the dam.
(3) Repeat the above two steps, until both the simulated pore pressure and deformation fit well with the observed data.

The optimized parameters obtained through back analysis are listed in Table 2, in which the bold characteristics represent that the values differ from the in-lab tested value.

3 Comparison of Analyzed and Observed Behavior

3.1 Modulus of Dam Shell and Overburden

The comparison of analyzed and observed settlements of the downstream dam shell on the maximum section (section 10 = 253.72 m) is shown in Fig. 3. In order to compare the settlement of the core wall and dam shell, the settlements of the core wall observed

Table 2. Inversion parameters of soil and rockfill materials

Material	Permeability		Shear strength			E-B model				
	k_0 cm/s	λ	C kPa	φ_0 (°)	$\Delta\varphi$ (°)	R_f	K	n	K_b	m
Gravelly clay	**1.0×10^{-7}**	40	127	**38.60**	/	**0.3**	**1105**	0.4	**695**	0.39
Rockfill I	/	/	/	**46.15**	**6.88**	0.78	**885.5**	0.34	**277.9**	0.29
Rockfill II	/	/	/	**44.98**	**5.04**	0.76	**866.6**	0.36	**296.8**	0.32
Filter I	5.0×10^{-3}	/	/	48.61	6.99	0.76	995	0.277	350	0.247
Filter II	5.0×10^{-2}	/	/	49.8	7.61	0.76	1166	0.327	412	0.266
Filter III	3.0×10^{-2}	/	/	50.14	7.44	0.74	1172	0.333	426	0.276
Transition	8.0×10^{-2}	/	/	50.95	8.03	0.79	1137	0.322	397	0.237
High plastic clay	**2.5×10^{-7}**	40	46	24	/	0.75	185	0.446	84.5	0.355
Overburden layer ①	0.5	/	/	48	7	0.8	**1430**	0.34	**728**	0.23
Overburden layer ②	0.3	/	/	46.9	6.5	0.77	**1398**	0.33	**715**	0.25
Overburden layer ③	0.3	/	/	47.9	4.1	0.79	**1124**	0.29	**585**	0.25

on section 10 = 256 are also marked in Fig. 3 with short dashed lines. For convenient, the analyzed results based on laboratory tested parameters are called as results of Theme 1, and back analyzed results are called as results of Theme 2 in following parts.

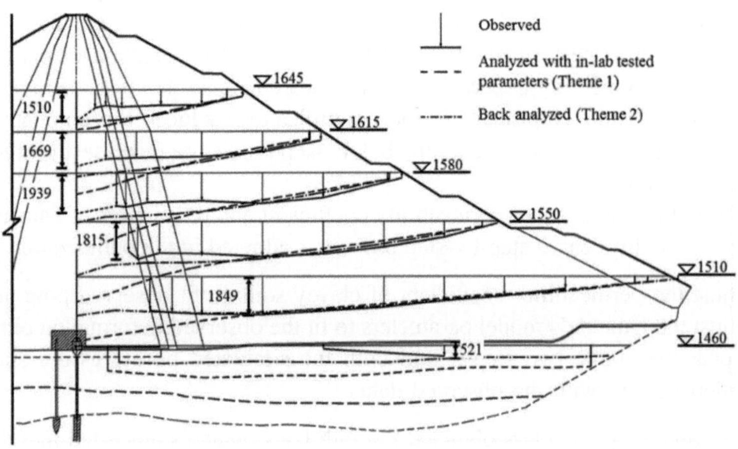

Fig. 3. Comparison of analyzed and observed settlements on the maximum section (Unit of settlements: mm, Unit of elevation: m)

It can be seen that the observed and back analyzed settlements in the filter and transition is smaller than in rockfill II. It can be inferred that the filter and transition is stiffer than rockfill II in-situ. In comparison of Tables 1 and 2, it can be found that inversion K and K_b of rockfill I and rockfill II are 30% less than laboratory tested values, and inversion φ_0 and $\Delta\varphi$ are 10% less than laboratory tested values, but the inversion parameters of filters and transitions material equals to the laboratory tested parameters.

It seems that the difference between in-situ and in-lab tested parameters of rockfill materials is related to particle size.

Scaling effect has been supposed to be an important cause for the difference of mechanical properties between in-situ rockfill and in-lab samples. Both in-situ and in-lab maximum particle sizes for all rockfill materials in Changheba dam are listed in Table 3. For coarse rockfill material (e.g. rockfill I and rockfill II in Fig. 1), the maximum in-situ particle size is near 900 mm, about 15 times of maximum particle size in tested samples, so there is obvious difference of parameters K K_b φ_0 and $\Delta\varphi$ between back analyzed and in-lab tested values. For fine rockfill material, such as filter material, the in-situ particle size is closed to in-lab samples, so there is little difference between back analyzed and in-lab tested parameters. For convenient, the following empirical equation is used to predict the in-situ parameters of rockfill material:

$$P_{in-situ} = P_{in-lab} \exp\left[-c_1\left(1 - \frac{D_{in-lab}}{D_{in-situ}}\right)\right] \tag{5}$$

where $P_{\text{in-situ}}$ and $P_{\text{in-lab}}$ denotes certain parameter of in-situ material and in-lab tested sample, $D_{\text{in-situ}}$ and $D_{\text{in-lab}}$ denotes characteristic particle size of in-situ material and in-lab tested sample respectively, and c_1 is an index that reflects the influence of scaling effect. According to back analyzed results, $c_1 = 0.4$ can be taken for K and K_b, and $c_1 = 0.1$ can be taken for φ_0 and $\Delta\varphi$ for rockfill I and rockfill II materials.

Table 3. In-situ and in-lab maximum particle sizes of rockfill materials

Material	In-situ maximum particle size mm	In-lab maximum particle size mm
Rockfill I	900	60
Rockfill II	900	60
Filter I	20	20
Filter II	80	60
Filter III	40	40
Transition	400	60

According to Fig. 3, the simulated settlement on the surface of the overburden in Theme 1 is much larger than observed. Accordingly, it can be seen in Tables 1 and 2, the inversion parameters K and K_b are 130% of the laboratory tested values. It can be inferred that the in-situ modulus of overburden is much higher than laboratory tested samples. Besides scaling effect, the long term consolidation history which cannot be simulated in laboratory might be another main cause for the difference between in-situ modulus and in-lab tested modulus of the overburden.

3.2 Behavior of the Gravelly Clay Core

Figure 4 shows the comparison of observed and analyzed settlements of the gravelly clay core. The monitored maximum settlement of the core is about 2134 mm at the elevation 1598.73 m. That might be not the real maximum settlement of the core,

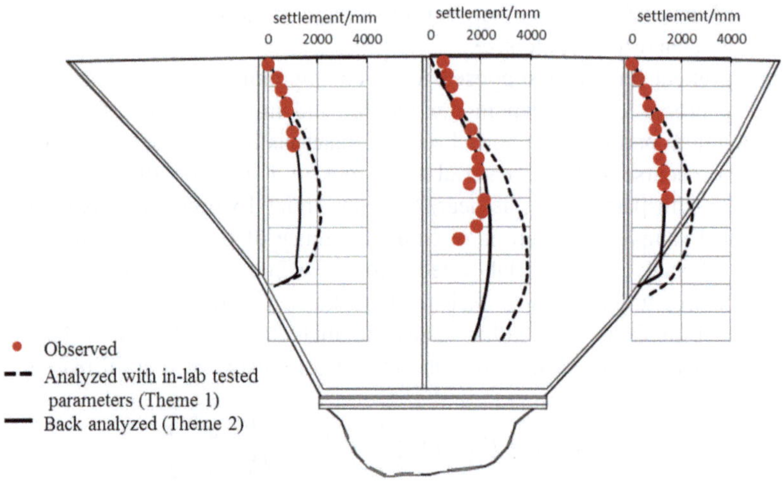

Fig. 4. Comparison of analyzed and observed settlements of the core wall

because some of the electromagnetic settlement instruments have been damaged. The back analyzed results show that the maximum settlement of the core wall is about 2400 mm, which is about 1% of the dam height, but less than 1% of the sum of dam height and overburden thickness. The simulated settlement based on in-lab tested parameters is much larger than observed settlements on most monitored points. That means the compressibility of the gravelly clay core is overestimated by in-lab tested parameters.

Figure 5 shows the comparison of observed and simulated pore pressure in the gravelly clay core on the maximum section. The observed maximum pore pressure in the core wall is about 1.8 MPa. The corresponding water head is about 3/4 of the dam height. It can be inferred that the permeability of the gravelly clay is very low, which make the dissipation of the excess pore pressure very slow. According to Tables 1 and 2, the back analyzed permeability coefficient is only 1/25 of the laboratory tested values. The permeability of the gravelly clay is greatly overestimated by in-lab test, so the simulated pore pressure in Theme 1 is much lower than observed.

The coexisting high pore pressure and small settlement is a special phenomenon occurred in the Changheba dam. The inversion results show that there is great difference between the in-situ mechanical behavior and laboratory tested behavior of gravelly clay. The influence of stress path may be an important factor. The actual stress path of gravelly clay core is constant stress ratio loading under a high stress level, but in conventional triaxial test the stress path is shear under constant minor stress. Researches show that, in constant stress ratio loading condition, the shear modulus of gravelly clay increase with shear strain; while in conventional triaxial test, the shear modulus decreases with shear strain [16]. It seems that the conventional triaxial test is not a good choice to determine the deformation parameters of the gravelly clay core during construction.

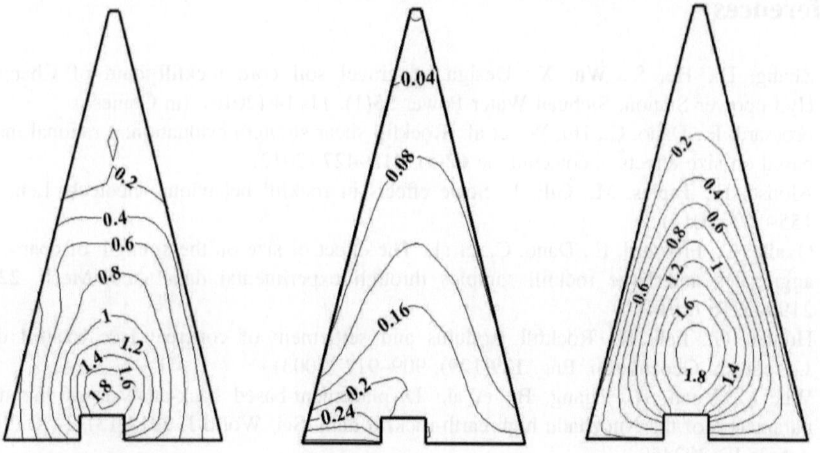

(a) Observed (b) Analyzed in Theme 1 (c) Analyzed in Theme 2

Fig. 5. Comparison of analyzed and observed pore pressure in the core wall (Unit: MPa)

4 Conclusions

The mechanical behavior of the Changheba ECRD has been studied through FE analysis. FE analysis based on laboratory tested parameters, and parameters back analysis are performed. The results show that there are differences between laboratory tested and in-situ properties of some dam materials. The primary findings of this study are summarized as follows:

The in-situ modulus and friction angle of blast rockfill material are smaller than laboratory tested values. The difference between the in-situ and laboratory tested parameters of rockfill materials is mainly caused by scaling effect, which depends on scaling ratio of particle size. So there are obvious difference for coarse rockfill materials, but little difference can be observed for fine rockfill materials. The in-situ modulus of overburden is much higher than laboratory tested samples. It might be due to the long term stress history and scaling effect.

The coexisting high pore pressure and small settlement is a special phenomenon occurred in the Changheba dam. The inversion parameters of gravelly clay are quite different from the in-lab tested parameters. The permeability coefficient of the gravelly clay is only 1/25 of the values obtained by in-lab seepage test. Back analyzed K and K_b are much larger and R_f is much smaller than in-lab tested values, which denotes that the in-situ modulus of gravelly clay core is much higher than in-lab test. The stress path in in-situ gravelly clay core is different from which in in-lab triaxial test is one of the main cause. Tests under constant stress ratio might be more suitable to evaluate the property of gravelly clay core during construction.

References

1. Zhang, D., He, S., Wu, X.: Design of gravel soil core rockfill dam of Changheba Hydropower Station. Sichuan Water Power **35**(1), 11–14 (2016). (in Chinese)
2. Frossard, E., Dano, C., Hu, W., et al.: Rockfill shear strength evaluation: a rational method based on size effects. Géotechnique **62**(5), 415–427 (2012)
3. Alonso, E., Tapias, M., Gili, J.: Scale effects in rockfill behaviour. Géotech. Lett. **2**(4), 155–160 (2012)
4. Ovalle, C., Frossard, E., Dano, C., et al.: The effect of size on the strength of coarse rock aggregates and large rockfill samples through experimental data. Acta Mech. **225**(8), 2199–2216 (2014)
5. Hunter, G., Fell, R.: Rockfill modulus and settlement of concrete face rockfill dams. J. Geotech. Geoenviron. Eng. **129**(129), 909–917 (2003)
6. Wu, Y., Yuan, H., Zhang, B., et al.: Displacement-based back-analysis of the model parameters of the Nuozhadu high earth-rockfill dam. Sci. World J. **2014**(13), 1–10 (2014). Article ID 292450
7. Won, M., Kim, Y.: A case study on the post-construction deformation of concrete face rockfill dam. Can. Geotech. J. **45**(6), 845–852 (2008)
8. Shen, Z., Zhang, W.: Analyses of effective stress and movement of Xiaolangdi sloping core rockfill dam. Hydro-Sci. Eng. **4**, 361–373 (1991). (in Chinese)
9. Duncan, J., Wong, K., Mabry, P.: Strength, stress-strain and bulk modulus parameters for finite element analysis of stresses and movements in soil masses. Report No. VCB/GT/78-02, University of California Berkeley (1978)
10. Leps, T.: Review of shearing strength of rockfill. J. Soil Mech. Found. Div. **96**(4), 1159–1170 (1970)
11. Taylor, D.: Fundamentals of Soil Mechanics. Wiley, New York (1984)
12. Tavenas, F., Jean, P., Leblond, P., et al.: The permeability of natural soft clays. Part II: permeability characteristics. Can. Geotech. J. **20**(4), 645–660 (1983)
13. Liu, X., Zhou, C., Zhang, L.: Application of neural network based on simulated annealing Gauss-Newton algorithm to seepage back analysis. Rock Soil Mech. **26**(3), 404–406 (2015)
14. Yu, Y., Zhang, B., Yuan, H.: An intelligent displacement back-analysis method for earth-rockfill dams. Comput. Geotech. **34**(6), 423–434 (2007)
15. Gang, M., Wei, Z., Chang, X.: A novel particle swarm optimization algorithm based on particle migration. Appl. Math. Comput. **218**(11), 6620–6626 (2012)
16. Chen, Z., Zhu, J., Liu, H.: Study on stress path tests of wide grading gravelly soil. J. Disaster Prev. Mitig. Eng. **30**(6), 614–619 (2010)

Load-Deformation Responses of Ballasted Rail Tracks: Laboratory and Discrete-Continuum Modelling

Ngoc Trung Ngo(✉), Buddhima Indraratna, and Cholachat Rujikiatkamjorn

University of Wollongong, Wollongong, Australia
trung@uow.edu.au

Abstract. The recent and rapid urbanization and frequent congestion of roads have led to more attention being focused on ballasted tracks for freight and commuter transport. The mechanisms of ballast degradation and deformation, the need for effective track confinement, understanding of interface behaviour, determining the dynamic bearing capacity of ballasted tracks require further insight to improve the existing design guidelines for future high speed commuter and heavier freight trains. The load-deformation behaviour of ballast under cyclic loads is measured in the laboratory using a novel large-scale Track Process Simulation Apparatus (TPSA). A novel coupling model based on discrete element method (DEM) and finite element method (FEM) is developed to predict the load-deformation responses of the ballast assembly considering the interaction of discrete ballast grains and continuum subgrade. In this coupled model, the discrete ballast grains are modelled by DEM and the subgrade domain is modelled as a continuum by FEM. The results indicate that significant settlements are observed during the initial load cycles, followed by gradually increased deformation, arriving at a steady value towards the end of tests. Contact force distributions, stress contours and corresponding broken bonds are captured.

Keywords: Railway ballast · Cyclic loading
Coupled discrete-continuum modelling

1 Introduction

Together with other countries, such as China, USA, Canada and India, Australia is playing a leading role in the development of heavy hauls that include 3–4 km long trains with axle loads exceeding 35 tonnes to optimise the efficiency of supply chains in the mining and agricultural sectors [1]. Ballast is an essential component of the rail track substructure, and it is commonly used for (a) distributing the wheel load from sleepers to the underlying subgrade, (b) maintaining track alignment, and (c) providing track drainage [2, 3]. Upon repeated train loads, the sharp corners of ballast aggregates break; this causes differential track settlement [4]. In addition, the ballast layer can be fouled by upward migration of subgrade clay fines and the downward migration of coal spilling from wagons [5–8], all of which seriously affect the drainage capacity of the track. Since ballasted tracks have minimum lateral support, the lateral confining

X. Shi et al. (Eds.): GSIC 2018, *Proceedings of GeoShanghai 2018 International Conference: Transportation Geotechnics and Pavement Engineering*, pp. 189–198, 2018.
https://doi.org/10.1007/978-981-13-0011-0_21

pressure must be increased to control lateral stability [9]. The performance of ballast is affected by the overall characteristics of a granular mass such as the particle size distribution, the void ratio, and the relative density [10, 11]. While the properties of individual grains of ballast such as size, shape, and angularity govern its degradation under traffic loading, deformation is also influenced by the magnitude of wheel (or axle) load, the number of load cycles, frequency (equivalent to train speed), and the impact loads [12, 13]. The magnitude of the impact loads depends on the type and nature of surface imperfections on the wheels and rails, as well as on the track's dynamic response.

In conventional track design, the ballast is often considered as an elastic medium where its degradation and associated plastic deformation is ignored. This issue is due to not understanding the complexity of particle degradation and fouling problems and not having a proper constitutive model. The constitutive behaviour of granular materials has been studied by various researchers using the discrete element modeling approach [14–16]. This approach to model the track granular media is promising as it can handle particles at micro-scale. In the last decades, the use of planar geosynthetic products such as geogrid, geotextile or geocomposite (bonding geogrid with geotextile) has been more popular as they are economical and increase track stability. Previous researches have indicated that geogrid reinforcement reduces the settlement and degradation of ballast [17–19] attributed to the interlocking effect that restricts the lateral movement of ballast. In addition, recent studies have shown that a geocell (three dimensional, polymeric cells, interconnected at the joint) can provide much better lateral track confinement than planar reinforcement [20].

Continuum approaches have been widely used to model granular material assemblies, and various conventional continuum constitutive models have been introduced to capture their stress-strain responses of granular materials. It is noted that due to the discrete nature of ballast particles, the continuum approach could not accurately simulate their micro-mechanical behaviour governed by the fabric anisotropy, contact orientations and strain localisations. Meanwhile the discrete element method (DEM) has been increasingly employed in the recent past as an alternative to the continuum-based methods for the study of granular materials. This paper presents a coupled DEM-FEM modelling approach to takes advantage of each numerical scheme to provide a realistic solution to model an integrated and layered ballasted track. The model is validated by extensive large-scale tests and presented in the following sections.

2 Experimental Investigations

A large-scale Track Process Simulation Apparatus (TPSA) can simulate a ballast assembly of: 800 mm × 600 mm × 600 mm was used to measure the deformation and degradation of ballast under cyclic loads, as shown Fig. 1. Ballast was selected from Bombo quarry, New South Wales, Australia, then cleaned and sieved following to the Australia Standards [21]. The particle size distribution of the ballast had an average particle size d_{50} = 38 mm, which is similar to current Australian practices. To simulate actual field conditions, a 150 mm-thick subballast and subgrade layer, made from coarse sand and gravel mixture (unit weight of 18.5 kN/m^3), was placed at the bottom of the TPSA. The ballast was then filled above the sub-ballast and compacted in every 50 mm-thick sub-layer to a unit weight of 15.3 kN/m^3, until a total height of the ballast layer achieved 300 mm. A lateral confining pressure of σ_{xx} = 10 kPa was applied in the direction parallel to sleepers to simulate low confinement given by the shoulder ballast in the field [22]. A cyclic load was applied via dynamic actuator with a maximum induced cyclic stress of σ_{yy} = 420 kPa, the frequency of f = 15 Hz, simulating a freight train (i.e. axle load: 30 tonnes; speed: 90 km/h). All tests were conducted up to 500,000 load cycles. The detailed experimental program and analysis of these tests were presented elsewhere by Indraratna et al. [3] and they concluded that the ballast showed a significant deformation within the first 100,000 load cycles, followed by a decreased rate of settlement up to 300,000 cycles, and then kept almost unchanged toward the end of tests. Ballast breakage was quantified after the completion of every test by comparing the differences of ballast particle size distribution curves before and after every test, using the ballast breakage index (BBI) proposed earlier by Indraratna et al. [23]. Some of the results of these tests are re-used in this study to calibrate and validate the combined DEM-FEM model.

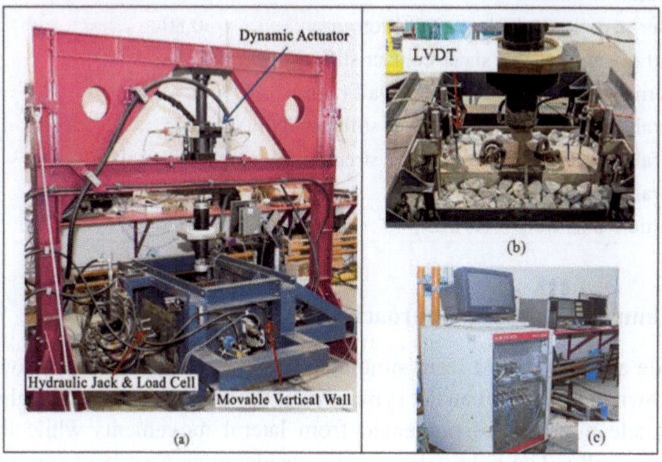

Fig. 1. (a) Track Process Simulation Apparatus – TPSA; (b) Ballast assembly; (c) Data logger instrumentations

3 Discrete-Continuum Modelling Approach

3.1 Discrete Element Method

The discrete element method (DEM) introduced by Cundall and Strack [24] has been increasingly applied to study the micromechanical behaviour of granular materials [25–28]. In DEM, particles are treated as rigid bodies and have a negligible contact area (i.e. at a point). A soft contact approach is commonly used at the contacts allowing the rigid particles to overlap one another at contact points and the extent of the overlap is related to the magnitude of contact forces governed by the force-displacement law.

The force-displacement law derives the contact force acting on two entities in contact to the relative displacement between them. In DEM, particles are spherical in three-dimension or circular in two-dimension; however it is also possible to create particles of arbitrary shape by connecting more spheres together at appropriate sizes and positions (i.e. clumps). Each clump composes of a number of overlapping spheres and acts as a rigid entity with a deformable boundary [29]. Furthermore, it is possible to bond particle together and model fracture of rock mass or breakage of granular materials when the bonds break, which was used by Indraratna et al. [3] to capture ballast breakage. Irregularly-shaped ballast particles were simulated in DEM by connecting of specified numbers of circular balls together to represent appropriate angularity and sizes. In the current DEM analysis, the material properties of the coupled DEM-FEM numerical model are given in Table 1. These values were determined based on the calibration with laboratory test data.

Table 1. Micromechanical parameters of ballast particles and walls used in DEM simulation.

Micro-mechanical parameters	Values
Radius of particle (m)	1.8×10^{-3}–16×10^{-3}
Inter-particle coefficient of friction	0.80
Particle normal and shear contact stiffness (N/m)	3.58×10^{8}
Normal and shear stiffness of wall (N/m)	3×10^{7}
Parallel bond normal and shear stiffness (N/m)	6.25×10^{10}
Parallel bond normal and shear strength (N/m^2)	5.78×10^{6}
Parallel bond radius multiplier	0.5
Particle unit weight (kN/m^3)	15.3

3.2 Continuum Modelling Approach

The subgrade and sub-ballast were simulated as a homogeneous layer having 150 mm depth, as shown in Fig. 2. Given the symmetry of the tracks, the left and right boundaries of the subgrade model were prevented from lateral movements while they allowed displacing vertically. The boundaries were considered as absorbent (viscous) to avoid spurious reflection of the cyclic wave. The bottom boundary was modelled as a pinned support (i.e. both the lateral and vertical displacements were restrained). The subgrade and sub-ballast were represented with a standard elasto-plastic Mohr Coulomb model where the model parameters (i.e. Young's modulus E, Poison's ratio v, cohesion c, friction angle ϕ, and dilatancy angle ψ) were determined from tests carried out on the subgrade soil. In the current coupled analysis, the subgrade having $E = 45$ MPa, $v = 0.32$, $c = 17$ kPa, $\phi = 14^0$, and $\psi = 4^0$ were used. It is noted that the DEM and FEM share the same geometrical boundaries where they are interfaced. Initially, a series of walls is generated in the DEM zone, where each wall at the interface corresponding to a single surface segment (continuum element) of the FEM zone [30].

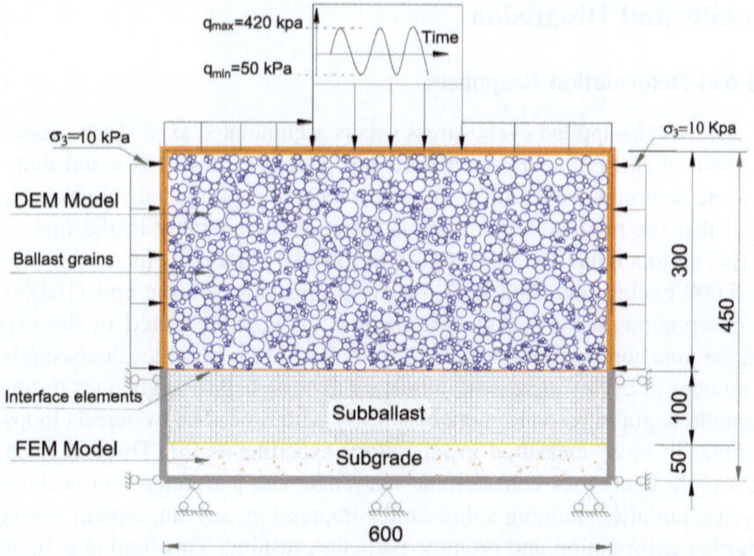

Fig. 2. Schematic diagram of a coupled DEM-FEM model (after Ngo et al. [22])

3.3 Coupled Discrete-Continuum Analysis

A coupled DEM-FEM that combines the discrete and continuum modelling approaches through the development of a force-displacement transmission mechanism at the ballast-subgrade interface is illustrated in Fig. 2; and the model dimensions represent the large-scale TPSA. Given that the movement of ballast in the longitudinal direction (along with the direction of train passage) is very small due to the confinement provided by embedded sleepers, the current analysis is carried out in an equivalent plane strain condition that is usually representative of long and straight tracks [31]. A ballast layer was modelled using the DEM, whereas subgrade layer was modelled by continuum approach. The interaction between ballast layer and sub-ballast was achieved through interface elements at the DEM and FEM boundary. Principally, the coupling between the DEM and FEM was achieved at the ballast-subgrade interface by: (i) applying the forces acting on the discrete particles as force boundary conditions to the continuum grids, and (ii) treating the continuum nodal displacements as velocity boundary conditions for the discrete elements. Subject to loading, the DEM zone (continuum mesh) deform in large strain; the displacements are transferred back to DEM, so that the walls in DEM zone moves in exactly the same way as the boundary segments of the FEM grid. The resulting wall forces, due to particles interacting with the walls, are transferred to FEM as applied grid-point forces. A detailed mathematical framework to assist the coupled model in transferring the forces and displacements between the two domains is presented earlier by Ngo et al. [22].

4 Results and Discussion

4.1 Load-Deformation Responses

Figure 3 shows the applied cyclic stress versus accumulated axial strain measured from the coupled DEM-FEM analysis at different loading cycles. It is noted that the axial strains were determined excluding any settlements of capping and subgrade layers. It is observed that the predicted axial strains increase considerably in the first 1000 load cycles (i.e. up to around 3% axial strain), followed by gradually increasing axial strains within 5,000 cycles, and then kept relatively unchanged to the end (10,000 cycles). These observations are in good agreement with those measured in the experiment. Indeed, the area confines by the cyclic (hysteresis) loops becomes increasingly smaller as the number of cycles increases, indicating that the ballast aggregates through cyclic densification begin to respond more elastically with time. The hysteresis loops are also very similar to those measured in laboratory experiments [3]. This indicates that the ballast sample undergoes considerable rearrangement and densification during initial load cycles, but after attaining a threshold compression, any subsequent loading would resist further deformation and promote particles crushing. This finding is in agreement with results presented by Lobo-Guerrero and Vallejo [32] where they observed that the ballast deformation considerably increased when the particle breakage was considered analysis.

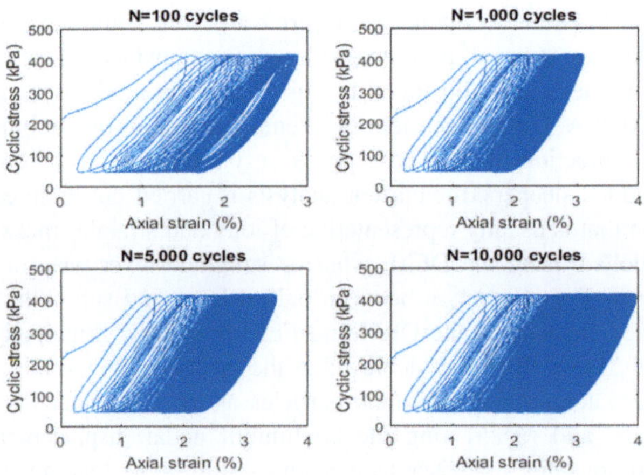

Fig. 3. Applied cyclic stress versus axial strain at different numbers of load cycles.

Applied loads transmit to the discrete ballast particles in the form of contact force-chains, where the pattern of force-fabric varies with the packing structure [33]. Figure 4 shows the inter-particle contact forces of the ballast specimens together with vertical stress contours at N = 10, 1000 and N = 5000 load cycles. Each contact force is represented at the contact point oriented in the direction of the force and with a thickness proportional to its intensity. It is seen that the majority of contact forces arrange vertically. Subjected to cyclic loads, there is a concentration of large contact forces occurring beneath the loading plate, and around wall edges, while a significant part of the applied load is still vertically transmitted to the underlying subballast and subgrade (Fig. 4b–c). It is also noted that the force distribution in the DEM region is always heterogeneous, where the maximum contact forces vary with load cycles. The compressive stress (σ_{yy}) in the subgrade is greater mainly around the interface area that is directly in contact with the aggregates, and decreases with depth as expected.

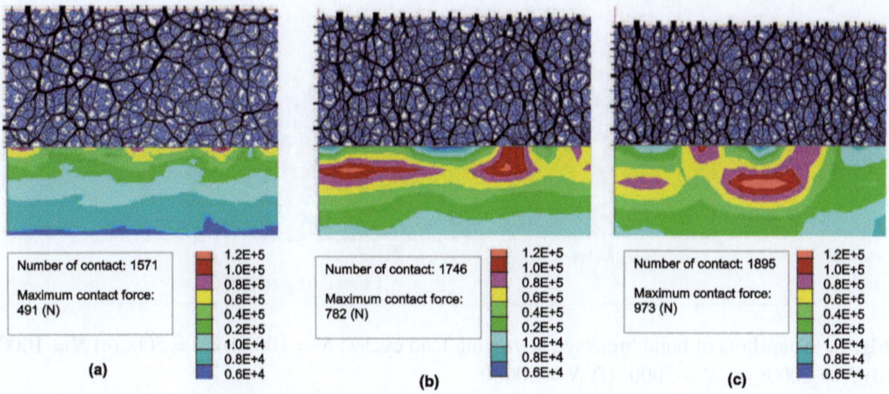

Fig. 4. Contact forces and stress contour (σ_{yy}) captured in discrete and continuum zones at different load cycles; (a) 10 cycles; (b) 1000 cycle; and (c) 5000 cycles (*after Ngo et al.* [22])

4.2 Ballast Breakage

In the current coupled DEM-FEM analysis, the breaks of bonds within a cluster of ballast are considered to represent ballast breakage. Figure 5 illustrates snapshots of the evolution of bond breakage at varying number of cyclic loads under a given load frequency of $f = 15$ Hz. The number and locations of bond breakages at different stages of cyclic loading, varying from 100 cycles to 10,000 cycles are presented in Figs. 5a–f. It is seen that, within the first 100 cycles, the majority of the bond breakage happens below the loading plate (i.e. sleepers) due to induced high contact forces, as shown in the Fig. 5a. With an increase in cyclic loads there is an increase in the bond breakage (Figs. 5b–f) and the re-arrangement of broken bonds (i.e. ballast get compacted and resisted from further degradation), which leads to more contact force distributions in the major principal stress direction as shown in Figs. 4b–c. It is seen that the evolution of broken bonds is very similar to the increased ballast breakage observed from the laboratory data [3]. This phenomenon clearly demonstrates that the formation of contact force distributions in the ballast assembly during cyclic loading is a dynamic process, significantly influenced by the breakage of the particles. Figure 5g illustrates typical locations and re-arrangement of ballast breakage during cyclic loading where bonds are broken and the corresponding particles are separating each other.

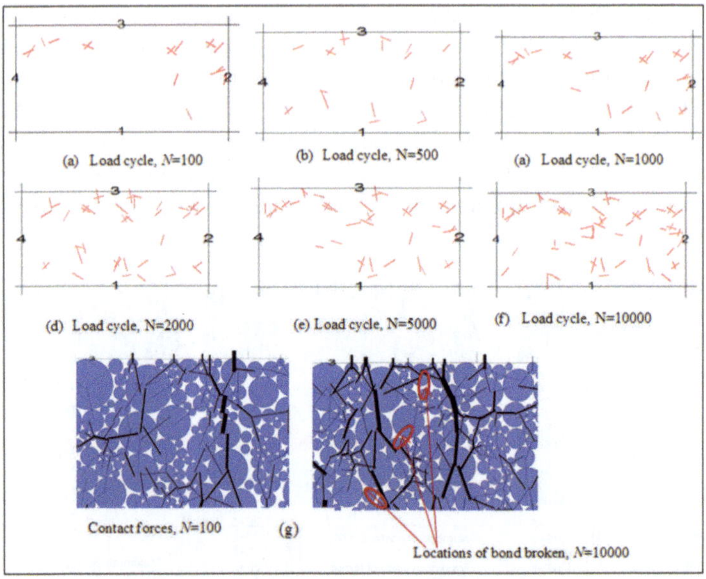

Fig. 5. Snapshots of bond breakage at varying load cycles: N = 100; (b) N = 500; (c) N = 1000; (d) N = 2000; (e) N = 5000; (f) N = 10000

5 Conclusions

A series of large-scale track process simulation apparatus (TPSA) tests were carried out to study the load-deformation responses of railway ballast under cyclic loading. A combined discrete-continuum model was introduced to study load-deformation responses of where the ballast aggregates were modelled by the discrete element method and the subgrade was modelled by finite continuum approach. Subgrade and subballast were modelled using continuum method, having a total thickness of 150 mm to simulate capping formation in the field. Results indicated that the ballast exhibited a significant axial strain in initial load cycles, followed by gradually increasing deformation and then remained relatively unchanged toward the end. Evolutions of contact force distributions and stress contours induced in the ballast at varying stages of cyclic loading were analysed and it confirmed that the stresses distributed non-uniformly across the ballast assembly where the maximum stresses occurring beneath the sleeper resulting in greater number of broken bond observed. The coupled DEM-FEM model will make an important intellectual contribution to advance knowledge, addressing the interface issues between the discrete and continuum approaches.

Acknowledgements. The Authors would like to acknowledge the Rail Manufacturing CRC, Australasian Centre for Rail Innovation (ACRI) Limited, and Tyre Stewardship Australia Limited (TSA) for providing the financial support needed to undertake this research (Project R2.5.1). The Authors are grateful to Mr. Alan Grant, Mr. Cameron Neilson, Mr Duncan Best and Mr. Ritchie McLean for their assistance in the laboratory.

References

1. Indraratna, B., Salim, W., Rujikiatkamjorn, C.: Advanced Rail Geotechnology - Ballasted Track. CRC Press, Taylor & Francis Group, London (2011)
2. Selig, E.T., Waters, J.M.: Track Geotechnology and Substructure Management. Thomas Telford, London (1994)
3. Indraratna, B., Ngo, N.T., Rujikiatkamjorn, C.: Deformation of coal fouled ballast stabilized with geogrid under cyclic load. J. Geotech. Geoenviron. Eng. **139**(8), 1275–1289 (2013)
4. Indraratna, B., Ngo, N.T., Rujikiatkamjorn, C.: Behavior of geogrid-reinforced ballast under various levels of fouling. Geotext. Geomembr. **29**(3), 313–322 (2011)
5. Tennakoon, N., Indraratna, B., Rujikiatkamjorn, C., Nimbalkar, S., Neville, T.: The role of ballast-fouling characteristics on the drainage capacity of rail substructure. Geotech. Test. J. **35**(4), 1–11 (2012)
6. Rujikiatkamjorn, C., Indraratna, B., Ngo, N.T., Coop, M.: A laboratory study of railway ballast behaviour under various fouling degree. In: The 5th Asian Regional Conference on Geosynthetics, pp. 507–514 (2012)
7. Tutumluer, E., Dombrow, W., Huang, H.: Laboratory characterization of coal dust fouled ballast behaviour. In: AREMA 2008 Annual Conference and Exposition, Salt Lake City, UT, USA (2008)
8. Ngo, N.T., Indraratna, B., Rujikiatkamjorn, C.: Micromechanics-based investigation of fouled ballast using large-scale triaxial tests and discrete element modeling. J. Geotech. Geoenviron. Eng. **134**(2), 04016089 (2017)
9. Lackenby, J., Indraratna, B., McDowell, G.R., Christie, D.: Effect of confining pressure on ballast degradation and deformation under cyclic triaxial loading. Geotechnique **57**(6), 527–536 (2007)
10. McDowell, G.R., Lim, W.L., Collop, A.C., Armitage, R., Thom, N.H.: Comparison of ballast index tests for railway trackbeds. Geotech. Eng. **157**(3), 151–161 (2008)
11. Erol, T., Yuanjie, X.: Gradation and packing characteristics affecting stability of granular materials: aggregate imaging-based discrete element modeling approach. Int. J. Geomech. **17** (3), 1–18 (2017). 04016064
12. Kaewunruen, S., Remennikov, A.M.: Dynamic properties of railway track and its components: a state-of-the-art review. In: Weiss, B.E. (ed.) New Research on Acoustics, pp. 197–220. Nova Science Publisher, Hauppauge (2008)
13. Powrie, W., Yang, L.A., Clayton, C.R.I.: Stress changes in the ground below ballasted railway track during train passage. Proc. Inst. Mech. Eng.: Part F: J. Rail Rapid Transit **221**, 247–261 (2007)
14. Bhandari, A., Han, J., Parsons, R.L.: Discrete element method investigation of geogrid-aggregate interaction under a cyclic wheel load. In: Geosynthetics Committee (AFS70) (2008)
15. Ngo, N.T., Indraratna, B., Rujikiatkamjorn, C.: A study of the geogrid-subballast interface via experimental evaluation and discrete element modelling. Granul. Matter **19**(3), 54 (2017)
16. McDowell, G.R., Harireche, O., Konietzky, H., Brown, S.F., Thom, N.H.: Discrete element modelling of geogrid-reinforced aggregates. Proc. ICE – Geotech. Eng. **159**(1), 35–48 (2006)
17. Indraratna, B., Hussaini, S.K.K., Vinod, J.S.: On the shear behaviour of ballast-geosynthetic interfaces. Geotech. Test. J. **35**(2), 1–8 (2012)
18. Biabani, M.M., Indraratna, B.: An evaluation of the interface behaviour of rail subballast stabilised with geogrids and geomembranes. Geotext. Geomembr. **43**(3), 240–249 (2015)

19. Indraratna, B., Nimbalkar, S.S., Ngo, N.T., Neville, T.: Performance improvement of rail track substructure using artificial inclusions – experimental and numerical studies. Transp. Geotech. **8**, 69–85 (2016)
20. Biabani, M.M., Indraratna, B., Ngo, N.T.: Modelling of geocell-reinforced subballast subjected to cyclic loading. Geotext. Geomembr. **44**(4), 489–503 (2016)
21. AS.289.6.2.2: Methods of testing soils for engineering purposes. Method 6.2.2: Soil strength and consolidation tests-determination of the shear strength of a soil-direct shear test using a shear box. Australian Standard (1998)
22. Ngo, N.T., Indraratna, B., Rujikiatkamjorn, C.: Simulation ballasted track behavior: numerical treatment and field application. Int. J. Geomech. **17**(6), 04016130 (2017)
23. Indraratna, B., Lackenby, J., Christie, D.: Effect of confining pressure on the degradation of ballast under cyclic loading. Geotechnique **55**(4), 325–328 (2005)
24. Cundall, P.A., Strack, O.D.L.: A discrete numerical model for granular assemblies. Geotechnique **29**(1), 47–65 (1979)
25. Tutumluer, E., Huang, H., Bian, X.: Geogrid-aggregate interlock mechanism investigated through aggregate imaging-based discrete element modeling approach. Int. J. Geomech. **12**, 391 (2012)
26. Ngo, N.T., Indraratna, B., Rujikiatkamjorn, C.: DEM simulation of the behaviour of geogrid stabilised ballast fouled with coal. Comput. Geotech. **55**, 224–231 (2014)
27. McDowell, G.R., Bolton, M.D.: On the micromechanics of crushable aggregates. Geotechnique **48**(5), 667–679 (1998)
28. Han, J., Bhandari, A., Wang, F.: DEM analysis of stresses and deformations of geogrid-reinforced embankments over piles. Int. J. Geomech. **12**(4), 340–350 (2011)
29. Itasca: Particle flow code in three dimensions (PFC3D). Itasca Consulting Group, Inc., Minneapolis (2014)
30. Indraratna, B., Ngo, N.T., Rujikiatkamjorn, C., Sloan, S.W.: Coupled discrete element-finite difference method for analysing the load-deformation behaviour of a single stone column in soft soil. Comput. Geotech. **63**, 267–278 (2015)
31. Ngo, N., Indraratna, B., Rujikiatkamjorn, C., Biabani, M.: Experimental and discrete element modeling of geocell-stabilized subballast subjected to cyclic loading. J. Geotech. Geoenviron. Eng. **142**, 0401510 (2016)
32. Lobo-Guerrero, S., Vallejo, L.E.: Crushing a weak granular material: experimental numerical analyses. Geotechnique **55**(3), 245–249 (2005)
33. Maeda, K., Sakai, H., Kondo, A., Yamaguchi, T., Fukuma, M., Nukudani, E.: Stress-chain based micromechanics of sand with grain shape effect. Granul. Matter **12**, 499–505 (2010)

Vehicular Impact Loading on the Barrier of a Retaining Wall System and Its Effect on the System Foundation

Yang An[1(✉)], Chin Leo[1], S. Liyanapathirana[1], and Henry Wong[2]

[1] School of Computing Engineering and Mathematics, Western Sydney University, Locked Bag 1797, Penrith, NSW 2751, Australia
e.an@westernsydney.edu.au

[2] Laboratoire de Tribologie et Dynamique des Systèmes (UMR-CNRS 5513), ENTPE, Université de Lyon, Lyon, France

Abstract. When a vehicle collides against a traffic barrier which is integrated into a retaining wall below, the impact load must be resisted by the barrier and the retaining wall system, and the concomitant effects will be eventually transmitted to the system foundation. Although the collision is a very short transient dynamic loading over an initially small impact zone, it has implications on the stability, hence design, of the barrier, the retaining wall system and its foundation. Little information, however, is available on how the impact loading transmits from the small impact zone over time to the broader structural system, and on how the effects of impact eventually interacts with the foundation soil. In this paper, numerical simulation of a Test Level 4 (TL-4, corresponding to regular traffic in Australian Standard (AS) 5100.2) vehicle crash on a 1.2 m high traffic barrier integrated into a 3 m high concrete retaining wall was performed and the effects of the impact loading on such an integral barrier-retaining wall system studied. The speed and impact angle of vehicle crash against the barrier in the model was calibrated so that the impact resulted in maximum loads in the transverse, longitudinal and vertical (downward) directions in the local area of impact which correspond to the design loads specified in AS 5100.2:2004. The soil-structure interactions at the base of the wall due to effects of the impact loading were then established for the calibrated vehicle collision, and results are discussed in this paper.

Keywords: Integral barrier-retaining wall system · Spread of transverse force Impact loading · Finite element model

1 Introduction

Retaining walls are a common feature of road and highway infrastructures. They exist in order to provide separation between the split-levels of a highway. On the high side of the highway where traffic normally travels, traffic safety barriers are also installed along the top of the retaining walls for vehicular protection. The impact load generated by the collision of an errant car or truck must generally be resisted by the barrier, retaining wall system and eventually the soil foundation.

X. Shi et al. (Eds.): GSIC 2018, *Proceedings of GeoShanghai 2018 International Conference: Transportation Geotechnics and Pavement Engineering*, pp. 199–206, 2018.
https://doi.org/10.1007/978-981-13-0011-0_22

The impact load due to vehicle collision is well known as a very short term transient dynamic load over an initially small impact zone. It is then spatially and temporally dispersed from the initial impact zone and time of collision to the broader structure. To understand the effects of the impact loading on the integral barrier-retaining wall system, it is necessary to establish what initial impact load is caused by the vehicle collision against the traffic barrier. Recommendations have been made in several local and international design codes. For instance, the ultimate transverse impact load for a regular performance level barrier according to Australian Standard 5100.2 (Bridge Design-Part 2: Design Loads) is 250 kN [1], and this load in Eurocode 1 is also specified as 250 kN [2]. However, in the United States, the AASHTO allowable stress design (ASD) bridge specifications [3] (which differs from the generally limit state-based approach of the other codes mentioned herein) recommend that the barrier with Test Level 4 (or TL-4, which is similar to a regular performance level barrier in Australia) needs to be designed for an equivalent static load of 44.5 kN. For the same vehicle impact, the AASHTO load and resistance factor design (LRFD) bridge specifications [4] recommend a design load of 240 kN. A further question that needs to be answered is "How does the impact load disperse within the wall and interacts with the foundation of the wall system?" For this question which has implications on the design of the wall system, most of design codes focus only on the localized impact loads on the traffic barrier and they are silent on the spread of impact loads and their interactions. Both AS5100.2 and AASHTO 2012 only recommend that the design loads shall be applied uniformly over the relevant specified contact lengths, i.e., 1.2 m and 1.22 m for the regular performance barrier in Australia and TL-4 barrier in America, respectively. Neither of them recommends how the impact loading spreads to the rest of the retaining wall system including its foundation.

In this study, TL-4 crash test [5] simulation corresponding to regular traffic performance in AS5100.2 was performed on a 1.2 m high barrier integrated into a 3 m high concrete retaining wall, using the finite element (FE) software Abaqus. The impact loading was created by the collision of a rigid truck against the traffic barrier. The FE model created was used to simulate the entire process of the collision, firstly, to capture the ultimate impact loads during the collision in three orthogonal directions: transverse, longitudinal and vertical. The speed and impact angle of vehicle collision in the model was calibrated so that the initial maximum impact forces at the contact zone in the transverse, longitudinal and vertical directions correspond to the design loads recommended in the Australian code. Following calibration, the dispersion of the effects of the impact loading from the impact zone to the rest of the wall system including the foundation of the system was simulated and studied. The normal force and shear resistance mobilized from the soil foundation in reaction to the impact loadings was then established and these results are discussed herein.

2 Simulation Model

A 3D FE model has been created using Abaqus/CAE in order to investigate the effects of impact loading on the integral barrier-retaining wall system,. There are three main parts included in the model as shown in Fig. 1. These are the integral barrier-retaining wall system (Fig. 1b), the soil backfill-foundation system (Fig. 1c) and the rigid truck

(Fig. 1d). Following simulations with various lengths of the wall system, it was established that a length of 30 m was sufficiently long in this instance to avoid boundary effects. Moreover, according to NCHRP report 350 [6], for impact testing, the length of a metal corrugated barrier should be longer than 30 m and that of a rigid barrier should be larger than 23 m. Consequently, a 3D model with a length of 30 m was used in this study.

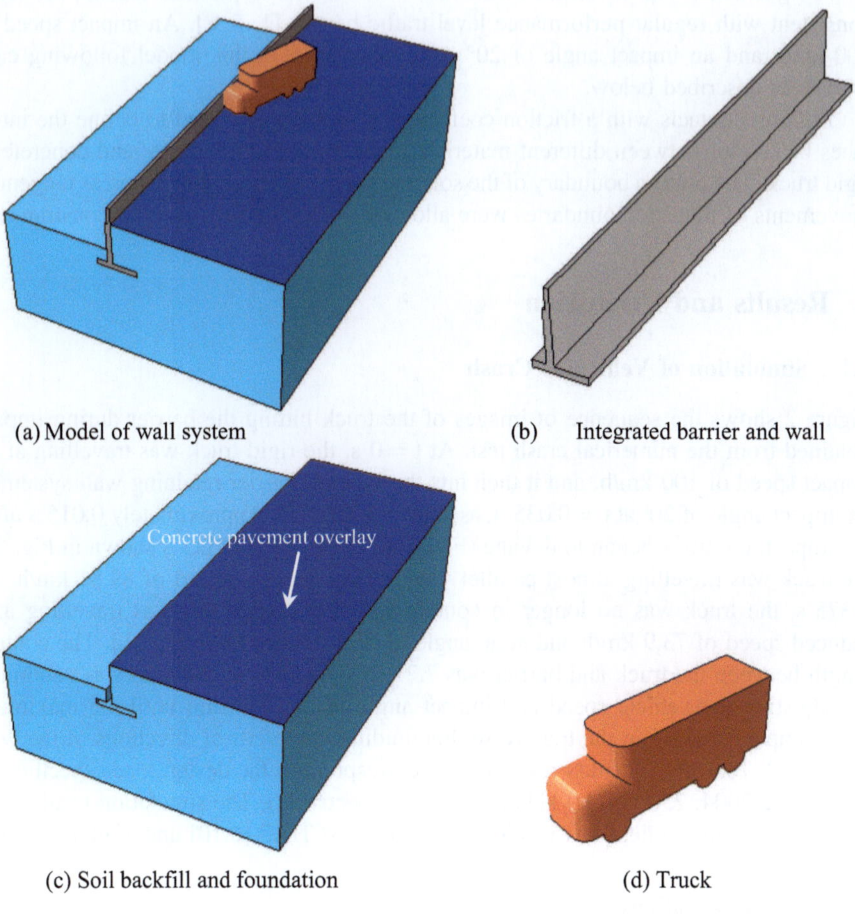

(a) Model of wall system (b) Integrated barrier and wall

(c) Soil backfill and foundation (d) Truck

Fig. 1. 3D finite element model

For the integral barrier-retaining wall system, the height of retaining wall was taken as 3 m and that of the barrier as 1.2-m. In the 3D FE model, the integral barrier-wall system was defined as a deformable part made out of reinforced concrete [7]. The solid element, C3D8, and Abaqus's damage plasticity material model were used to capture the mechanical behaviour of the reinforced concrete. The reinforcement within the concrete was modeled as steel bars and simulated using the beam element, B31. The material behaviour was considered as elastic-plastic.

The soil backfill-foundation system includes two components, (i) soil mass (comprising backfill and foundation soil) and (ii) rigid pavement which overlays the soil backfill. The pavement is made of concrete with a thickness of 100 mm. The shell-element S4R and concrete damage plasticity material model were used to represent this pavement. Soil mass was modelled with the solid-element C3D8R and the material behaviour was simulated using the Mohr-Coulomb model. The soil mass was assumed to have a friction angle of 35° and a small value of cohesion of 3 kPa.

A rigid truck with a weight of 22 tons was created to produce the impact loads consistent with regular performance level traffic barrier [1, 5, 8]. An impact speed of 100 km/h and an impact angle of 20° have been used in this model following calibration as described below.

Friction contacts with a friction coefficient of 0.5 [9] was used to define the interfaces behaviour between different materials including soil to concrete and concrete to rigid truck. The bottom boundary of the soil mass was held immobile whereas tangential movements of the side boundaries were allowed, corresponding to roller constraints.

3 Results and Discussion

3.1 Simulation of Vehicular Crash

Figure 2 shows the sequence of images of the truck hitting the barrier during impact obtained from the numerical crash test. At t = 0 s, the rigid truck was travelling at an impact speed of 100 km/h, and it then hits the integral barrier-retaining wall system at an impact angle of 20º at t = 0.035 s, as shown in Fig. 2a. Approximately 0.015 s after the impact, the truck began to deviate (Fig. 2b). Then, at t = 0.325 s shown in Fig. 2c, the truck was travelling almost parallel with the barrier at a speed of 89.81 km/h. At 0.375 s, the truck was no longer in contact with the barrier and was travelling at a reduced speed of 73.9 km/h and at an angle of 5.71° as shown in Fig. 2d. The contact length between the truck and barrier was 7.21 m. The vehicle collision was calibrated by adjusting the vehicle speed and impact angle of the collision until the maximum initial impact loading in the transverse, longitudinal and vertical directions of the wall (279.7 kN, 78.4 kN, 71.0 kN respectively) correspond to the design loads specified in AS 5100.2:2004; 250 kN, 80 kN and 80 kN respectively). The simulation results also match well with the field test results provided by NCHRP [6,10] and Winkler [11].

3.2 Impact Loadings

The impact loading generated by the collision of the truck can be obtained from the FE model directly, as shown in Fig. 3. The impact loading includes three components as shown in Fig. 3a. They are the transverse force, F_T, longitudinal force, F_L, and vertical force, F_V, respectively. For the transverse force, F_T, shown in Fig. 3b, the impact force increases rapidly at the beginning of the collision (t = 0.035 s) and reaches the ultimate value of 279.7 kN at t = 0.05 s when the truck begins to deviate. Then, as the truck begins to deviate, the impact transverse force starts to decrease. When the truck's position is parallel to the barrier (t = 0.325 s), the impact force is almost negligible.

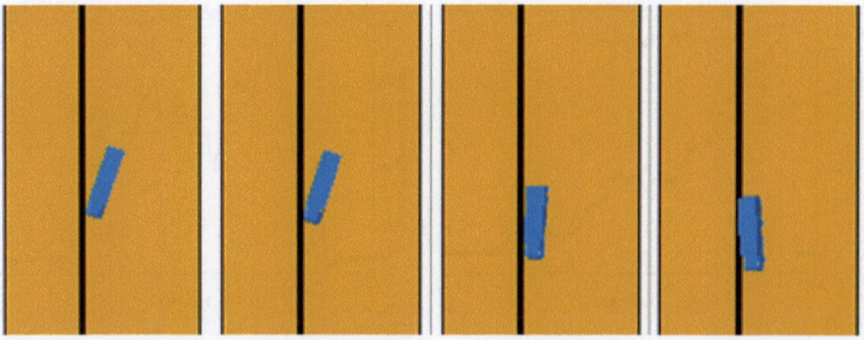

(a) t = 0.035 sec (b) t = 0.050 sec (c) t = 0.325 sec (d) t = 0.375 sec

Fig. 2. Sequence of images during impact

The whole impact lasts about 0.4 s. Similar evolutions can be observed for the longitudinal vertical forces as shown in Figs. 3c and d. However, the maximum longitudinal and vertical forces do NOT occur at the same time.

3.3 Interactions at Stem of the Wall During Collision

When a vehicle hits a barrier, the integral barrier-retaining wall system may fail structurally or due to sliding and/or overturning. The interactions at stem of the wall due to the impact loading would be of interest and must be checked. Figure 4 shows the longitudinal distribution of the driving force, F_D, and driving overturning moment, M_{OT}, at the stem of the wall due to the impact loading. The maximum driving force at the stem occurs at 0.05 s, which is the same time as when the maximum transverse force occurs. The maximum value of driving force is 111.3 kN and the location is at where the impact occurs. Meanwhile, the maximum driving overturning moment also occurs at the same time and same location. The maximum value of driving overturning moment is 321.4 kN.m.

3.4 Soil-Structure Interactions at Foundation

When a vehicle hits a barrier, the integral barrier-retaining wall system may fail due to sliding and/or overturning. Thus, to study the stability of such systems in response to an impact, it is essential to investigate the effects of impact loading on the mobilized forces from soil foundation. As shown in Fig. 5d, the mobilized forces include normal force N, moment M, and shear force S. The spatial distributions of these mobilized forces per unit length along the longitudinal direction of the integral barrier-wall system at different times, obtained from the numerical simulations, are shown in Figs. 5a, b and c.

These figures show that the maximum mobilized forces from the soil foundation occur at 0.175 s. The maximum values of the normal force, moment and shear force are 43.0 kN/m, 119.7 kNm/m and 8.0 kN/m, respectively. As discussed in the above, the ultimate impact loading occurs at t = 0.05 s, which is approximately 0.125 s earlier

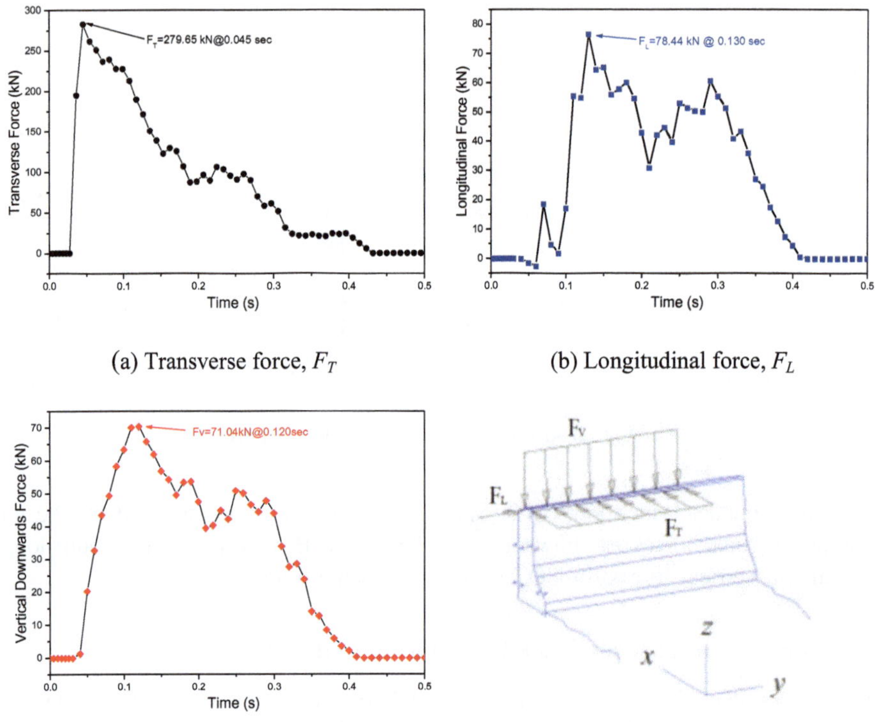

(a) Transverse force, F_T

(b) Longitudinal force, F_L

(c) Vertical downwards force, F_V

(d) Three components of impact loading

Fig. 3. Impact loading time history curves

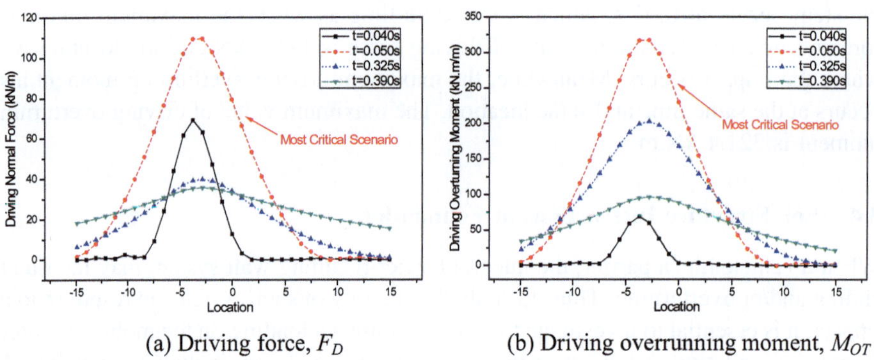

(a) Driving force, F_D

(b) Driving overrunning moment, M_{OT}

Fig. 4. Longitudinal distributions of driving force and moment

than the maximum mobilized forces have occurred. This delay indicates the finite speed of dispersion of the impact loading in the integral barrier-retaining wall. It is clear that the stability analysis of the barrier-wall system need to take into account the dispersion of the impact loading although current literature does not yet provide sufficient guidance on how it should be dealt with.

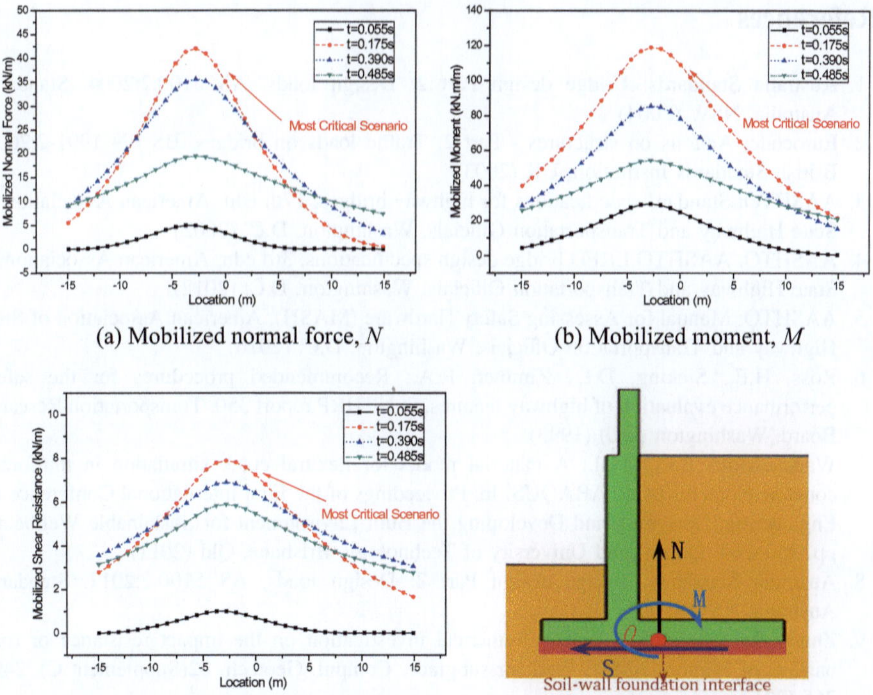

(a) Mobilized normal force, *N*

(b) Mobilized moment, *M*

(c) Mobilized shear resistance, *S*

(d) Three components of mobilized forces

Fig. 5. Longitudinal distributions of mobilized forces

4 Conclusions

This study investigated the effects of a localised impact loading on the integral barrier-retaining walls system. A finite element model was created to perform a numerical crash test on a 1.2-m-high barrier integrated with a 3-m-high reinforced concrete retaining wall. The vehicle collision creates initial impact loading which includes transverse, longitudinal and vertical forces at the contact zone. The maximum values for these correspond to the design loads specified in the Australia Standard, AS5100.2:2004 for regular traffic. The effects of impact loading on the mobilized forces of the soil foundation have also been discussed. Due to the fact that the maximum effects of impact loading on different parts of the wall system take place at different times, it seems that a systematic and rational way to incorporate these effects into design of the wall system in general, and the system foundation in particular, requires further investigation.

Acknowledgement and Disclaimer. The content of this paper is motivated by a project sponsored by the Roads and Maritime Services (RMS), Australia. However, the views of this paper do not represent the opinion of RMS and are entirely those of the authors alone.

References

1. Australia Standards: Bridge design Part 2: Design loads, AS 5100.2:2004. Standards Australia, NSW (2004)
2. Eurocode: Actions on structures - Part 2: Traffic loads on bridges, BS EN 1991-2:2003. British Standards Institution, UK (2003)
3. AASHTO: Standard specifications for highway bridges, 17th edn. American Association of State Highway and Transportation Officials, Washington, D.C. (2002)
4. AASHTO: AASHTO LRFD bridge design specifications, 3rd edn. American Association of State Highway and Transportation Officials, Washington, D.C. (2012)
5. AASHTO: Manual for Assessing Safety Hardware (MASH). American Association of State Highway and Transportation Officials, Washington, D.C. (2016)
6. Ross, H.E., Sicking, D.L., Zimmer, R.A.: Recommended procedures for the safety performance evaluation of highway features, in NCHRP report 350. Transportation Research Board, Washington (DC) (1993)
7. Wahalathantri B.L., et al.: A material model for flexural crack simulation in reinforced concrete elements using ABAQUS. In: Proceedings of the First International Conference on Engineering, Designing and Developing the Built Environment for Sustainable Wellbeing. pp. 260–264. Queensland University of Technology, Brisbane, Qld (2011)
8. Australia Standards: Bridge design Part 2: Design loads, AS 5100.2:2017. Standards Australia, NSW (2017)
9. Zhang, Z., Chen, Y., Liu, H.: Numerical investigation on the impact resistance of road barriers of Micropile-MSE Wall for subgrade. Comput. Geotech. **82**(Supplement C), 249–265 (2017)
10. Kim, K.M., et al.: Full-Scale impact test of four traffic barriers on top of an instrumented MSE wall. J. Geotech. Geoenvironmental Eng. **136**(3), 431–438. (2010)
11. Gutkowski, R.M., Winkler, D.J.: Simplified impact testing of traffic barrier systems. pp. 1–143, North Dakota State University, Fargo, N.D. (2003). MPC Report

Dynamic Response Modeling of High-Speed Railroad Ballastless Track Over Pile-Raft Foundation

Yuanjie Xiao[1](✉), Liuxin Chen[1], Kunye Zhou[2], and Erol Tutumluer[3]

[1] School of Civil Engineering, Ministry of Education Key Laboratory for Heavy Haul Railway Engineering Structures, National Engineering Laboratory for High Speed Railway Construction, Central South University, Changsha 410075, Hunan, China
yjxiao@csu.edu.cn

[2] School of Civil Engineering, Tongji University, Shanghai 200092, China

[3] Department of Civil and Environmental Engineering, University of Illinois Urbana-Champaign, Urbana IL 61801, USA

Abstract. Ballastless tracks have been extensively used in newly built high-speed railroads in China. Most of their foundations are constructed without pile improvement; however, the pile-raft design concept has been increasingly used as a rapid yet effective construction technique to reinforce the foundations. Limited attention has been paid to the behavior of this foundation system under dynamic loads either in the laboratory trials or field applications. This study presented the results of displacement responses obtained from three-dimensional (3D) dynamic finite element simulations of a typical high-speed railroad ballastless track over the pile-raft foundation. Both the soft ground improved by the pile-raft technique and the unreinforced ground, as control, were modeled and compared to quantify the effectiveness of the ground improvement in mitigating the detrimental effects of dynamic impact loads induced by high-speed trains. Other features of the 3D finite element models included the simulation of the moving train at different speed levels, the use of nonlinear constitutive models for track foundation geomaterials, and the consideration of the pile-soil interaction. The results of this study could lead to further research in this area as guidelines to design piled raft foundations economically.

Keywords: High speed railroad · Ballastless track · Displacement responses
Finite element analysis · Pile-raft foundation · Soil-pile interaction

1 Introduction

The use of ballastless slab tracks has prevailed in newly built high speed railroads (HSR) in China and other places. The function of track foundations is to transmit the train loads to subgrade soils in a safe yet cost-effective manner, as well as to ensure operational stability. In soft soil areas where HSR embankments are constructed, technical concerns related to post-construction settlement, long-term stability, and train-induced vibration of the track foundation remain to be tackled (e.g., Dieterman

X. Shi et al. (Eds.): GSIC 2018, *Proceedings of GeoShanghai 2018 International Conference: Transportation Geotechnics and Pavement Engineering*, pp. 207–217, 2018.
https://doi.org/10.1007/978-981-13-0011-0_23

and Metrikine [1], Kaynia et al. [2], Mashus and Kaynia [3], Gao et al. [4], Shih et al. [5]). Recently, pile-supported slab track has been widely constructed as a practical solution for railway track construction over soft soils in Europe and China. The pile-supported slab track system consists of pile foundation, cross beams and platforms. This system has been applied to high-potential collapse or above soft ground embankment system [6]. In Europe, the pile-supported slab track system has been applied to a high embankment on peat and clay [7]. To reinforce the foundations and ensure the serviceability of the surrounding structures, the pile-raft design concept has been increasingly adopted and been regarded as an effective yet rapid reinforcement technique (e.g. [8–14]). The rigid platform provided by the raft is intended for distributing loads and housing piles, while the piles are designed to share the loads transmitted from superstructures by interacting with surrounding soils. The piled raft foundation system is highly effective in restraining settlement of track superstructure; however, it has frequently been found to be damaged by dynamic loads, such as moving train or earthquake loads. Previous researchers (e.g., Eekelen et al. [15], Sinha and Hanna [16]) have developed design methods and performed analyses to address stability and settlement issues of piled raft embankments. However, most of these studies focused on the behavior of piled raft embankments under static loading rather than transient loading (e.g., dynamic train loading). knowledge on their dynamic behaviors under transient loading of moving vehicles is rather limited in the literature. Therefore, it is necessary to perform in-depth study on dynamic responses of slab track system supported by the piled raft foundation under the application of moving loads.

In this study, a three-dimensional (3D) finite-element model of a piled raft foundation was developed to simulate the case of a piled raft embankment in a typical cross-sectional design of a pre-selected high-speed railroad ballastless track. The model accounts for pile-to-pile, raft-to-pile, pile-to-soil, and raft-to-soil interactions. The developed 3D numerical model was first validated with available data in the literature. After model validation, dynamic responses of the pile-raft supported slab track system subjected to a high-speed train were simulated for two scenarios: soft ground improved by piled raft foundation, and untreated soft ground. The effectiveness of the ground improvement in mitigating detrimental vibrations induced by high-speed trains was evaluated and compared in time domain. The results of this study could lead to further research in this area as guidelines to design piled raft foundations economically.

2 Numerical Model

2.1 Model Description

A 3D numerical model was developed using the commercial program Abaqus® to simulate the engineering application of a piled raft embankment in the high-speed railway slab track system. Figure 1(a) presents a schematic sketch of the cross-sectional layout of the slab track system to be modeled in this study. As shown in Fig. 1(a), the piled raft embankment of the high-speed railroad ballastless track consists of a raft and a group of piles that are embedded in the soft soil. The piles selected for ground improvement are cast-in-place concrete large diameter pipe piles. Principles of this ground improvement method can be found in reference [17]. Based on its technical specification [17], the following dimensions were selected: pile length is 20 m; outer pile diameter is 0.5 m; and the piles are arranged in a rectangular pattern with pile spacing of 2.4 m. The ground consists of a thick soft clay layer which is underlain by a stiff clay layer, and a rigid base or bedrock is assumed below this stiff clay layer. There is a fill layer on the soft soil layer, which serves as a platform for pilling. The upper structure is a single-lane ballastless slab track embankment. Full description of the model geometry is shown in Fig. 1(a). In this study, the length of pile-supported slab track system is 25.95 m, and the selected train has five carriages with a total length of 100 m. The loading of a moving train was represented approximately as a series of equally distributed vertical stress, as shown in Fig. 2. Note that the areas of the distributed loads are so small as compared to the model dimensions that they appear as if they were point loads in Fig. 2. The intensity of the distributed load p was calculated from the static axle load F acting on the contact area ΔS, i.e., $p = F/\Delta S$. The values of both F and ΔS were selected from the literature [18] to be 72.5 kN and 115.8 mm^2, respectively; thus, the intensity p was calculated as 6×108 N/m^2. The train axle loads used in the finite element model is shown in Fig. 2 along with the track structure. In this study, the top surface of the rail was referred to as loading face, while the axle loads of the five-carriage train were applied to the rail simulated by eight-node solid elements. A VDLOAD subroutine was developed using Microsoft Visual Studio® in Abaqus® to simulate the moving axle loads. Each of the axle loads became active when it reached the predetermined reference points, which was associated with train speed and time. As mentioned previously, different travel speed values (i.e., v = 60, 80, 100, and 120 m/s) were modeled in this study.

By taking advantage of symmetry, only one half of the finite element model was built as shown in Fig. 1(b). Its total length and width are 25.95 m and 20.6 m. The model was meshed using 8-node linear hexahedral elements with reduced integration and hour-glass control. The symmetry boundary was restrained from the translation in X-direction, and fixed boundary was used at the bottom to represent the bedrock. Absorbing boundaries were applied using infinite elements of type CIN3D8. In the vicinity of the piled raft and the loads, fine meshes were used with a gradual, smooth transition of element sizes from fine to coarse elsewhere. The maximum size of the hexahedral elements was 3 mm. The total number of elements used in the analysis was approximately 60575. Lumped mass scheme was used to discretely represent the continuous distribution of mass within each finite element. Rayleigh damping was used to represent energy-dissipating mechanisms in the simulation model, in which the mass and stiffness proportional damping constants were selected to provide slight damping ratios of 2–4% in the frequency range of 3–50 Hz. In the analysis, the time period

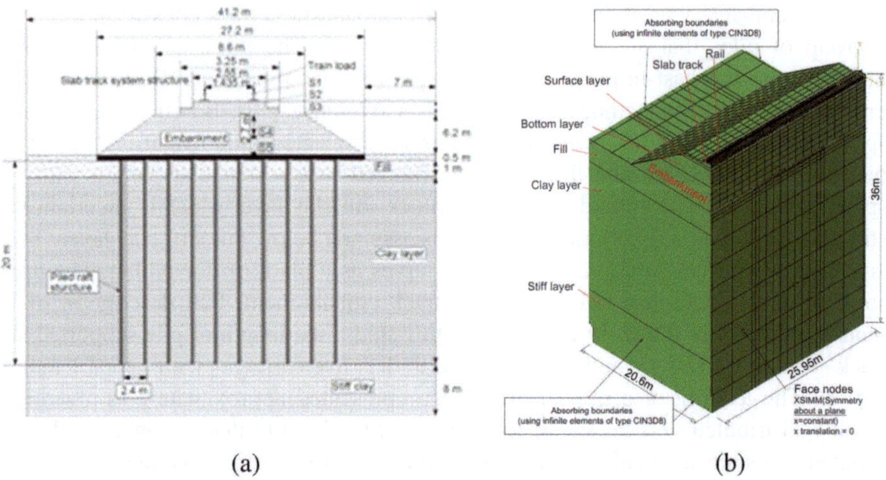

Fig. 1. Model details: (a) Schematic cross-sectional layout of a typical High-speed railroad ballastless track over piled raft embankment and (b) Illustration of the geometry of the developed 3d finite element model

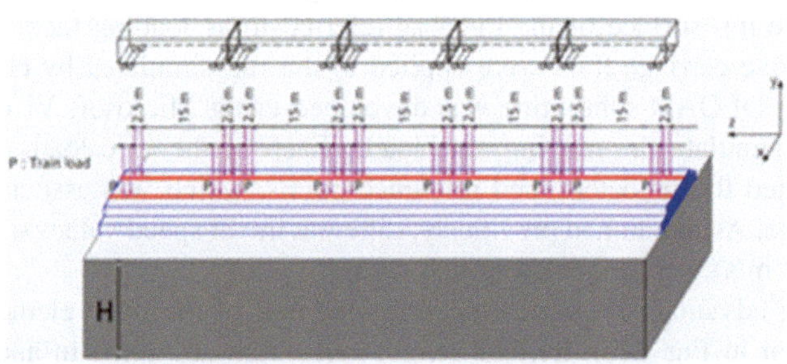

Fig. 2. Illustration of loading conditions in the developed 3D finite element model

corresponding to train speed of 60, 80, 100, and 120 m/s was computed as 2.44, 1.83, 1.46, and 1.22 s, respectively. The time step size used in the simulations is 2.0 × 10-5 s, which is small enough to warrant the solution accuracy and the numerical stability of the explicit integration. The finite element program performed dynamic analyses in the time domain using an explicit method. Furthermore, to simulate the interactions between the soil and the pile-raft, the node-to-surface and surface-to-surface contact options were examined. It was decided to implement the surface-to-surface contact due to its high level of accuracy, while the penalty type frictional constraint method was used to formulate the stiffness of contact surfaces. The friction coefficient of the pile-soil interface was taken as 0.466 as a baseline reference value in the model.

The discrete supporting points of the rail beams were spaced at an interval of 0.65 m that reflects the actual fastening systems. The fastening system was modeled as the Cartesian type connecting element in the Interaction module. The function of the fastening system was represented by the stiffness and damping coefficient values in the X, Y, and Z directions. The soft soil were modeled using a linear elastic-perfectly plastic model with Mohr–Coulomb failure criterion. The Mohr–Coulomb model requires five parameters: effective cohesion, c, friction angle, dilatancy angle, effective Young's modulus, E, and Poisson's ratio ρ. The parameters used in the analyses are summarized in Tables 1 and 2. Note that v_s and v_R are shear and Rayleigh wave speeds, respectively. The piled raft structure modeled in this study represents those commonly used in real-world high-speed railroad lines, of which the close-up view is shown in Fig. 3 with specific dimensions.

Table 1. Geometric and mechanical properties of slab track components

Material	Modulus (GPa)	Poisson's ratio	Thickness (m)	Dimensions (m)	Density (kg/m^3)
Slab	35.5	0.167	–	6.45 × 2.55 × 0.2	2450
CA Mortar layer	7	0.167	–	25.95 × 2.55 × 0.03	1800
Supporting layer	27	0.167	–	25.95 × 2.55 × 0.3	2350
Surface layer of embankment	0.18	0.25	0.4	–	2000
Sub-embankment	0.11	0.25	2.3	–	1800
Embankment	0.06	0.2	3.5	–	1700
Raft	40	0.2	–	27.2 × 12 × 0.5	2500
Pile	40	0.2	–	Φ0.5 × L20	2500

2.2 Model Validation

Fu et al. reported finite element analysis results of a ballasted track over piled raft foundation [19]. Due to similar scope, the results by [19] were employed in this study for model validation purpose. The geometry and the material properties used in the validation analyses are same with those in [19]. Figure 4 presents a summary of the validation results, i.e., the calculated vertical displacement contours at a specific time step, as well as the comparison of vertical displacement curves obtained by the developed finite element model against those by [19]. It can be noted that although the peak values were slightly different due to the use of assumed parameter values missing in the reference by [19], overall there is a reasonably good agreement between those two results. Therefore, the developed finite element simulation approach for piled raft foundation was validated.

Table 2. Mechanical parameters of foundation geomaterials used in finite element analyses

Element	E/MPa	μ	ρ (kg·m^{-3})	Vs (m·s^{-1})	V_R (m·s^{-1})	Damping		c/MPa
						α	β	
Soft clay	10	0.35	1500	–	–	–	–	10
Stiff clay	50	0.3	1800	103.4	95.9	1.4	0.00024	–
Fill	35	0.3	1800	86.5	80.2	1.4	0.00024	–

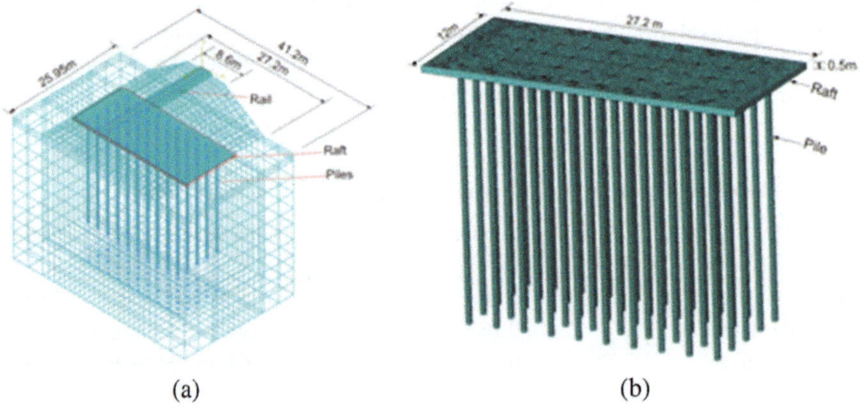

Fig. 3. Piled raft structure in the developed 3d finite element model: (a) perspective view and (b) close-up view

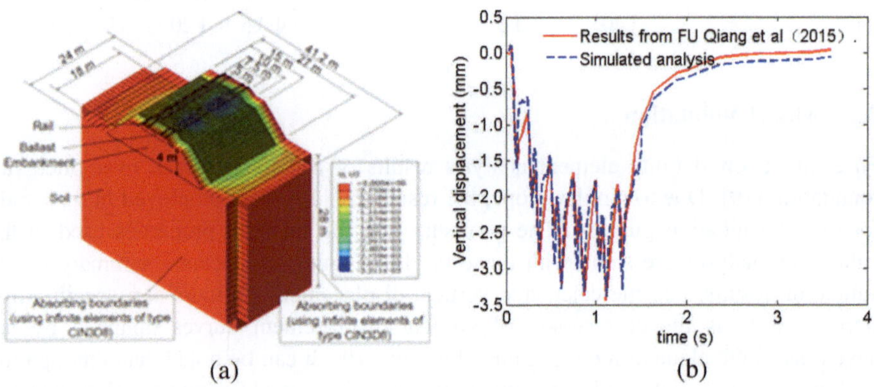

Fig. 4. Model validation results: (a) vertical displacement contours of rail top surface and (b) comparison of model-predicted time history of vertical displacement of rail top surface against that reported by [19]

3 Results and Analysis

To better understand the vibration mitigation of piled raft foundation, time histories of vertical displacement at the top of rail for train speeds of 60, 80, 100, and 120 m/s are presented. It can be known that the vertical displacement reaches a peak value when the wheel sets arrive, which explains the horn shaped distribution for two cases: soft ground improved by piled raft foundation, and untreated soft ground. Each carriage has two bogies spaced at a distance of 17.5 m and each bogie has two wheel sets which is 2.5 m from each other. The distance of bogies between two adjacent carriages are 7.5 m center-to-center. Five carriages are simulated, and accordingly there are 20 peak values presented. The dynamic response induced by the passing two wheel sets of the same bogie and two bogies between two adjacent carriages were overlapped respectively because of short distance.

As for the ground improved by piled raft, the peak displacement is reduced at each of the train speeds studied as compared to that without ground improvement. In particular, the peak displacement is reduced by 47.8% from 2.72 mm to 1.42 mm at 60 m/s, by 45.0% from 2.60 mm to 1.43 mm at 80 m/s, by 42.6% from 2.65 mm to 1.52 mm at 100 m/s, and by 45.0% from 2.71 mm to 1.49 mm at 120 m/s, respectively. Note that peak displacement values corresponding to improved ground exhibit insignificant change with increasing train speed. This is probably due to the use of identical loading conditions among different speed scenarios. In fact, the dynamic wheel load increases with increasing train speed, which has to be taken into account in future studies.

For the case where the ground is untreated, the peak displacement values under the loads of each bogie are considerably different. For the speed levels of 60 m/s and 80 m/s shown in Fig. 5(a) and (b), the peak displacement values under the middle wheel sets are lower than those under other sets. Whereas for the speed levels of 100 m/s and 120 m/s, the peak displacement values induced by middle sets of loads increase drastically to the maximum. Such a displacement increase is attributed to the superposition and dissipation of vibration waves due to passing train loads. After ground improvement by piled raft foundation, peak displacements under each bogie load are approximately maintained at the same level, thus mitigating most of the adverse vibration effects.

Figure 6 compares the vertical displacement time histories at five typical locations denoted as S1 to S5 (see Fig. 1(a)) for the ground with piled raft improvement and not in order to investigate the propagation and attenuation of vibration wave in geotechnical material (e.g. embankment) and the reinforcement effect, taking the case of train speed 100 m/s as an example. It is easily understood that the displacements at five different observation locations decrease with the increase of the distance from the track center vertically and horizontally. Accordingly, before the piled raft consolidation, it can be obtained that the peak displacement at location S1 is about 2.8 mm which is biggest among these observation points. The peak displacement of S2 besides the bottom of the rail is 2.3 mm, which is a bit lower than that of S1. The maximum

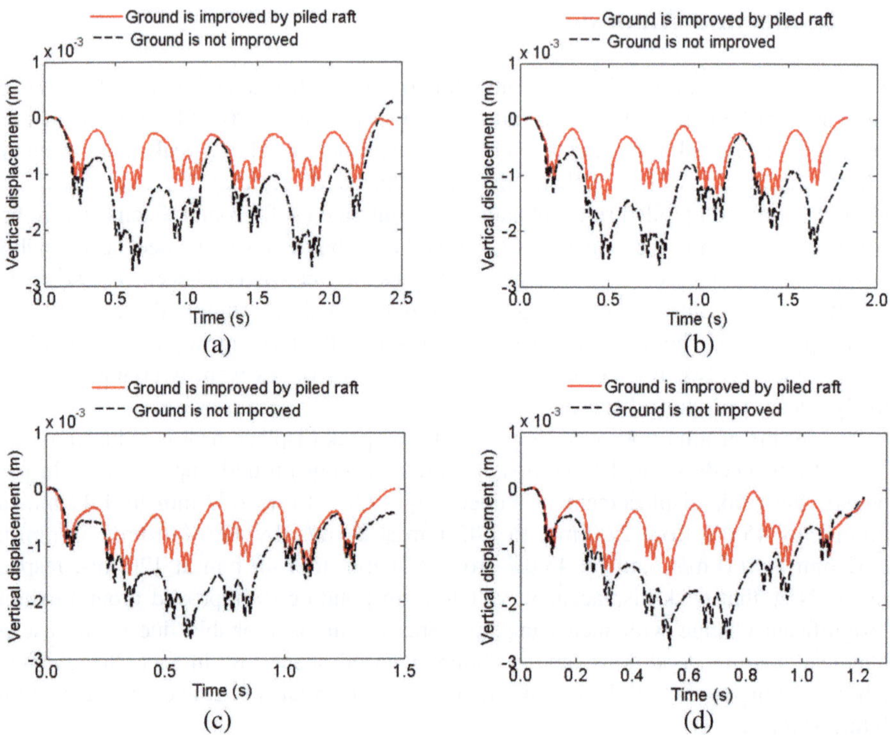

Fig. 5. Time history of vertical displacement at the top of rail for different train speed: (a) v = 60 m/s; (b) v = 80 m/s; (c) v = 100 m/s; (d) v = 120 m/s

displacement of S4 (3.45 m or so away from the top of rail vertically) and S5 (about 6.95 m away from the top of rail in the vertical direction) are approximately 2.2 mm and 2.1 mm respectively, both of which are higher than that of the location S3. The horizontal and vertical distance of S3 are 3.58 m and 0.75 m or so respectively, perfectly reflecting that the vertical propagation of vibration is the main direction, when it comes through these mediums with damping.

Another discovery is that the degree of wave superimposed effect increases with the increase of the distance away from the track center, weakening horn shaped distribution. After using piled raft foundation, the propagation character of wave is not changed, however the bearing capacity and dynamic character of geotechnical material (e.g. embankment) are improved. Thus, Fig. 6(b) shows similar displacement pattern, like horn shape for the five observation points, however, the vibration at five different observed points are effectively controlled and reduced considerably up to about 50%. The settlement curve of the point on the raft surface, namely S5 is from below to above that of S3 after soil improvement, mainly because of the function of piled raft. Moreover, the piled raft foundation evens the peak displacements under each bogie load, which mitigate most of the adverse vibration effects.

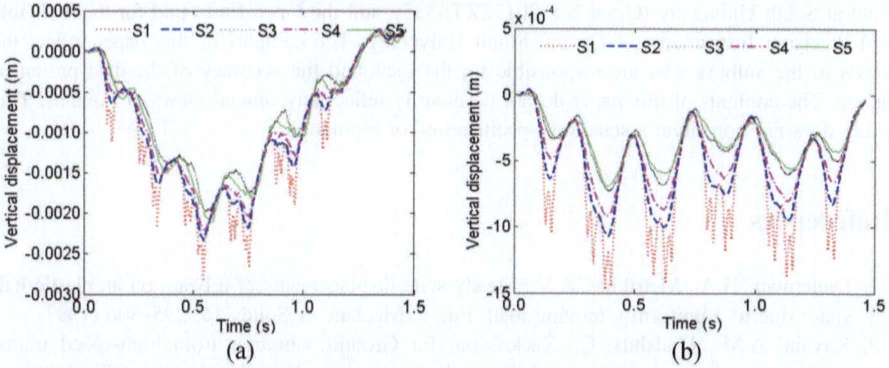

Fig. 6. Time histories of vertical displacement of top rail surface in the case of train speed 100 m/s for (a) untreated foundation and (b) piled raft foundation

4 Summary and Conclusions

A 3D finite-element model was developed in this study to examine the effectiveness of piled raft foundation in high-speed railroad ballastless track in terms of mitigating train-induced vibrations in soft soil areas, as well as to assess the effect of the governing parameters on the performance of piled raft foundations. The Mohr-Coulomb plasticity model provided an excellent basis for analyzing the stress-strain of soil continnum constitutive law, and the software program Abaqus®, which is based on solid mechanics principles, was used successfully to simulate the stated problem. The following conclusions can be drawn for the problems examined in this study:

- A three-dimensional finite element model was developed to simulate the dynamic response of ballastless track-subgrade-ground system subjected to moving loads caused by high speed trains. The vibration waves induced on the rail spread out into the embankment, and were mitigated by the pile-raft reinforcement. The vibration displacement of railway system can reflect the superposition of vibration, which obviously shows horn-like distribution.
- After ground improvement by piled raft foundation, peak displacements under each bogie load are approximately maintained at the same level and reduced at all considered train speeds, thus mitigating most of the adverse vibration effects.
- The vertical displacement time histories at five typical locations denoted as S1 to S5 (see Fig. 1(a)) for the ground with piled raft improvement and not perfectly reflect that the vertical propagation of vibration is the main direction, when it comes through these mediums with damping. Another discovery is that the degree of wave superimposed effect increases with the increase of the distance away from the track center, weakening horn shaped distribution.

Acknowledgements. The authors acknowledge the financial support from National Key R&D Program of China under Grant No. 2017YFB1201204, the National Natural Science Foundation of China under Grant No. 51508577, the Graduate Student Autonomous Exploration Project of

Central South University (Grant No. 2017ZZTS525), and the Open-End Fund for the Valuable and Precision Instruments of Central South University. The contents of this paper reflect the views of the authors who are responsible for the facts and the accuracy of the data presented herein. The contents of this paper do not necessarily reflect any official views or policies. This paper does not constitute a standard, specification, or regulation.

References

1. Dieterman, H.A., Metrikine, A.V.: Steady state displacements of a beam on an elastic half space due to a uniformly moving load. Eur. J. Mechan. A Solids **16**, 295–306 (1997)
2. Kaynia, A.M., Madshus, C., Zackrisson, P.: Ground vibration from high-speed trains: prediction and countermeasures. J. Geotech. Geoenviron. Eng. **126**(6), 531–537 (2000)
3. Mashus, C., Kaynia, A.M.: High speed railway lines on soft ground: dynamic behavior at critical train speed. J. Sound Vib. **231**(3), 689–701 (2000)
4. Mashus, C.: Modeling, monitoring and controlling the behavior of embankments under high speed train loads. In: Correia, A.G., Brandl, H., (eds.) Geotechnics for Roads, Rail Tracks and Earth Structures, pp. 225–238. Balkema, Rotterdam (2001)
5. Banimahd, M., Woodward, P.K.: Numerical study of train speed effect on railway track response. In: Proceedings of the 9th International Conference of Railway Engineering, London (2007)
6. Xiao, J.H., Wang, B.L., Wang, Z.D., Yang, L.C., Gong, Q.M.: Differential Settlement of Subgrade and its Control for High Speed Railway. Tongji University Press (2015)
7. Wang, C., Wang, B., Guo, P., et al.: Experimental analysis on settlement controlling of geogrid-reinforced pile-raft-supported embankments in high-speed railway. Acta Geotech. **15**(9), 867–878 (2015)
8. Zheng, G., Jiang, Y., Han, J., Liu, Y.: Performance of cement fly ash-gravel pile-supported high-speed railway embankments over soft marine clay. Mar. Georesour. Geotech. **29**(2), 145–161 (2011)
9. Hong, X., Qiang, L., Jiang-dong, D., Ming, H.: Test study of reinforcing railway subgrades by using concrete compacted base-enlarged piles and geogrids. J. Geotech. Mech. **29**(8), 2157–2162 (2008)
10. Xu, X.: Study on Field Measurement and Full-Scale Model Test on Dynamic Response of High-Speed Railway Subgrade. Zhejiang University (2014)
11. Bhasi, A., Rajagopal, K.: Study of the effect of pile type for supporting basal reinforced embankments constructed on soft clay soil. Indian Geotech. J. **43**(4), 344–353 (2013)
12. Lixibin, W.: The numerical analysis of pile-supported embankment deformation mechanism induced by high-speed train load. J. Rock Mech. Geotech. Eng. **33**(Suppl 2), 232–237 (2011)
13. Guido, V.A., Chang, D.K., Sweeney, M.A., et al.: Comparison of geogrid and geotextile reinforced earth slabs. Can. Geotech. J. **23**(4) (2011)
14. Van Eekelen, S.J.M., Bezuijen, A., Lodder, H.J., et al.: Model experiments on piled embankments: Part I. Geotext. Geomembr. **32**, 6981 (2012)
15. Lai, H.-J., Zheng, J.-J., Zhang, J.: DEM analysis of "soil" arching within geogrid-reinforced and unreinforced pile-supported embankments. Comput. Geotech. **61**, 13–23 (2014)
16. Sinha, A., Hanna, A.M.: 3D numerical model for piled raft foundation. Int. Geomech. **17**(2), 04016055 (2017)

17. Xiao, Q., Huang, B., Xu, H., et al.: Analysis of steady rolling contact creep characteristics of high speed wheel/rail under different running conditions. Chin. Railway Sci. **36**(3), 81–87 (2015)
18. Sinha, A., Hanna, A.M.: 3D numerical model for piled raft foundation. Int. J. Geomech. 04016055 (2016)
19. Fu, Q., Liu, H., Ding, X., Zheng, C.: Numerical investigation of piled raft foundation in mitigating embankment vibrations induced by high-speed trains. J. Central South Univ. **22** (11), 4434–4444 (2015)

Vibration Characteristics of Discretely-Supported Floating Slab Track Under Moving Load

Qiang Huang[1,2], Hongwei Huang[1,2(✉)], and Dongmei Zhang[1,2]

[1] Department of Geotechnical Engineering, Tongji University, Shanghai 200092, China
huanghw@tongji.edu.cn

[2] Key Laboratory of Geotechnical and Underground Engineering, Tongji University, Ministry of Education, Shanghai 200092, China

Abstract. Floating slab track (FST), as an effective way to reduce the train-induced vibration, has been widely used in the practice. The FST with discrete support is presented in the paper, consisting of an upper Euler-Bernoulli accounting for the rail and a lower layer free short beam accounting for the floating slab. Dynamic response of the discretely-supported FST under a moving load is determined using mode superposition method combined with Runger-Kutta numerical method. Rail deflection and force in the slab bearing are chosen to evaluate the vibration characteristics of FST. The dynamic response of FST is significantly influenced by the slab length. When the floating slab is not long (less than about 10 m), the rail deflection and force in the slab bearing decrease rapidly as slab length increases. However, the response becomes stable if the slab is too long (more than about 30 m). The vibration-isolation performance of FST is investigated by a parametric study. It is found that a stiffer rail pad or floating slab contributes to reduce the rail deflection and inner force of slab bearing at the same time while heavier slab is adverse in vibration attenuation, and softer slab bearing is effective to reduce the inner force of slab bearing but induces more rail deflection. A possible way to improve the vibration-isolation performance and train moving comfort of FST is to increase the slab length, stiffness of rail pad and flexural stiffness of slab as well as moderate decrease of the stiffness of slab bearing.

Keywords: Floating slab track · Mode superposition method · Frequency
Slab length · Rail deflection · Slab bearing force

1 Introduction

To reduce the train-induced vibration and re-radiated noise from underground railways in the urban rail transit system, one effective means is to introduce floating slab track [1–3]. The floating slab track (FST), which basically consists of concrete slabs supported on resilient elements such as rubber bearing or steel spring, has been used on modern rail transit system for years. The design of FST is to reduce the vibration transmitted to the supporting foundation and surrounding areas by inserting the floating

X. Shi et al. (Eds.): GSIC 2018, *Proceedings of GeoShanghai 2018 International Conference: Transportation Geotechnics and Pavement Engineering*, pp. 218–226, 2018.
https://doi.org/10.1007/978-981-13-0011-0_24

slabs between the rail and foundation. Simple mass-spring models were often used to reveal the isolation performance of FST and is assumed to isolate the vibration for frequency above $\sqrt{2}$ times the natural frequency of the system. A series of researches have been conducted on the vibration characteristics of FST. Cui and Chew [4] analyzed the effectiveness of FST with continuous or discrete support using receptance method by modelling floating slabs as mass layer. Hussein and Hunt [5] used double-beam system to model the FST with continuous slabs and obtained the analytical solution of FST under moving oscillating load, then Hussein and Hunt [6] further studied the response of FST with discontinuous slabs under moving oscillating load, in which each floating slab was modelled as discrete beam supported by continuous slab bearing. Li and Wu [7] compared the responses of three FST models by assuming the slab as mass, infinite beam and free beam, respectively. Kuo et al. [8] also studied the vibration characteristics of discretely-supported FST, in which the vehicle-track interaction was considered. In this paper, the FST with discontinuous slab and discrete support is taken into account, the slabs are modelled as free beam and the effect of slab length is carefully considered. Subsequently, a parametric study is conducted to analyze the effects of varying factors on the vibration-isolation performance of FST. Finally, a feasible method to improve vibration isolation of FST as well as not sacrificing the train moving comfort is proposed.

2 Analytical Equations of FST

In reality, the slab usually has finite length from less than one meter to dozens of meters, and the slab vibrates in its own mode shapes that may affect the vibration isolation performance of FST [7]. Therefore, the slab with finite length is modelled as a free beam, and the FST is modelled as shown in Fig. 1. For simplicity, the support spacing below the rail and slab is the same, equal to 0.6 m.

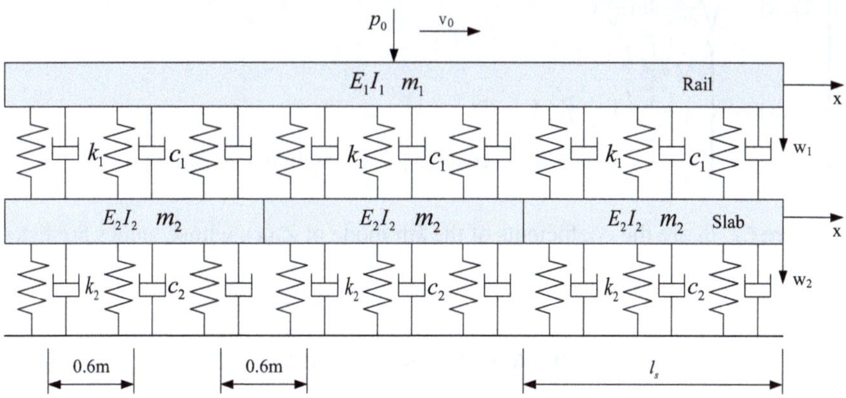

Fig. 1. Long beam-short beam model

The motion equations of long-short beam model are given by

$$\begin{cases} E_1 I_1 \frac{\partial^4 w_1}{\partial x^4} + m_1 \frac{\partial^2 w_1}{\partial t^2} + \sum_{i=1}^{N} F_{ri} = p_0 \delta(x - vt) \\ E_2 I_2 \frac{\partial^4 w_{2m}}{\partial x^4} + m_2 \frac{\partial^2 w_{2m}}{\partial t^2} + \sum_{i=1}^{NP} F_{si} = \sum_{i=1}^{NP} F_{ri} \end{cases} \quad (1)$$

Where $w_1(x)$ is the rail deflection, w_{2m} is the deflection of the mth slab, F_{ri}, F_{si} is the corresponding force in the rail pad and slab bearing, given by

$$\begin{cases} F_{ri} = k_1(w_1\delta(x - x_i) - w_{2m}\delta(x - (m-1)l_s)) + c_1(\dot{w}_1\delta(x - x_i) - \dot{w}_{2m}\delta(x - (m-1)l_s)) \\ F_{si} = k_2 w_{2m}\delta(x - x_i + (m-1)l_s) + c_2\dot{w}_{2m}\delta(x - x_i + (m-1)l_s) \end{cases} \quad (2)$$

There are n slabs in the FST, each slab is supported by NP slab bearings, so the number of total slab bearing or rail pad is $N = n \times NP$. The length of each slab is l_s, and the calculated length of rail is $l_0 = n \times l_s$. To facilitate the solution of partial differential equations in Eq. (1), mode superposition method is adopted. The deflection expression for the rail and free short beam are as follows [9, 10]:

$$\begin{cases} w_1(x,t) = \sum_{k=1}^{nmr} Y_k(x) q_k(t) \\ w_{2m}(x,t) = \sum_{k=1}^{nms} Z_k(x) \phi_{km}(t) \; (m = 1, 2, \cdots n) \end{cases} \quad (3)$$

Where the total number of modes for the rail and slab is *nmr and nms*, respectively; $Y_k(x)$ and $Z_k(x)$ are the shape function of the rail and slab, $q_k(x)$ and $\Phi_{km}(t)$ are the corresponding kth mode time coordinate of the rail and the mth slab; the shape function $Y_k(x)$, $Z_k(x)$ are given by

$$\begin{cases} Y_k(x) = \sqrt{\frac{2}{m_1 l_0}} \sin \frac{k\pi}{l_0} x \\ Z_k(x) = \begin{cases} \sqrt{\frac{1}{m_2 l_s}} \; k = 1 \\ \sqrt{\frac{3}{m_2 l_s}} \left(1 - \frac{2x}{l_s}\right) \; k = 2 \\ \sqrt{\frac{1}{m_2 l_s}} (\cosh(\beta_k x) + \cos(\beta_k x) - C_k(\sinh(\beta_k x) + \sin(\beta_k x))) \; k \geq 3 \end{cases} \end{cases} \quad (4)$$

Where C_k, β_k are the coefficients of the kth mode of $Z_k(x)$, whose values are listed in Table 1.

Table 1. Values of C_k and β_k

Mode k	1	2	3	4	5	6
C_k	–	–	0.9825	1.00077	0.999966	1.0000
$\beta_k l_s$	0	0	4.7304	7.8532	10.9956	$(2k-3)\frac{\pi}{2}$

As a result, the Eq. (1) is transformed into the following form:

$$\begin{cases} \ddot{q}_k(t) + \sum_{i=1}^{N} c_1 Y_k(x_i) \sum_{k=1}^{nmr} Y_k(x_i)\dot{q}_k(t) + \sum_{i=1}^{N} k_1 Y_k(x_i) \sum_{k=1}^{nmr} Y_k(x_i) q_k(t) + \frac{E_1 I_1}{m_1}\left(\frac{k\pi}{l_0}\right)^4 q_k(t) - \\ \sum_{i=1}^{N} c_1 Y_k(x_i) \sum_{k=1}^{nms} Z_k(x_i - (m-1)l_s)\dot{\phi}_{km}(t) - \sum_{i=1}^{N} k_1 Y_k(x_i) \sum_{k=1}^{nms} Z_k(x_i - (m-1)l_s)\phi_{km}(t) = p_0 Y_k(vt) \\ \ddot{\phi}_{km}(t) + \sum_{i=1}^{NP} (c_1 + c_2) Z_k(x_i - (m-1)l_s) \sum_{k=1}^{nms} Z_k(x_i - (m-1)l_s)\dot{\phi}_{km}(t) \\ + \sum_{i=1}^{NP} (k_1 + k_2) Z_k(x_i - (m-1)l_s) \cdot \sum_{k=1}^{nms} Z_k(x_i - (m-1)l_s)\phi_{km}(t) + \frac{E_2 I_2}{m_2}\beta_k^4 \phi_{km}(t) \\ - \sum_{i=1}^{NP} c_1 Z_k(x_i - (m-1)l_s) \sum_{k=1}^{nmr} Y_k(x_i)\dot{q}_k(t) - \sum_{i=1}^{NP} k_1 Z_k(x_i - (m-1)l_s) \sum_{k=1}^{nmr} Y_k(x_i) q_k(t) = 0 \end{cases} \quad (5)$$

Equation (5) can be expressed in the form of standard matrix equations, in the form of matrix ($nmr + n \times nms$) $\times$ ($nmr + n \times nms$). The generalized displacement vector X is $(q_1\ q_2\ \ldots\ q_{nmr}\ \Phi_1\ \ldots\ \Phi_{1nms}\ \Phi_2\ \ldots\ \Phi_{2nms}\ \Phi_{n1}\ \ldots\ \Phi_{nnms})^T$.

3 Model Validation

The aforementioned analytical equations of FST models can be solved by Runger-Kutta numerical method. To validate the accuracy of mode superposition method and Runger-Kutta method in calculating dynamic response of FST, an example is presented for model validation. The example is from the FST with the continuous support in Hussein and Hunt [5]. The FST model is a double infinite Euler beam model, whose analytical equations can be given by

$$\begin{cases} \ddot{q}_k(t) + \frac{c_1}{m_1}\dot{q}_k(t) - \frac{c_1}{\sqrt{m_1 m_2}}\dot{\phi}_k(t) + \left(\frac{k_1}{m_1} + \frac{E_1 I_1}{m_1}\left(\frac{k\pi}{l_0}\right)^4\right) q_k(t) - \frac{k_1}{\sqrt{m_1 m_2}}\phi_k(t) = p_0 Y_k(vt) \\ \ddot{\phi}_k(t) + \frac{c_1 + c_2}{m_2}\dot{\phi}_k(t) - \frac{c_1}{\sqrt{m_1 m_2}}\dot{q}_k(t) + \left(\frac{k_1 + k_2}{m_2} + \frac{E_2 I_2}{m_2}\left(\frac{k\pi}{l_0}\right)^4\right)\phi_k(t) - \frac{k_1}{\sqrt{m_1 m_2}} q_k(t) = 0 \end{cases} \quad (6)$$

The parameters of FST model are listed in Table 2. Take load velocity v_0 as 20 m/s and frequency as 10 Hz for example, the calculated rail deflection and force in rail pad are compared in Figs. 4 and 5, respectively.

Table 2. Parameters of the FST [5]

Rail				Slab			
E_1I_1/Nm^2	m_1/kgm^{-1}	k_1/kNm^{-2}	c_1/Nsm^{-2}	E_2I_2/Nm^2	m_2 kgm^{-1}	k_2/kNm^{-2}	c_2/Nm^{-2}
1e7	100	4e4	6.3e3	1.43e9	3500	5e4	4.18e4

As seen in Fig. 2, the results from mode superposition agree quite well with the theoretical results from Hussein and Hunt [6]. Therefore, mode superposition method is valid and accurate to determine the dynamic response of floating slab track.

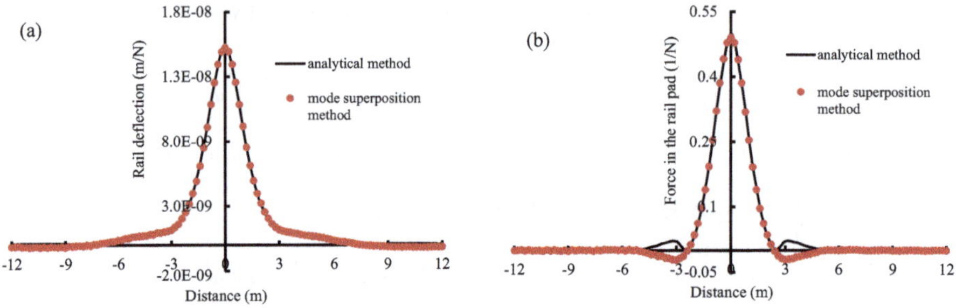

Fig. 2. Model validation: (a) rail deflection; (b) force in the rail pad

4 Vibration of FST with Different Slab Length

To have a further knowledge of vibration characteristics of FST, the dynamic responses of FST under different slab length are carefully investigated. The parameters of FST are from the real track of Taipei rail transit system (Kuo et al. 2008), as listed in Table 3. The applied load p_0 is assumed as a unit load.

Table 3. Parameters of FST [8]

Rail				Slab			
E_1I_1/Nm^2	m_1/kgm^{-1}	k_1/kNm^{-2}	c_1/Nsm^{-2}	E_2I_2/Nm^2	m_2/kgm^{-1}	k_2/kNm^{-2}	c_2/Nsm^{-2}
6.14e6	60.34	1.0e8	3e4	6.615e7	1640	5.8e6	3.5e4

The vibration characteristics of FST with different slab length is analyzed first, the slab length ranges from 0.6 m to 36 m. In the paper, the rail deflection and force in the slab bearing are chosen to evaluate the dynamic response of FST. The rail deflection is related to the train moving comfort, and the inner force of slab bearing is related to the environmental vibration level. The rail deflection and inner force of slab bearing are the deflection and the force at the midpoint of FST. The response results are shown in Figs. 3 and 4.

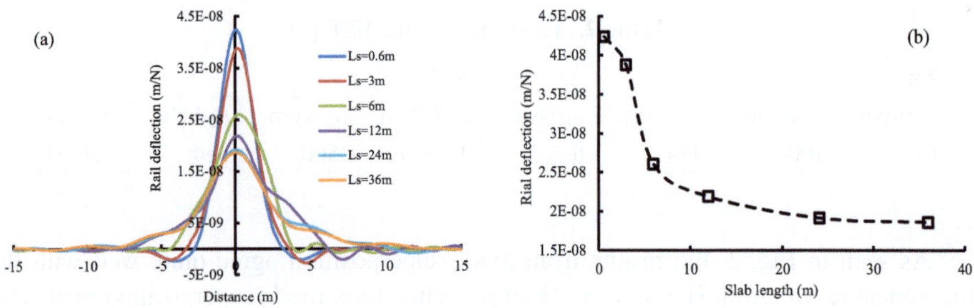

Fig. 3. The effect of slab length on rail deflection: (a) rail deflection distribution; (b) maximum rail deflection

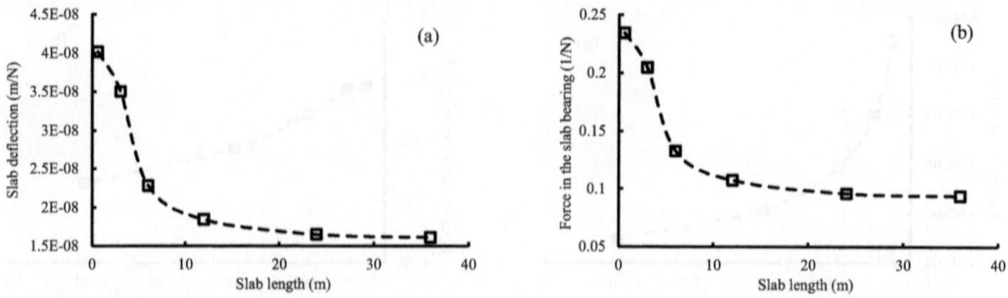

Fig. 4. The effect of slab length: (a) maximum slab deflection; (b) force in the slab bearing

As seen in the figures, the response of FST is significantly influenced by the slab length. When the slab length is not very long, less than 10 m in terms of the current track parameters, the rail deflection, slab deflection as well as the inner force of slab bearing decrease rapidly as slab length increases. However, as the slab length continues to increase, decrease of the deflection or force in the slab bearing becomes slower and slower, and approaches a stable value if the slab length is too long (more than 30 m fro example). In another word, increasing the slab length is an effective way to reduce the vibration level, but overlong slab length is not rational.

5 Parametric Study

It is well known that FST effectively reduces the force transmitted to the foundation, however, it also induces more rail deflection, which reduces the train moving comfort as a result. In this section, parametric study is conducted to investigate the vibration-isolation performance of FST with a slab length of 3 m. The effects of stiffness of rail pad, slab bearing, slab mass and flexural stiffness of slab on rail deflection and the force in the slab bearing is clarified.

5.1 The Effect of Stiffness of Rail Pad

The effect of stiffness of rail pad is shown in Fig. 5. It is seen that increasing rail pad stiffness can effectively reduce the rail deflection and slab bearing force at the same time, especially for the rail deflection. Therefore, a stiffer rail pad contributes to the vibration attenuation and train moving comfort increase, which agrees with the conclusion of Kuo et al. [8].

5.2 The Effect of Slab Mass

The effect of slab mass is shown in Fig. 6. It is assumed that increasing the slab mass dose not induce the change of flexural stiffness of slab. It can be seen that increasing slab mass is proved to be unfavorable in vibration isolation. A lighter slab is much better than the heavier slab.

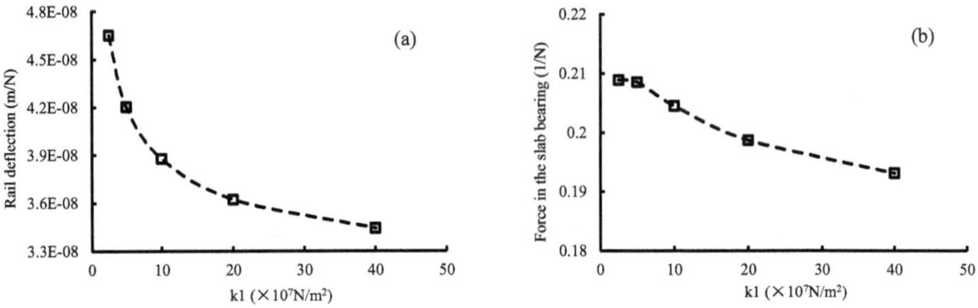

Fig. 5. The effect of stiffness of rail pad: (a) maximum slab deflection; (b) force in the slab bearing

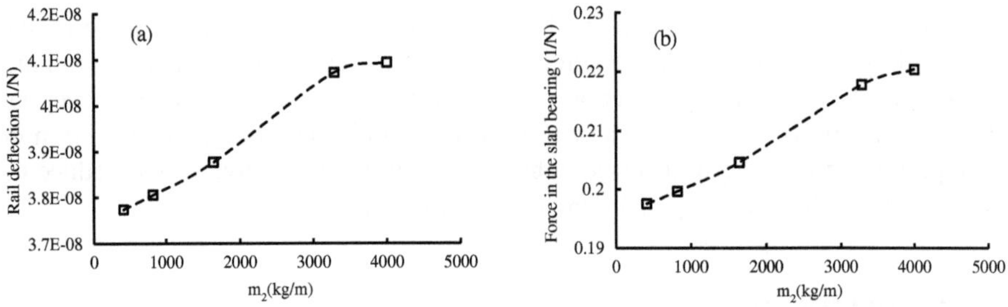

Fig. 6. The effect of slab mass: (a) maximum slab deflection; (b) force in the slab bearing

5.3 The Effect of Flexural Stiffness of Slab

The effect of flexural stiffness of slab is shown in Fig. 7. When the height of slab increases, the flexural stiffness of slab increases correspondingly, and the increase of flexural stiffness is more evident than the increase of slab mass. Figure 7 shows that increasing the flexural stiffness of slab is effective in reducing rail deflection and slab bearing force, which counteracts the adverse effect of slab mass increase to some extent. Therefore, it is feasible to increase the flexural stiffness of slab to improve the train-isolation performance of FST.

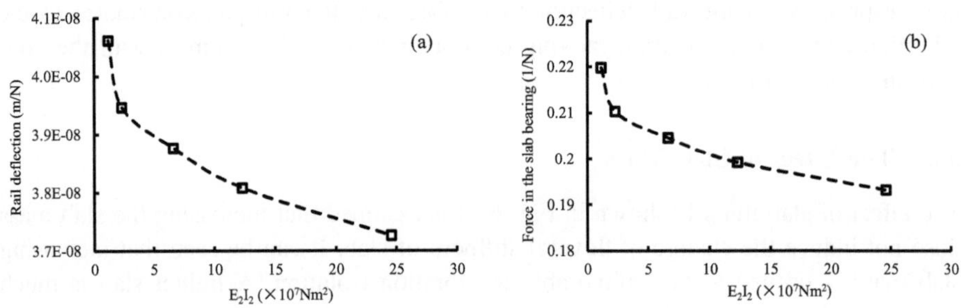

Fig. 7. The effect of flexural stiffness of slab: (a) maximum slab deflection; (b) force in the slab bearing

5.4 The Effect of Flexural Stiffness of Slab

Similarly, the effect of stiffness of slab earing is shown in Fig. 8. It is found the effects of slab bearing stiffness on rail deflection and force in the slab bearing are different. Increasing the stiffness of slab bearing results in evident reduction of rail deflection, but induces more force in the slab bearing. As a result, the vibration reduction performance of FST is achieved by sacrificing the train moving comfort when slab bearing stiffness increases. From the analysis above, in can conclude that there are several methods to decrease the rail deflection, such as increase of rail stiffness, increase of flexural stiffness of the slab.

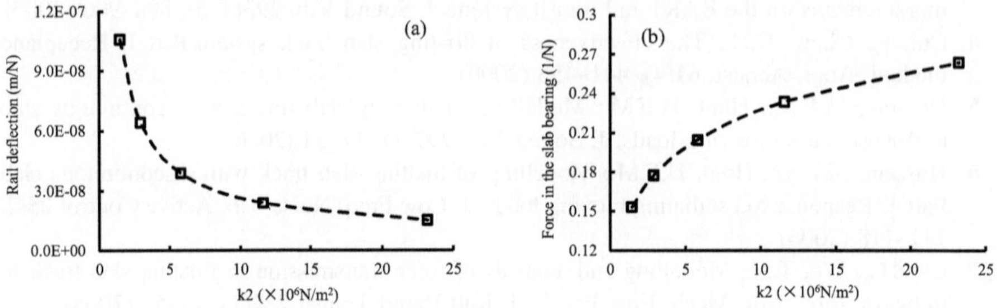

Fig. 8. The effect of stiffness of slab bearing: (a) maximum slab deflection; (b) force in the slab bearing

6 Conclusion

In the paper, the dynamic response of discretely-supported FST under moving load is studied using mode superposition method. The effect of slab length is analyzed and the vibration-isolation performance of FST under different parameters is investigated. Based on the above analyses, some conclusions are obtained as follow:

(1) Dynamic response of FST is significantly influenced by the length of floating slab, when the slab length is not long (less than about 10 m), increasing slab length can greatly reduce the rail deflection and force in the slab bearing; when the slab length is long (more than 30 m for example), increasing slab length is no longer effective in improving the vibration isolation performance of FST.

(2) Parametric study reveals that increasing stiffness of rail pad or flexural stiffness of slab can effectively reduce the rail deflection and force in the slab bearing; increasing the stiffness of slab bearing can decrease the rail deflection but induce larger slab bearing force; increasing slab mass is adverse in vibration isolation.

(3) According to the parametric study, a possible method to improve vibration-isolation performance of FST is to increase the stiffness and rail pad and increase the flexural stiffness of slab, combined with a moderate decrease of stiffness of slab bearing, thereby the adverse effect of slab mass increase and larger rail deflection due to decrease of slab bearing can be counteracted.

Acknowledgement. This study is substantially supported by Natural Science Foundation of China (NSFC Grant No. 51538009), the support is gratefully acknowledged. The authors also wish to appreciate the reviewers' kind comments on this paper.

References

1. Wilson, G.P., Saurenman, H.J., Nelson, J.T.: Control of ground-borne noise and vibration. J. Sound Vib. **87**(2), 339–350 (1983)
2. Nelson, J.T.: Recent development in ground-borne noise and vibration control. J. Sound Vib. **193**, 367–376 (1996)
3. Saurenman, H., Philips, F.: In-service tests of the effectiveness of vibration control measurements on the BART rail transit system. J. Sound Vib. **293**(3–5), 888–900 (2006)
4. Cui, F., Chew, C.H.: The effectiveness of floating slab track system-Part I. Receptance method. Appl. Acoust. **61**(4), 441–453 (2000)
5. Hussein, M.F.M., Hunt, H.F.M.: Modelling of floating-slab tracks with continuous slabs under oscillating moving loads. J. Sound Vib. **297**(1), 37–54 (2006)
6. Hussein, M.F.M., Hunt, H.F.M.: Modelling of floating-slab track with discontinuous slabs Part 1. Response to oscillating moving loads. J. Low Freq. Noise Vib. Active Control **25**(2), 111–118 (2006)
7. Li, Z.G., Wu, T.X.: Modelling and analysis of force transmission in floating-slab track for railways. Proc. Inst. Mech. Eng. Part F: J. Rail Rapid Transit **222**(1), 45–57 (2008)
8. Kuo, C.M., Huang, C.H., Chen, Y.Y.: Vibration characteristics of floating slab track. J. Sound Vib. **317**(3), 1017–1034 (2008)
9. Young, D.: Vibration of rectangular plates by the Ritz method. J. Appl. Mech. **17**, 448–453 (1950)
10. Zhai, W.M., Cai, Z.: Dynamic interaction between a lumped mass vehicle and a discretely supported continuous rail track. Comput. Struct. **63**(5), 987–997 (1997)

The Effect of Soil Non-linearity on Mixed Traffic Railway Lines: Passenger vs Freight Loads

Kaitai Dong[1], David P. Connolly[2(✉)], Omar Laghrouche[1], Peter K Woodward[2], Carlton Ho[3], and Pedro Alves Costa[4]

[1] Heriot Watt University, Edinburgh, UK
[2] University of Leeds, Leeds, UK
d.connolly@leeds.ac.uk
[3] University of Massachusetts, Amherst, USA
[4] University of Porto, Porto, Portugal

Abstract. To add additional capacity to railway networks, freight services might be added to lines that have previously only be used for passenger services. Existing ballasted lines may have mixed subgrade conditions and thus the effect of increased axle loads on track behavior is unclear. Typically, such cases will result in elevated track deflections in comparison to passenger vehicles. As a result, the supporting subgrade experiences higher strain levels, which can fall into the large strain range. The related non-linear subgrade behavior plays an important role in track response but is challenging to model. As a solution, this paper presents a new semi-analytical numerical model, where the track is simulated analytically and allows for 1D wave propagation. The ground is modelled using a non-linear equivalent thin-layer finite element formulation. This allows for the subgrade stiffness to be updated in an iterative manner with minimal computational effort. A case study is presented to show that modest increases in axle load can have a marked effect on track deflections.

Keywords: Railway freight · Railroad · Non-linear soil

1 Literature Review

With the aim of adding additional capacity to existing railway lines, it may be desirable to add freight services to tracks that have previously only be used for passenger services. Some of these lines may have relatively low subgrade stiffness's and thus the effect of increased axle loads on track behavior is unclear.

To investigate and predict the track performance and ground response under various train loads and speeds, a number of modelling techniques have been proposed. The approaches include analytical models (Krylov 1995; Degrande and Lombaert 2001; Takemiya and Bian 2005), semi-analytical models (Sheng et al. 1999; Madshus and Kaynia 2000; Sheng et al. 2003; Kaynia et al. 2000; Thompson 2008; Triepaischajonsak and Thompson 2015). There are also numerical models: 2.5D models (Yang et al. 2003; P. Alves Costa et al. 2012; Pedro Alves Costa et al. 2010) and fully 3D models using

X. Shi et al. (Eds.): GSIC 2018, *Proceedings of GeoShanghai 2018 International Conference: Transportation Geotechnics and Pavement Engineering*, pp. 227–236, 2018.
https://doi.org/10.1007/978-981-13-0011-0_25

finite element (FE) and possibly boundary element (BE) theories (Hall 2003; Kouroussis et al. 2011; Arlaud et al. 2015; El Kacimi et al. 2013).

For freight trains, the dominant frequency components of the vibration are within 4-30 Hz (Jones and Block 1996). In order to study the vibrations induced by the freight trains, both dynamic and quasi-static generation mechanism, a track response model combined with transfer functions from sleeper to ground was utilized by (Jones and Block 1996). Another numerical model was proposed for the studies of longitudinal dynamics of the trainset (Belforte et al. 2008). On-site tests can be costly (Jones 1994), meaning theoretical models are often used to examine the track performance and ground response from freight trains.

In modelling the ground vibrations from railways, linear elastic models of the soil are commonly used, because strains are small. Nonetheless, when axle loads increase and/or the train speed gets close to the critical velocity, the track deflections increase and non-linear soil response occurs (Madshus and Kaynia, 2000) (Pedro Alves Costa et al. 2010). To simulate this non-linear behavior, soil stiffness' can be artificially reduced (Madshus and Kaynia 2000; Kaynia et al. 2000). Alternatively, using an automated, equivalent non-linear approach, the shear modulus can be adjusted based on the maximum effective octahedral shear strain in each soil element. Then it can be updated element by element until a tolerance requirement is met (Pedro Alves Costa et al. 2010).

Since the supporting non-linear ground behavior plays a key role when modelling the vibrations generated by the freight trains, this paper provides a robust and efficient semi-analytical to model non-linear soil effects. The track is modelled analytically and allows for 1D wave propagation. The soil is modelled using a non-linear equivalent thin-layer method (TLM). The soil stiffness is updated in an iterative manner to simulate the non-linear behavior of the soil with the minimum computational effort.

2 Numerical Model Development

Freight trains subject railway tracks to heavy axle loads which result in elevated strains within the supporting subgrade. Large strains cause non-linear soil behavior, resulting in reduced support stiffness. Modelling non-linear soil behaviour is computationally intensive and thus difficult to include in a sensitivity analysis. Therefore, to reduce computational requirements, a thin-layer finite element model was developed, and then combined with an equivalent non-linear procedure. This soil model was then efficiently coupled with a track model that permitted 1D wave propagation.

2.1 Track Model

Ballasted track was modelled considering, rail, railpad, sleeper and ballast components, as shown in Fig. 1. One dimensional wave propagation was considered in the ballast and the track was coupled to the soil using an equivalent spring, using the approach outlined in (Dieterman and Metrikine 1996).

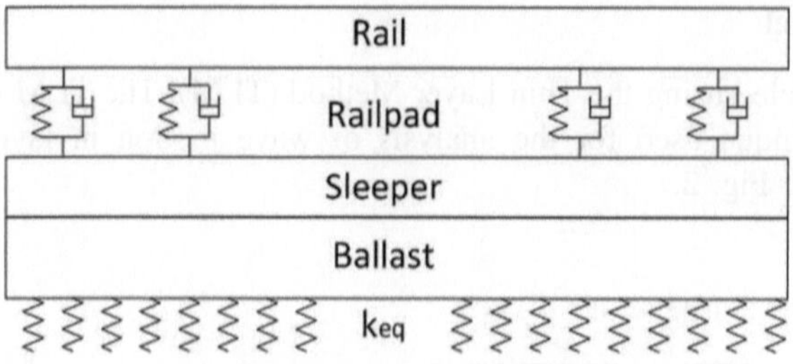

Fig. 1. Analytical ballasted track model layout

$$\begin{bmatrix} EI_r k_1^4 + k_p^* - \omega^2 m_r & -k_p^* & 0 \\ -k_p^* & k_p^* + \frac{2\omega E_b^* b\alpha}{tan\left(\frac{\omega h}{C_p}\right) C_p} - \omega^2 m_s & \frac{-2\omega E_b^* b\alpha}{sin\left(\frac{\omega h}{C_p}\right) C_p} \\ 0 & \frac{-2\omega E_b^* b\alpha}{sin\left(\frac{\omega h}{C_p}\right) C_p} & \frac{2\omega E_b^* b\alpha}{tan\left(\frac{\omega h}{C_p}\right) C_p} + k_{eq} \end{bmatrix} \begin{Bmatrix} \tilde{u}_r(k_1,\omega) \\ \tilde{u}_s(k_1,\omega) \\ \tilde{u}_{bb}(k_1,\omega) \end{Bmatrix} = \begin{Bmatrix} \tilde{P}(k_1,\omega) \\ 0 \\ 0 \end{Bmatrix} \quad (1)$$

Where EI_r is the bending stiffness of the rail; m_r is the mass of rails per meter; m_s is the equivalent distributed mass of sleepers; k_p^* is the complex stiffness of the railpad; k_{eq} is the equivalent stiffness of the ground; E_b^* is the Young's modulus of the ballast; C_p is the compression wave speed in the ballast; h is the ballast layer height; α is the a dimensional parameter, taken as 0.5; b is the half-width of the track.

The ballasted track model includes the coupling between the track and the soil, i.e., the complex equivalent stiffness of the ground k_{eq}. It was suggested by (Steenbergen and Metrikine 2007) that the equivalent stiffness can be calculated using the ratio between the load and average displacement along the track-soil interface. Therefore, the equivalent stiffness can be mathematically represented in the wavenumber-frequency domain by the formula:

$$\tilde{k}_{eq}(k_1,\omega) = \frac{2\pi}{\int_{-\infty}^{+\infty} \tilde{u}_{zz}^G(k_1,k_2,0,\omega) \frac{\sin(k_2 b)^2}{(k_2 b)^2} dk_2} \quad (2)$$

Where u_{zz} is the Green's function of vertical displacement of the ground in the wavenumber-frequency domain, and k_1 and k_2 are the Fourier images of coordinate x and y, respectively. The Green function is computed by the Haskell-Thompson approach (Sheng et al. 1999).

2.2 Soil Model

The soil is modeled using the Thin-Layer Method (TLM). The TLM is a semi-discrete numerical technique used for the analysis of wave motion in layered media. It is illustrated in the Fig. 2.

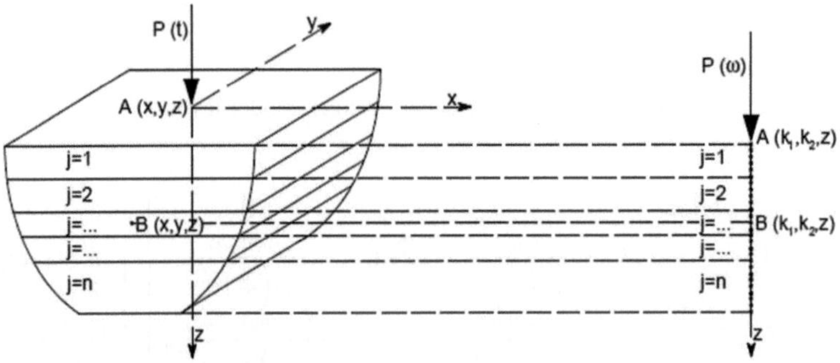

Fig. 2. Schematic diagram of Thin-Layer Method modelling process (Pedro Alves Costa, 2011)

It is worth noting that:

- The thickness of the thin layers was computed as $h = \frac{wavelength}{8} = \frac{2\pi}{8k_{max}}$, where k_{max} is the maximum wavenumber defined
- Quadratic elements were used for the soil model, as demonstrated in the Fig. 4
- After obtaining the displacement of each node, the strain/stress field inside the layer was then calculated using Eqs. (3) and (4)

$$\{\boldsymbol{\varepsilon}\} = [\mathbf{B}]\{\mathbf{u}\} \tag{3}$$

$$\{\boldsymbol{\sigma}\} = [\mathbf{D}]\{\boldsymbol{\varepsilon}\} = [\mathbf{D}][\mathbf{B}]\{\mathbf{u}\} \tag{4}$$

Where $[\mathbf{B}] = [\mathbf{B}_1 \quad \mathbf{B}_2 \quad \mathbf{B}_3]$ and

$$[\mathbf{B}_i] = \begin{bmatrix} ik_1N_i & 0 & 0 \\ 0 & ik_2N_i & 0 \\ 0 & 0 & \frac{\partial N_i}{\partial z} \\ ik_2N_i & ik_1N_i & 0 \\ 0 & \frac{\partial N_i}{\partial z} & ik_2N_i \\ \frac{\partial N_i}{\partial z} & 0 & ik_1N_i \end{bmatrix} \tag{5}$$

$$\begin{aligned} N_1(\xi) &= \frac{1}{2}\xi^2 - \frac{1}{2}\xi \\ N_2(\xi) &= 1 - \xi^2 \\ N_3(\xi) &= \frac{1}{2}\xi^2 + \frac{1}{2}\xi \end{aligned} \tag{6}$$

- The inverse Fourier Transform was used to convert the results from the wavenumber-frequency domain back to the time-space domain.

2.3 Equivalent Non-linear Model

For freight trains lines, the supporting subgrade is likely to experience high levels of strain. This can result in soil stiffness degradation, thus increasing the track displacements and causing track deterioration. To simulate this, a non-linear equivalent model, based on an iterative stiffness updating procedure, was used. This model was well-suited to the discretized nature of the TLM method and summarized using the following steps:

(1) Assume low strain properties for all elements
(2) Compute strain time histories and determine the maximum effective octahedral shear strain values for all elements
(3) Use stiffness degradation curves, as shown in the Fig. 3, and calculated maximum effective octahedral strains to obtain the new stiffness for all elements
(4) Use the same procedure to compute the new damping values for all elements
(5) Repeat steps 2–4 until the established tolerance is met for all elements.

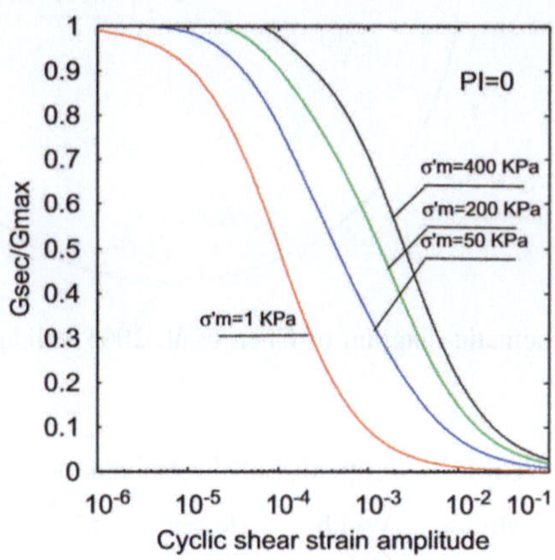

Fig. 3. Modulus reduction curves for non-plastic soil (Pedro Alves Costa et al. 2010)

3 Model Validation

(Chen et al. 2005) proposed an analytical approach that calculates the stresses in the ground using the equivalent stiffness on the basis of the model of an Euler beam resting on the half space subjected to a moving load. In order to validate the TLM model for the ground response, same case was studied and the stresses in the soil compared against Chen et al. 2005's simulation result.

3.1 Model Description

As depicted in the Fig. 4, the train-embankment-ground model contains an Euler beam resting on top of the half-space with a concentrated moving force acting on the beam. The half-space was modelled as thin layers and the coupling between the embankment and ground was represented by the equivalent stiffness. Assuming the load is at the centre of the embankment at the beginning, then it will move along the central line with a certain speed. The stresses generated by the contact force between the embankment and ground were calculated at 2 m depth below the loading point.

Key embankment and ground properties related to the validation are listed in the Tables 1 and 2 respectively. Aside from the listed parameters, the load speed was 30 m/s and the amplitude of the point load used in the simulation was 160 kN, without consideration of the irregularity of the contact surface.

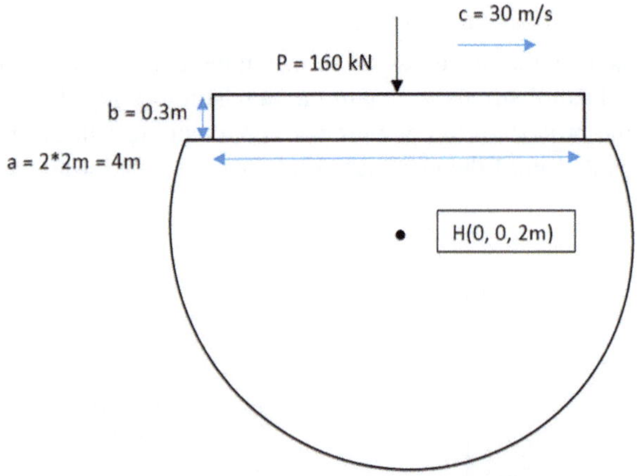

Fig. 4. Schematic diagram of Chen et al. 2005 validation model

Table 1. Properties of the embankment

Density (kg/m^3)	Young's modulus (MPa)	Width (m)	Height (m)	Mass (kg)	Second moment of area (m^4)
1900	30000	4	0.3	2280	0.009

3.2 Simulation Result

Since the direction of the point load is vertical and forces on other directions are not considered, shear stress τ_{xy} and τ_{yz}, for soil elements beneath the moving point load, were zero. Therefore, apart from normal stresses σ_{xx}, σ_{yy} and σ_{zz}, only $\tau_{xz}(=\tau_{zx})$ was analysed. In this validation case, the dynamic stresses generated by the moving point load were computed using the TLM model and the response points for comparison were chosen at 2 m depth directly underneath the motion line of the moving load.

Table 2. Properties of the ground

Shear modulus (MPa)	Poisson ratio	Density (kg/m^3)	Secondary wave speed (m/s)
10	0.45	1800	74.54

Figure 5 reveals good agreement for all dynamic stresses between Chen et al. 2005 and TLM's simulation result is found. Moreover, for a given soil, the strain can be calculated as $\varepsilon = \sigma/E$, where E is the Young's modulus of the soil layer. Therefore, the TLM is also applicable for the calculation of strains. This validation manifests that the TLM model is able to accurately describe the ground response induced by the moving load and also enables to predict the response in the soil at any given point.

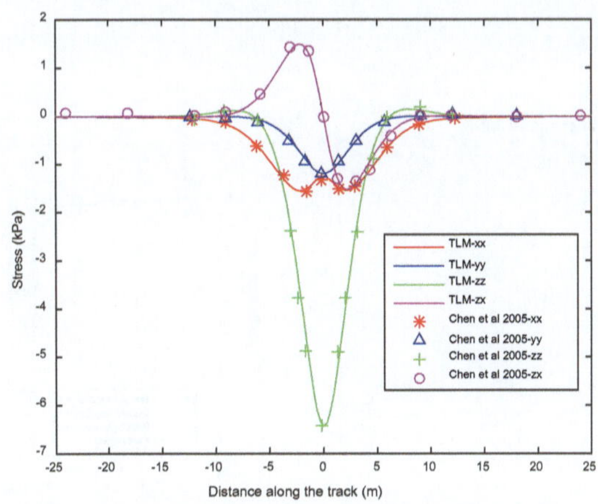

Fig. 5. Comparisons of the dynamic stresses of an element with 2 m depth underneath the moving load

4 Analysis and Results

Simulations were run to determine the effect of adding 25 tonne fright axle loads to a previous passenger-only (17 tonne) ballasted line, with the aim of determining increases in track displacement and soil strain. To do so, the following track properties were assumed: $m_r = 120\ kg/m$, $m_s = 490\ kg/m$, $k_p^* = 5 \times 10^8\ N/m^2$, $E_b^* = 125\ MPa$, $h = 0.35\ m$, $b = 2.5$ m. The soil was modelled as a homogenous half-space using the following properties: $density = 2000\ kg/m^3$, $Young's\, modulus = 25MPa$, $Poissons\ ratio = 0.35$, $damping = 0.03$. The stiffness degradation profile was the same as that shown previously. Train speed for both the passenger and freight axle loads was 26 m/s.

Figure 6 (left) shows the variation of strain versus depth within the soil. It is observed that the maximum strain level is found approximately 1 m below the ground surface and decays rapidly with depth. Correspondingly, Fig. 6 (right) shows maximum strain levels and their corresponding effect on soil stiffness. After the first

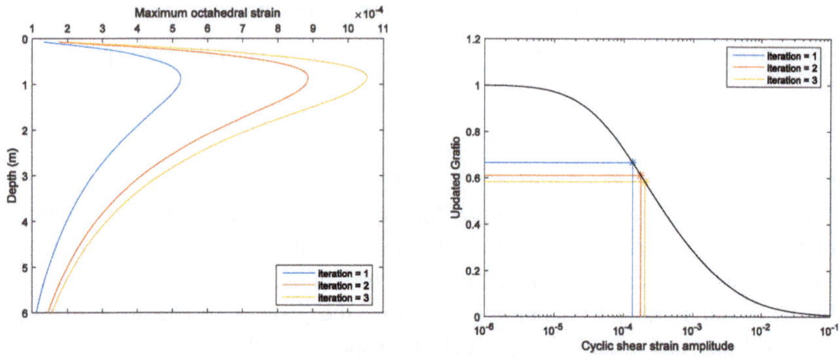

Fig. 6. Left: Octahedral strain vs soil depth, Right: Soil stiffness degradation during freight train passage

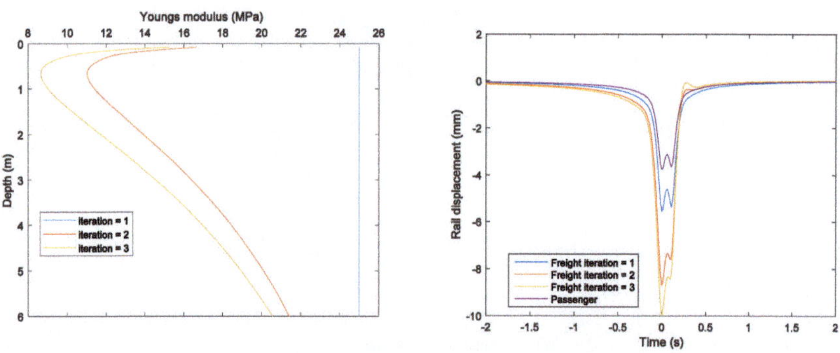

Fig. 7. Left: Young's modulus reduction with depth, Right: Track displacements

iteration, the soil drops to 67% of its original stiffness and by the third (and final) iteration, it has reached a value of 59%.

The resulting reduction in stiffness (Young's modulus) with depth is shown in Fig. 7 (left). For iteration 1, stiffness is constant with depth, however after strain updating, the subsequent iterations show large variations with depth, and are all lower than the starting value, particularly near the soil surface. For the passenger train, track displacements are 3.7 mm, however for the freight train, the linear value is 5.5 mm displacement, and the non-linear (iteration 3) is 8.4 mm. Therefore, it can be seen that the soil behavior is significantly non-linear, and that traditional linear analysis would greatly underestimate track deflections. This would result in much faster loss of track geometry and require frequent tamping. In addition, it is interesting to note that as the soil stiffness decreases, dynamic effects become more prevalent, with iteration 3 displacements appearing less symmetric than iteration 1.

5 Conclusions

Under certain circumstances, it may be desirable to run freight trains on ballasted track originally designed for passenger services. In such cases, the track may have a relatively low subgrade stiffness, meaning the effect of freight axles loads can lead to non-linear behavior. To determine the effect of increased axle loads in such cases, an equivalent non-linear numerical model was developed, capable of quickly assessing soil stresses and strains, and resulting track displacements. The model was validated and then used to assess the behavior of freight axle loads on a low stiffness ballasted line. It was shown that the track displacements have the potential to become high, due to non-linear stiffness reduction and the resulting dynamic amplification.

References

Alves Costa, P.: Vibrações Do Sistema Via-Maciço Induzidas Por Tráfego Ferroviário. Modelação Numérica E Validação Experimental. (2011)

Alves Costa, P., Calçada, R., Silva Cardoso, A.: Track-ground vibrations induced by railway traffic: In-situ measurements and validation of a 2.5D FEM-BEM model. Soil Dyn. Earthq. Eng. **32**(1), 111–128 (2012)

Alves Costa, P., Calçada, R., Silva Cardoso, A., Bodare, A.: Influence of soil non-linearity on the dynamic response of high-speed railway tracks. Soil Dyn. Earthq. Eng. **30**(4), 221–235 (2010)

Arlaud, E., Costa D'Aguiar, S., Balmes, E.: Validation of a reduced model of railway track allowing long 3D dynamic calculation of train-track interaction. In: Computer Methods and Recent Advances in Geomechanics - Proceedings of the 14th International Conference of International Association for Computer Methods and Recent Advances in Geomechanics, IACMAG 2014, (September), pp. 1193–1198 (2015)

Belforte, P., Cheli, F., Diana, G., Melzi, S.: Numerical and experimental approach for the evaluation of severe longitudinal dynamics of heavy freight trains. Veh. Syst. Dyn. **46** (SUPPL.1), 937–955 (2008)

Chen, Y.M., Wang, C.J., Chen, Y.P., Zhu, B.: Characteristics of stresses and settlement of ground induced by train. In: Environmental Vibrations: Prediction, Monitoring, Mitigation and Evaluation: Proceedings of the International Symposium on Environmental Vibrations, pp. 33–42 (2005)

Degrande, G., Lombaert, G.: An efficient formulation of Krylov's prediction model for train induced vibrations based on the dynamic reciprocity theorem. J. Acoust. Soc. Am. **110**(3), 1379–1390 (2001)

Dieterman, H.A., Metrikine, A.V.: The equivalent stiffness of a half-space interacting with a beam. Critical velocities of moving load along the beam. Eur. J. Mech. A. Solids **15**, 67–90 (1996)

El Kacimi, A., Woodward, P.K., Laghrouche, O., Medero, G.: Time domain 3D finite element modelling of train-induced vibration at high speed. Comput. Struct. **118**, 66–73 (2013)

Hall, L.: Simulations and analyses of train-induced ground vibrations in finite element models. Soil Dyn. Earthq. Eng. **23**(5), 403–413 (2003)

Jones, C.J.: Use of numerical models to determine the effectiveness of anti-vibration systems for railways. Proc. Inst. Civ. Eng. Transp. **105**(1), 43–51 (1994)

Jones, C.J.C., Block, J.R.: Prediction of ground vibration from freight trains. J. Sound Vib. **193** (1), 205–213 (1996)

Kaynia, A.M., Madshus, C., Zackrisson, P.: Ground vibration from high-speed trains: prediction and countermeasure. J. Geotech. Geoenviron. Eng. **126**(Igvc), 531–537 (2000)

Kouroussis, G., Gazetas, G., Anastasopoulos, I., Conti, C., Verlinden, O.: Discrete modelling of vertical track-soil coupling for vehicle-track dynamics. Soil Dyn. Earthq. Eng. **31**(12), 1711–1723 (2011)

Krylov, V.V.: Generation of ground vibrations by superfast trains. Appl. Acoust. **44**(2), 149–164 (1995)

Madshus, C., Kaynia, A.M.: High-speed railway lines on soft ground: dynamic behaviour at critical train speed. J. Sound Vib. **231**(3), 689–701 (2000)

Sheng, X., Jones, C.J.C., Petyt, M.: Ground vibration generated by a harmonic load acting on a railway track. J. Sound Vib. **225**(1), 3–28 (1999)

Sheng, X., Jones, C.J.C., Thompson, D.J.: A comparison of a theoretical model for quasi-statically and dynamically induced environmental vibration from trains with measurements. J. Sound Vib. **267**(3), 621–635 (2003)

Steenbergen, M.J.M.M., Metrikine, A.V.: The effect of the interface conditions on the dynamic response of a beam on a half-space to a moving load. Eur. J. Mech. A. Solids **26**(1), 33–54 (2007)

Takemiya, H., Bian, X.: Substructure simulation of inhomogeneous track and layered ground dynamic interaction under train passage. J. Eng. Mech. **131**(7), 699–711 (2005)

Thompson, D.: Railway Noise and Vibration: Mechanisms, Modelling and Means of Control. Elsevier, Amsterdam (2008)

Triepaischajonsak, N., Thompson, D.J.: A hybrid modelling approach for predicting ground vibration from trains. J. Sound Vib. **335**, 147–173 (2015)

Yang, Y.B., Hung, H.H., Chang, D.W.: Train-induced wave propagation in layered soils using finite/infinite element simulation. Soil Dyn. Earthq. Eng. **23**(4), 263–278 (2003)

Investigation of the Effect of Track on the Dynamic Stress in Saturated Soils Induced by the Moving Train Loads in a Shield Metro Tunnel

Honggui Di[1,2](✉), Shunhua Zhou[2], Zhe Luo[1], Chao He[2], and Xiaohui Zhang[2]

[1] The University of Akron, Akron, OH 44325, USA
hdi@uakron.edu
[2] Key Laboratory of Road and Traffic Engineering of the Ministry of Education, Tongji University, Shanghai 201804, China

Abstract. This study proposed an improved tunnel model for evaluating the dynamic stress in saturated soils subjected to moving train loads. The lining and grouting layer of the shield tunnel are modeled as a Flügge double cylindrical shell. The saturated soils are modeled as a poroelastic medium using the Biot's theory. The infinite rail and track slab are simplified as Euler–Bernoulli beams, and the track slab is coupled to the tunnel via one line of uniform supports. The rail, track slab, tunnel and soil medium are coupled by the compatibility of force and displacement at the contact surface. The developed tunnel model is validated against an existing tunnel model. Based on this model, the effect of railway track on the train-induced dynamic stress in the saturated soils is investigated by the proposed model. The model reveals that the track has significant influence on the magnitude and stress path of the dynamic stress in saturated soils.

Keywords: Shield tunnel · Saturated soil · Dynamic stress

1 Introduction

The excessive tunnel settlement in the metro system of the Yangtze River delta region in China has currently impeded the operation; as a result, the annual maintenance cost has considerably increased [1, 2]. The field monitoring data reveal that the train loading is one of the major components of the total tunnel settlement [3]. In the analysis of traffic load-induced settlement of soft soil foundation, the proper evaluation of dynamic stress in soils is the prerequisite for the research approaches using cyclic triaxial tests and empirical formulas [4].

The existing approaches to evaluating the dynamic response of underground railways mainly include analytical (or semi-analytical) modeling approaches and numerical modeling approaches. The modeling of underground railways using full three-dimensional finite element method (FEM) is shown to be computationally expensive; as a result, some investigators adopted 2.5D FEM to improve the computational efficiency [5, 6]. In the conventional FEM approaches, the discretization of soil

X. Shi et al. (Eds.): GSIC 2018, *Proceedings of GeoShanghai 2018 International Conference: Transportation Geotechnics and Pavement Engineering*, pp. 237–251, 2018.
https://doi.org/10.1007/978-981-13-0011-0_26

and structure and the truncation of the infinite domain by an artificial boundary can cause considerable simulation error. In this regard, the 2.5D FE-BE model was developed, in which the soil layers at the far ends are modeled with BEM [7, 8]. A general 2.5D FE-BE for the computation of the dynamic interaction between a layered soil and structures with a longitudinally invariant geometry was also proposed [9]. In the 2.5 D FE-BE model, the discretization is performed over an arbitrary tunnel cross-section, while solutions in the tunnel longitudinal direction are determined by Fourier transform. The computational efficiency can also be improved by incorporating the Floquet transform that accounts for periodicity in the tunnel direction [10]. In another effort, the 2.5D finite/infinite element approach was proposed, in which the wave propagation properties of the infinite far-end field are modeled by the infinite elements [11]. The finite difference method was also used as an alternative to compute the surface vibration induced by an underground train [12]. Numerical models, such as coupled FE–BE models, can account for different forms of tunnel cross-sections (e.g., rectangular tunnel) and refined modeling; however, a significant computational effort is required.

Other research efforts in this field are focused on developing tunnel models using analytical or semi-analytical approaches. The embedded Euler-Bernoulli beam can be used to study the ground vibration induced by a constant, harmonic, or a stationary random loading inside a tunnel [13]. The PiP model [14] is a three-dimensional semi-analytical model for calculating the vibration of an underground railway tunnel buried in an elastic full space. The tunnel lining was conceptualized as a thin cylindrical shell surrounded by soils with infinite radial extent. The PiP model was further enhanced by incorporating a railway track, simulating scenarios, such as two parallel circular tunnels, a double-deck circular tunnel and a layered elastic half-space [15–18]. The dynamic responses of a tunnel in a poroelastic full-space subjected to moving harmonic load and train loads were also analyzed using the improved PiP model [19]. Although the PiP model is only suitable for a tunnel with a circular cross-section, it is computationally efficient. Recently, a benchmark solution to the vibrations due to a moving point load in a tunnel embedded in an elastic half-space was also proposed [20]. However, the aforementioned research primarily focused on the soil displacement generated by the underground railways.

Several studies have examined the stress-state variation in soils generated by train movements of ground railways and the influence range of the train-load-induced stress of the ground surrounding a shield tunnel [21, 22]. Note that previous research simply treated soils as single-phase elastic or visco-elastic media. However, pore-water pressure is an important factor that should be considered in the modeling that analyzes train-induced dynamic stress in saturated soil. In addition, there is little research on modeling the grouting layer in shield tunnels. To further address the effect of pore water of soils and the grouting layer on the dynamic stress in soils, an improved tunnel model based on PiP was proposed to evaluate the dynamic stress of the saturated soils under the moving train load [23]; however, the track was not taken into account in the simulation.

In this paper, an improved 3D track-tunnel-soil model for evaluating the dynamic stress in the saturated soils is developed by incorporating the track structure into the previous tunnel-soil model [23]. Based on the proposed model, the effect of the track on the dynamic stresses in the soils generated by the moving train loads is investigated. The simulation results can provide a reference for the analysis of the stress state variation and tunnel settlement in soft soils induced by metro train loads.

2 Vehicle-Track-Tunnel-Soil Model

2.1 Track-Tunnel-Soil Model

In this section, the solution of a track-tunnel-soil system is derived. The lining and grouting layer of the shield tunnel are modeled as a Flügge double cylindrical shell of infinite length, as shown in Fig. 1. The soils are modeled as a saturated poroelastic medium using Biot's theory. The infinite rail and track slab are simplified as Euler–Bernoulli beams, and the track slabs are coupled to the tunnel via one line of uniform supports. The vehicle, rail, track, tunnel and soil medium are coupled by the compatibility of force and displacement conditions at the contact surface.

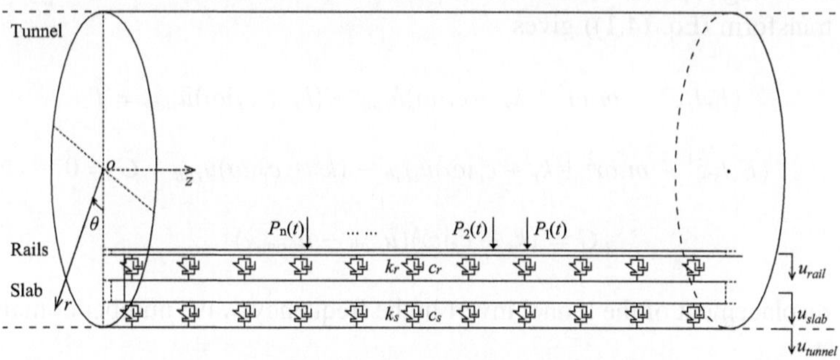

Fig. 1. Track-tunnel-soil coupled model

The governing equation of the rail beam and track slab beam subjected to an external force $f(z, t)$ is given by [24]

$$E_rI_r\frac{\partial^4 u_{rail}}{\partial z^4} + m_r\frac{\partial^2 u_{rail}}{\partial t^2} + k_r(u_{rail} - u_{slab}) + c_r(\frac{\partial u_{rail}}{\partial t} - \frac{\partial u_{slab}}{\partial t}) = f \tag{1}$$

$$E_sI_s\frac{\partial^4 u_{slab}}{\partial z^4} + m_s\frac{\partial^2 u_{slab}}{\partial t^2} - k_r(u_{rail} - u_{slab}) - c_r(\frac{\partial u_{rail}}{\partial t} - \frac{\partial u_{slab}}{\partial t}) + G = 0 \tag{2}$$

$$G = k_s(u_{slab} - u_{tunnel}) + c_s(\frac{\partial u_{rail}}{\partial t} - \frac{\partial u_{tunnel}}{\partial t}) \tag{3}$$

where E_rI_r and E_sI_s are the bending stiffness of the rail beam and track slab beam, respectively; u_{rail}, u_{slab} and u_{tunnel} are the displacements of the rail, track slab and tunnel invert, respectively; m_r and m_s are the mass of rail beam per unit length and the track slab beam per unit length, respectively; k_r and k_s are the stiffness of the rail pads and stiffness of the track slab bearings, respectively; c_r and c_s are the damping of the rail pads and the track slab bearings, respectively; and G is the interaction force between the track slab and the tunnel invert.

The coupling of the track-tunnel-soil model is solved in the frequency and wavenumber domain. The double Fourier transforms with respect to time t to frequency ω and with respect to coordinate z (along the tunnel axis) to wavenumber ξ are adopted; they are expressed as:

$$\tilde{f}(\xi,\omega) = \int_{-\infty}^{\infty}\int_{-\infty}^{\infty} f(z,t)\,\mathrm{e}^{-i(\xi z+\omega t)}\,dzdt \tag{4.1}$$

$$f(z,t) = \frac{1}{4\pi^2}\int_{-\infty}^{\infty}\int_{-\infty}^{\infty} \tilde{f}(\xi,\omega)\,\mathrm{e}^{i(\xi z+\omega t)}\,d\xi d\omega \tag{4.2}$$

where $i = \sqrt{-1}$, and the tilde "$\sim$" on the uppercase coefficients indicates that the quantities are in the frequency-wavenumber domain.

Transforming Eqs. (1–3) into the frequency and wavenumber domain by using the Fourier transform (Eq. (4.1)) gives

$$(E_r I_r \xi^4 - m_r\omega^2 + k_r + c_r\mathrm{i}\omega)\tilde{u}_{rail} - (k_r + c_r\mathrm{i}\omega)\tilde{u}_{slab} = \tilde{f} \tag{5}$$

$$(E_s I_s \xi^4 - m_s\omega^2 + k_r + c_r\mathrm{i}\omega)\tilde{u}_{slab} - (k_r + c_r\mathrm{i}\omega)\tilde{u}_{rail} + \tilde{G} = 0 \tag{6}$$

$$\tilde{G} = (k_s + c_s\mathrm{i}\omega)(\tilde{u}_{slab} - \tilde{u}_{tunnel}) \tag{7}$$

The displacement of the tunnel invert in the frequency-wave number domain can be written as

$$\tilde{u}_{tunnel} = \tilde{H}(\xi)\tilde{G} \tag{8}$$

where $\tilde{H}(\xi)$ is the frequency-response function of $H(z)$ (for the response of the uncoupled tunnel invert to a unit point load at z = 0) in the wavenumber domain.

The solution to the dynamic response of the tunnel-soil system under a unit moving harmonic point load was developed in the previous work [23], where $\tilde{H}(\xi)$ is given by

$$\tilde{H}(\xi) = \sum_{n=0}^{\infty} \tilde{U}_{rn}(r,\xi,\omega)\cos n\theta\Big|_{r=a,\theta=0} \tag{9}$$

Form Eqs. (5)–(9), the following are obtained:

$$\tilde{u}_{rail} = (\zeta_3(1+\zeta_4\tilde{H}) + \zeta_4)\tilde{f}/\zeta_5 \tag{10}$$

$$\tilde{u}_{slab} = \zeta_2(1+\zeta_4\tilde{H})\tilde{f}/\zeta_5 \tag{11}$$

$$\tilde{u}_{tunnel} = \zeta_2\zeta_4\tilde{H}\tilde{f}/\zeta_5 \tag{12}$$

$$\tilde{G} = \zeta_2\zeta_4\tilde{f}/\zeta_5 \tag{13}$$

where

$$\begin{aligned}
&\zeta_1 = E_rI_r\xi^4 - m_r\omega^2 + k_r + c_r\mathrm{i}\omega, \zeta_2 = k_r + c_r\mathrm{i}\omega \\
&\zeta_3 = E_sI_s\xi^4 - m_s\omega^2 + k_r + c_r\mathrm{i}\omega, \zeta_4 = k_s + c_s\mathrm{i}\omega \\
&\zeta_5 = \zeta_1\zeta_3 - \zeta_2^2(1 + \zeta_4\tilde{H}) + \zeta_1\zeta_4
\end{aligned} \tag{14}$$

By using the inverse Fourier transform (Eq. (4.2)) in Eqs. (10–13), the displacements of the rail, track slab and tunnel invert in the time and space domain can be obtained as

$$\begin{Bmatrix} u_{rail} \\ u_{slab} \\ u_{tunnel} \end{Bmatrix} = \frac{1}{4\pi^2}\int_{-\infty}^{+\infty}\int_{-\infty}^{+\infty}\begin{Bmatrix} \tilde{u}_{rail} \\ \tilde{u}_{slab} \\ \tilde{u}_{tunnel} \end{Bmatrix} \mathrm{e}^{\mathrm{i}\omega t}\mathrm{e}^{\mathrm{i}\xi z} d\omega d\xi \tag{15}$$

To obtain the dynamic response at some radius R elsewhere in the soils, the FRFs should be determined. The FRFs in the wavenumber domain can be written as

$$\begin{cases} \tilde{H}_r = \sum_{n=0}^{\infty} \tilde{U}_{rn}(r,\xi,\omega)\cos n\theta\big|_{r=R,\theta=\phi} \\ \tilde{H}_\theta = \sum_{n=0}^{\infty} \tilde{U}_{\theta n}(r,\xi,\omega)\sin n\theta\big|_{r=R,\theta=\phi}, \\ \tilde{H}_z = \sum_{n=0}^{\infty} \tilde{U}_{zn}(r,\xi,\omega)\cos n\theta\big|_{r=R,\theta=\phi} \end{cases} \begin{cases} \tilde{H}_{\tau_{rr}} = \sum_{n=0}^{\infty} \tilde{\tau}_{rrn}(r,\xi,\omega)\cos n\theta\big|_{r=R,\theta=\phi} \\ \tilde{H}_{\tau_{rz}} = \sum_{n=0}^{\infty} \tilde{\tau}_{rzn}(r,\xi,\omega)\cos n\theta\big|_{r=R,\theta=\phi} \\ \tilde{H}_{\tau_{zz}} = \sum_{n=0}^{\infty} \tilde{\tau}_{zzn}(r,\xi,\omega)\cos n\theta\big|_{r=R,\theta=\phi} \\ \tilde{H}_{P_f} = \sum_{n=0}^{\infty} \tilde{P}_{fn}(r,\xi,\omega)\cos n\theta\big|_{r=R,\theta=\phi} \end{cases} \tag{16}$$

Similarly, the dynamic response of the soils in the frequency-wave number domain can be written as

$$\begin{cases} \tilde{u}_r = \tilde{H}_r\tilde{R} \\ \tilde{u}_\theta = \tilde{H}_\theta\tilde{R}, \\ \tilde{u}_z = \tilde{H}_z\tilde{R} \end{cases} \begin{cases} \tilde{\tau}_{rr} = \tilde{H}_{\tau_{rr}}\tilde{R} \\ \tilde{\tau}_{rz} = \tilde{H}_{\tau_{rz}}\tilde{R} \\ \tilde{\tau}_{zz} = \tilde{H}_{\tau_{zz}}\tilde{R} \\ \tilde{P}_f = \tilde{H}_{P_f}\tilde{R} \end{cases} \tag{17}$$

Thus, the dynamic response in the soils in the time and space domain can be obtained by using the inverse Fourier transform (Eq. (4.2)):

$$\begin{Bmatrix} u_r \\ u_\theta \\ u_z \end{Bmatrix} = \frac{1}{4\pi^2} \int_{-\infty}^{+\infty} \int_{-\infty}^{+\infty} \begin{Bmatrix} \tilde{u}_r \\ \tilde{u}_\theta \\ \tilde{u}_z \end{Bmatrix} \mathrm{e}^{\mathrm{i}\omega t} \mathrm{e}^{\mathrm{i}\xi z} d\omega d\xi,$$
$$\begin{Bmatrix} \tau_{rr} \\ \tau_{rz} \\ \tau_{zz} \\ P_f \end{Bmatrix} = \frac{1}{4\pi^2} \int_{-\infty}^{+\infty} \int_{-\infty}^{+\infty} \begin{Bmatrix} \tilde{\tau}_{rr} \\ \tilde{\tau}_{rz} \\ \tilde{\tau}_{zz} \\ \tilde{P}_f \end{Bmatrix} \mathrm{e}^{\mathrm{i}\omega t} \mathrm{e}^{\mathrm{i}\xi z} d\omega d\xi \tag{18}$$

2.2 Train Load Model

If the train load F is a series of moving axle loads P_n at a constant speed v_0 (Fig. 2), then the load F can be written as

$$F = \sum_{n=1}^{N} F_n(x - vt)\mathrm{e}^{i\omega_0 t}$$
$$F_n(x - vt) = P_{n1}\delta(x - vt + \sum_{k=1}^{n-1} l_k + l_d) + P_{n2}\delta(x - vt + w_a + \sum_{k=1}^{n-1} l_k + l_d)$$
$$+ P_{n3}\delta(x - vt + w_a + w_b + \sum_{k=1}^{n-1} l_k + l_d) + P_{n4}\delta(x - vt + 2w_a + w_b + \sum_{k=1}^{n-1} l_k + l_d) \tag{19}$$

where N is the carriage number of the train, P_{nj} is the jth wheel load of nth carriage, l_k is the length of train carriage, w_a is the distance between two wheel sets of a bogie, w_b is the distance between the second wheel set and the third wheel set, and l_d is the distance between the first wheel set and the observation point.

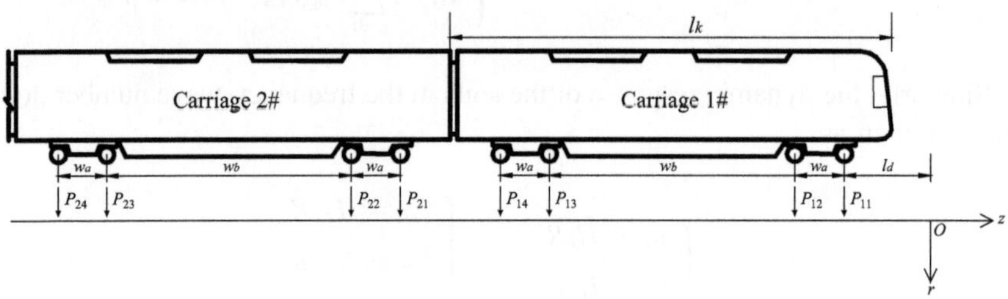

Fig. 2. Train load model

Using the Fourier transform (Eq. (4.1)), the train load in the frequency-wavenumber domain can be written as

$$\tilde{F} = 2\pi\delta(\omega + \xi v - \omega_0)Q(\xi)$$
$$Q(\xi) = \sum_{n=1}^{N}[P_{n1} + P_2 e^{i\xi w_a} + P_{n3}e^{i\xi(w_a+w_b)} + P_{n4}e^{i\xi(2w_a+w_b)}]e^{i\xi(\sum_{k=1}^{n-1} l_k + l_d)} \tag{20}$$

Subsequently, the dynamic response to the moving train load in the time-space domain can be obtained by performing the inverse Fourier transform (Eq. (4.2)).

3 Numerical Results and Discussion

3.1 Validation Against Existing PiP Model

To validate the presented model through comparing with the PiP model [15], the improved model was first degraded. Given the parameters of grouting layer $h_2 = 0$, the double shell degrades to a single shell. If the Biot's parameters $\alpha = 0$, $M = 0$, $b = 0$, $m = 0$ and $\rho_f = 0$ are assumed, then the porous medium degrades into a single-elastic medium. The other parameter values of the tunnel and the soils are shown in Tables 1 and 2. The parameter values of the track are obtained from reference [15]. After the degradation, the dynamic response in the soils at $z = 0$, $r = 30$ m, and $\theta = 120°$ under a unit stationary time-harmonic point load acting on the $z = 0$ m, $r = 2.75$ m, and $\theta = 0°$ is calculated. The calculated soil displacement is compared with those obtained by [15], as shown in Fig. 4. The results from two models are found to be very consistent (Fig. 3).

Table 1. Parameters for the tunnel lining in the track-tunnel-soil model

Parameter	Value
Modulus of elasticity of the lining layer, E_1	5×10^{10} Pa
Poisson's ratio of the lining layer, v_1	0.3
Density of the lining layer, ρ_1	2500 kg/m^3
Inner radius of the lining layer, R_0	2.75 m
Thickness of the lining layer, h_1	0.25 m
Loss-factor damping for the lining, β_t	0.03

Table 2. Parameters for the soils in the track-tunnel-soil model

Parameter	Value
Lamé constant of the bulk material, λ	1.4×10^9 Pa
Lamé constant of the bulk material, μ	191×10^6 Pa
Density of the soil grains, ρ_s	2600 $_{kg}$/m^3
Density of the fluid, ρ_f	0.0001 $_{kg}$/m^3
Porosity of the materials, n_0	0.4
Biot's parameter accounting for the compressibility of the materials, α	0.0001
Biot's parameter accounting for the compressibility of the materials, M	0.0001 Pa
Parameter that accounts for the internal friction between the solid skeleton and the pore fluid, b	0.0001 N.s/m^4
Loss-factor damping for the soil, β_s	0.06

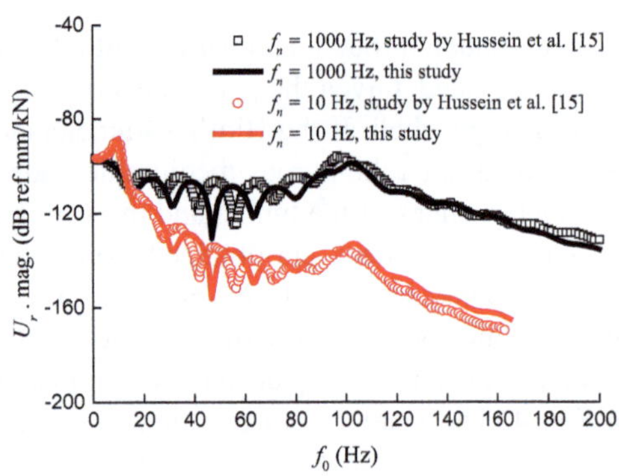

Fig. 3. Comparison between this study and the existing tunnel model [15]

3.2 Numerical Results

To investigate the effect of the railway track on the train-induced dynamic stress in the saturated soils, three scenarios are simulated using the proposed track-tunnel-soil model: (1) simulation without railway track, (2) simulation with a directly fixed slab track and (3) simulation with a floating slab track. Specifically, for the tunnel-soil model without considering the track structure, the train load is simplified as a series of moving wheel loads acting directly on the tunnel invert. For the simulation with a directly fixed slab track, the stiffness of the slab bearings is $k_s = \infty$. Tables 3, 4, 5 and 6 show the parameters for saturated soils, grouting layer, tunnel lining and railway track used in this study. The details of the parameter values of the grouting layer are discussed and documented in [25]. The parameter values of the tunnel lining and saturated soils are taken from [19]. The parameter values of the floating slab track and metro vehicle are chosen as those in [15, 23].

Table 3. Parameters for the tunnel lining and the grouting layer in the track-tunnel-soil model

Parameter	Value
Modulus of elasticity of the lining layer, E_1	5×10^{10} Pa
Poisson's ratio of the lining layer, v_1	0.3
Density of the lining layer, ρ_1	2500 kg/m^3
Inner radius of the lining layer, R_0	2.75 m
Thickness of the lining layer, h_1	0.35 m
Modulus of elasticity of the grouting layer, E_2	5.5×10^8 Pa
Poisson's ratio of the grouting layer, v_2	0.2
Density of the grouting layer, ρ_2	2500 kg/m^3
Thickness of the grouting layer, h_2	0.2 m
Loss-factor damping for the lining and the grouting layer, β_{tunnel}	0.03

Table 4. Parameters for the saturated soils in the track-tunnel-soil model

Parameter	Value
Lamé constant of the bulk material, λ	3×10^7 Pa
Lamé constant of the bulk material, μ	2×10^7 Pa
Density of the soil grains, ρ_s	2600 kg/m^3
Density of the fluid, ρ_f	1000 kg/m^3
Porosity of the materials, n_0	0.4
Biot's parameter accounting for the compressibility of the materials, α	1.0
Biot's parameter accounting for the compressibility of the materials, M	5×10^9 Pa
Parameter that accounts for the internal friction between the solid skeleton and the pore fluid, b	10^8 N.s/m^4
Loss-factor damping for the soil, β_{soil}	0.04

Figure 4 shows the dynamic stress response of the soil below the tunnel invert when subjected to moving train loads. The estimated amplitudes of the dynamic stress components (τ_{rr}, P_f, and τ_{rz}) of the soil at the position of the train load for the three scenarios are different. The pulses caused by the wheel loads are reduced because of the buffer of the track slab, which leads to a more uniform load distribution along the longitudinal direction. Compared with the model without tracks (floating slab and directly fixed slab), the dynamic stress components (τ_{rr}, P_f and τ_{rz}) simulated by modeling with tracks are lower in magnitude. The stress curve in a floating slab track scenario is smoother than that in a directly fixed slab track scenario.

Table 5. Parameters for the track in the vehicle-track-tunnel-soil model

Parameter	Value
Bending stiffness of the rail beam, E_rI_r	1×10^7 Pa.m^4
Mass of rail beam per unit length of track, m_r	100 kg/m
Stiffness of the rail pads, k_r	4×10^7 N/m^2
Damping of the rail pads, c_r	6.3×10^3 N.s/m^2
Bending stiffness of the slab, E_sI_s	1.43×10^9 Pa.m^4
Mass of slab per unit length of track, m_s	3.5×10^3 kg/m
Stiffness of the slab bearings, k_s	5×10^7 N/m^2
Damping of the slab bearings, c_s	4.18×10^4 N.s/m^2
Loss-factor damping for the rail, β_r	0.02
Loss-factor damping for the slab, β_s	0.05

Table 6. Parameters for the train in the track-tunnel-soil model

Parameter	Value
Carriage number of the metro train, N_c	6
Wheel load, P_{j}	80 kN
Distance between two wheel sets of a bogie, w_{a}	2.5 m
Distance between the second wheel set and the third wheel set, w_{b}	13.2 m
Length of the train carriage, l_{k}	22.8 m
Distance between the first wheel set and the observation point, l_{d}	0 m

The stress state for a soil element under the path of moving load ($\theta = 0°$) is further investigated in this study. As shown in Fig. 2, two soil elements, Point O under the grouting layer (r = 3.2 m, z = 0 m) and Point M below the grouting layer (r = 6.2 m, z = 0 m), are selected for the analysis. When a carriage moves from point A to point B at a speed of v = 16.7 m/s, the change in the stress state of the soil elements at O and M is computed with the improved model.

Figure 5 shows the dynamic stress with time and stress path for the soil element at Point O induced by multiple moving loads for poroelastic soils. As shown in Fig. 5(b), the stress path for the poroelastic soils is asymmetrical about the axis $\tau_{rz} = 0$. Compared with the tunnel with a directly fixed slab track, the maximum normal stress difference ($\tau_{rr} - \tau_{zz}$) and maximum shear stress level ($2\tau_{rz}$) in the soils are lower than those for the tunnel with a floating slab track. The dynamic stress state for the soil element at Point M (3 m below O) induced by multiple moving loads for poroelastic

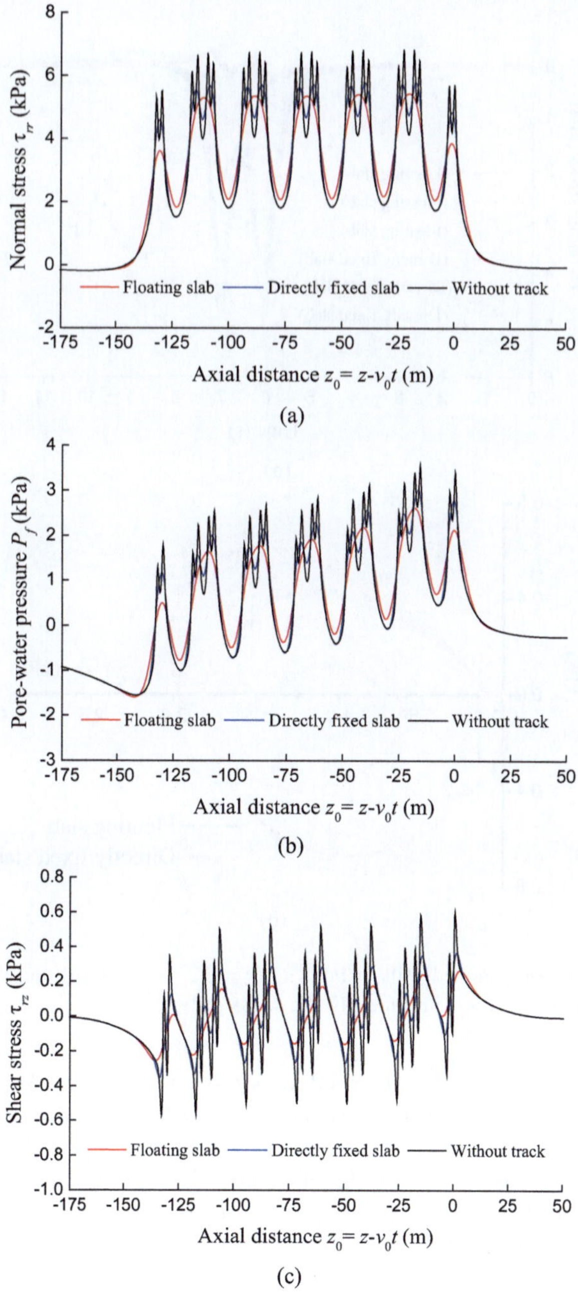

Fig. 4. Influence of track type on the dynamic stresses in the saturated soils: (a) normal stress τ_{rr}; (b) pore-water pressure P_f; (c) shear stress τ_{rz}

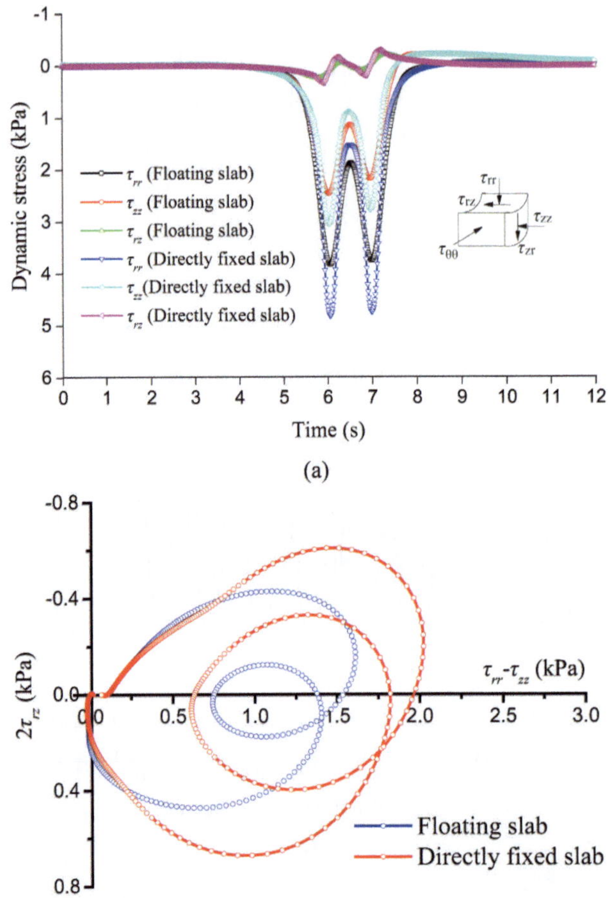

(b)

Fig. 5. Stress state of saturated soils at Point O (r = 3.2 m, z = 0 m) under multiple moving loads: (a) the dynamic stress with time; (b) the stress path

soil is further shown in Fig. 6. The resulting stress levels at Point M (in Fig. 6(a)) are lower than the stress levels at Point O (in Fig. 5(a)). Unlike the stress path for Point O in Fig. 5(b), the stress path for the soil element at the deeper location, Point M, is roughly symmetrical at axis τ_{rz} = 0 (in reference to Fig. 6(b)). From Figs. 4, 5 and 6, it can be concluded that the dynamic responses of the tunnel-soil system can be more reasonably simulated and evaluated by considering the track structure and the supports.

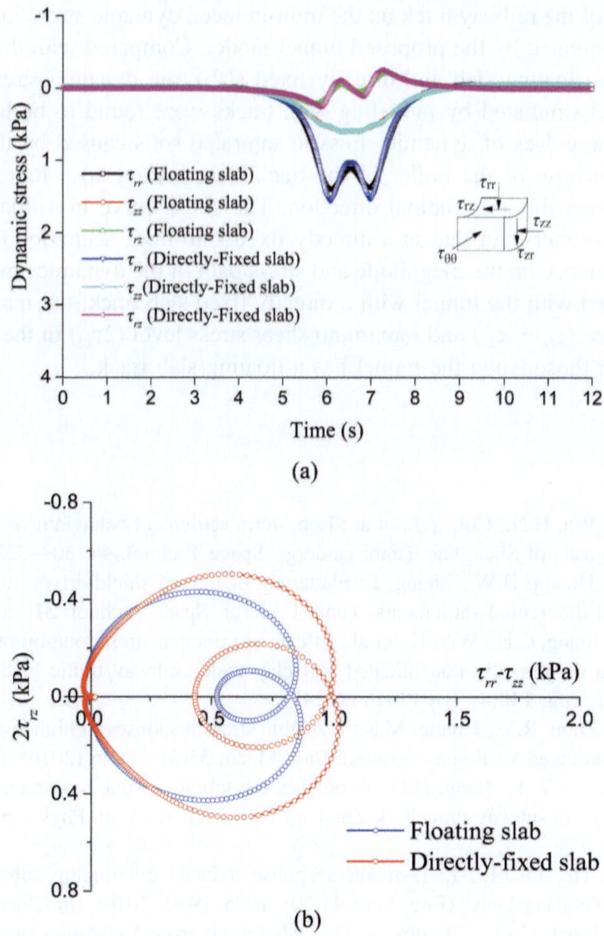

Fig. 6. Stress state of saturated soils at Point O (r = 6.2 m, z = 0 m) under multiple moving loads: (a) the dynamic stress with time; (b) the stress path

4 Conclusions

In this study, an improved tunnel model was developed to evaluate the dynamic stresses in the saturated soil. In the model, the lining and grouting layer of the shield tunnel are modeled as a Flügge double cylindrical shell. The saturated soils are modeled as a poroelastic medium using the Biot's theory. The infinite rail and track slab are simplified as Euler–Bernoulli beams, and the track slab is coupled to the tunnel via one line of uniform supports. The rail, slab, tunnel and soil medium are coupled by the compatibility of force and displacement at the contact surface. The reliability of the developed approach was validated by comparing the results obtained with the results of an existing model.

The effect of the railway track on the train-induced dynamic stress in saturated soils was then investigated by the proposed tunnel model. Compared with the tunnel model without tracks (floating slab and directly fixed slab), the dynamic stress components ($\tau_{\rm rr}$, $P_{\rm f}$ and $\tau_{\rm rz}$) simulated by modeling with tracks were found to be less in terms of magnitude. The pulses of dynamic stress in saturated soils caused by the wheel loads are reduced because of the buffer of the track slab, leading to a more uniform load distribution along the longitudinal direction. The stress curve in a floating slab track scenario is smoother than that in a directly fixed slab track scenario. The track has a significant influence on the magnitude and stress path of the dynamic stress in saturated soils. Compared with the tunnel with a directly fixed slab track, the maximum normal stress difference ($\tau_{rr} - \tau_{zz}$) and maximum shear stress level ($2\tau_{rz}$) in the saturated soils are lower than those when the tunnel has a floating slab track.

References

1. Shen, S.L., Wu, H.N., Cui, Y.J., et al.: Long-term settlement behaviour of metro tunnels in the soft deposits of Shanghai. Tunn. Undergr. Space Technol. **40**, 309–323 (2014)
2. Huang, X., Huang, H.W., Zhang, J.: Flattening of jointed shield-driven tunnel induced by longitudinal differential settlements. Tunn. Undergr. Space Technol. **31**, 20–32 (2012)
3. Xiao, J.H., Juang, C.H., Wei, K., et al.: Effects of principal stress rotation on the cumulative deformation of normally consolidated soft clay under subway traffic loading. J. Geotech. Geoenviron. Eng. **140**(4), 04013046 (2014)
4. Qian, J.G., Zhou, R.Y., Huang, M.S.: Dynamic stress responses to high-speed moving load on elastic saturated semi-space ground. Eng. Mech. **33**(3), 39–46 (2016). (in Chinese)
5. Bian, X.C., Jin, W.F., Jiang, H.G.: Ground-borne vibrations due to dynamic loadings from moving trains in subway tunnels. J. Zhejiang Univ.-Sci. A (Appl. Phys. Eng.) **13**(11), 870–876 (2012)
6. Gao, G.Y., He, J.F., Li, J.: Dynamic response induced by running subway in saturated ground. J. Zhejiang Univ. (Eng. Sci.) **44**(10), 1925–1930 (2010). (in Chinese)
7. Sheng, X., Jones, C.J.C., Thompson, D.J.: Modelling ground vibration from railways using wavenumber finite and boundary-element methods. Math. Phys. Eng. Sci. **461**, 2043–2070 (2005)
8. He, C., Zhou, S.H., Di, H.G., et al.: A 2.5-D coupled FE–BE model for the dynamic interaction between saturated soil and longitudinally invariant structures. Comput. Geotech. **82**, 211–222 (2017)
9. François, S., Schevenels, M., Galvín, P., et al.: A 2.5D coupled FE–BE methodology for the dynamic interaction between longitudinally invariant structures and a layered halfspace. Comput. Methods Appl. Mech. Eng. **199**(23–24), 1536–1548 (2010)
10. Degrande, G., Clouteau, D., Othman, R., et al.: A numerical model for ground-borne vibrations from underground railway traffic based on a periodic finite element-boundary element formulation. J. Sound Vib. **293**(3–5), 645–666 (2006)
11. Hung, H.H., Yang, Y.B.: Analysis of ground vibrations due to underground trains by 2.5D finite infinite element approach. Earthq. Eng. Eng. Vib. **9**(3), 327–335 (2010)
12. Nejati, H.R., Ahmadi, M., Hashemolhosseini, H.: Numerical analysis of ground surface vibration induced by underground train movement. Tunn. Undergr. Space Technol. **29**, 1–9 (2012)

13. Metrikine, A.V., Vrouwenvelder, A.C.W.M.: Surface ground vibration due to a moving train in a tunnel: two-dimensional model. J. Sound Vib. **234**(1), 43–66 (2000)
14. Forrest, J.A., Hunt, H.E.M.: A three-dimensional model for calculation of train-induced ground vibration. J. Sound Vib. **294**(4/5), 678–705 (2006)
15. Hussein, M.F.M., Hunt, H.E.M.: A numerical model for calculating vibration from a railway tunnel embedded in a full-space. J. Sound Vib. **305**, 401–431 (2007)
16. Hussein, M.F.M., François, S., Schevenels, M., et al.: The fictitious force method for efficient calculation of vibration from a tunnel embedded in a multi-layered half-space. J. Sound Vib. **333**, 6996–7018 (2014)
17. Kuo, K.A., Hunt, H.E.M., Hussein, M.F.M.: The effect of a twin tunnel on the propagation of ground-borne vibration from an underground railway. J. Sound Vib. **330**(25), 6203–6222 (2011)
18. Clot, A., Arcos, R., Romeu, J., et al.: Dynamic response of a double-deck circular tunnel embedded in a full-space. Tunn. Undergr. Space Technol. **59**, 146–156 (2016)
19. Zeng, C., Sun, H.L., Cai, Y.Q., et al.: Analysis of three-dimensional dynamic response of a circular lining tunnel in saturated soil to harmonic loading. Rock Soil Mech. **35**(4), 1147–1157 (2014). (in Chinese)
20. Yuan, Z.H., Boström, A., Cai, Y.Q.: Benchmark solution for vibrations from a moving point source in a tunnel embedded in a half-space. J. Sound Vib. **387**, 177–193 (2017)
21. Wang, C.J., Chen, Y.M.: Stress state variation and principal stress axes rotation of ground induced by moving loads. Chin. J. Rock Mechan. Eng. **26**(8), 1698–1704 (2007). (in Chinese)
22. Gong, Q.M., Xu, Y., Zhou, S.H.: Dynamic response analysis of tunnel foundation by vehicle vibration in metro. China Railway Sci. **26**(5), 47–50 (2005). (in Chinese)
23. Di, H.G., Zhou, S.H., He, C., et al.: Three-dimensional multilayer cylindrical tunnel model for calculating train-induced dynamic stress in saturated soils. Comput. Geotech. **80**, 333–345 (2016)
24. Yuan, Z.H., Cai, Y.Q., Zeng, C.: Dynamic response of track system and underground railway tunnel in saturated soil subjected to moving train loads. Chin. J. Rock Mech. Eng. **34**(7), 1470–1479 (2015). (in Chinese)
25. Zhang, Y., Ying, Z.Z., Xu, Y.F.: Analysis on three dimensional ground surface deformations due to shield tunnel. J. Rock Mech. Eng. **21**(3), 388–392 (2002). (in Chinese)

Three-Dimensional Shakedown Solutions for Railway Structure Under Train Loads

Yan Zhuang[1(✉)] and Kangyu Wang[2]

[1] Key Laboratory for RC and PRC Structures of Education Ministry, Southeast University, No. 2 Sipailou, Nanjing 210096, Jiangsu, China
zhuangyan4444@hotmail.com

[2] Key Laboratory of Ministry of Education for Geomechanics and Embankment Engineering, Geotechnical Research Institute, Hohai University, 1 Xikang Road, Nanjing 210098, China

Abstract. Shakedown analysis is a robust approach to solve the strength problem of a structure under cyclic or repeated loading, e.g. railway structures subject to rolling and sliding loads. A three-dimensional analytical shakedown solution is developed in this paper for the strength analysis of railway structure under multiple wheel loads, in which wheel/rail contact load is idealized as a Hertz distribution with circle contact area. In order to apply the Melan static shakedown theorem to the stability analysis of railway structures, an analytical solution of the elastic stress distribution in a single layer half-space under moving train loads is derived. Meanwhile, the residual stress distribution in the railway structure is simplified. Finally, the shakedown limit can be obtained in a direct way. Parametric study show that the shakedown limit significantly decreases with the increase of frictional coefficient from 0.0 to 1.0 and the Poisson's ratio has little effect on the shakedown limit. The obtained results give a useful reference for the engineering design of the railway structure.

Keywords: Shakedown analysis · Railway structure · Multiple Hertz loads

1 Introduction

The evaluation of railway performance requires proper assessment of the permanent deformation and fatigue under moving traffic loads. A particularly effective way of gaining insight into the combination of loads at which the structure shakedown is to evaluate the shakedown limit loads using either the incremental elastic-plastic analysis and shakedown analysis. The former is based on the calculation of the full history of stress-strain curve, and therefore involves in a very large computational effort. Shakedown analysis on the other hand is a direct approach to calculate the limit load of an elastoplastic structure through searching the critical failure point, and is based on two fundamental shakedown theorems, i.e. the lower bound theorem [1] and the upper bound theorem [2]. The shakedown limit load is used since the 1960's for the rational design of metallic contacts such as rails, roller bearings, and traction drives [3], and has been successfully used to in the field of pavement engineering and the design of pavements against excessive rutting [4–10].

X. Shi et al. (Eds.): GSIC 2018, *Proceedings of GeoShanghai 2018 International Conference: Transportation Geotechnics and Pavement Engineering*, pp. 252–261, 2018.
https://doi.org/10.1007/978-981-13-0011-0_27

The problem of railway structure deterioration is similar to that of pavement degradation, however, owing to the complexity of the railway structure, numerical techniques are required to obtain the shakedown limit load [11–14]. However, the majority of these studies focused on the rolling contact fatigue between wheels and rails. Less attention has been paid to the performance of the railway substructure, despite the fact that the performance of the components of the substructure under cyclic rail traffic loads can be critical for the cost of track maintenance.

This paper aims to develop a three-dimensional (3D) shakedown solution for the railway structure under multiple wheel loads, in which each of the wheel/rail contact load is presented as a 3D Hertz load distribution. For a homogenous single layered subgrade, the analytical elastic stress distribution under multiple Hertz loads is derived. Together with the residual stress formulation, the shakedown limit multiplier can then be found based on the Melan's lower-bound shakedown theorem. Finally, a parametric study is performed, to investigate the effect of problem parameters on the shakedown limit of the railway subgrade.

2 Static Shakedown for a Half-Space Under Multiple Hertz Loads

2.1 Problem Definition

The geometry and the loading distribution of one carriage in the plan view are presented in Fig. 1. The carriage contains 4 pairs of wheels and therefore 4 wheel-loads acts on each side of the rail. In the paper, a train considered in the model has five carriages and has the length of 125 m. Actually, for the railway structure, a series of moving wheel loads are acting on the surface of two parallel rails, and then transformed to the soil mass, which results in a more complicated load distribution in the railway substructure. In order to simplified the problem in, a homogenous single layered subgrade is assumed herein, in which the wheel loads are applied directly onto the surface of the subgrade along two parallel lines as shown in Fig. 1, ignoring the redistribution of the stress from the rails and sleepers.

The effect of rolling and sliding contact on the bearing capacity of transportation structures (i.e. pavements and railways) is a challenging task for engineers. The rolling and sliding contact in the wheel-rail interface is of particularly important, which can be simplified as two forces transferring from the wheel to the rails: total normal force P applied along the vertical direction due to the wheel rolling, and the total shear force Q applied along the travel direction due to sliding (Fig. 1). Due to its efficiency and simplicity, the Hertz contact theory which leads to an elliptical contact area and a semi-ellipsoid contact pressure distribution in the contact region, has been widely used in the rail-wheel contact problems [13, 15, 16].

Considering the contact surface of the Hertz load to be circular with radius a allows calculating the normal and shear pressure p and q as:

$$p = \frac{3P}{2\pi a^3}\left(a^2 - x^2 - y^2\right)^{1/2}$$
$$q = \frac{3Q}{2\pi a^3}\left(a^2 - x^2 - y^2\right)^{1/2} \tag{1}$$

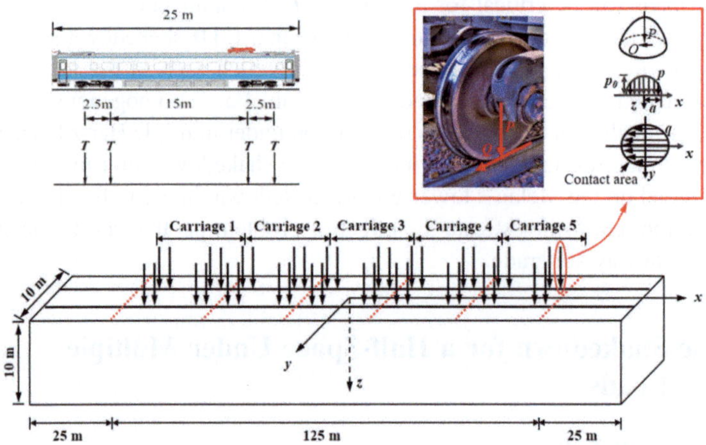

Fig. 1. Railway subgrade subjected to the axle load of a five-carriage train

The load distribution results in a maximum compressive pressure $p_0 = 3P/(2\pi a^2)$ at the center of the loaded area ($x = y = z = 0$). The rolling friction coefficient μ between wheel and rail may be assumed to be constant, with the slip between the wheel and the rail being fully developed, so that the shear force is proportional to the normal force, as:

$$Q = \mu P \tag{2}$$

2.2 Elastic Stress Field of 3D Hertz Loads

One of the key works for shakedown analysis is to get the explicit expression of the elastic stress field. Different with the pavement, the railway structure subjects to multiple wheel loads rather than a single Hertz load. The analytical solution of the elastic stress field at any point (x, y, z) in the half space subjected to only one 3D Hertz load was obtained by Hamilon [17]. For the problem considering multiple loads (the number of train loads is 40), the elastic stress fields at any point (x, y, z) can be determined as the sum of those induced by 3D Hertz loads located at the surface of the half-space (x_i, y_i, 0), in which i means the ith 3D Hertz load, and i = 1 − 40.

Assuming $x_i' = x - x_i, y_i' = y - y_i$, the elastic stress field at any point (x, y, z) due to the normal loads P is given as follows.

$$
\begin{aligned}
\left(\sigma_{xx}^{e}\right)_{p} &= \sum_{i=1}^{40} \frac{p_0}{a}\Bigg[(1+\upsilon)z\phi + \frac{1}{r^2}\Bigg\{-N\left(x_i'^2 + 2\upsilon y_i'^2\right) - \frac{Mx_i'^2 za}{S} \\
&\quad + \frac{y_i'^2 - x_i'^2}{r^2}\left[(1-\upsilon)Nz^2 - \frac{1-2\upsilon}{3}(NS + 2AN + a^3) - \upsilon Mza\right]\Bigg\}\Bigg] \\
\left(\sigma_{zz}^{e}\right)_{p} &= \sum_{i=1}^{40} \frac{p_0}{a}\left[-N + \frac{azM}{S}\right] \\
\left(\sigma_{xz}^{e}\right)_{p} &= \sum_{i=1}^{40} \frac{p_0}{a}\left[-z\left(\frac{x_i'N}{S} - \frac{x_i'zH}{G^2 + H^2}\right)\right]
\end{aligned} \tag{3}
$$

And the elastic stress field due to the tangential shear force Q is given as

$$
\begin{aligned}
\left(\sigma_{xx}^{e}\right)_{q} &= \sum_{i=1}^{40} \frac{q_0}{a}\Bigg[\frac{ax_i'M}{r^4}\Bigg\{\left(\frac{3}{2} - \frac{2x_i'^2}{r^2}\right)(S\upsilon - 2A\upsilon + z^2) + \frac{x_i'^2 z^2}{S} + \frac{7\upsilon r^2}{4} - 2\upsilon x_i'^2 + r^2\Bigg\} \\
&\quad + \frac{x_i'zN}{r^4}\Bigg\{\left(\frac{3}{2} - \frac{2x_i'^2}{r^2}\right)\left[-\frac{S}{6}(1-2\upsilon) - \frac{A}{3}(1-2\upsilon) - \frac{1}{2}(z^2 + 3a^2)\right] \\
&\quad + \frac{a^2 x_i'^2}{S} - \frac{\upsilon r^2}{4} - \frac{7r^2}{4}\Bigg\} - x_i'\left(\frac{\upsilon}{4} + 1\right)\phi + \frac{4a^3 x_i' z}{3r^4}\left(\frac{3}{2} - \frac{2x_i'^2}{r^2}\right)(1-2\upsilon)\Bigg] \\
\left(\sigma_{zz}^{e}\right)_{q} &= \sum_{i=1}^{40} \frac{q_0}{a}\left[\frac{x_i'zN}{2r^2}\left(1 - \frac{r^2 + z^2 + a^2}{S}\right)\right] \\
\left(\sigma_{xz}^{e}\right)_{q} &= \sum_{i=1}^{40} \frac{q_0}{a}\Bigg[\frac{3z\phi}{2} + \frac{azM}{r^2}\Bigg\{1 + \frac{x_i'^2}{r^2} - \frac{x_i'^2}{S}\Bigg\} + \frac{N}{r^2}\Bigg\{-\frac{3}{4}(S + 2A) + z^2 - \frac{3}{4}a^2 - \frac{1}{4}r^2 + \frac{z^2}{2}\left(\frac{1}{2} - \frac{2x_i'^2}{r^2}\right)\Bigg\}\Bigg]
\end{aligned} \tag{4}
$$

where

$$
\begin{aligned}
A &= r^2 + z^2 - a^2 & S &= \left(A^2 + 4a^2 z^2\right)^{1/2} \\
r^2 &= x^2 + y^2 & M &= \left(\frac{S+A}{2}\right)^{1/2} \\
N &= \left(\frac{S-A}{2}\right)^{1/2} & \phi &= \tan^{-1}\left(\frac{a}{M}\right) \\
G &= M^2 - N^2 + zM - zN & H &= 2MN + aM + zN
\end{aligned} \tag{5}
$$

The total elastic stresses at any point (x, y, z) in the half-space are the sum of the solutions of normal loads and shear loads.

$$
\sigma_{ij}^{e} = \left(\sigma_{ij}^{e}\right)_{p} + \left(\sigma_{ij}^{e}\right)_{q} \tag{6}
$$

2.3 Residual Stress Field Under Moving Train Load

For the railway structure under moving train loads, the residual stress field, which is different with that in pavement, may be more complicated owing to the complexity of structural geometry and load distribution. Calculation of the shakedown limit requires the establishment of a residual stress field in the half-space defined by the plane where external loads are acting viz. the stresses resulting when the magnitude of the load is sufficient to induce plastic deformations. As a result, the problem can be rationally simplified to find the most critical plane that influence the most on the shakedown limit. Provided that the half-space is homogenous and every cross-section perpendicular to the travel directions experienced the same stress history, the residual stress fields are therefore independent of the travel direction. For the 3D railway problem considered here, Yu [6] assumed that under a moving three-dimensional Hertz pressure distribution, the most critical plane is one of the xz plane defined by y = const. On these planes, the only critical non-zero residual stress that may increase the shakedown limit would be σ^r_{xx} as a function of y and z. In the y-direction, the residual stress σ^r_{yy} may well exist, as a function of z.

The total stress at any vertical plane xz (y = constant) can be expressed as follows:

$$\begin{aligned} \sigma_{xx} &= \lambda\sigma^e_{xx} + \sigma^r_{xx} \\ \sigma_{zz} &= \lambda\sigma^e_{zz} \\ \sigma_{xz} &= \lambda\sigma^e_{xz} \end{aligned} \tag{7}$$

2.4 Lower-Bound Shakedown Solution

For the soil materials, the Mohr-Coulomb yield criterion is often used. Based on the Melan shakedown theorem, the total stress of the elastic stress and the residual stress has to lie within the Mohr-Coulomb failure surface. Since σ_{yy} is an intermediate principle stress, the above requirement then leads to the following expression [6].

$$f = \left(\sigma^r_{xx} + M\right)^2 + N \leq 0 \tag{8}$$

where

$$\begin{aligned} M &= \lambda\sigma^e_{\mathrm{xx}} - \lambda\sigma^e_{zz} + 2\tan\varphi\left(c - \lambda\sigma^e_{zz}\tan\varphi\right) \\ N &= 4\left(1 + \tan^2\varphi\right)\left[\left(\lambda\sigma^e_{xz}\right)^2 - \left(c - \lambda\sigma^e_{zz}\tan\varphi\right)^2\right] \end{aligned} \tag{9}$$

In order to make the Eq. (8) satisfied at any condition, it must have

$$N \leq 0 \Rightarrow \lambda \leq \frac{c}{\left|\sigma^e_{xz}\right| + \sigma^e_{zz}\tan\varphi} \tag{10}$$

By searching for the maximum value of $\left|\sigma^e_{xz}\right| + \sigma^e_{zz}\tan\varphi_n$ through the whole half-space, it can provide the shakedown multiplier defined as λ_I.

The shakedown condition (10) is based on the assumption that the first term of the Eq. (8) is zero, and therefore it gives a maximum value to the exact lower-bound shakedown limit. The residual stress is calculated by substituting λ_I into the Eq. (8) may not fulfill the equilibrium condition. According to Eq. (8), the residual stress σ^r_{xx} at any point in the half-space must be either of the two roots of $f = 0$, i.e., $-M_i - \sqrt{-N_i}$ (smaller root) and $-M_i + \sqrt{-N_i}$ (larger root) in which N_i is always negative. For the system independent of the travel direction x, the possible residual stress σ^r_{xx} at any depth z is unique and has to lie between two critical residual stresses of $\max_x\left(-M_i-\sqrt{-N_i}\right)$ (referred to as the maximum smaller root) and $\min_x\left(-M_i+\sqrt{-N_i}\right)$ (referred to as the minimum larger root). It should be noted that at a given load parameter, two critical residual stress fields can be obtained by calculating the maximum smaller root and the minimum larger root at each point independently. For the load parameter at or within the shakedown limit, these critical residual stress fields constitute a region which contains the real residual stress field.

$$\max_x\left(-M_i-\sqrt{-N_i}\right) \le \sigma_{xx}^r \le \min_x\left(-M_i+\sqrt{-N_i}\right) \tag{11}$$

Then, the Melan shakedown theorem can be mathematically expressed for a half-space under moving traffic loads as follows.

$$\lambda_{\text{sd}} = \max \lambda \text{ s.t.} \begin{cases} \sigma_{xx}^r = \max_x\left(-M_i-\sqrt{-N_i}\right) \text{ or } \sigma_{xx}^r = \min_x\left(-M_i+\sqrt{-N_i}\right) \\ f\left(\sigma_{xx}^r\left(\lambda\sigma^e\right),\ \lambda\sigma^e\right) \le 0 \\ f\left(\sigma_{xx}^r\left(\lambda\sigma^e\right)\right) \le 0 \end{cases} \tag{12}$$

By searching the maximum load multiplier that satisfies the above formulation, the shakedown limit λ_{sd} can then be obtained.

2.5 Validation of the Shakedown Solution

In order to verify the validity of the proposed shakedown solution, this section will get some results for only single 3D Hertz load (i.e., n = 1) on the surface of the half-space. The half space was assumed as the cohesive material obeying the Mohr-Coulomb criterion with the elastic modulus E = 100 MPa, the Poisson ratio v = 0.3, the frictional angle φ = 25°, and the cohesion c = 100 kPa.

The elastic stress field in the half-space due to the single 3D Hertz load is firstly analyzed, in which the elastic stress is normalized by the maximum compressive pressure p_0, as shown in Figs. 2 and 3. As discussed above (see Eq. 10), the elastic stress components that govern the shakedown of the railway structure are the shear stress σ_{xz} and the vertical normal stress σ_{zz}. Moreover, the critical point for the shakedown lies on a vertical xz plane aligned with the direction of travel. Therefore, the elastic stresses σ_{xz} and σ_{zz} are presented along the direction of travel (x direction). As expected, the elastic stresses at the surface of the half-space are distributed within the circular contact area with the radius a, while transferring to a relatively large area as the thickness increases. The maximum normal vertical stress σ_{zz} at the surface of the half-space equals to the external stress p_0, and is independent on the friction coefficient μ, while the shear stress σ_{xz} at the surface of the half-space is a function of the external stress p_0 as formulated in Eq. 2. For the half-space at a thickness of z/a = 2, the maximum absolute elastic stresses increase with the rise in the friction coefficient due to the increase of the shear force (Eq. 10), with the critical point moves right along the travel direction (Fig. 3).

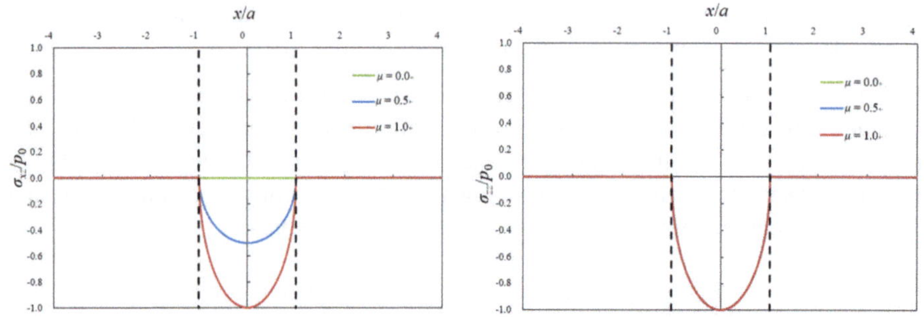

(a) Shear stress σ_{xz} (b) Normal stress σ_{zz}

Fig. 2. Elastic stresses on the half-space surface due to Hertz loads along the travel direction

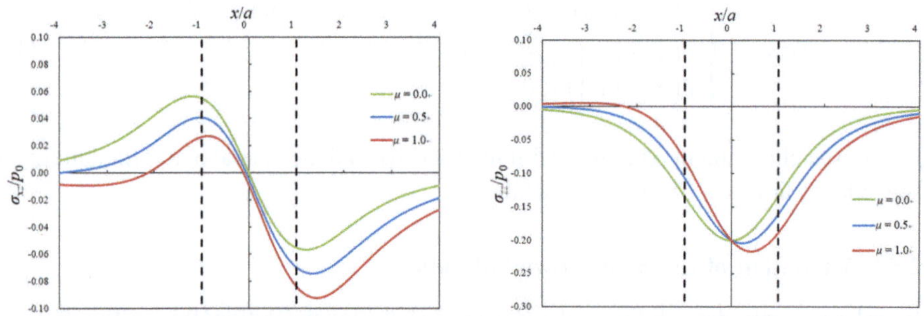

(a) Shear stress σ_{xz} (b) Normal stress σ_{zz}

Fig. 3. Elastic stresses in the thickness of z/a = 2 due to Hertz loads along the travel direction

The normalized shakedown limit $\lambda_{sd}p_0/c$ is then compared with some previous research works [18, 19], as shown in Fig. 4, where p_0 is the maximum compressive pressure of the 3D Hertz load and c is the cohesion of the material. The results of the present shakedown limit are found to be good consistence with the upper-bound shakedown limit of Ponter et al. [18] using the Hertz load, especially for the case of the frictional coefficient $\mu = 0$ and $\mu \geq 0.3$. The difference is significant for the case of $\mu = 0.2$, where the failure mechanism changes from alternating plasticity to incremental collapse. The numerical lower-bound shakedown limit of [19] for uniform load distribution is also presented in Fig. 4, which exhibits the similar trend with the present result. It shows that the numerical result of [19] is generally lower than the present shakedown solution for $\mu \leq 0.6$. The comparison in Fig. 4 indicates that the proposed shakedown solution can well capture the shakedown limit.

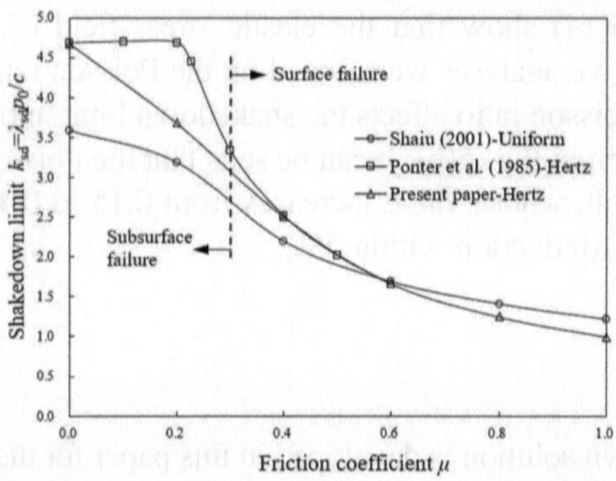

Fig. 4. Validation of the shakedown solution

3 Parametric Study

In order to find how material properties influence the shakedown limit of railway structures under moving train loads, this section will give some results with various friction angle and Poisson ratio of materials, as shown in Fig. 5. From Fig. 5(a), it can be seen that the shakedown limit increases significantly with the rise of the friction angle of materials, especially when the frictional coefficient is relatively small (i.e., $\mu < 0.4$). The increase of the shakedown limit is approximately 82% when the friction angle rises from 0° to 45° for the frictional coefficient $\mu = 0$. The influence of the friction angle is negligible for the frictional coefficient larger than 0.6. For the small values of μ, the failure of structures occurs below the surface of structures. However, for the larger values of μ, the failure of structures occurs at the surface of structures. For the failure at the surface, the shear stress is the critical stress component. Consequently, the frictional coefficient has a significant effect on the shakedown limit. With the rise of the frictional coefficient from 0.0 to 1.0, the normalized shakedown limit rapidly decreases around 90%.

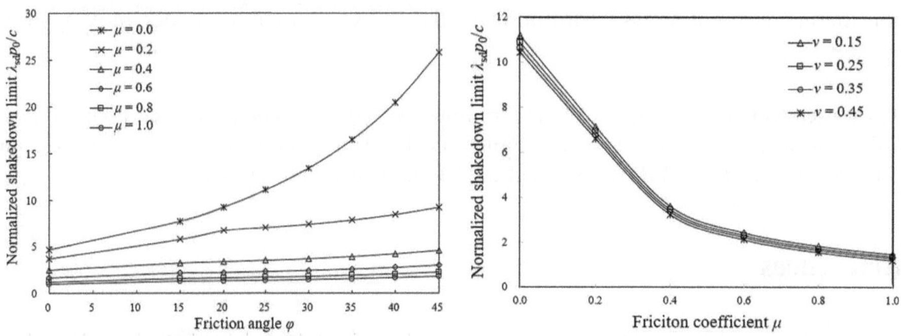

(a) Effect of friction angle

(b) Effect of Poisson's ratio

Fig. 5. Results of parametric study on normalized shakedown limits of railway subgrade

Equations (3) and (4) show that the elastic stress field σ_{xx}^{e} is a function of the Poisson ratio. The above analyses were based on the Poisson ratio $\nu = 0.3$. In order to investigate how the Poisson ratio affects the shakedown limit, more results with various Poisson ratios are given in Fig. 5(b). It can be seen that the Poisson ratio has little effect on the shakedown limit, whose value increases from 0.15 to 0.45 slightly reduces the shakedown limit load (reduction within 5%).

4 Conclusions

A analytical shakedown solution is developed in this paper for the strength analysis of a single layered railway subgrade under multiple wheel loads, in which the wheel/rail contact load is idealized as a Hertz distribution with circle contact area. The shakedown analysis is based on the Melan static shakedown theorem and then the lower bound to the shakedown limit is obtained. Through the analytical and numerical study of the stress history in a half-space under moving train loads, a self-equilibrated critical residual stress distribution is proposed. And the elastic stress distribution is analytically derived for a single layered railway structure under wheel loads. Then, the shakedown limit is obtained by means of an optimization technique. The developed method can overcome the numerical instability and too much computation of the traditional incremental elastic-plastic analysis for the shakedown problem of structures under cyclic loading. The method provides a powerful tool for the engineering design of railway structures.

The developed shakedown solution is successfully applied to the single layered railway subgrade, in which five carriages with a total of 40 wheel loads are considered. Parametric study show that the shakedown limit significantly decreases with the increase of frictional coefficient from 0.0 to 1.0 and the Poisson's ratio has little effect on the shakedown limit. It is found that the failure mechanism of railway structures moves from the subsurface to the surface with the rise of the friction angle materials. Actually, the application of the developed method in railway structures ignores the redistribution of the stresses transferring from the rails and sleepers, and is therefore a simplification situation. The developed shakedown solution can be further extended considering the real stress distribution in the railway structures under moving train loads.

Acknowledgements. The financial support from the National Natural Science Foundation of China (Grant No. 51478166), Qing Lan Project, the Fundamental Research Funds for the Central Universities (Grant No. 2015B17814 and 2016B20614) is acknowledged.

References

1. Melan, E.: Theorie Statisch Unbestimmter Tragwerke aus idealplastischem Baustoff. Sitzungsbericht der Akademie der Wissenschaften (Wien) Abt IIA **195**, 145–195 (1938)
2. Koiter, W.T.: General theorems for elastic–plastic solids. In: Sneddon, I.N., Hill, R. (eds.) Progress in Solid Mechanics, pp. 165–221. North Holland, Amsterdam (1960)

3. Johnson, K.L.: Contact Mechanics. Cambridge University Press, Cambridge (1985)
4. Sharp, R.W., Booker, J.R.: Shakedown of pavements under moving surface loads. J. Transp. Eng. **110**(1), 1–14 (1984)
5. Boulbibane, M., Weichert, D.: Application of shakedown theory to soils with non-associated flow rules. Mech. Res. Commun. **24**(5), 516–519 (1997)
6. Yu, H.S.: Three-dimensional analytical solutions for shakedown of cohesive-frictional materials under moving surface loads. Proc. R. Soc. A: Math. Phys. Eng. Sci. **461**, 1951–1964 (2005)
7. Li, H.X.: Kinematic shakedown analysis under a general yield condition with non-associated plastic flow. Int. J. Mech. Sci. **52**, 1–12 (2010)
8. Zhao, J., Sloan, S.W., Lyamin, A.V., Krabbenhøft, K.: Bounds for shakedown of cohesive-frictional materials under moving surface loads. Int. J. Solids Struct. **45**(11), 3290–3312 (2005)
9. Wang, J., Yu, H.S.: Three-dimensional shakedown solutions for anisotropic cohesive-frictional materials under moving surface loads. Int. J. Numer. Anal. Meth. Geomech. **38**(4), 331–348 (2014)
10. Zhuang, Y., Wang, K.Y.: Three-dimensional shakedown analysis of ballasted railway structures under moving surface loads with different load distributions. Soil Dyn. Earthq. Eng. **100**, 296–300 (2017)
11. Van, K.D., Maitournam, M.H.: Rolling contact in railways: modelling, simulation and damage prediction. Fatigue Fract. Eng. Mater. Struct. **26**(10), 939–948 (2003)
12. Bocciarelli, M., Cocchetti, G., Maier, G.: Shakedown analysis of train wheels by Fourier series and nonlinear programming. Eng. Struct. **26**, 455–470 (2004)
13. Taraf, M., Zahaf, E.H., Oussouaddi, O., Zeghloul, A.: Numerical analysis for predicting the rolling contact fatigue crack initiation in a railway wheel steel. Tribol. Int. **43**, 585–593 (2010)
14. Zhao, X., Li, Z.: A three-dimensional finite element solution of frictional wheel–rail rolling contact in elasto-plasticity. Proc. Inst. Mech. Eng. Part J: J. Eng. Tribol. **229**(1), 86–100 (2015)
15. Ekberg, A., Bjarnehed, H., Lundbéan, R.: A fatigue life model for general rolling contact with application to wheel/rail damage. Fatigue Fract. Eng. Mater. Struct. **18**(10), 1189–1199 (1995)
16. Conrado, E., Foletti, S., Gorla, C., Papadopoulos, I.V.: Use of multiaxial fatigue criteria and shakedown theorems in thermo-elastic rolling–sliding contact problems. Wear **270**(5), 344–354 (2011)
17. Hamilton, G.M.: Explicit equations for the stresses beneath a sliding spherical contact. Proc. Inst. Mech. Eng. Pt. C: Mech. Eng. Sci. **197**, 53–59 (1983)
18. Ponter, A.R.S., Hearle, A.D., Johnson, K.L.: Application of the kinematical shakedown theorem to rolling and sliding point contacts. J. Mech. Phys. Solids **33**(4), 339–362 (1985)
19. Shiau, S.H.: Numerical methods for shakedown analysis of pavements under moving surface loads. University of Newcastle (2001)

The Stress Distribution in Airport Runway Subgrade Subjected to Airplane Moving Loads

Fei Ruan[1] and Xilin Lü[1,2(✉)]

[1] Department of Geotechnical Engineering, Tongji University, 1239 Siping Road, Shanghai 200092, China
xilinlu@tongji.edu.cn

[2] Key Laboratory of Geotechnical and Underground Engineering of Ministry of Education, Tongji University, Shanghai 200092, China

Abstract. The calculation of stress distribution in subgrade subjected to an airplane load is crucial to the design of airport runway. Utilizing the Boussinesq's solution and integration method, this paper analyzes the stress variation in a soil element of subgrade caused by the passage of a moving airplane. The obtained results show the stress variation as the airplane moves on the ground surface; the development of the heart-shaped stress path in the deviator stress space is clearly shown. In the case of multiple soil layers ground covered with a rigid pavement, the finite element method is adopted to calculate the stress distribution. Comparing the analytical solution with numerical results, the theoretical model can capture the stress distribution in the vertical direction. Analysis indicates that the upper rigid pavement can significantly reduce the maximum stress in subgrade. Therefore, it is proven an effective way to control the settlement of subgrade by enhancing pavement.

Keywords: Airport runway · Subgrade · Moving load · Stress path Heart-shaped

1 Introduction

The prediction of settlement of soft subgrade under repeated moving traffic loading condition is a significant problem in transportation engineering (Chai and Miura 2002). The difficulty for precise calculation comes from the complicated stress path of soil during the traffic loading process (Ishihara 1983, 1996). Previous results showed that the stress path demonstrated principal stress rotations and magnitude changes in the deviator stress (Hight et al. 1983; Cai et al. 2016). Qian et al. (2012) found that larger soil deformation would be induced in a dynamic hollow cylinder test with traffic load induced stress path generated than conventional triaxial tests. The deformation subjected to a stress path with both cyclic principal stress rotation and magnitude change exceeds that subjected to a common cyclic stress (Du et al. 2016; Qian et al. 2017). Consideration of the real stress path caused by moving traffic loads is important to the calculation of the subgrade settlement and design of embankment.

Using the Boussinesq's solution, Ishihara (1996) analyzed the development of stress induced by earthquake, wave, and traffic loads; the obtained heart-shaped stress

X. Shi et al. (Eds.): GSIC 2018, *Proceedings of GeoShanghai 2018 International Conference: Transportation Geotechnics and Pavement Engineering*, pp. 262–269, 2018.
https://doi.org/10.1007/978-981-13-0011-0_28

path under traffic moving loads is distinct from that which is in the other situations. Powrie et al. (2007) indicated the complexity of stress changes in a soil element of subgrade when a train passes. By finite element analysis, the stress changes during train passage were studied, and the effects of soil element location and elastic parameters were investigated. Zhou et al. (2014) adopted a saturated poroelastic model to investigate the dynamic stress response induced by traffic moving loads, and the influence of speed and material parameters were obtained. Based on theoretical method, Lu et al. (2016) adopted the Mindlin's solution to compute the stress of subgrade when subjected to a traffic load. Wei et al. (2017) introduced amplification coefficients to account for the effect of moving speed on the wheel load induced stresses. All these studies contribute to the determination of stress distribution in a subgrade. However, a simpler solution for the problem is still required for easy usage in engineering design.

This paper studied the influence of airplane moving load on the stress state of subgrade based on the quasi-static analysis. The influence of pavement was considered in the Boussinesq's solution by using the pressure between pavement and subgrade, and the effects of speed and time effect were not considered. The stress development and generation of a heart-shaped stress path caused by the passage of an airplane are obtained by the integration of Boussinesq's solution. The effect of the contrast between the rigid pavement and soil layer on the stress distribution in subgrade is investigated by analytical and numerical methods. The comparison indicates the strong reducing effect on the stress amplitude by the rigid pavement above the subgrade.

2 The Stress State of Subgrade Under Moving Loads

To avoid the complexity of loading condition yet without the loss of geniality, the loading of airplane tires was considered as a rectangular distributed load. As shown in Fig. 1a), the ground is assumed to be a homogeneous semi-infinite elastic space. The moving traffic load p is uniformly distributed with a length of l and a width of b. The point $M(d, 0, 0)$ is the center of the load area, and the load moves from minus to plus infinity along the X-coordinate on the ground surface. On a differential area $dxdy$, we can treat $pdxdy$. as a concentrated vertical force, and then the stresses of soil element subjected to it can be obtained according to the Boussinesq's elastic solution. The calculated stresses are as follows

$$d\sigma_z = \frac{3pdxdy}{2\pi}\frac{z^3}{R^5} \tag{1}$$

$$d\sigma_x = \frac{3pdxdy}{2\pi}\left\{\frac{x^2 z}{R^5} + \frac{1-2v}{3}\left[\frac{1}{R(R+z)} - \frac{(2R+z)x^2}{R^3(R+z)^2} - \frac{z}{R^3}\right]\right\} \tag{2}$$

$$d\sigma_y = \frac{3pdxdy}{2\pi}\left\{\frac{y^2 z}{R^5} + \frac{1-2v}{3}\left[\frac{1}{R(R+z)} - \frac{(2R+z)y^2}{R^3(R+z)^2} - \frac{z}{R^3}\right]\right\} \tag{3}$$

$$d\tau_{xy} = \frac{3pdxdy}{2\pi}\left[\frac{xyz}{R^5} + \frac{1-2v}{3}\frac{(2R+z)xy}{R^3(R+z)^2}\right] \tag{4}$$

$$d\tau_{xz} = \frac{3pdxdy}{2\pi}\frac{xz^2}{R^5} \tag{5}$$

$$d\tau_{yz} = \frac{3pdxdy}{2\pi}\frac{yz^2}{R^5} \tag{6}$$

where $r = \sqrt{x^2+y^2}$, $R = \sqrt{x^2+y^2+z^2}$ and v is Poisson's ratio.

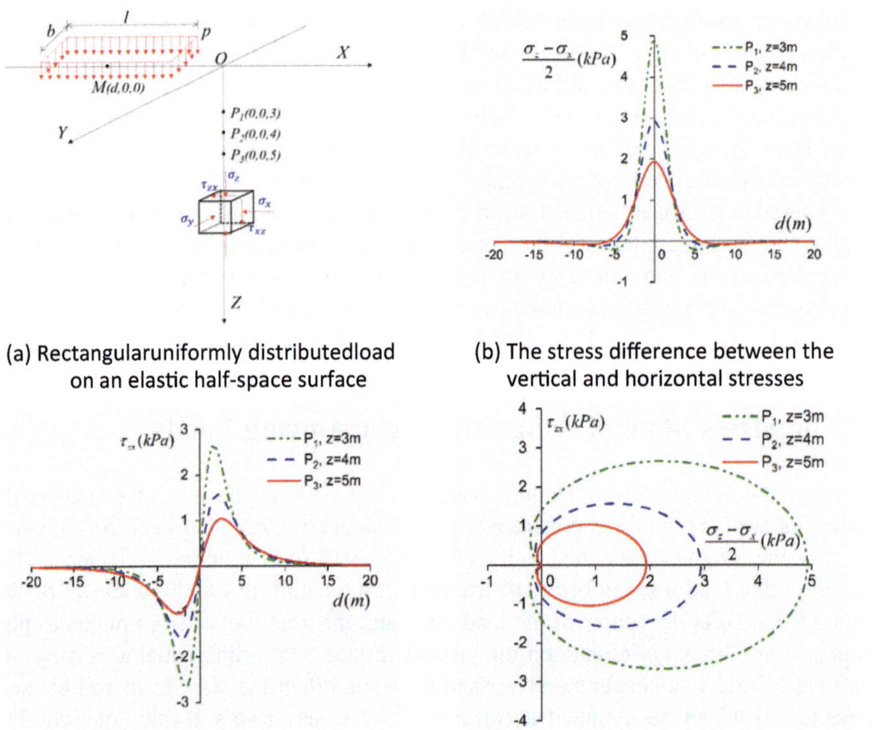

(a) Rectangularuniformly distributedload on an elastic half-space surface

(b) The stress difference between the vertical and horizontal stresses

(c) The shear stress

(d) The stress path in a deviator stress space

Fig. 1. Stresses of soil elements induced by an uniformly distributed moving load

Then, the stresses due to the uniform load can be obtained by integrating the Eqs. (1) to (6)

$$\begin{aligned}
\sigma_z &= \int_{-b/2}^{b/2}\int_{-l/2}^{l/2} d\sigma_z, \sigma_x = \int_{-b/2}^{b/2}\int_{-l/2}^{l/2} d\sigma_x, \sigma_y = \int_{-b/2}^{b/2}\int_{-l/2}^{l/2} d\sigma_y \\
\tau_{xz} &= \int_{-b/2}^{b/2}\int_{-l/2}^{l/2} d\tau_{xz}, \tau_{yz} = \int_{-b/2}^{b/2}\int_{-l/2}^{l/2} d\tau_{yz}, \tau_{xy} = \int_{-b/2}^{b/2}\int_{-l/2}^{l/2} d\tau_{xy}
\end{aligned} \tag{7}$$

Since the six components of stress are obtained, the general shear stress can be calculated by Eq. (8):

$$q = \frac{1}{\sqrt{2}}\left[\left(\sigma_x - \sigma_y\right)^2 + \left(\sigma_y - \sigma_z\right)^2 + \left(\sigma_z - \sigma_x\right)^2 + 6 \times \left(\tau_{xy}^2 + \tau_{yz}^2 + \tau_{zx}^2\right)\right]^{\frac{1}{2}} \quad (8)$$

Three points of different depths under the origin as shown in Fig. 1a) are chosen to display stress states and how they evolve as the soil element goes deeper. To express the results by numerical data, we set l, b, p and μ to be 2 m, 1 m, 100 kPa and 0.3 respectively. Calculated results demonstrated in Figs. 1 and 2 indicate how the deviator stress $(\sigma_z - \sigma_x)/2$ and the shear stress τ_{xz} change in the course of passage of a rectangular load on the ground surface, and how the stress (general deviator stress here) distributes when the load center arrives at the origin. The deviator stress increases as the loads move closer and then decreases when the loads move farther, developing sharply near the origin but gently away from the origin. Most of all, Fig. 2(d) clearly illustrates the generation of the heart-shaped profile of stress path in a deviator stress space under moving loads. The profiles of the heart-shaped path shrink when the soil depth increases, as the stresses disperse gradually in deeper subgrade. Figure 2 shows the reduction of the stress with the increasing distance from the load center and meanwhile the dispersion along the depth by comparing its two sub-graphs.

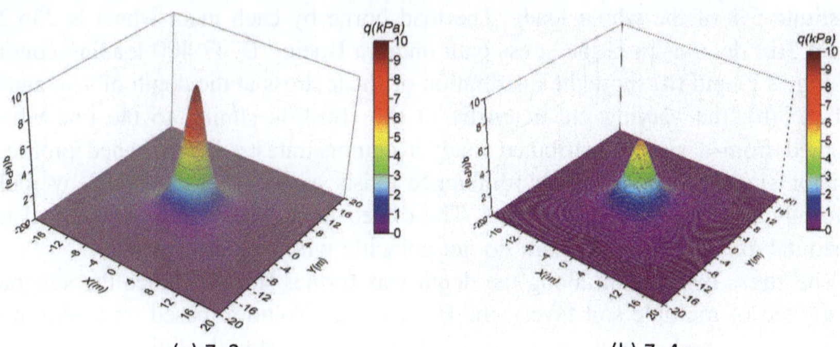

(a) Z=3m (b) Z=4m

Fig. 2. Distribution of shear stress (distributed load center locates the original point)

3 Case Study - Shanghai Pudong International Airport Runway

The stress distribution in the subgrade of the second runway in Shanghai Pudong International Airport runway is further studied. The Airport has been put into use since 1999. The second runway is 3,800 m in length and 60 m in width. Table 1 lists the material parameters for the pavement and soil layers.

The Boussinesq's solution was firstly adopted to analyze the stress distribution in the subgrade under airplane loading condition, and the typical load was selected according to the Boeing B747-400. The distribution of the tire pressure for Boeing

Table 1. Soil properties of Pudong Airport runway foundation.

Foundation		Average thickness (m)	Water content (%)	Elastic modulus (MPa)	Poisson ratio
Pavement layers	Concrete structure	0.44	–	30000	0.15
	Gravel cushion	0.56	–	5000	0.2
Soil layers	Hydraulic fills (fine sand)	2.64	–	100	0.2
	Clayey silt and silty clay	1.76	34.5	16.1	0.3
	Sandy silt	5.8	30	34.5	0.3
	Muddy clay	11.18	50.1	6.4	0.45
	Clay	8.31	37.8	11.2	0.4
	Silty clay	10.95	35.1	15.0	0.35

B747-400 is shown in Fig. 3(a). To simplify analysis, the loads were treated as four uniformly distributed rectangular loads which are plotted in dashed rectangles and transmitted by 16 wheels. The loads caused by two nose wheels were ignored, for they constitute 5% of the whole loads. The load borne by each main wheel is 236.2 kN. Figure 3(b) demonstrates the stress path under a Boeing B747-400 loading condition, and Fig. 3(c) and (d) show the distribution of shear stress at the depth of 4 m and 5 m. In Fig. 3(b), the varying characteristic of the stress is similar to the one which is obtained from a single distributed load; it demonstrates a heart-shaped profile in a deviator stress space. However, difference exists as the stress induced by multiple uniformly distributed loads overlaps. The dotted lines, which are symmetrical to the horizontal axis of the upper part, do not coincide with the lower part.

The stress distribution along the depth was further studied. Since the subgrade is constituted of multiple soil layers, the Boussinesq's solution based on a semi-infinite elastic space is no longer appropriate. Therefore, we resort to the finite element analysis to solve the problem. The finite element model, as shown in Fig. 4, is a half of the runway foundation for the symmetry between airplane loads and its moving direction. The length, width, and depth of the model are 50 m, 50 m, and 40 m, respectively. The adopted materials parameters are listed in Table 1. Airplane moving load is applied as a uniformly distributed pressure on the ground surface; constraints of three directions are imposed on the bottom and symmetric constraint on the symmetrical side. The C3D8 (8-node linear brick) element was used to generate mesh and the mesh was constituted of 69,300 elements.

To provide comparison, the stress distribution in subgrade was also analyzed by homogeneous assumption, and the adopted material parameters were weighted mean values from each layer. The obtained results are shown in Fig. 3(c) and (d). The stress at the center-line of the two neighboring rectangular loads reaches the maximum value. The soil elements under this center-line were selected to demonstrate the variation of

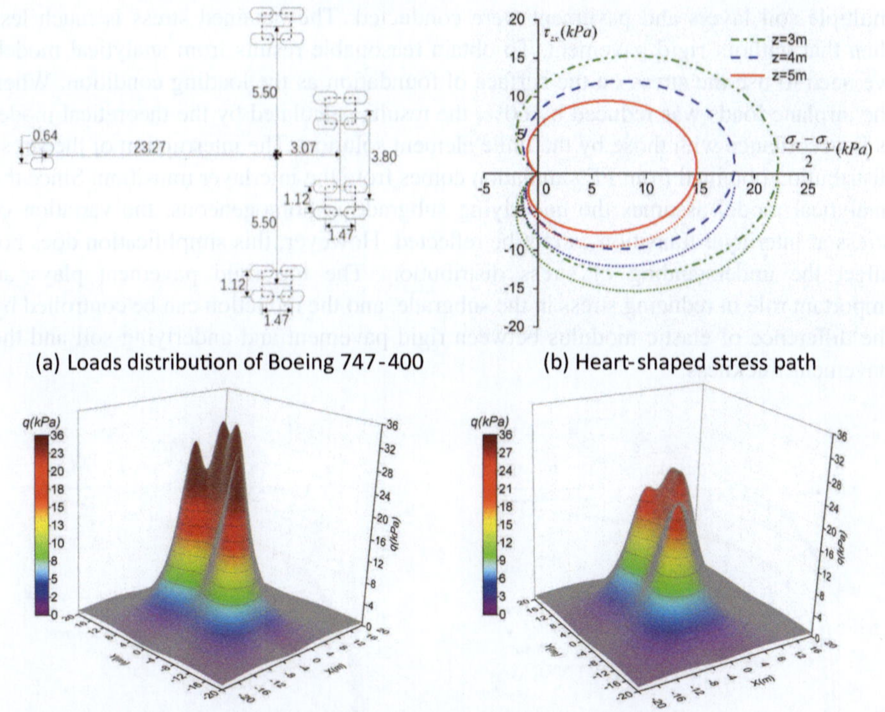

(a) Loads distribution of Boeing 747-400 (b) Heart-shaped stress path

(c) Distribution of shear stress when Z=4m (d) Distribution of shear stress when Z=5m

Fig. 3. Airplane moving loads and stress states of soil elements in subgrade at different depth

Fig. 4. Finite element mesh of the runway foundation

stress with depth. As shown in Fig. 5(a) and (b), the obtained results from FE simulation and those from analytical solution are very close. To consider the effect of concrete pavement on the stress distribution in subgrade, finite element analysis with multiple soil layers and pavement were conducted. The obtained stress is much less than that without rigid pavement. To obtain reasonable results from analytical model, we need to use the stress on the surface of foundation as the loading condition. When the airplane loads was reduced by 60%, the results calculated by the theoretical model is in accordance with those by the finite element solution. The interruption of the stress distribution obtained from FE simulation comes from the interlayer transition. Since the analytical model assumes the underlying subgrade is homogeneous, the variation of stress at interlayer transition cannot be reflected. However, this simplification does not affect the understanding of stress distribution. The top rigid pavement plays an important role in reducing stress in the subgrade, and the reduction can be controlled by the difference of elastic modulus between rigid pavement and underlying soil and the pavement thickness.

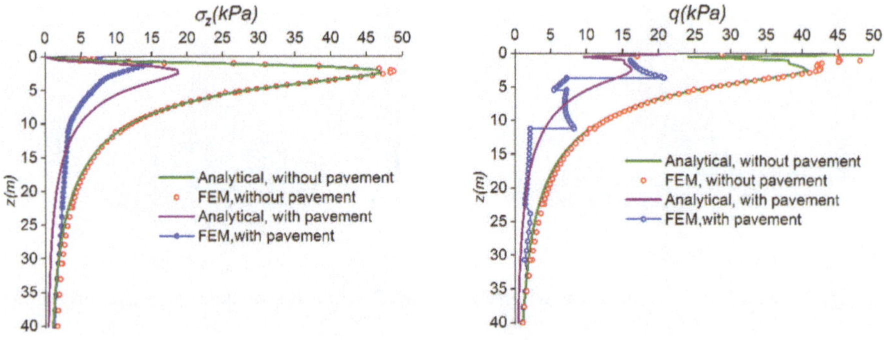

(a) Vertical stress along the depth (b) General shear stressalong the depth

Fig. 5. Stress distribution in subgrades with and without rigid pavement

4 Conclusions

Soil elements of the subgrade subjected to moving airplane loads undergo a complicated stress path. Using analytical solution and finite element simulation, this paper analyzes the stress development during the passage of moving airplane loads. The generation of the heart-shaped stress path is clearly shown. Conclusions can be made as follows.

A quasi-static method was employed in the theoretical and numerical analyses, and the airplane load was considered as a uniformly distributed quasi-static load. When the load moves from minus to plus infinity along a direction on the surface ground, stress undergoes a process from zero to a highest value and then decreases to zero again. By plotting the deviator stress versus shear stress, a stress path of heart-shaped profile is obtained. Unlike the symmetrical heart-shaped profile of stress path subjected to a rectangular moving load, the stress path under airplane moving loading condition shows an eccentric profile.

The rigid pavement on the surface reduces the stress magnitude in the subgrade, which is useful for the settlement control. The difference of the elastic modulus between pavement and underlying soil and the thickness are the most important factors of influence on the reduction of stress in the subgrade. One can infer that the settlement of subgrade can be effectively controlled by a well-designed pavement.

Acknowledgements. The financial supports by National Science Foundation of China (NSFC through grant No. U1433123) is gratefully acknowledged.

References

Cai, Y.Q., Guo, L., Jardine, R.J., Yang, Z., Wang, J.: Stress–strain response of soft clay to traffic loading. Géotechnique **67**(5), 446–451 (2016)

Chai, J.C., Miura, N.: Traffic-load-induced permanent deformation of road on soft subsoil. J. Geotech. Geoenviron. Eng. **128**(11), 907–916 (2002)

Du, Z.B., Qian, J.G., Huang, M.S.: Experimental study on deformation behavior of saturated soft clay under traffic loading considering effect of principal stress rotation. Chin. J. Rock Mechan. Eng. **35**(5), 1031–1040 (2016)

Hight, D.W., Gens, A., Symes, M.J.: The development of a new hollow cylinder apparatus for investigating the effects of principal stress rotation in soils. Géotechnique **33**(4), 355–383 (1983)

Ishihara, K.: Soil response in cyclic loading induced by earthquakes, traffic and waves. Soils Found. **32**(3), 25–26 (1983)

Ishihara, K.: Soil Behaviour in Earthquake Geotechnics. Oxford Engineering Science (1996)

Lu, X.L., Fang, H., Zhang, J.F.: A theoretical solution for long-term settlement of soft subgrade induced by traffic loading. Rock Soil Mech. **37**, 435–440 (2016)

Powrie, W., Yang, L.A., Clayton, C.R.I.: Stress changes in the ground below ballasted railway track during train passage. Proc. Inst. Mech. Eng. Part F: J. Rail Rapid Transit **221**(2), 247–262 (2007)

Qian, J.G., Guo, S.B., Huang, M.S., Zhang, J.F.: Experimental study on responses of saturated clay to traffic loading. In: Miura, S., et al. (eds.) Advances in Transportation Geotechnics 2 (2012)

Qian, J.G., Du, Z.B., Yin, Z.Y.: Cyclic degradation and non-coaxiality of soft clay subjected to pure rotation of principal stress directions. Acta Geotechnica, 1–17 (2017). https://doi.org/10.1007/s11440-017-0567-8

Wei, X., Wang, G., Wu, R.Z.: Prediction of traffic loading-induced settlement of low-embankment road on soft subsoil. Int. J. Geomech. **17**(2), 06016016 (2017)

Zhou, R.Y., Qian, J.G., Huang, M.S.: Dynamic stress responses to traffic moving loading in the saturated poroelastic ground, vol. 240, pp. 117–125. Geotechnical Special Publication, ASCE (2014)

Strength Properties and Behavior of Fresh and Abraded Railway Ballasts in Unsaturated Conditions

Andrew K. Rohrman(✉) and Carl L. Ho

University of Massachusetts, Amherst, MA 01003, USA
arohrman@umass.edu

Abstract. Railway ballast is an important component of the railroad substructure, providing resistance to forces, distribution of loads to reduce stress on the subgrade, and drainage of water away from the track. To better understand the strength properties and behavior that govern its ability to perform these tasks, laboratory testing can be performed on ballast under various conditions. In this study, 10-inch (25.4 cm) diameter triaxial specimens are prepared using an abraded ballast most closely matching AREMA #4 gradation, taken from track located in Shelburne Falls, Massachusetts. Each specimen is prepared to the same density, with variation in water content and confining pressures tested. Tests are performed in the dry condition, field capacity condition, and at 50% of field capacity. For each of these water contents, confining pressures of 5 psi (34.5 kPa), 10 psi (68.9 kPa), and 15 psi (103.4 kPa) are used. From these tests, friction angle, Poisson's ratio, and Modulus of Elasticity are determined, as well as stress-strain and volumetric strain behavior. Furthermore, these results are compared to tests performed previously at the University of Massachusetts Amherst. This series of tests were performed under the same exact conditions, but instead used a highly angular, freshly quarried Connecticut granite. Comparison of these two materials under the same conditions provides an understanding of how ballast behavior and strength properties may change over time.

Keywords: Railroad · Ballast · Triaxial · Unsaturated · Angularity

1 Introduction

Ballast is the gravel-sized granular material placed below the tracks in the railway substructure. This material is typically selected to have high angularity and shear strength, with a uniform gradation. These properties allow for the ballast to serve some of its primary functions, which are to resist applied forces and retain the position of the track, distribute loads and reduce pressure to the underlying subgrade, and provide drainage of water away from the track [1].

Triaxial testing has been used extensively in past research [2–6] to determine strength properties and strain behavior, which is important in understanding ballast performance. Specifically, triaxial testing can determine friction angle, shear strength, Poisson's ratio, and Young's modulus of granular materials. It can also be used to determine the axial strain behavior and volumetric strain behavior.

X. Shi et al. (Eds.): GSIC 2018, *Proceedings of GeoShanghai 2018 International Conference: Transportation Geotechnics and Pavement Engineering*, pp. 270–278, 2018.
https://doi.org/10.1007/978-981-13-0011-0_29

This research uses triaxial tests to investigate an abraded ballast which has been worn over time from repeated traffic loading on an actual track. These results are compared to tests previously performed at the University of Massachusetts Amherst under the same testing conditions [7]. Comparison of these results will provide an understanding of the differences in behavior between fresh ballast and a naturally abraded ballast.

2 Laboratory Testing

This study was conducted with used ballast taken from a track in Shelburne Falls, Massachusetts. It is unknown how long the ballast was in place, but it is moderately abraded from repeated traffic loading over many years. Based on visual inspection, the ballast most likely originates from a trap rock such as basalt. Grain size distribution tests indicate that the ballast likely met AREMA #4 [8] gradations at time of placement.

The previously performed investigation at the University of Massachusetts tested a highly angular Connecticut granite ballast prepared to meet AREMA #4 gradation. This material was taken directly from a quarry, and thus is representative of a fresh, newly placed ballast [7].

Water content measurements and consolidated drained triaxial (CDIC) tests were performed on the material used in this study. The following sections further detail the processes used to perform each of these tests. The processes outlined matched those from the previous investigation as closely as possible to reduce the number of variables between them. This allows for direct comparisons of the material behavior to be made.

2.1 Water Content Measurements

Tests were performed at three water contents: dry, 50% field capacity, and field capacity. The field capacity of the ballast is determined by saturating a sample of ballast and allowing it to fully drain. The water content is then immediately measured in general accordance with ASTM D2216-10 [9]. This value was then used to determine the amount of water added to the triaxial specimens. After test completion, water content measurements were taken using the entire triaxial specimens. This ensured that all tests met the desired water contents.

2.2 Triaxial Testing Methods

Drained triaxial compression tests (CIDC) were performed in general accordance with ASTM D7181-11 [9]. The triaxial cell at the University of Massachusetts Amherst accommodates 10-inch (254 mm) diameter specimens. Each specimen is prepared by compacting the ballast in eight 2.5-inch (38.1 mm) layers to create a 20-inch (508 mm) tall specimen of approximately uniform density. Figure 1 presents an example of a completed specimen within the cell. When water is required for a test, it is added from the top and 16 h are allowed to pass to encourage full permeation of the water into the soil.

The triaxial tests were run with no water (dry), at 50% field capacity (50FC), and at field capacity (FC). For each water content, three tests are performed, each at a different

Fig. 1. 10-inch (255 mm) diameter specimen triaxial setup

confining pressure so that Mohr-Coulomb failure properties can be determined. Since ballast is only confined by its self-weight from the track shoulders, low confining pressures of 5 psi (34.5 kPa), 10 psi (68.9 kPa), and 15 psi (103.4 kPa) are used to be representative of the field case.

The triaxial specimens are loaded using an Instron actuator on a self-reacting frame. A constant strain rate of 0.04 in/min is applied to the specimen. The load and axial strain are measured by the actuator. Conventional volumetric strain measurements cannot be used since the specimens are not saturated. Instead, circumferential string potentiometers are placed at the center and quarter sections of the specimen. The measured radial changes are then used to estimate the volumetric strain using the Stain Invariant Method.

3 Results and Discussion

3.1 Friction Angle and Ultimate Drained Strength

The ultimate drained strength (s_d) of each test is determined by finding the maximum point reached on the stress path. By grouping three stress paths at different confining pressures but at the same water content, the Mohr-Coulomb friction angle (φ') can be determined for the material at each water content. Table 1 presents the determined values for the ultimate drained strength and friction angle for the two ballast materials. The ultimate strength values are also presented graphically in Fig. 2 so that trends based on confining pressure and water content can be more easily identified.

These data show that the ultimate drained strength of the abraded ballast is slightly greater than the Connecticut Granite in every case except for the 10 and 15 psi tests in the dry condition. Similarly, the friction angles for the abraded ballast are higher than the granite except for in the dry condition.

Table 1. Drained shear strength and friction angle values determined for Connecticut granite and a naturally abraded ballast using triaxial tests

Water condition	Confining pressure	Conn. granite		Abraded	
		φ'	s_d	φ'	s_d
	psi	degrees	kPa	degrees	kPa
Dry	5	47.4	121.4	46.5	129.3
	10		209.0		184.5
	15		264.1		252.8
50FC	5	45.1	119.5	47.7	122.2
	10		167.0		196.1
	15		236.2		282.7
FC	5	44.7	104.7	47.4	114.7
	10		183.1		201.9
	15		218.9		273.2

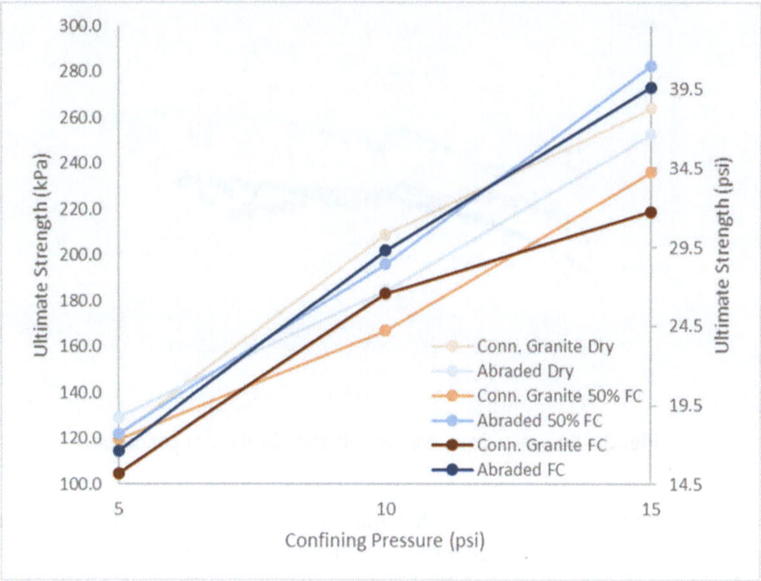

Fig. 2. Ultimate strength recorded for each test, showing trends with confining pressure and moisture content

These results are significant for several reasons. Generally, ballast is selected for high angularity so that there is more interlock, which should result in greater strengths. However, the abraded ballast which has a lower overall angularity shows higher strengths in most cases. This seems to indicate that the slightly broader gradation of the abraded ballast is a greater contributor to strength than the angularity. It is also possible that these differences occur because the ballast come from different parent rock, granite and basalt, despite being prepared to the same specifications.

Regarding friction angle, the determined values for the Connecticut granite decreases with water content as might be expected. The abraded ballast, alternatively, exhibits an increase in friction angle at 50% field capacity followed by a slight decrease. The results overall seem to indicate that the strength increases with increased water content for the abraded ballast, which is not typically expected.

3.2 Strain Behavior

This section will examine both the stress-strain behavior and the volumetric strain behavior. Figure 3 presents the stress-strain curves for all tests in the dry condition as an example of the behavior typically seen across all tests. For ease of interpretation, the abraded ballast tests are shown in shades of blue and the Connecticut granite tests are in shades of orange.

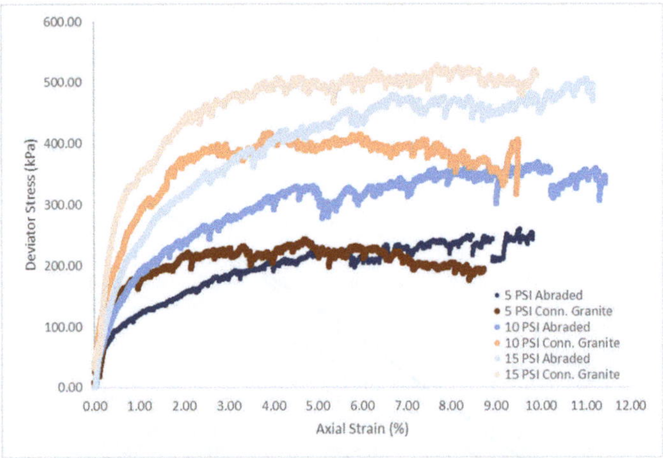

Fig. 3. Stress-strain curves for all tests in the dry condition.

It is clear from this data that each ballast material behaves differently. For each confining pressure, the Connecticut granite achieves much higher stresses at lower axial strains. Upon reaching approximately 2% strain, the Connecticut granite has reached a point where the curve flattens out, with no considerable gain in strength as the test continues. The abraded ballast, in contrast, is much slower to build strength. The stress-strain curve is more gradual, with the curves beginning to flatten out closer to 6% axial strain. It is beyond this point where both materials behave the same, approximately maintaining the same stress until the end of the test. These stress-strain behaviors are also observed in each of the tests performed at the 50% field capacity and field capacity conditions.

The specimens tested are not saturated and therefore the volumetric strain cannot be directly measured. Therefore, string potentiometers are used to measure radial changes throughout the test and the Strain Invariant method is used to estimate the volume.

Figure 4 presents the volumetric strain curves determined by this method for the tests performed in the dry condition and in the field capacity condition. Again, the abraded ballast tests are shown in shades of blue and the Connecticut granite tests are in shades of orange. Note that in this analysis, geotechnical engineering conventions are used rather than solid mechanics convention, meaning that positive strains indicate contraction, while negative strains indicate dilation.

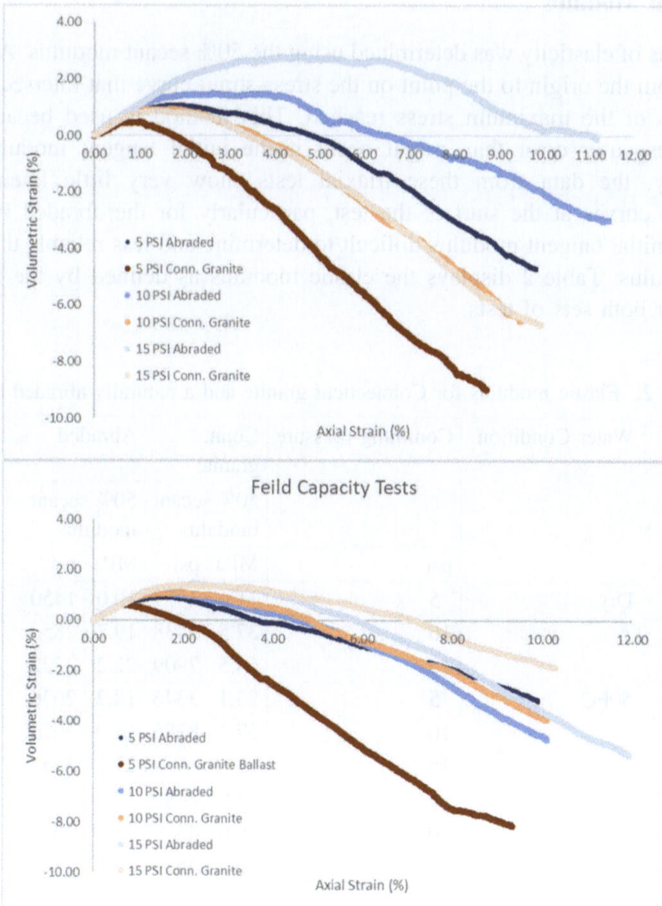

Fig. 4. Volumetric strain curves for all Conn. granite and abraded ballast tests in the dry condition and field capacity condition.

These data indicate two distinct behaviors from the two materials. In the dry condition, the abraded ballast contracts significantly before dilation begins, and the extent of this contraction increases with confining pressure. The Connecticut granite contracts to a much lesser extent and is less influenced by the confining pressure. Additionally, the Connecticut granite dilates more than the abraded ballast does.

The introduction of water, however, changes this behavior entirely. At field capacity, the Connecticut granite shows higher contraction and less dilation as confining pressures increase. The abraded ballast, in contrast, exhibits nearly the same volumetric behavior during each test, regardless of confining pressure. The tests performed at 50% of field capacity exhibit volumetric strain behavior that falls somewhere between the two plots above, but it is more similar to the field capacity tests.

3.3 Secant Modulus

The modulus of elasticity was determined using the 50% secant modulus. A secant line is drawn from the origin to the point on the stress-strain curve that intersects the value that is 50% of the maximum stress reached. This method is used because there is generally less user error than might occur if the initial tangent modulus is used. Additionally, the data from these triaxial tests show very little linearity in the stress-strain curves at the start of the test, particularly for the abraded ballast. This makes the initial tangent modulus difficult to determine and less reliable than the 50% secant modulus. Table 2 displays the elastic modulus as defined by the 50% secant modulus for both sets of tests.

Table 2. Elastic modulus for Connecticut granite and a naturally abraded ballast

Water Condition	Confining pressure	Conn. granite		Abraded	
		50% secant modulus		50% secant modulus	
	psi	MPa	psi	MPa	psi
Dry	5	31.6	4579	10.0	1450
	10	37.5	5438	19.7	2852
	15	54.5	7909	22.2	3222
50FC	5	23.1	3346	18.2	2636
	10	57.1	8286	14.0	2023
	15	–	–	27.3	3955
FC	5	24.0	3480	10.7	1554
	10	30.0	4350	17.6	2559
	15	33.3	4833	24.0	3480

Is clear that the Connecticut Granite gives consistently higher modulus values than the abraded ballast. This indicates that more angular ballast is stiffer than the abraded ballast. This behavior conforms with the volumetric strain curves, which show much lower levels of contraction for the angular ballast. Together, these sets of data indicate that the highly angular Connecticut granite can bear higher applied loads with less material deformation.

These data also indicate that the secant modulus, in most cases, increases with increasing confining pressure. This result is in agreement with previous investigations performed on other ballast materials. The effect of water content on the modulus is less clear, with scattered results given from this series of tests. The secant modulus results are plotted in Fig. 5 to show this scatter.

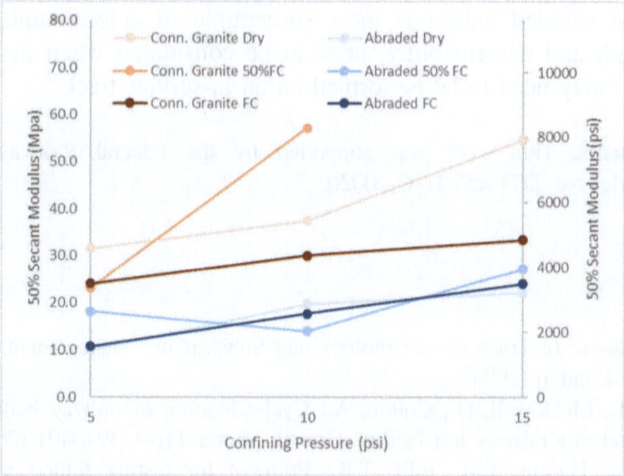

Fig. 5. 50% Secant modulus of all tests on Connecticut granite and abraded ballasts

It might be expected that the addition of water would decrease the stiffness of the material, resulting in a lower modulus. This holds true for the Connecticut granite in the dry and field capacity states, however the half field capacity condition produces scattered results that do not conform to this behavior. The tests on the abraded ballast show that the field capacity and dry tests produce nearly the same results for each confining pressure. Like the Connecticut granite, the half of field capacity tests produce behavior inconsistent with the other tests.

4 Conclusions

Ballasted railway track will become abraded over time due to repeated traffic loading, and this change in the particle shape can change the overall behavior and characteristics of the material. The results of this study show that for most cases tested, a naturally abraded ballast achieves higher strengths and generally higher friction angles than a fresh ballast. This indicates that either the broader gradation of this material is a greater contribution to the strength than the angularity of the particles, or that the differing rock types result in different strengths even though the specimens are prepared in the same manner and to the same density and void ratio. However, the highly angular Connecticut granite achieves a higher modulus of elasticity for each case. The volumetric strain behavior of the two materials is also distinctly different, with the Connecticut granite exhibiting less initial contraction than the abraded ballast in the dry condition. However, in the partially saturated conditions tested, both materials have similar volumetric strain behaviors. The results from this investigation show that the biggest difference in the behavior of these two materials is their deformability when under load. While a naturally abraded ballast may achieve higher strengths than the angular ballast, the additional strength is not very significant. Instead, the volumetric strain behavior and secant modulus results together indicate a significant difference in behavior that shows that the abraded ballast is more susceptible to deformation. Both of these aspects, strength and deformability, need to be considered when determining when ballast renewal may need to be performed on an in-service track.

Acknowledgements. This work was supported by the Federal Railway Administration (FRA) under order No. DTFR53-16-C-00020.

References

1. Selig, E., Waters, J.: Track Geotechnology and Substructure Management. Thomas Telford Publications, London (1994)
2. Aursudkij, B., McDowell, G., Collop, A.: Cyclic loading of railway ballast under triaxial conditions and in a railway test facility. Granul. Matter **11**(6), 391–401 (2009)
3. Ebrahimi, A., Tinjum, J.M., Edil, T.B.: Protocol for testing fouled railway ballast in large-scale cyclic triaxial equipment. Geotech. Test. J. **35**(5), 796–804 (2012)
4. Holtz, W.G., Gibbs, H.J.: Triaxial shear tests on pervious gravelly soils. J. Soil Mech. Found. Div. **82**, 1–22 (1956). American Society of Civil Engineers – Proceedings
5. Indraratna, B., Ionescu, D., Christie, H.D.: Shear behavior of railway ballast based on large-scale triaxial tests. J. Geotech. Geoenviron. Eng. **124**(5), 439–449 (1998)
6. Suiker, A.S., Selig, E.T., Frenkel, R.: Static and cyclic triaxial testing of ballast and subballast. J. Geotech. Geoenviron. Eng. **131**(6), 771–782 (2005)
7. Kashani, H.F.: Evaluating the influence of breakdown fouling and moisture content on the mechanical and electromagnetic properties of ballasted railroad track. Ph.D. dissertation, Department of Civil and Environmental Engineering, University of Massachusetts Amherst (2016)
8. American Railway Engineering and Maintenance-of-Way Association, Manual for Railway Engineering, Lanham, MD
9. American Society for Testing and Materials, Annual Book of Standards. Philadelphia, PA

Blasting Vibration Study of a Deep Buried Tunnel Entrance

Jiang-Bo Xu(✉), Chang-Gen Yan, and Hang Yuan

School of Highway, Chang'an University, Xi'an 710064, Shaanxi, China
xujiangbo@yeah.net

Abstract. Taking the Luo-Sha tunnel as the research site, we monitored the blasting vibration in a tunnel portal and performed time-frequency analysis on the monitoring data. During the time domain analysis, the maximum horizontal radial component of the blasting vibration velocity is greater than the maximum horizontal tangential component, and the maximum horizontal tangential component is greater than the maximum vertical component; this result is the opposite of the result that we obtained from the blasting vibration research on a phyllite tunnel portal. With the distance increasing, especially in the tunnel portal location, the blasting vibration velocity shows a significant increase, and the slope surface of the tunnel portal shows significant vibration amplification. In the frequency domain analysis, as the distance increased, the high frequency region in the tunnel portal was magnified, whereas the low frequency region gradually weakened. In the hard rock tunnel, the high frequency region is obviously increased, whereas the low frequency region is reduced; this behaviour was the opposite of that of a soft rock tunnel.

Keywords: Luo-Sha tunnel · Time domain analysis
Frequency domain analysis

1 Introduction

With the proposed "Belt and Road" blueprint, our country's infrastructure construction will certainly experience a rapid development period; such development will involve the construction of many tunnels. However, during the tunnel excavation process, blasting construction is inevitable; such blasting will cause considerable damage to the surrounding rock and lining. In certain special parts, especially in the tunnel's entrance and exit, the damage to the surrounding rock is not only related to the construction progress but also to the tunnel's post-construction maintenance.

Both domestic and foreign scholars have studied the effect of blasting on tunnels. Blair (2015) studied the relationship between free face and blasting vibration in theory and verified their results using dynamic finite element analysis. Based on the Jinping-II hydropower project, Yang et al. (2014) studied the safety distance between the secondary lining and the blasting tunnel face and evaluated the different stages of second lining concrete using numerical simulation. Xia et al. (2014) studied the influence of steps and ravines' on blasting seismic waves using both theoretical analysis and field tests. Zhang et al. (2014) studied the attenuation result of blasting vibration in an open

X. Shi et al. (Eds.): GSIC 2018, *Proceedings of GeoShanghai 2018 International Conference: Transportation Geotechnics and Pavement Engineering*, pp. 279–288, 2018.
https://doi.org/10.1007/978-981-13-0011-0_30

bench using dynamic finite element software and presented a method to protect the constructions around blasting areas in two aspects: peak velocity and energy attenuation. Blair (2014) studied the relationship between blasting vibration and the charge length, the detonation velocity, and the layered medium. Segarra et al. (2015) studied the characteristics of certain coupling methods under blasting vibration using field tests. Gao et al. (2011) conducted research on tunnel blasting excavation and surface vibration monitoring, taking the exit section excavation of the Huolangyu tunnel project in the Mi-xing road as the research object. According to the monitoring data of the blasting vibration in a shallow-buried tunnel during construction, Lin et al. (2006) obtained the variation result of the ground vibration velocity in Tai-Shan granite via fitting and analysed the vibration frequency spectrum and energy distribution characteristics of blasting vibration in a shallow-buried tunnel using the method of wavelet analysis. Xu et al. (2007) studied the amplification effect of blasting vibration during the excavation of a tunnel via a field test in the Ren-he-chang shallow-buried tunnel blasting excavation in the Yuhuai Railway. Yun et al. (2012) studied the amplification factors of blasting vibration velocity in the entrance of New Pagoda Mountain tunnel and checked the stability of the building and slope at the entrance of tunnel. Huang and Wu (2012), Huang et al. (2014) conducted a series of research studies on the tunnel portal with focus on the eccentric compression and slope stability problems in the tunnel portal. Nateghi (2011) predicted the blasting's influence level on the surrounding concrete structure using the monitoring data of site blasting. Mohamad et al. (2012) compared the ground vibrations predicted from an empirical formula and an analytical programme with the real data and observed that the ANN method produced more accurate prediction than the empirical formula. Yan et al. (2014) studied the stress-strain relationship of soil mass to calculate the safety factor of a slope in natural conditions, as well as with blasting vibration, to determine the most endangered sliding surface and summarized the varying patterns of slope stability under the influence of blasting vibration. Li et al. (2015) studied the adverse effects caused by blasting during the construction of highway tunnels. Jian-min et al. (2016) performed experiments on blasting vibration at the Zuangou Limestone open-pit mine to investigate the influence of the rock structure planes on the particle peak velocity, the frequency characteristics and the energy distribution of blasting vibration signals.

The above-described research considered the different entrances and exits of a tunnel from different perspectives; however, these studied did not consider the blasting vibration effect on a deep tunnel (especially frequency spectrum and amplitude analysis of the blasting vibration). Based on this background, this study involves the blasting vibration test in a deep buried tunnel portal and analysis of the blasting vibration data.

2 Overview of the Project

Luo-Sha tunnel is located between Bashan Brook and Scissors River in Dangchang County. The geological map is shown in Fig. 1. The uniaxial compression strength is 450–500 kPa. The geological details are as follows.

Fig. 1. Topographic map of the Luo-Sha tunnel

2.1 Formation Lithology

The tunnel is primarily glutenite and mudstone. The bed rock in the entrance of the tunnel is bare, and the exit end of the tunnel, the slope and the top of the slope are covered by slope deposit fine (coarse) breccia, proluvial coarse breccia and gravelly soil of Quaternary Holocene and eolian deposit sandy loess of Late Pleistocene. The surface is covered by landslide deposit fine (coarse) breccia.

2.2 Geological Structure

The geo-tectonics of construction site belongs to the Qinghai-Xizang eta-type structure. Affected by tectonism, the folds and faults are well-developed, and the geological structure is very complicated.

Affected by regional tectonics, the folds are well-developed, and the structure is relatively simple within the scope of the construction site.

The maximum depth of the tunnel body is approximately 696 m with a minimum ground elevation of 1850 m and a maximum elevation of 2610 m. The tunnel has country roads connected with the 212 State Road. The mileage of the tunnel between the origin and the terminal is 8001 m from DK259(km) + 843(m) to DK267 + 844. The route in the tunnel has downhill grades of approximately 12.8‰, 13‰ and 12.8‰ from the entrance.

The primary stratums through which the tunnel passes are tertiary conglomerate sandstone, conglomerate sandstone with mudstone, mudstone with conglomerate and sandstone with mudstone. The entrance, gully, slope, and top of the slope are covered by the quaternary Holocene diluvial silty clay, fine breccia soil, landslide accumulation of gravel soil and diluvial gravel soil that is distributed on the entrance of the inclined shaft. The IV level of the surrounding rock of the tunnel is 7004 m, which represents 87.5% and the V level of the surrounding rock is 997 m, which represents 12.5%. The tectonic structure of the Luo-Sha tunnel belongs to the Qinghai-Tibet eta-type tectonic system, which was affected by tectonism; thus, the fold fracture is well-developed, and the geological structure is highly complicated.

The mileage of the entrance of Luo-Sha tunnel is DK259 + 846, and the mileage of blasting tunnel face is DK267 + 843. When the sensors were laid, monitoring point 5 was at the entrance of the Luo-Sha tunnel, and monitoring point 4, monitoring point 3, monitoring point 2 and monitoring point 1 were set successively from the entrance to the blasting tunnel face in steps of 90 m. The mileage marker of each point is shown in Table 1.

Table 1. Mileage markers of the monitoring points

Number of monitoring points	5	4	3	2	1
Mileage	DK259 + 846	DK259 + 936	DK260 + 026	DK260 + 116	DK260 + 206

3 Monitoring Data Analysis

We could analyse the blasting earthquake wave's characteristic, propagation result and its influence on soft rock tunnel via the blasting vibration test, thereby effectively controlling the blasting work to prevent or reduce the destruction of the deep soft rock tunnel. The essence of the blasting vibration test is to monitor the vibration rule (the time-displacement/velocity/acceleration curves) of the medium point to obtain the fundamental characteristics of deep soft rock tunnel during blasting vibration via analysis of the waveform graph. The field monitoring system uses an acceleration sensor to translate the vibration parameter into a current signal and later translates the current signal into a voltage signal via a signal conditioner. Next, vibration data are collected by the acquisition system. Finally, the vibration signal is analysed using the analysis software in a computer.

3.1 Time Domain Analysis of Blasting Vibration

The vibration of blasting earthquake wave is a type of random vibration, and its vibration signal is formed by the overlaying of simple harmonic waves with different frequencies. Time domain analysis is primarily used to study the vibration velocity and acceleration during the process of vibration. To study the surrounding rock rupture during blasting process, we must analyse the whole vibration records that had been measured to determine the nature of the surrounding rock in the tunnel.

The duration of blasting vibration and the vibration's propagation velocity in the surrounding rock are important parameters to study blasting vibration. The longer the blasting vibration continues, the greater harm to the structures would be. The study of the propagation velocity of blasting vibration in surrounding rock contributes to analysis of the circumstance of the surrounding rock already excavated, quantifying the ranks of it and analysing the broken circumstance of the front surrounding rock thereby predicting it by using engineering analogy method.

The first column in the Table 2 is the field monitor date, the second column is the arrival time on the initial wave reaching the 1 sensor, the third column is the arrival

Table 2. Statistical table of blasting earthquake velocity in Luo-Sha tunnel

Date (2013)	Time of the wave's arriving at the 1 sensor (ms)	Time of the wave's arriving at the 5 sensor (ms)	Time lag between the 1 sensor and 5 sensor (ms)	Blasting wave propagation velocity (m/s)
August 7	1678	1768	70.00	5142.86
August 8	287	325	58.00	6206.90
August 9	1311	1383	62.00	5806.45
August 10	1069	1140	71.00	5070.42
August 11	702	768	66.00	5454.55
August 12	811	874	63.00	5714.29
August 13	1054	1134	70.00	5142.86
August 16	299	372	73.00	4931.51
August 24	666	724	58.00	6206.90
August 28	2074	2158	84.00	4285.71
August 29	2011	2050	69.00	5217.39
August 30	1378	1447	69.00	5217.39
September 1	2022	2061	59.00	6101.69

time on the initial wave reaching the 5 sensor, the 4th column is time difference between the 1 sensor and 5 sensor, the 5th column is the vibration propagation velocity; the distance between the 1 sensor and 5 sensor divided by the time difference is the propagation velocity, which is an important index for the surrounding rock in front of the tunnel face.

Through the weighted average of the blasting propagation velocities in the last column of the above table, we can determine the propagation velocity of blasting earthquake wave in Luo-Sha tunnel is 5422 m/s. The quantization of the propagation velocity of blasting vibration has significance for us to know the front surrounding rock and has reference value on guiding construction.

3.2 Blasting Vibration Frequency Spectral Analysis

The blasting earthquake wave can break up into waveforms with different frequency spectrums. The contents of waveforms with various frequencies have large differences, and we can transfer the waveforms into energies to denote it via spectral analysis, thereby confirming the dominant frequency and the frequency band.

(1) **The basic theory of spectral analysis.** Any periodic waveform can be denoted by being transferred into Fourier series; thus, the blasting vibration can also be transferred as a combination of many simple harmonic vibration waves with discrete frequencies. The test point is a very complicated motor process, and the vibrational frequency is composed of successive frequencies from zero to infinity; however, the spectrum is not composed of discrete frequencies, i.e., the frequency spectrum is continuous.

Function $f(t) \in L^2(R)$ is continuous; thus, its Fourier expression is

$$F(\omega) = \int_{-\infty}^{+\infty} f(t)e^{-i\omega t}dt$$

This function's complex formula is $F(\omega) = A(\omega) - iB(\omega) = R(\omega)^{-i\varphi(\omega)}$.

Thus, the range frequency spectrum of Fourier function $f(t)$ is $R(\omega) = \sqrt{[A(\omega)]^2 + [B(\omega)]^2}$.

The dominant frequency and the domain of dominant frequencies are important parameter to study the performance of blasting vibration. In the spectrogram, the frequency corresponding to the maximum frequency spectrum amplitude is the main frequency of this waveform, and the frequency region, for which the frequency spectrum component is larger than the maximum amplitude for 0.707 times, is the domain of the dominant frequency.

(2) **Power frequency spectrum analysis.** $x(t)(t = 1, 2, \cdots, N-1, N)$ is a discrete-time series, the Fourier expression of which is

$$X(t) = \sum_{i=1}^{N-1} x(i) \exp(\frac{-j2\pi ti}{N}) \; (t = 0, 1, 2, \cdots, N-1)$$

Its power spectral density function is

$$G_x(t) = \frac{2\Delta t}{N} \left| \sum_{T=0}^{N-1} x(t) \exp(\frac{-j2\pi ti}{N}) \right|$$

The purpose of power frequency spectrum analysis is to extract the vibration characteristics of vibration signal, judge the concentrated domain frequency of the waveform's vibration energy, reach a scientific blasting scheme and reduce the blasting vibration.

There are 15 acquisition channels in Figs. 9–16; every sensor can gather three directions' blasting earthquake waves, i.e., five sensors have 15 channels. The first channel represents the frequency vertical component of the 1 monitor sensor, the second channel represents the frequency horizontal tangential component, and the third channel represents the frequency horizontal radial component. The 4th channel represents the frequency vertical component of the 2 monitor sensor, the 5th channel represents the frequency horizontal tangential component of the 2 monitor sensor, and the 6th channel represents the frequency horizontal radial component of the 2 monitor sensor. Similarly, the 7–9 channels represent the three components of the 3 monitor sensor, the 10–12 channels represent the three components of the 4 monitor sensor, and the 13–15 channels represent the three components of the 5 monitor sensor. In the plot of Fig. 9, the abscissa indicates the acquisition channels, and the ordinate corresponds to the frequency spectrum (Figs. 2, 3, 4, 5, 6, 7, 8, and Table 3).

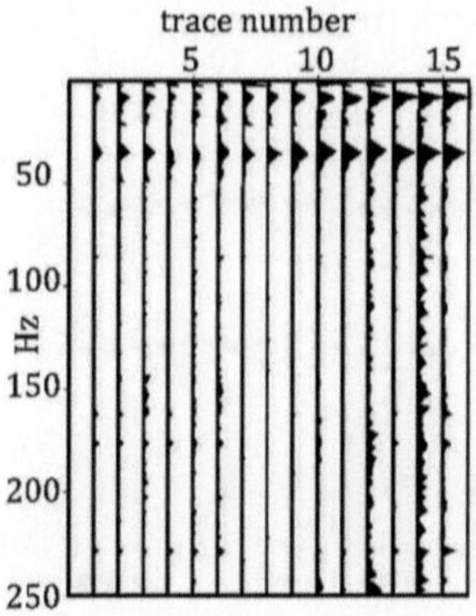

Fig. 2. Frequency spectrogram of Luo-Sha tunnel on August 7

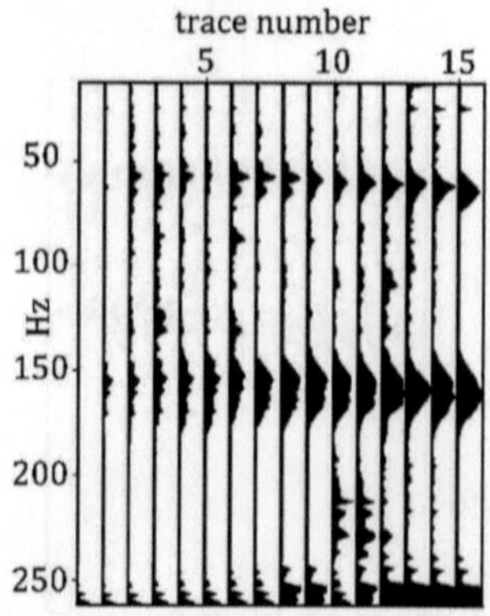

Fig. 3. Frequency spectrogram of Luo-Sha tunnel on August 8

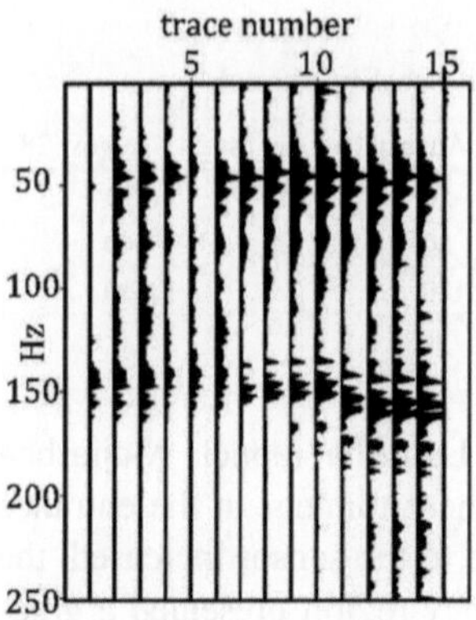

Fig. 4. Frequency spectrogram of Luo-Sha tunnel on August 10

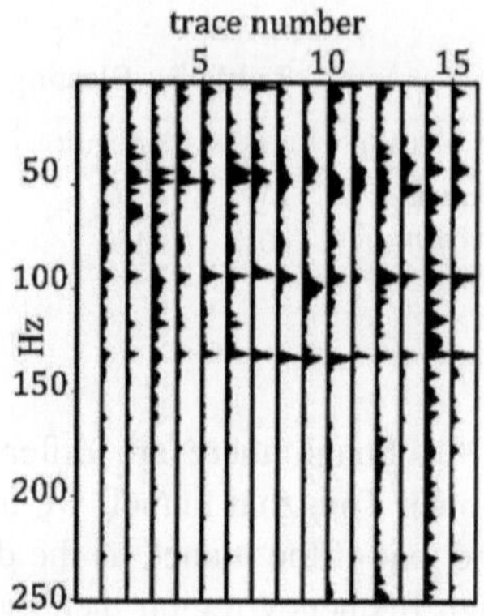

Fig. 5. Frequency spectrogram of Luo-Sha tunnel on August 12

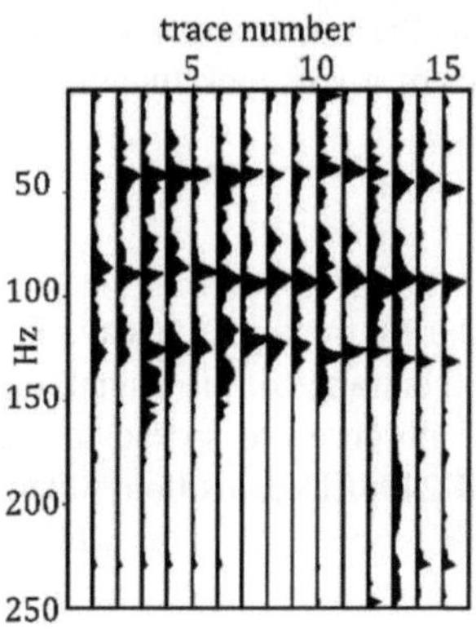

Fig. 6. Frequency spectrogram of Luo-Sha tunnel on August 13

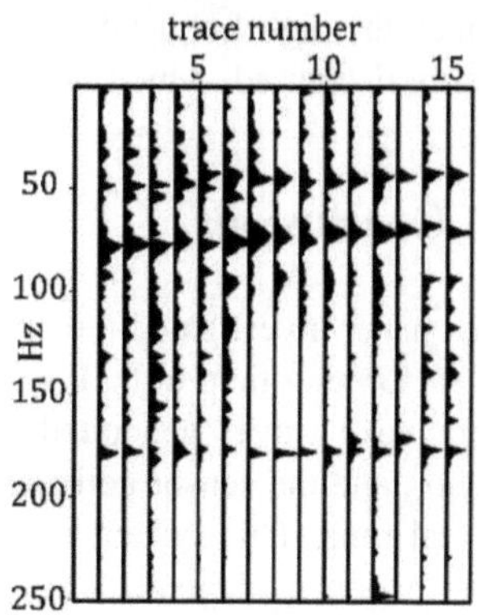

Fig. 7. Frequency spectrogram of Luo-Sha tunnel on August 16

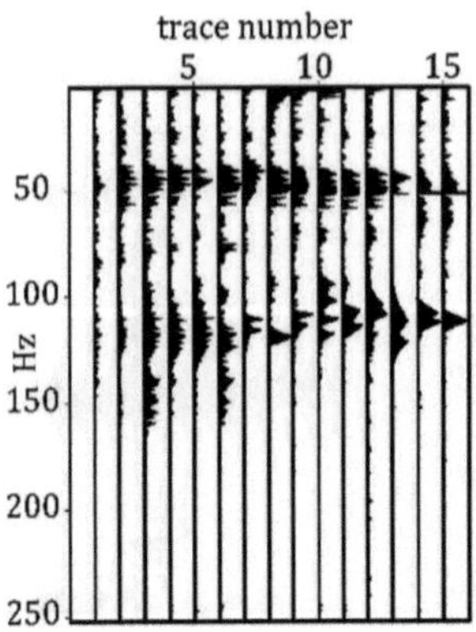

Fig. 8. Frequency spectrogram of Luo-Sha tunnel on August 24

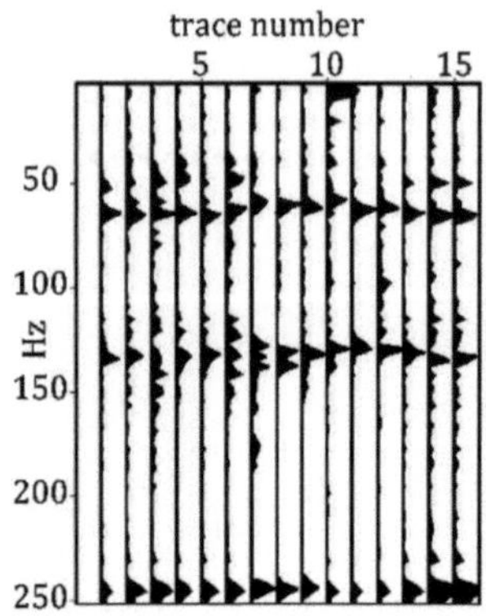

Fig. 9. Frequency spectrogram of Luo-Sha tunnel on August 28

Table 3. Blasting domain frequency spectrum of Luo-Sha tunnel

Date(2013)	August 7	August 8	August 10	August 12	August 13	August 16	August 24	August 28
Domain frequency (HZ)	15	60	50	50	50	50	50	70
	40	150	150	100	90	80	120	140
		250		140	130	180		250

Although there are differences in lithology among Luo-Sha tunnel, Majiashan tunnel, Tongzhai tunnel, we obtained the same result in these tunnels: at the entrance and exit of the tunnel, as the distance from the tunnel face to the sensor increased, the high frequency region increased, whereas the low frequency region presented a gradually weakening trend. Because the monitoring points are far from the tunnel face, the distinction of the domain frequency spectrum obtained during the blasting earthquake wave is not pronounced. Nevertheless, we can still determine that in the hard rock tunnel, the high frequency region clearly increased, whereas the low frequency region clearly decreased, which was the opposite of the behaviour of a soft rock tunnel.

4 Conclusions

This paper investigated the tunnel face blasting executed during the tunnel excavation by performing field tests. Laying the LC114 M sensor at the entrance of the tunnel, the earthquake waves generated by tunnel face blasting were observed to spread to the sensor, and the sensor delivered the obtained data to the SE2404EI acquisition system in real time. This system realized the collection and transmission of data, as well as information management of the data during the blasting process of the tunnel construction. The study's results can be outlined as follows.

1. In the tunnel portal, during blasting, the blasting vibration velocity increases, and the slope surface exhibits a vibration amplification effect in the tunnel face.
2. According to the analysis of the blasting vibration velocity component, the maximum peak velocities monitored by sensors in three different directions were determined to be different. In the same tunnel, the maximum peak velocities with different directions should have definite rules.
3. The higher the velocity at which the blasting earthquake wave propagates, the better the rock's integrity is. The quantization of the propagation velocity of blasting vibration provides information on the front surrounding rock and has reference value for guiding the construction process.
4. In the Luo-Sha tunnel entrance, as the distance from the tunnel face to the sensor increases, the high frequency region increases, whereas the low frequency region exhibits a gradually weakening property.
5. In the hard rock tunnel, the high frequency region grew, whereas the low frequency region shrank; this behaviour was the opposite of that of a soft rock tunnel.

5 Conflict of Interest

The authors declare that there are no conflicts of interests regarding the publication of this paper.

Acknowledgements. This work was partially supported by the National Key R&D Program of China (Nos. 2016YFC0802203), Guangxi Key Laboratory of Geomechanics and Geotechnical Engineering (15-KF-05), Zhejiang Collaborative Innovation Center for Prevention and Control of Mountain Geologic Hazards (PCMGH-2017-Z-01) and the Xi'an Railway Fund (Nos. 2016GG010).

References

Blair, D.P.: Blast vibration dependence on charge length, velocity of detonation and layered media. Int. J. Rock Mech. Min. Sci. **65**, 29–39 (2014)

Blair, D.P.: The free surface influence on blast vibration. Int. J. Rock Mech. Min. Sci. **77**, 182–191 (2015)

Gao, W.X., Yan, P.C., Li, Z.X., Wei, P.W., Hou, B.H.: Blasting excavation and vibration effects of shallow tunnel excavation. Chin. J. Rock Mech. Eng. **30**(S2), 4153–4157 (2011)

Huang, Z.Q., Wu, L.: Effects of blasting vibration on stability of bedding rock slope at tunnel entrance. Blasting **29**(1), 110–118 (2012)

Huang, Z.Q., Wu, L., Xing, X.K.: Optimized design of large span tunnel entrance blasting under complicated geological condition. Blasting **31**(1), 69–74 (2014)

Lin, C.M., Yang, L.D., Cui, J.H.: Study on the vibration characteristic of the shallow-tunneling blasting. Chin. J. Undergr. Space Eng. **2**(2), 276–279 (2006)

Nateghi, R.: Prediction of ground vibration level induced by blasting at different rock units. Int. J. Rock Mech. Min. Sci. **48**, 899–908 (2011)

Segarra, P., Sanchidrian, J.A., Castedo, R., Lopez, L.M., del Castillo, I.: Performance of some coupling methods for blast vibration monitoring. J. Appl. Geophys. **112**, 129–135 (2015)

Xu, H.L., Zhang, J.C., Ren, H.J.: Experimental study on magnifying effects of vibration resulting from driving blasting in tunnel. Chin. J. Undergr. Space Eng. **3**(4), 609–616 (2007)

Xia, X., Li, H.B., Niu, J.T., Li, J.C., Liu, Y.Q.: Experimental study on amplitude change of blast vibrations through steps and ditches. Int. J. Rock Mech. Min. Sci. **71**, 77–82 (2014)

Yun, Y.F., Wang, F., Wu, D.L., Xu, C.M.: Stability of the slope and house of the new Baota tunnel entrance. J. Xi'an Univ. Sci. Technol. **32**(2), 204–213 (2012)

Yang, J.H., Lu, W.B., Zhao, Z.G., Yan, P., Chen, M.: Safety distance for secondary shotcrete subjected to blasting vibration in Jingping- II deep-buried tunnels. Tunn. Undergr. Space Technol. **43**, 123–132 (2014)

Zhang, Y.J.: Study on attenuation law of open-pit bench blasting vibration. In: International Symposium on Safety Science and Technology. Procedia Engineering, vol. 84, pp. 868–872 (2014)

Mohamad, E.T., Armaghani, D.J., Noorani, S.A., et al.: Prediction of flyrock in boulder blasting by using artificial neural network. Electron. J. Geotech. Eng. **17**, 2585–2595 (2012)

Yan, Y., Zhang, Y., Huang, C.: Impact of blasting vibration on soil slope stability. Electron. J. Geotech. Eng. **19**, 6559–6568 (2014)

Li, Z., Li, Q., Cui, M., et al.: Blasting analysis of structural vibration to highway tunnels. Electron. J. Geotech. Eng. **20**(3), 909–918 (2015)

Jian-Min, Z., Xu-Guang, W., Ming-Sheng, Z., et al.: The effects of rock structure planes on the propagation of blasting seismic wave—a case study. Electron. J. Geotech. Eng. **11**(7), 4587–4599 (2016)

Experimental Studies on Strength Performance of Subgrade Soil Mixed with Bottom Ash and Coir Fiber

G. Karthikeyan, S. Karthic, and Sreevalsa Kolathayar(✉)

Department of Civil Engineering, Amrita School of Engineering, Amrita Vishwa Vidyapeetham, Amrita University, Coimbatore, India
sreevalsakolathayar@gmail.com

Abstract. The traffic loads on pavements are ultimately transferred to the subgrade and the performance of the pavement largely depends upon properties of sub grade soil. It becomes necessary to stabilize weak soils for infrastructural development such as for the construction of pavements. As the natural resources are depleting, limiting the use of natural resources and enhancing the use of locally available materials and waste materials is very important for sustainable development. The present study aims to evaluate the possible utilization of bottom ash (BA) which is a by-product obtained from thermal power plant, along with natural coir fiber as reinforcing agent for a locally available weak clayey soil in Southern India. Various strength parameters like California Bearing ratio (CBR), unconfined compression strength (UCS), Split tensile strength were estimated for soil-BA-coir fiber mixes in different percentages. A small amount of cement (5%) is also added to promote pozzolanic reaction. It was observed that the engineering properties of the soil depend mainly on the amount of coir and BA as well as curing period. This experimental investigation recommends the use of coir fiber and BA in combined form to get maximum benefits in soil stabilization. The results of experiments are presented and the suitability of using coir fiber and bottom ash are discussed in detail.

Keywords: Soil subgrade · Bottom ash · Coir fiber · California Bearing Ratio Unconfined Compressive Strength · Tensile strength

1 Introduction

Soil has been used as a construction material from time immortal. For most of the construction purpose, be it river embankments, dams or road pavements, the stability of sub grade soil plays a major role. Being poor in mechanical properties, it has been putting challenges to civil engineers to improve its properties depending upon the requirement which varies from one site to another. One such problem commonly faced is the damage of the road pavements constructed above soft soils. One common solution for such problem is excavation of large volume of soft soil and replacing it with harder soil or chemical treatment of soil to stabilize it which seems to be costly. Instead of using chemicals, abundantly available residues like bottom ash and natural coir fiber provide a wide range of application in soil stabilization. Thermal power plants

X. Shi et al. (Eds.): GSIC 2018, *Proceedings of GeoShanghai 2018 International Conference: Transportation Geotechnics and Pavement Engineering*, pp. 289–298, 2018.
https://doi.org/10.1007/978-981-13-0011-0_31

across the world face acute problem of disposing of the ash produced. It has become a major concern for the safe disposal of the ash produced without affecting the environment. Thus, several ways were proposed to utilize the ash produced instead of disposing it to the environment. One such method is by adding it to the sub grade and improving the geotechnical properties of the sub grade soil. Present study explores the suitability of using Bottom ash and natural coir fiber to improve strength characteristics of soil in terms of Compaction tests, Unconfined Compressive Strength (UCC), Split Tensile Strength Tests and California Bearing Ratio (CBR).

Rajakumar (2015) explored utilizing bottom ash in stabilizing expansive sub grade soil with experiments conducted to study the changes in soil properties such as plasticity characteristics, free swell index, pH etc., using bottom ash of varying percentage such as 10%, 20%, 30%, 40% and 50%. They reported that CBR increased until 20% addition of bottom ash and then decreased for every 10% addition of bottom ash whereas USC increased for increase in bottom ash until 40% and then decreased for 50%. Hastuty (2017) studied on the use of bottom ash and cement in stabilizing clay soil, by adding bottom ash at varying levels in clay soil with 2% cement. The results showed that compared to 2% cement and 9% bottom ash at 14 day curing and 1% cement 10% bottom ash at 14 day curing, an increase in compressive strength for 2% mixture was found. Lopez (2015) studied about the bearing capacity of bottom ash and its mixture with soil. Their study showed that soils can be improved by adding bottom ash contents that vary from 15% to 40% of the soil weight, thus improving their load-bearing capacity and reducing their plasticity. CBR improved to about 24% with percentage of 40% added ash. There were several studies on use of natural fibers as soil reinforcement and prominent studies on use of natural fibers in soil is listed in Table 1.

Table 1. Studies on soil reinforcement using natural fibers

Fiber type	Literature
Coir	Rowell et al. (2000), Babu and Vasudevan (2008), Subaida et al. (2009), Chauhan et al. (2008), Ramesh et al. (2010)
Sisal	Mishra et al. (2004), Ghavami et al. (1999), Prabhakar and Sridhar (2002)
Palm	Marandi et al. (2008), Jamellodin et al. (2010), Ahmed et al. (2010), Salehan and Yaacob (2011)
Jute	Aggarwal and Sharma (2010)
Flax	Segetin et al. (2007), Chean and Morgan (2009)
Barley straw	Key (1988), Bouhicha et al. (2005)

Yadav (2016) performed various tests and evaluated the compaction and strength behavior of the cement-stabilized, and sodium hydroxide treated coir fiber reinforced clay-pond ash mixtures. The results showed that the inclusion of fiber and pond ash in cementations material showed an increase in unconfined compression strength, split tensile strength and axial strain at failure. The incorporation of treated coir fiber reduced the stiffness, rate of loss of post-peak strength and changed the behavior of the composite from brittle to ductile. Dang (2016) studied the behavior of expansive soils stabilized with hydrated lime and bagasse fibers. He investigated the influences of bagasse fibers on the engineering behavior of expansive soil with varying proportions of randomly distributed bagasse fibers of 0.5%, 1.0%, and 2.0% which were added to expansive and lime hydrated expansive soils. The results showed that bagasse fiber reinforcement blended with hydrated lime showed increase in compressive strength with increase in curing time and additives contents whereas the linear shrinkage of stabilized expansive soils decreased with increasing hydrated lime and bagasse fiber proportions and curing periods. Thus studies showed that expansive soils can be successfully stabilized by combination of hydrated lime and bagasse fibers. Anggraini (2015) in his paper discussed about the effects of coir fiber on tensile and compressive strength of lime treated soft soil. The results revealed that both tensile and compressive strength increased with addition of lime, coir fiber and its curing time. Maximum value of both indirect tensile strength (ITS) and unconfined compressive strength (UCS) was gained at 1% coir content. The fibers efficiently obstructed the further development of cracks and prevented the samples from complete failure.

Lekha (2015) discussed about the effects of including randomly spaced Arecanut coir into laterite soil. The test results indicated that the improvement in characteristics of the soil cement coir mixtures were functions of coir dosage, soil type and curing days. Fauzi et al. (2010) studied the utilization of fly ash and bottom ash in improving the sub grade materials in highway construction. The strength gain in stabilization mainly depended on two factors: fly ash and bottom ash and water content. The fly ash and bottom ash stabilization increased the CBR values substantially for the mixtures tested. Peter (2014) presented an investigation on the behavior of soft soil stabilized with varying percentages of coir pith (0–3%) and coir fiber (0–1%) by carrying out Standard Proctor, Static Tri axial test and California Bearing Ratio (CBR) tests. They reported that stabilization with coir waste had a significant effect on the compaction, Elastic modulus as well as CBR characteristics.

The present study explores the suitability of using Bottom Ash and Coir Fiber in improving the strength behaviour of weak soil through experimental investigations in terms of Compaction tests, Unconfined Compression Strength tests, California Bearing Ratio Tests and Split Tensile Strength tests.

2 Materials

2.1 Soil

Soil sample investigated is an excavated clayey soil obtained from a construction site at Periyanaickenpalayam, near Coimbatore, Tamil Nadu, India. The soil properties are listed in Table 2.

2.2 Bottom Ash

Bottom ash is used as stabilizing agent which was obtained from Bottom ash Neyveli Lignite Corporation NLC, Neyveli. The specific gravity of the bottom ash was found to be 2.43. Bottom ash is a non-combustible residue, used in the production of many

Table 2. Soil Properties.

Parameter	Value
Liquid limit (%)	63.67
Plastic limit (%)	31.72
Plasticity index (%)	31.65
Optimum moisture content (%)	28.73
Maximum dry density (g/cc)	1.42
Free swell index (%)	77.78
Specific gravity	2.33
Soil classification	CH group, i.e., inorganic clay of high compressibility

concrete blocks for the constructing walls. Suitability of these materials for the stability of subgrade is demonstrated in this paper. The bottom ash has a specific gravity of 2.43, bulk density of 1.21 g/cc and fine content 4.65.

2.3 Fibers

Natural coir fibers are used as reinforcement in soil for the entire project work. Diameter of the coir fiber was found to be 0.0125 cm by using screw gauge. The length of the fiber used in this study was 5 mm.

2.4 Cement

The cement used was a OPC cement by Mahagold Cements, which was available in the construction Material laboratory of Amrita School of Engineering.

3 Experimental Programme

Experimental tests was done on the virgin soil sample mixed with different percentages of waste bottom ash, coir fibers and cement. The experiments performed include Compaction Tests, California Bearing Ratio tests, Unconfined Compression tests, Split Tensile Tests. The optimum percentage of bottom ash added to stabilize the soil was obtained by compaction tests results and was found to be 30%. Coir fiber of length 5 mm was used to reinforce the soil. The fiber content to be used in the soil was varied in different percentages of 0.5, 1, and 1.5 by weight.

3.1 Preparation of Specimens and Testing Procedure

Air dry soil samples were taken to compute the OMC of the sample using the values obtained from light compaction test. The relation between moisture content and maximum dry density for virgin soil and soil mixed with bottom ash to stabilize the soil was obtained. CBR values of virgin soil and soil mixed with different percentage of bottom ash were obtained by experimentally investigating California Bearing Ratio tests and optimum percentage of bottom ash was determined from the experimental results; optimum percentage of bottom ash was found out from optimum moisture content and dry density. Then the optimum percentage of bottom ash and soil with different percentage of coir fiber was varied with fixed percentage of cement in addition. All the samples were cured for 7, 14 and 28 days. Split Tensile tests and Unconfined Compressive Strength tests experiments were done on cylindrical specimens of diameter 39 mm and height 78 mm.

4 Results and Discussions

4.1 Compaction Tests

The soil samples were mixed with 0%, 10%, 20%, 30% and 40% of bottom ash to experimentally investigated and find out the optimum moisture content and maximum dry density of soil by light compaction according to IS: 2720 (Part V11) - 1980. The values of OMC and MDD for varying percentage of Bottom Ash are presented in Table 3.

Table 3. OMC and MDD for varying percentage of bottom ash CBR results on soil sample with 30% BA and 5% cement

Bottom ash in %	Optimum moisture content in %	Maximum dry density in g/cc
0	28.73	1.42
10	23.73	1.53
20	21.12	1.56
30	18.82	1.64
40	3.73	1.54

4.2 California Bearing Ratio Tests

The tests were conducted according to IS: 2720 (Part 16) - 1987. The specimens were heavily compacted and tests were performed for different mix percentages of bottom ash, soil, and cement and coir fiber. CBR results for the virgin soil and with 30% of Bottom ash and 5% of cement were presented in the Table 4.

Table 4. CBR Results on soil sample with 30% BA and 5% cement

Samples penetration	Virgin soil	Virgin soil + 30% of bottom ash and 5% of cement
CBR @ 2.5 mm	2.78	33.27
CBR @ 5 mm	2.24	34.77

Various proportions of bottom ash, soil, fiber and cement were experimentally investigated to find out the CBR results. The optimum mix of soil with bottom ash was found from compaction test. This experiment was done to evaluate the strength of sub grade mainly responsible for pavements and roads.

Figure 1 shows the variation of CBR value with different proportions of coir fiber with cement, soil and bottom ash. For 0.5%, 1% and 1.5% coir fibers the unsoaked CBR results were determined as 48%, 65% and 50% respectively. For soaked CBR the results were determined as 26%, 39% and 31% respectively. CBR soaked and unsoaked results shows that the CBR value increases to maximum addition of 1% coir fibers beyond increase in percentage of coir fibers shows decrease in strength for both cases of soaked and unsoaked.

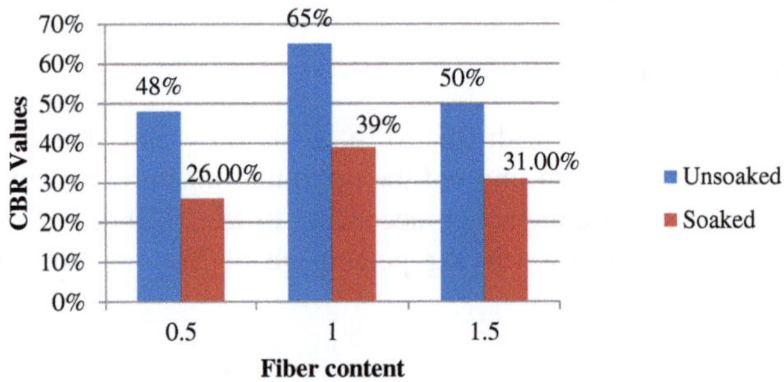

Fig. 1. Variation of CBR with different proportions of coir fiber

4.3 Unconfined Compressive Strength Tests

The specimens were prepared with varying quantity of Bottom Ash like 0%, 10%, 20%, 30% and 40% to find out the required density and optimum moisture content. The values for UCS with varying quantity of Bottom Ash are presented in Table 5.

Table 5. Variation of UCS with % BA

Bottom ash (%)	Unconfined compressive strength (KPa)
0	163.39
10	169.69
20	170
30	174.95
40	107.01

The rate of strain was 0.5 to 1 percent per minute. The samples mixed with different percentages of coir fibers, bottom ash, soil, and cement as follows

Soil
Soil + 30% Bottom ash
Soil + 30% Bottom ash + 0.5% of coir fibers
Soil + 30% Bottom ash + 1% of coir fibers
Soil + 30% Bottom ash + 1.5% of coir fibers

The tests were done for the samples with certain mix proportions having 30% of bottom ash, soil, 5% of cement and varying percentages of fiber (0%, 0.5%, 1%, and 1.5%). Samples were cured properly for 7 days, 14 days and 28 days respectively. The experimental values for the samples made with 90% soil, 30% bottom ash, 5% cement and different percentages of fiber (0%, 0.5%, 1%, and 1.5%) are shown in Fig. 2. It was clear that by varying different percentage of coir fibers the value of UCS increases. The average UCC value for virgin soil was 164 KN/mm^2. From the experimental results it is clear that by reinforcing the coir fiber into soil there was an increase in strength when compare to the samples without coir fibers.

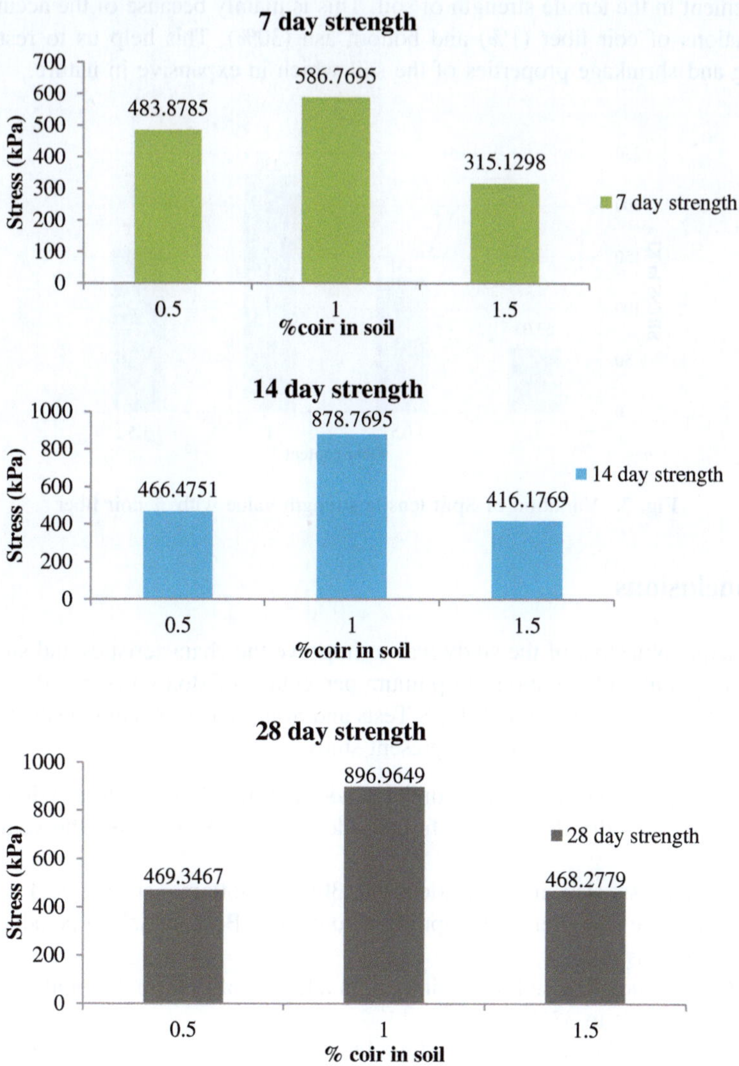

Fig. 2. Variation of UCS with % coir fiber [7 day strength, 14 day strength, 28 day strength (Top to bottom)]

We can infer from Fig. 2 showing UCS of various mix proportions, tested after 7, 14 and 28 days, that strength of soil increases up to an optimum percentage of coir fiber (1%) and with further increase in coir fiber, a decreases in strength of soil is observed.

4.4 Split Tensile Strength Tests

FigVarious combinations of soil samples were prepared and tests were carried out to according to ASTM D3967-16 for determining the tensile strength of soil specimens. Split tensile test results (Fig. 3) shows that the 1% of coir fiber mixes have 3.65 times the tensile strength of pure soil sample. This indicates that there is a massive improvement in the tensile strength of soil. This is mainly because of the accurate mix combinations of coir fiber (1%) and bottom ash (30%). This help us to restrain the swelling and shrinkage properties of the soil which in expansive in nature.

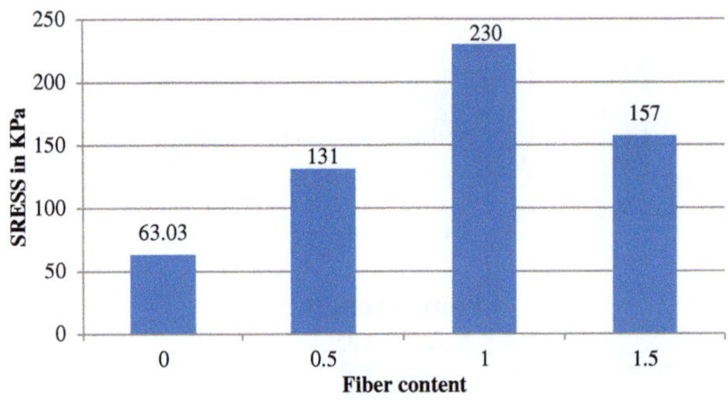

Fig. 3. Variation of Split tensile strength value with % coir fiber

5 Conclusions

The principle objective of the study was to improve the characteristics and strength of weak clayey soil with addition of optimum percentage of Bottom Ash and coir fibers. Based on the test results of CBR, UCS Tests and Split Tensile Strength tests, following conclusions were derived from the present study.

- From compaction tests, it was found that addition of 30% of Bottom Ash to the soil gave maximum dry density and further addition of BA decreased the value of dry density.
- From UCS tests also, it was evident that 30% of Bottom Ash inhibited maximum strength to the soil. Hence the optimum content of Bottom Ash to be added to the soil is fixed as 30%.
- The CBR value increased by 11 times with addition of Bottom Ash and 5% cement. (From 3% CBR to 33% CBR).
- The California Bearing Ratio was found to increases up to 65% with addition of 1% of coir fibers, whereas the increase in CBR for 0.5% and 1.5% of coir fiber, were 48% and 50% respectively.

- The Unconfined Compressive Strength (in all cases: 7 day, 14 day and 28 day curing) was also found to increase with addition of coir fiber up to 1% and further addition of coir fiber showed a decrease in strength.
- Similar trend was observed in case of Split Tensile Strength as well. Hence it can be concluded that the optimum percentage of coir fiber to be added to the soil is 1%, to arrive at better performance in terms of compression, tension and CBR.

To summarize, the weak soil mixed with optimum percentage of Bottom Ash and coir fiber gives 6 to 8 fold increase in strength compared to that of raw soil. The global annual production of coir fiber is about 350,000 metric tons. Yet, even in the world's top two producers, India and Sri Lanka, which account for about 90% of global coir fiber production, combined, this renewable resource is underutilized; local coir mills process only a fraction of the available husks, which accrue more or less year round as a waste during coconut processing. Present study adds a market value for coir and envisages the effective use of coir fiber in soil stabilization along with bottom ash, another waste product from thermal power plants.

References

Rajkumar, C.: Experimental study on utilization of bottom ash to stabilize expansive soil subgrades. Int. J. ChemTech Res. (IJCRGG) **8**(3), 997–1005 (2015)

Hastuty, I.P.: A study of the effectiveness of the use of cement and bottom ash towards the stability of clay in terms of UCT value. Procedia Eng. **171**, 484–491 (2017)

Lopez, L.E.: Bearing capacity of bottom ash and its mixture with soil. Soils Found. **55**(3), 529–535 (2015)

Yadav, J.S.: Behavior of cement stabilized treated coir fiber-reinforced caly-pond ash mixtures. J. Build. Eng. **8**, 131–140 (2016)

Dang, L.C.: Behavior of expansive soils stabilized with hydrated lime and bagasse fibers. In: International Conference on Transportation Geotechnics (ICTG 2016), pp. 658–665 (2016)

Anggraini, V.: Effects of coir fiber on tensile and compressive strength of lime treated coir fiber. Measurement **59**, 372–381 (2015)

Lekha, B.M.: Evaluation of lateritic soil stabilized with Arecanut coir for low volume pavements. Transp. Geotech. **2**, 20–29 (2015)

Fauzi, A., Nazmi, W.M., Fauzi, U.J.: Subgrades stabilization of Kuantan clay using fly ash and bottom ash. In: 8th International Conference on Geotechnical and Transportation Engineering (2010)

Peter, L.: Laboratory investigation in the improvement of sub grade characteristics of expansive soil stabilized with coir waste. Transp. Res. Procedia **17**, 558–566 (2014)

Rowell, M., Han, S., Rowell, S.: Characterization and factors effecting fiber properties. Nat. Polym. Agr. Compos. 115–134 (2010)

Babu, S., Vasudevan, K.: Strength and stiffness response of coir fiber-reinforced tropical soil. J. Mater. Civ. Eng. **20**, 571–577 (2008). ASCE

Subaida, A., Chandrakaran, E., Sankar, N.: Laboratory performance of unpaved roads reinforced with woven coir. Geotext. Geomembr. **27**, 204–210 (2009)

Chauhan, S., Mittal, S., Mohanty, B.: Performance evaluation of silty sand subgrade reinforced with fly ash and fiber. Geotext. Geomembr. **26**, 429–435 (2008)

Ramesh, N., Krishna, V., Mamatha, V.: Compaction and strength behavior of lime-coir fiber treated black cotton soil. Geomech. Eng. **2**, 19–28 (2010)

Mishra, S., Mohanty, K., Drzal, T., Misra, M., Hinrichsen, G.: A review on pineapple leaf fibers, sisal fibers and their biocomposites. Macromol. Mater. Eng. **289**, 955–974 (2004)

Ghavami, K., Filho, R., Barbosa, P.: Behaviour of composite soil reinforced with natural fibers. Cem. Concr. Compos. **21**, 39–48 (1999)

Prabakara, J., Sridhar, R.: Effect of random inclusion of sisal fiber on strength behavior of soil. Construct. Build. Mater. **16**, 123–131 (2002)

Marandi, M., Bagheripour, H., Rahgozar, R., Zare, H.: Strength and ductility of randomly distributed palm fibers reinforced silty-sand soils. Am. J. Appl. Sci. **5**, 209–220 (2008)

Jamellodin, Z., Talib, Z., Kolop, R., Noor, N.: The effect of oil palm fibre on strength behaviour of soil. In: 3rd SANREM Conferences, Kota Kinabalu, Malaysia (2010)

Ahmad, F., Bateni, F., Azmi, M.: Performance evaluation of silty sand reinforced with fibers. Geotext. Geomembr. **28**, 93–99 (2010)

Salehan, I., Yaacob, Z.: Properties of laterite brick reinforced with oil palm empty fruit bunch fibers. Pertanika J. Sci. Technol. **19**, 33–43 (2011)

Aggarwal, P., Sharma, B.: Application of jute fiber in the improvement of subgrade characteristics. In: Proceedings of International Conference on Advances in Civil Engineering, Trabzon, Turkey (2010)

Segetin, M., Jayaraman, K., Xu, X.: Harakeke reinforcement of soil–cement building materials: manufacturability and properties. Build. Environ. **42**, 3066–3079 (2007)

Cheah, J., Morgan, T.: UKU: concept to construction using flax-fibre reinforced stabilised rammed earth. In: 11th International Conference on Non-conventional Materials and Technologies; 6–9 September 2009, Bath, UK (2009)

Key, L.: Straw as an erosion control mulch, a technical report from US Agriculture Department, Portland, Oregon, No. 49 (1988)

Bouhicha, M., Aouissi, F., Kenai, S.: Performance of composite soil reinforced with barley straw. Cem. Concr. Compos. **27**, 617–621 (2005)

Use of Industrial Byproducts to Improve Unsuitable Soils for Road Construction in Sri Lanka

Udeni P. Nawagamuwa[1(✉)], A. G. Shamil Shabry[2], Kanapathippillai Ainkaran[3], D. L. S. Prasad[4], and H. K. P. Madushanka[1]

[1] University of Moratuwa, Moratuwa, Sri Lanka
udeni@uom.lk
[2] National Water Supply and Drainage Board, Ratmalana, Sri Lanka
[3] Road Development Authority, Battaramulla, Sri Lanka
[4] KDA Weerasinghe & Co. (Pvt) Ltd., Nugegoda, Sri Lanka

Abstract. When an unsuitable soil is come across in areas with limited resources, engineers must add transportation cost to the overall cost if a suitable soil has to be hauled from a long distance. In such instances, improvement of the existing soils with locally available materials would be the most beneficial solution. Most of the rural roads in Sri Lanka are now under a phase lift although with lot of constraints. Road Development Authority is practicing a guideline developed consisting of the grading of soil, Atterberg limits, Proctor density and California Bearing Ratio (CBR) to select a suitable subgrade material. However, in most of the cases, soils are rejected due to low CBR, although the other conditions are satisfactory. This study discusses the improvements of such rejected soils from distinct places of Sri Lanka with quarry dust, lime stone dust, paddy husk ash, fly ash and copper slag. Basic soil tests such as particle size analysis, Atterberg limits, Proctor compaction and CBR tests were conducted on unimproved and improved soils. A considerable increase in the CBR (in the range of 10–20 times) was observed with the addition of industrial byproducts to improve rejected soils. Mixing is proposed with a traditional conveyor belt mechanism. Use of industrial byproducts in the infrastructure development is to be encouraged with the objective in preserving environment by reducing utilization of new resources and excessive dumping of by products to the environment.

Keywords: CBR · Industrial byproducts · Roads in Sri Lanka

1 Introduction

Soil is a primary engineering material for road construction and maintenance (Biggs and Mahony 2004). Constructing and maintaining properly functioning transportation infrastructure is a very resource intensive activity. Substantial amounts of materials and natural resources are required and they consume proportionately large amounts of energy and fuel (Correia et al. 2016). Traditional road construction consumes soil, stone aggregates, sand, bitumen, cement etc. and their future mining is declining

X. Shi et al. (Eds.): GSIC 2018, *Proceedings of GeoShanghai 2018 International Conference: Transportation Geotechnics and Pavement Engineering*, pp. 299–308, 2018.
https://doi.org/10.1007/978-981-13-0011-0_32

gradually due to limited natural resources. Also, the public pressure against mining of natural materials is immense while cost of extracting good quality natural material is increasing. Available soils in the proposed construction area sometimes do not possess the requirements imposed by the consultants to be used in road works. Hauling materials from faraway areas will create additional cost and environmental damage due to longer distance transportation. Considering these factors, engineers always look for alternative materials for road construction, and industrial byproducts which have proven their capacity to replace some portion of the natural resources.

Several ground improvement methods are available; however, decisions should be supported by comprehensive analyses of the carbon footprint, life cycle and cost studies, and energy consumption analyses of each of the candidate methods in order to determine the one that proves to be the most sustainable (Correia et al. 2016). Soil stabilization with industrial by-products has become an established practice for improving the engineering characteristics of problematic soils (Ahnberg et al. 2003; Indraratna 1996; Ingles and Metcalf 1972; James et al. 2008; Jones 1990; Porbaha et al. 2000; Wilkinson et al. 2006).

In recent years, applications of industrial by-products have been considered in road construction with great interest in many industrialized and developing countries. Use of these materials in road making is based on technical, economic, and ecological criteria (Sen and Mishra 2010). Materials such as fly-ash from thermal power plants and other coal fired industries, blast furnace slag from steel industries, cement kiln dust from cement related industries, phosphogypsum from phosphatatic fertilizer industries, and many other solid wastes have already proved to be useful for road construction in many countries (Sen and Mishra 2010). Swell control potential of fly ash can be probed in order to further reinforce it as a universal pozzolanic additive in soil stabilization. Permeability and compressibility performance of lime/cement stabilized soils with fly ash, rice husk ash, and ground granulated blast furnace slag can be taken up to complement existing literature (James and Pandian 2016). Fly ash is an effective agent for chemical and/or mechanical stabilization of soils. Blast furnace slag has been used as a cementitious binder in road construction.

After the 30-year civil war, Sri Lanka is now at a rapid rate of infrastructure development. Therefore, there is a huge need of construction materials for those projects, especially, soils for the rural roads and national highways. However, due to locality issues of geology, soil types are varying in various places of the country. It was observed that in some areas, soils do not satisfy the requirements imposed by ICTAD (2009). These conditions consist of grading curves, Atterberg limits, maximum dry density and California bearing ration (CBR). Although the soils do satisfy most of the mentioned requirements, they lack in providing a sufficient CBR value for sub base applications. This study aims to identify such soils in different localities which have come across during road projects and to propose improvements in soils with the addition of industrial byproducts.

2 Methods

2.1 Site Description

Four localities, Addalaichenai area in Ampara district, Jaffna Peninsula, Thissa-Sithulpawwa road in Hambantota and Bodagama-Hambegamuwa-Kalthota road in Monaragala are shown in Fig. 1. Soil samples were collected form those areas for possible improvements. According to Moormann and Panabokke (1961), Panabokke (1971) and Cooray (1984), Addalaichenai consists of noncalcic brown soils with solodized solonetz, Jaffna peninsula with calcite red yellow latsols with solodized solonetz, Thissa-Sithulpawwa with reddish brown earth with solodized solonetz and Bodagama-Hambegamuwa-Kalthota with reddish brown earth.

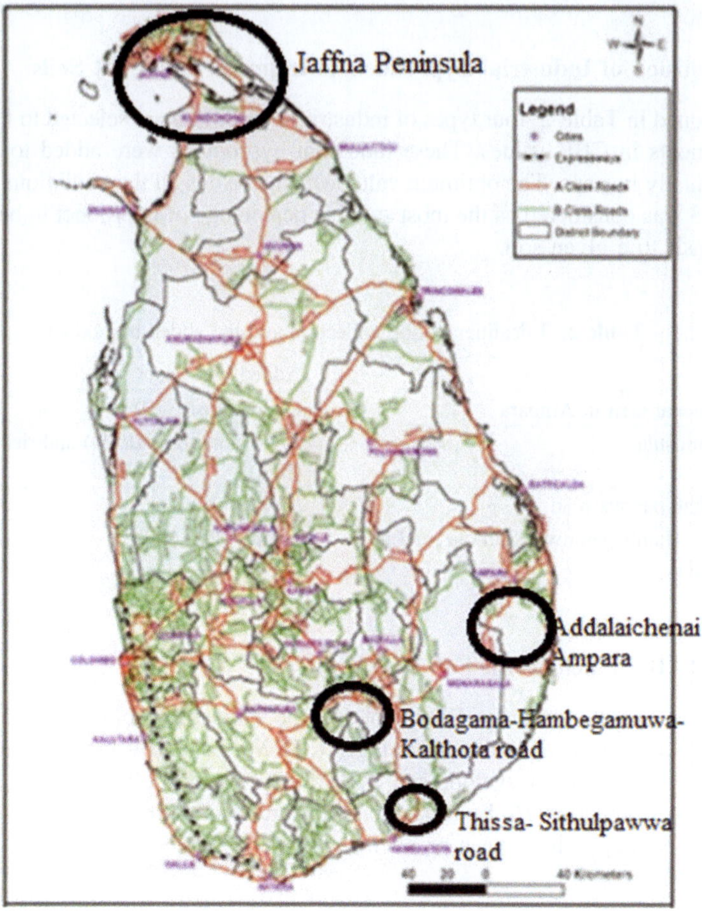

Fig. 1. Four selected sites in the national road network in Sri Lanka (http://www.rda.gov.lk)

2.2 Standards Maintained in the Laboratory

Table 1 provides the standards and specifications maintained during laboratory investigations.

Table 1. Testing standards followed during the laboratory tests

Test	Testing method
Particle size distribution	ASTM D6913 - 04e1
Plastic limits (PL) (%), Liquid limits (LL) (%)	ASTM D4318 - 10e1
Maximum dry density (kgm^{-3})	ASTM D698 - 12e2
Specific gravity	ASTM D854 - 14
California Bearing Ratio (4 days soaked at 98% MDD) (%)	ASTM D1883

2.3 Inclusion of Industrial Byproducts to Improve Low CBR Soils

As mentioned in Table 2, four types of industrial byproducts are selected to find out the improvements in CBR values. These industrial byproducts were added to respective soils gradually in steps. The optimum value which satisfies all the conditions stipulated in Table 3 was considered as the most suitable percentage of byproduct to be added on weight basis to a given soil.

Table 2. Localities of soil collection and the added byproduct

Soil	Added byproduct
Addalaichenai area in Ampara	Quarry dust (QD)
Jaffna peninsula	Lime stone dust (LSD) and rice husk ash (RHA)
Thissa-Sithulpawwa road	Fly ash (FA)
Bodagama-Hambegamuwa-Kalthota road in Moneragala	Copper slag (CS)

2.4 ICTAD Guidelines for Sub-base Soils

Road Development Authority (RDA) follows the standard specification for construction and maintenance of roads and bridges (SSCM) issued by the Institute for construction Training and Development (ICTAD 2009). From SSCM and Highways schedule rates prepared (16) by RDA, conditions mentioned in Table 3 for sub base in the road construction have been practiced.

3 Results

3.1 Particle Size Analysis

Soils were mixed with industrial byproducts on w/w basis from 0 to 50%. Optimum proportion, which provides the satisfactory performances on the stipulated conditions mentioned in Table 3, was selected. Those selected proportions are provided in Table 4. Following identifications proposed in Table 4 are used in this paper for different proportions of improved soils for easy reference.

Table 3. Stipulated conditions for sub-base soils (RDA guidelines adopted from ICTAD 2009)

Maximum dry density	$\geq$ 1650 kg/m^3
California Bearing Ratio (at 98% of MDD)	$\geq$ 20%
Liquid limit	$\leq$ 40%
Plasticity index	$\leq$ 15%

Table 4. Identification of unimproved and improved soils

Unimproved soil		Improved soil with industrial byproducts		Optimum w/w percentage
Addalaichenai area in Ampara	AD	QD	ADI	AD 75% - QD 25%
Jaffna peninsula	JF	LSD and RHA	JFI	JF80% - LSD5% - RHA15%
Thissa-Sithulpawwa road	TS	FA	TSI	TS90% - FA10%
Bodagama-Hambegamuwa-Kalthota road in Moneragala	MO	CS	MOI	MO65% - CS35%

Grading requirements stipulated in ICTAD (2009) were checked with the four soil types as shown in Figs. 2 and 3 shows the grading of selected proportions on soils improved with the industrial byproducts. Table 4 summarizes the optimum w/w percentages of soils improved with the byproducts added in gradual steps. Financial benefits were too considered for selecting the optimum proportion for JF sample as JF80% - LSD15% - RHA5% combination too provides a high satisfactory CBR value while satisfying other conditions.

3.2 Atterberg Limits

Plasticity behavior was compared before and after improvement of soils as given in the Fig. 4. All the unimproved and improved soils do satisfy the ICTAD (2009) requirements. As some soils (especially after improvements) did not demonstrate workable limits using standard Casagrande apparatus, cone penetration test was used as proposed by Wasti (1987).

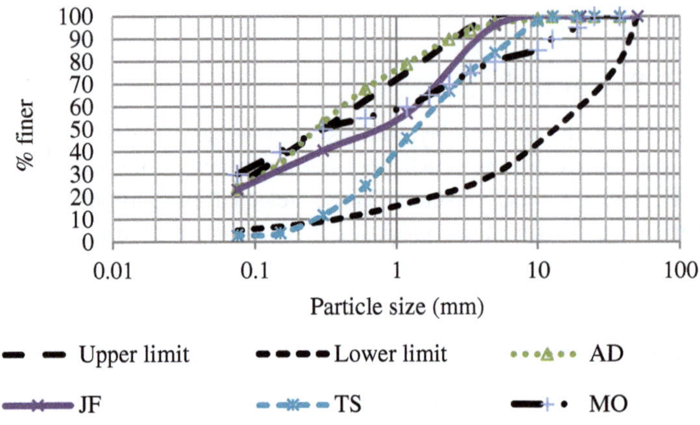

Fig. 2. Grading curves of soils before improvement

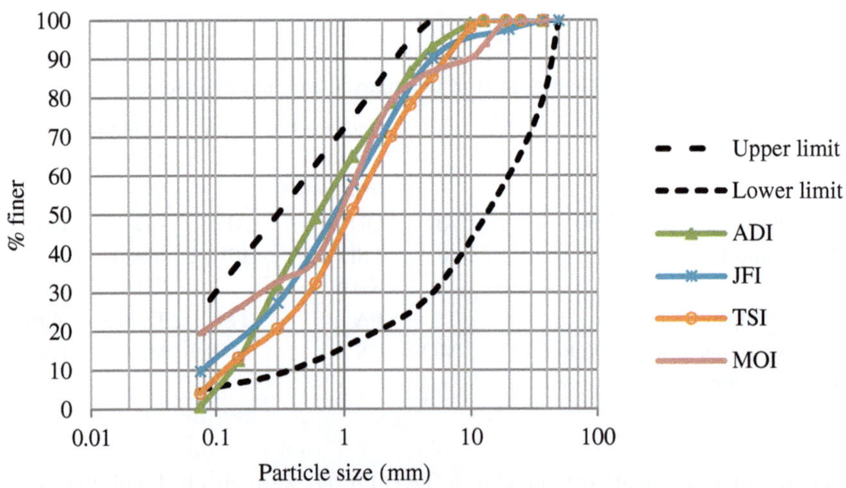

Fig. 3. Soils improved with industrial byproducts

3.3 Compaction Characteristics of Soils Before and After Improvement

Standard Proctor compaction test was conducted on the soils before and after improvement and found to be well satisfactory against the specified guidelines and the results are illustrated in Fig. 5. MOI shows a relatively higher maximum dry density (MDD) mainly because of the additive, copper slag, is of very high specific gravity around 3.5.

3.4 Improvement of CBR Values

Summary of the improved California Bearing Ratio values are presented in Fig. 6 with respect to their maximum dry density values obtained from Fig. 5. Although MOI

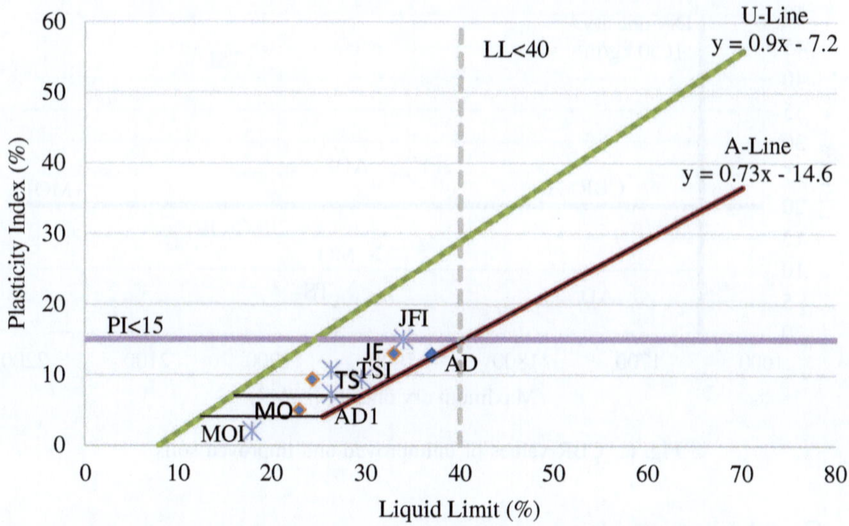

Fig. 4. Atterberg limits of unimproved and improved soils

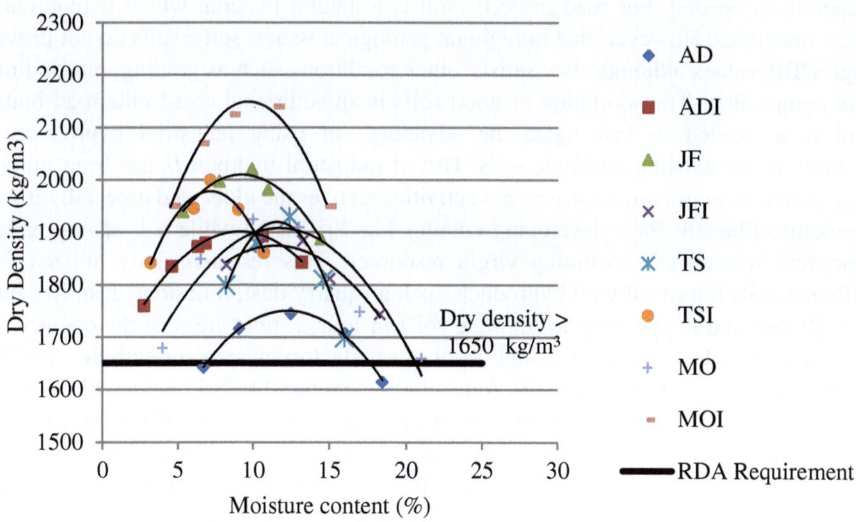

Fig. 5. Compaction characteristics of unimproved and improved soils

exhibits the highest MDD, it doesn't improve its CBR value considerably. This is probably due to its neutral or non-pozzolanic behavior and hence couldn't contribute much to CBR value.

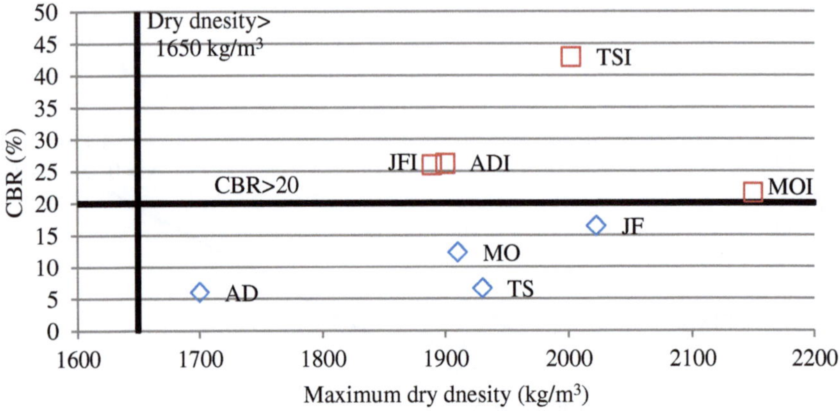

Fig. 6. CBR values of unimproved and improved soils

4 Conclusion

Due to new infrastructure projects in Sri Lanka, a considerable amount of construction materials is needed. For road projects, soil is a natural material which is required in great quantities. However, due to regional geological issues, some soils do not provide high CBR values although they satisfy other conditions such as grading, plastic limits and compaction. Transportation of good soils is an additional cost to the road budget and it is needed to investigate the advantage of using industrial byproducts in improving the existing available soils. Use of industrial byproducts has been immensely practiced in ground improvement activities all over the globe and especially it is an economical benefit for a developing country like Sri Lanka while it is saving natural resources instead of consuming virgin resources. This research study utilized four different soils improved with byproducts such as quarry dust, limestone dust, rice husk ash, fly ash and copper slag to improve soils in four distinct areas of the country.

Laboratory tests indicated a high level of satisfactory agreement with the specified conditions given in ICTAD (2009). A developing country like Sri Lanka will be greatly benefitted by such findings. These promising results with high CBR values of soils after improvement, encourage the construction engineers to apply such techniques with lot of monitoring. Large scale mixing needs a conveyor belt type mixing arrangement fixed locally. This will further enhance the income of the local residents in multiple ways.

Acknowledgement. Staff members at Soil Mechanics Laboratory, University of Moratuwa, Engineers in the regional offices of the Road Development Authority, Sri Lanka and KDA Weerasinghe Co. (Pvt) Ltd. are acknowledged for allowing the authors to carry out this research.

References

Ahnberg, H., Johansson, S.E., Pihl, H., Carlsson, T.: Stabilizing effects of different binders in some Swedish soils. Ground Improv. **7**(1), 9–23 (2003)

ASTM D1883: Edition, 1 March 2016 Standard Test Method for California Bearing Ratio (CBR) of Laboratory-Compacted Soils (2016)

ASTM D4318 - 10e1: Standard Test Methods for Liquid Limit, Plastic Limit, and Plasticity Index of Soils (2010)

ASTM D6913 - 04e1: Standard Test Methods for Particle-Size Distribution (Gradation) of Soils Using Sieve Analysis (2009)

ASTM D698 - 12e2: Standard Test Methods for Laboratory Compaction Characteristics of Soil Using Standard Effort (12 400 ft-lbf/ft^3 (600 kN-m/m^3)) (2014)

ASTM D854 - 14: Standard Test Methods for Specific Gravity of Soil Solids by Water Pycnometer (2014)

Biggs, A.J.W., Mahony, K.M.: Is soil science relevant to road infrastructure? In: 13th International Soil Conservation Organization Conference (ISCO 2004), Conserving Soil and Water for Society: Sharing Solutions, Brisbane, July 2004

Cooray, P.G.: An Introduction to the Geology of Sri Lanka, 2nd edn., 340 p., National Museum of Sri Lanka Publication, Colombo (1984)

Correia, A.G., Winter, M.G., Puppala, A.J.: A review of sustainable approaches in transport infrastructure geotechnics. In: Transportation Geotechnics. Elsevier (2016). http://dx.doi.org/10.1016

Highways Schedule Rates: Ministry of Higher Education and Highways Sri Lanka, January 2016

ICTAD: ICTAD Specifications for Construction and Maintenance of Roads and Bridges, 2nd edn. ICTAD Publication No. SCA/5, June 2009

Indraratna, B.: Utilization of lime, slag and fly ash for improvement of a colluvial soil in New South Wales, Australia. Geotech. Geol. Eng. **14**(3), 169–191 (1996)

Ingles, O.G., Metcalf, J.B.: Soil Stabilization: Principles and Practice. Butterworths, Sydney (1972)

James, J., Pandian, P.K.: Industrial wastes as auxiliary additives to cement/lime stabilization of soils. Adv. Civ. Eng. **2016**, 17 p. (2016). Article ID 1267391. http://dx.doi.org/10.1155/2016/1267391

James, R., Kamruzzaman, A.H.M., Haque, A., Wilkinson, A.: Behavior of lime slag treated soils. Proc. Inst. Civ. Eng. Ground Improv. **161**(4), 207–216 (2008)

Jones, L.W.: Interference Mechanisms in Waste Stabilization/Solidification Processes. US Army Corps of Engineers Waterways Experiment Station, Vicksburg, MMS, USA. United States Environmental Protection Agency EPA/ 600/2-89/067 (1990)

Moormann, F.R., Panabokke, C.R.: Soils of Ceylon, a New Approach to the Identification and Classification of the Most Important Soil Groups of Ceylon. Government Press, Ceylon (1961)

Panabokke, C.R.: Soil map of Sri Lanka, Land use division, Irrigation Department of Sri Lanka (1971)

Porbaha, A., Shibuya, S., Kishida, T.: State of the art in deep mixing technology. Part III: geomaterial characterization. Ground Improv. **4**(3), 91–110 (2000)

Sen, T., Mishra, U.: Usage of industrial waste products in village road construction. Int. J. Environ. Sci. Dev. **1**(2) (2010). ISSN 2010-0264

Wasti, Y.: Liquid and plastic limits as determined from the fall cone and the casagrande methods. Geotech. Test. J. GTJODJ **10**(1), 26–30 (1987)

Wilkinson, A., Haque, A., Kodikara, J.: Factors affecting the geotechnical response of cementitiously stabilized soils in a range of geo-environments. In: Advances in Unsaturated Soil, Seepage, and Environmental Geotechnics, ASCE Geotechnical Special Publication No. 148. ASCE, Reston, pp. 323–328 (2006)

http://www.rda.gov.lk

Pavement Structures and Materials

Application of Gyratory Compaction for Determining the Target Values for Pavement Subgrade Compaction

Jiake Zhang(✉), Jianming Ling, Jingsong Qian, and Chen Jin

Tongji University, Shanghai 201800, China
zhjiake@tongji.edu.cn

Abstract. Pavement subgrade compaction consist of determining the "optimum" moisture content and maximum dry unit weight from laboratory Proctor tests, and then specify an acceptable relative compaction value and moisture content limits in the field. Although this approach has been widely accepted, it does not provide the mechanistic properties of the compacted material (i.e., strength or stiffness). Design parameters for pavements, slopes, foundations retaining structures, etc. typically rely on strength or modulus/compressibility of the compacted materials. Thus, this research applies the gyratory compaction with an integrated pressure distribution analyzer (PDA) sensor kit for determining the mechanistic based target values for pavement subgrade compaction. The results presented herein linked the moisture-density-engineering properties of the gyratory compaction samples, which showed the feasibility of using gyratory compaction for determining the mechanistic based target values for pavement subgrade compaction.

Keywords: Gyratory compaction · Target value · Subgrade compaction

1 Introduction

Moisture content and dry unit weight are the key parameters for the target value determination of in-situ roller compaction. Current standard compaction control consist of determining the "optimum" moisture content and maximum dry unit weight from Proctor test and then a minimum relative compaction value and moisture content range is specified for acceptance during construction. However, the performance of compacted fill materials relies on the mechanistic-based parameter values, such as modulus, strength or stiffness etc. Thus, using the mechanistic properties of the compacted material as the quality control operation would be more directly relate to the design and also provide perhaps more value for ensuring quality as a final product.

This work studied the laboratory gyratory compaction coupled with a pressure distribution analyzer (PDA) as a method to link water content and dry unit weight to mechanistic parameters such as shear strength and unconfined compressive strength of the compacted samples. Researchers using gyratory compaction on soils and demonstrated that gyratory compaction provided more comprehensive information of the compacted materials than impact compaction [1, 2]. Lee et al. (2007) found that field compacted specimens yield similar strength and modulus values compared with gyratory

X. Shi et al. (Eds.): GSIC 2018, *Proceedings of GeoShanghai 2018 International Conference: Transportation Geotechnics and Pavement Engineering*, pp. 311–318, 2018.
https://doi.org/10.1007/978-981-13-0011-0_33

compacted soils using 600 kPa vertical pressure [3]. Recently, Li et al. (2015) demonstrated that the gyratory compactor with pressure distribution analyzer (PDA) facilitates characterizing unique moisture-density-energy-strength relationships for waste materials [4]. Guler et al. (2000) developed the PDA to capture the applied pressure on sample during gyratory compaction. It mainly includes three 9 kN load cells, and a data recording system that is used to download data from the PDA [5]. The resultant ram force (*R*) and the average eccentricity (*e*) are obtained from the PDA output data, which can be used to calculate the shear resistance of the compacted specimens.

The purpose of this research was to correlate the mechanistic-based parameters which obtained from the gyratory compaction with light weight deflectometer (LWD) modulus, and explore the feasibility of using gyratory compaction for target value determination of subgrade compaction.

2 Research Methodology

2.1 Compaction Test

Gyratory compaction was conducted using a AFGB1A Brovold gyratory compactor with pressure distribution analyzer (PDA) (Fig. 1). ASTM D3387-2011 was followed to produce the compacted specimens [6]. Materials were compacted with selected vertical stresses (σ_o) ranging from 100 to 600 kPa at a constant rate of 30 gyrations per minute with the gyration angle set at 1.25°. PDA was placed above the sample to capture the pressure distribution across the sample during compaction. It provides the resultant force (*R*) and the eccentricity (*e*). With *R* and *e*, the frictional resistance or shear resistance ($_G$) of the compacted materials can be calculated using Eq. (1) [7]:

$$\tau_G = \frac{R \bullet e}{A \bullet H} \tag{1}$$

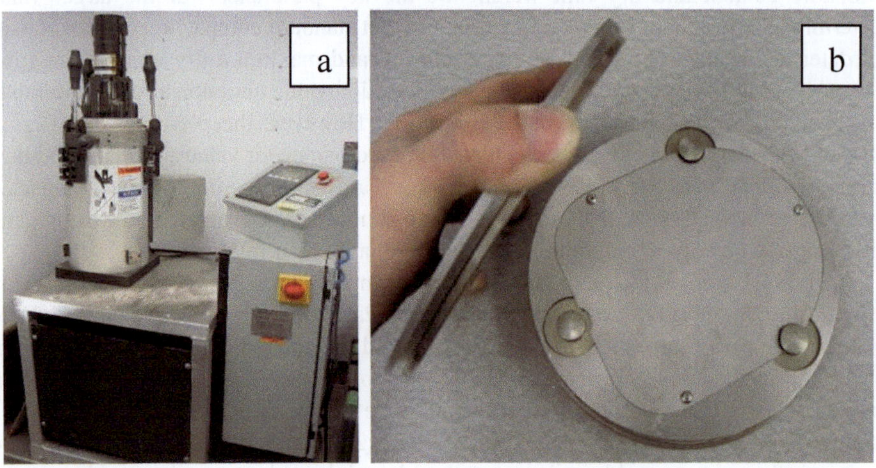

Fig. 1. AFGB1A gyratory compactor (a) and pressure distribution analyzer (b)

where, R = resultant force, e = eccentricity, A = sample cross-sectional area, and H = sample height at any gyration cycle. The relationships between $_{G}$ and s_u, and E_{LWD} were explored in this research.

Both standard Proctor and modified Proctor tests were conducted on the materials that were sampled from the field in accordance with ASTM D698 [8] and ASTM D1557 [9], respectively.

2.2 Light Weight Deflectometer (LWD) Test

LWD tests were performed in the laboratory on the gyratory compacted specimens. The compacted specimens were extruded from the gyratory mold and tested with two different boundary conditions as shown in Fig. 2, which are considered as "stiff" and "rigid" boundary condition in this paper. LWD tests were conducted in accordance with the manufacture recommendations [10], and the LWD modulus can be determined using Eq. (2):

$$E_{LWD} = \frac{(1 - v^2)\sigma_o r}{d_o} \times F \quad (2)$$

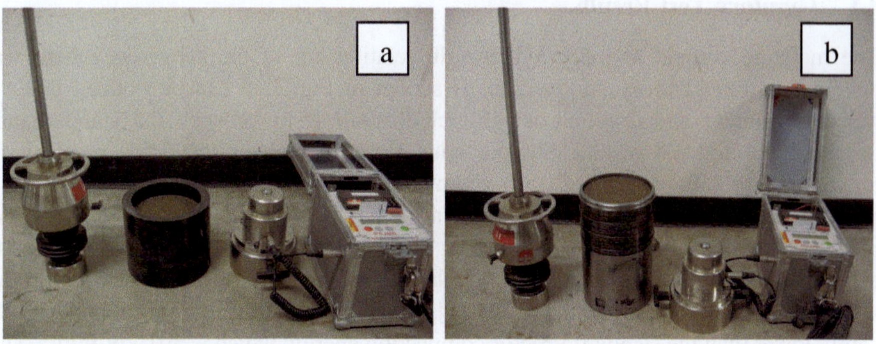

Fig. 2. LWD testing on gyratory compacted specimens with two boundary conditions: (a) stiff boundary; (b) rigid boundary

where, E_{LWD} = elastic modulus (MPa), d_o = measured deflection (mm), v = Poisson's ratio (assumed to be 0.4), σ_o = applied stress (MPa), r = radius of the plate (mm) and f = shape factor that depends on the stress. Also, in-situ LWD tests were conducted in the field of sampling the materials.

3 Materials

The subgrade material can be classified as CL (lean clay sand) based on the USCS classification, and the soil index properties are summarized in Table 1.

Table 1. Summary of the material properties for test material

Parameter/Material	Subgrade
Material description	Sandy lean clay
Maximum dry unit weight (kN/m^3) and optimum moisture content (%)	
Standard Proctor	17.2 (17.7)
Modified Proctor	19.3 (12.8)
Gravel Content (%) (>4.75 mm)	3
Sand Content (%) (4.75 mm – 75 μm)	30
Silt Content (%) (75 μm – 2 μm)	38
Clay Content (%) (<2 μm)	29
Liquid Limit, LL (%)	39
Plasticity Index, PI	19
USCS	CL

4 Results and Discussion

4.1 Gyratory Test Results

The moisture content and dry unit weight relationship of the subgrade, which was developed from gyratory compaction, is shown in Fig. 3. It can be noticed that the moisture content and dry unit weight relationship is in between the standard and modified Proctor test results. Sample with moisture content of 15.5% reached the maximum dry unit weight of 18.0 kN/m^3, when compacted with 300 kPa vertical pressure at 100 gyrations, and the sample achieved the standard Proctor maximum dry unit weight (17.2 kN/m^3) at 18 gyrations. The gyratory compaction energy applied to achieve the same maximum dry unit weight from standard Proctor was 133 $kN\text{-}m/m^3$, which can be calculated from Eq. (3) [4]. Thus, the gyratory compaction is significantly lower than the standard Proctor compaction energy, which is 591 $kN\text{-}m/m^3$. This suggests that gyratory compaction is a more energy efficient method for compacting the subgrade materials relative to impact compaction. Also, standard or modified Proctor compaction is the "energy per volume" controlled method and the same compaction energy is applied to the materials which is in the controlled size mold. On the other hand, gyratory compaction is the "energy per mass" controlled method [11], which means all compaction energy is applied on compacting the material.

$$E_{gyratory} = \frac{p_{vertical} \times A \times (H_o - H_N) + 4 \times \omega \times \sum_0^N \tau_G V_i}{V_N} \quad (3)$$

Where $E_{gyratory}$ = gyratory compaction energy (kJ/m^3); p_{vertical} = vertical applied pressure (kPa); A = area of specimen (m^2); H_0 = initial specimen height (m); H_N= final specimen height after compaction; τ_G = measured shear stress during compaction (kPa); ω = gyration angle (radians); and V_i = volume of specimen (m^3).

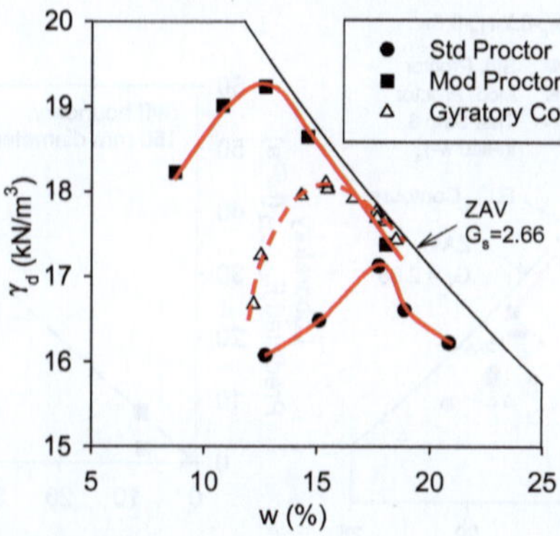

Fig. 3. Comparison of impact compaction and gyratory compaction

4.2 Target Value Determination

The approach of developing w-γ_d-E_{LWD} contours is useful in determining target E_{LWD} values for a specified moisture range and target minimum density in situ. Gyratory compacted specimens were prepared and tested using 100-mm diameter plate Zorn LWD under "stiff" and "rigid" boundary conditions. PDA shear resistance was measured during gyratory compaction process. Multiple regression analysis was performed by incorporating w, γ_d, and interaction terms as indicated in Eq. (4). Statistical significance of each variable was assessed based on p- and t-values. The selected criteria for identifying the significance of a parameter included: p-value < 0.05 = significant, < 0.10 = possibly significant, > 0.10 = not significant, and t-value < −2 or > + 2 = significant. The p-value indicates the significance of a parameter and the t-ratio value indicates the relative importance.

$$E_{LWD} \text{ or } \tau = b_0 + b_1 w + b_2 \gamma_d + b_3 w^2 + b_4 \gamma_d^2 + b_5 w \gamma_d \tag{4}$$

Where, b_o = intercept, b_1, b_2, b_3, b_4, and b_5 = regression coefficients.

Contours of $E_{LWD\text{-}Z1.0}$ and τ_G can be developed by applying the w, γ_d in the equation, which are presented from Figs. 4, 5 and 6 for "stiff" boundary model and "rigid" boundary model. The contours indicate that increasing w and decreasing γ_d generally decreases the E_{LWD} and τ values. Also the predicted values are compared with the actual in-situ measurements for similar material, and in-situ measurements that are in the range of moisture and density of laboratory samples are considered for the comparison, which only include two measurements.

Multiple regression model with rigid boundary condition is used to compare in-situ w-γ_d measurements at E_{LWD} test locations in Fig. 6. It combines all the data points collected from the same site at different locations, and soils were sampled from three locations to perform the standard proctor test and modified proctor test. Since all soils

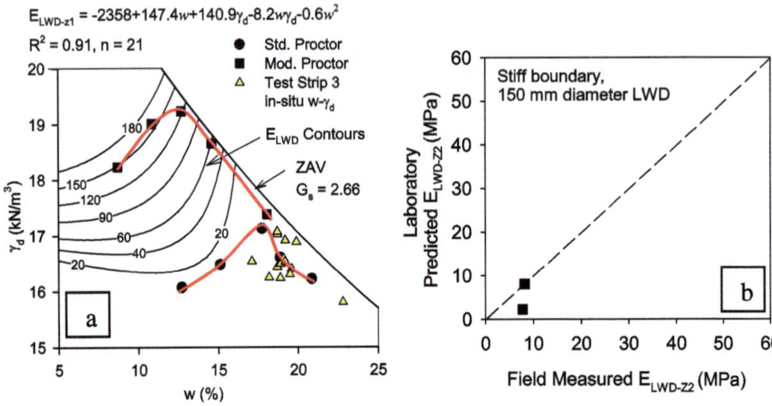

Fig. 4. Comparison between in situ LWD measurements and laboratory predicted LWD target values ("stiff" boundary model): (a) developed E_{LWD} contour; (b) comparison of predicted E_{LWD} and field measured E_{LWD}

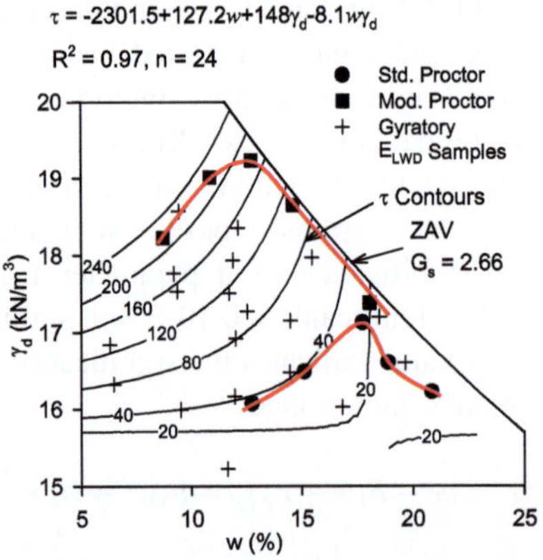

Fig. 5. Shear resistance contours in relationship with moisture and dry unit weight

exhibited similar compaction trend, this work combined all the data points as shown in Fig. 6a. The in-situ measurements showed w values in the range of 65% to 140% of w_{opt}. The predicted values are compared with actual in-situ measurement values which combined all the data points from multiple test strips as shown in Fig. 6b. The regression relationships indicate positive relationships and predicted E_{LWD} values are about 0.6 times the actual in-situ E_{LWD} measurements with the poor regression parameter of 0.22 due to the wide range of E_{LWD} values.

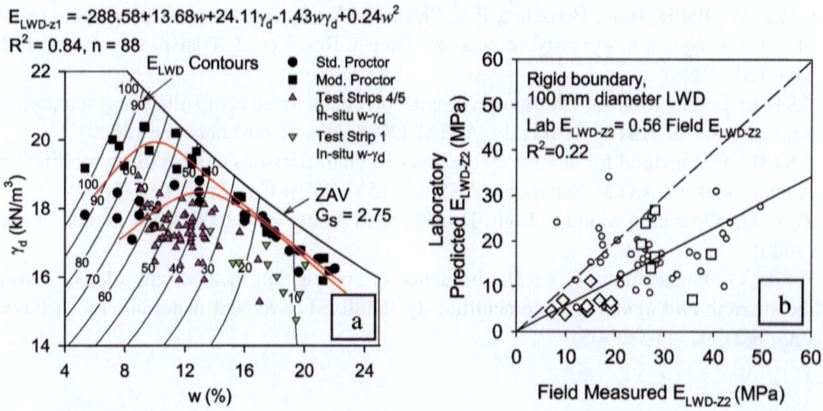

Fig. 6. Comparison between in situ LWD measurements and laboratory predicted LWD target values ("rigid" boundary model): (a) developed E_{LWD} contour; (b) comparison of predicted E_{LWD} and field measured E_{LWD}

5 Conclusions

Based on the tests results and analysis, the following conclusions can be drawn:

- Gyratory compaction is a more energy efficient method for subgrade compaction compared with impact compaction;
- Compared with impact compaction, gyratory compaction with PDA can provide the mechanistic based parameter that may be able to use as the compaction control parameter;
- Results from this research shown the feasibility of using gyratory compaction for subgrade compaction target value determination.

References

1. Coyle, H.M., West, E.C.: Laboratory compaction of a silty clay to simulate field density curves. M.S. thesis, Massachusetts Institute of Technology, MA (1956)
2. McRae, J.L.: Gyratory Testing Machine Technical Manual for Bituminous Mixtures, Soils, and Base Course Materials. Engineering Developments Company Inc., Vicksburg (1965)
3. Lee, K., Prezzi, M., Kim, M.: Subgrade design parameters from samples prepared with different compaction methods. J. Transp. Eng. **133**(2), 82–89 (2007)
4. Li, C., White, D.J., Vennapusa, P.: Moisture-density-strength-energy relationships for gyratory compacted geomaterials. Geotech. Test. J. **38**(4), 1–13 (2015)
5. Guler, M., Bahia, H.U., Bosscher, P.J., Plesha, M.E.: Device for measuring shear resistance of hot-mix asphalt in gyratory compactor. Transp. Res. Rec. **1723**, 116–124 (2000)
6. ASTM.: Standard test method for compaction and shear properties of bituminous mixtures by means of the U.S. corps of engineers gyratory testing machine (GTM), Annual book of ASTM Standards, ASTM D3387, West Conshohocken (2011)

7. Guler, M., Bahia, H.U., Bosscher, P.J., Plesha, M.E.: Device for measuring shear resistance of hot-mix asphalt in gyratory compactor. Transp. Res. Rec. J. Transp. Res. Board **1723**(1), 116–124 (1996)
8. ASTM.: Test method for laboratory compaction characteristics of soils using standard effort. Annual book of ASTM Standards, ASTM D698, West Conshohocken (2000)
9. ASTM.: Test method for laboratory compaction characteristics of soils using modified effort. Annual book of ASTM Standards, ASTM D1557, West Conshohocken (2000)
10. Zorn, G.: Operating Manual: Light Drop-Weight Tester ZFG2000. Zorn Stendal, Germany (2000)
11. White, G., Gnanendran, C.T.: The influence of compaction method and relative density on the strength and modulus of cementitiously stabilized pavement materials. Int. J. Pavement Eng. **6**(2), 97–110 (2005)

Impact of Distressed Modulus of Existing Concrete Pavement on Performance of Unbonded Concrete Overlay

Gauhar Sabih(✉) and Rafiqul A. Tarefder

Department of Civil Engineering, University of New Mexico, Albuquerque, NM, USA
gsabih@unm.edu

Abstract. Unbonded concrete overlay (UBCO) is used to improve the performance of the existing cracked jointed plain concrete pavement (JPCP). Unbonded JPCP overlay involves placement of an interlayer of hot mix asphalt (HMA), which acts as a separation layer between the existing concrete pavement and the JPCP overlay. One of the factors that affect the performance of unbonded JPCP overlays is existing pavement condition and the severity of damage of existing pavement immensely affects the performance of unbonded overlay. The severity of damage is described by the distressed elastic modulus thus, accurate determination of the distressed elastic modulus of the existing concrete pavement is essential for predicting the accurate performance of the unbonded overlay. This study focuses on analyzing the impact of distressed modulus by conducting simulations in the AASHTOWare pavement ME design software version 2.3 and evaluating the predicted performance of JPCP overlay for two different climatic regions. The results indicated that the distressed modulus of existing concrete pavement significantly affects the transverse cracking and pavement roughness of JPCP overlay while it has a minimal impact on joint faulting.

Keywords: Unbonded concrete overlay · Transverse cracking
Joint faulting · International roughness index · Distressed

1 Background

Unbonded concrete overlay (UBCO) is a layer of concrete placed over an existing concrete pavement to improve the performance of the existing pavement. The main purpose is to preserve and extend pavement service for years beyond the original design life. There are various design factors, which need to be considered for effective designing of concrete overlays. These include future design life, traffic volume, climate, existing pavement condition, joint details and properties of the pavement materials. Out of these factors, the existing pavement condition is a major factor which regulates the overlay performance. The existing condition of concrete pavement can be described by the distressed elastic modulus, which is a function of the level of damage to the existing concrete pavement.

X. Shi et al. (Eds.): GSIC 2018, *Proceedings of GeoShanghai 2018 International Conference: Transportation Geotechnics and Pavement Engineering*, pp. 319–326, 2018.
https://doi.org/10.1007/978-981-13-0011-0_34

Unbonded concrete overlay acts like an independent pavement resting on a stable base, which is the existing concrete pavement. Various phenomenon responsible for the behavior of new concrete pavement are also valid for concrete overlays. Selezneva et al. [1] identified the material characteristics of concrete including strength, CTE and ultimate shrinkage as the key design factors that affect the structural performance of continuously reinforced concrete pavements. Sabih and Tarefder [2] investigated the effects of variability of mechanical and thermal properties of concrete on mechanistic-empirical performance predictions of concrete pavements. They found that concrete properties such as elastic modulus, flexural strength and coefficient of thermal expansion significantly affect the performance of concrete pavement over the pavement design life.

There is not much research found on impact of existing pavement condition on performance of unbonded overlay and this study focuses on the effects of distressed elastic modulus of existing jointed plain concrete pavement (JPCP) on the performance of unbonded JPCP overlays. The Pavement ME Design procedure considers the elastic modulus of existing concrete pavement as an input for subsequent design of JPCP overlay.

1.1 Existing Pavement Condition

The condition of existing concrete pavement can be described by the elastic modulus of concrete, which can be obtained by coring and laboratory testing from an intact slab. Since existing concrete pavements have various levels of damage/distress, ascertaining a distressed elastic modulus is a difficult task. For a concrete pavement without any signs of damage, distress and cracks, the elastic modulus value obtained from the laboratory is reasonable. However, when the existing pavement is distressed/cracked, the distressed elastic modulus cannot be obtained from the laboratory tests of the intact cores as this will provide the elastic modulus as if the slab is intact. Thus, the intact elastic modulus needs to be converted to distressed modulus.

2 Performance Indicators and Prediction Models

The performance indicators for unbonded JPCP overlays are joint faulting, transverse cracking and pavement roughness (measured as international roughness index, IRI) as per the pavement ME design.

2.1 Transverse Cracking Model

This parameter is calculated as percent of slabs with transverse cracks. Transverse slab cracking predictions are calculated from a set of equations as follows [3]:

$$\log(N_{allowable}) = C_1(\frac{MOR}{\sigma_{PCC}})C_2 \quad (1)$$

$$Crack = \frac{100}{1 + C_4 * (\frac{N_{applied}}{N_{allowable}})^{C_5}} \quad (2)$$

where MOR = Modulus of rupture of the concrete; σ = Critical stress in the slab; $N_{applied}$ = Applied number of load applications; $N_{allowable}$ = Allowable number of load applications; C_1, C_2, C_4, C_5 = Calibration coefficients.

2.2 Joint Faulting Model

Transverse joint faulting is the differential elevation across the transverse joint. The mean faulting of all transverse joints in a pavement section is the parameter predicted by the ME design software. The primary faulting prediction model is shown in Eq. 3 [3].

$$FaultMax_0 = C_{12} \cdot \delta_{curl} \cdot [\log(1 + C_5 \cdot 5^{Erod}) \cdot \log(\frac{P_{200} \cdot Wetdays}{P_s})]^{C6} \quad (3)$$

where $FaultMax_0$ = Initial maximum mean transverse joint faulting; δ_{curl} = Maximum mean monthly slab corner upward deflection due to temperature curling; Ps = Overburden on subgrade; P_{200} = Percent subgrade material passing #200 sieve; *Wet Days* = Average annual number of wet days; *C 5, 6, 12* = Calibration coefficients.

2.3 Pavement Roughness Model

Pavement roughness is generally defined as an expression of irregularities in the pavement surface that adversely affect the ride quality of a vehicle. Roughness is typically quantified using international roughness index (IRI) and it is used to define a characteristic of the longitudinal profile of a traveled wheel-track [4]. IRI prediction model is as follows:

$$IRI = IRI_{ini} + C_1 \cdot Crack + C_2 \cdot Spall + (C_3 \cdot Fault \cdot 5280/JSP) + C_4 \cdot SF \quad (4)$$

Where IRI = Predicted IRI; IRI_{ini} = Initial smoothness measured as IRI; *Crack* = Percent slabs with transverse cracks (all severities); *SPALL* = Percentage of joints with spalling (medium and high severities); *Fault* = Total joint faulting cumulated; *SF* = Site factor; $C_{1,\ 2,\ 3,\ 4}$ = Calibration coefficients; *JSP* = Transverse joint spacing.

3 Objective

The objective of this study is to compare the existing methods of transforming the intact elastic modulus of existing JPCP into the distressed elastic modulus and analyze the impact of distressed elastic modulus on the mechanistic-empirical performance of unbonded concrete overlays.

4 Determination of Distressed Modulus of Existing JPCP

For this study, the intact modulus of existing JPCP was assumed as constant with a value of 28.94 GPa (4.2 E6 psi). This value was obtained from an average of the back calculated elastic modulus values from FWD test results as conducted by Glover and Stanley [5]. The distressed elastic modulus was determined according to the three methods available in the literature [5–7]. The summary of the distressed elastic modulus of existing JPCP is tabulated in Table 1.

Table 1. Summary of distressed elastic modulus of existing JPCP

Adjustment factor (C)	Condition of damage	Detailed condition	Distressed modulus (GPa)
USACE method			
1	Good	Pavement in good condition	28.9
0.75	Moderate	Pavements exhibit initial cracking due to load	21.7
0.35	Severe	Pavements exhibit progressive cracking	10.1
ME design method			
0.58	Good	Less than 10% slabs cracked	16.8
0.32	Moderate	10 to 50% slabs cracked	9.2
0.13	Severe	More than 50% slabs cracked	3.7
Glover and stanley method			
0.92	Intact	No cracking	26.5
0.6	Low severity	Slab is broken into two pieces	17.3
0.45	Medium	Slab is broken into three pieces	13
0.35	High severity	Slab is broken into four pieces	10.1

5 Simulation Methodology

Entire simulation work was conducted in Pavement ME design software version 2.3, which is the latest tool for pavement design and analysis. The design inputs required by Pavement ME design for design of unbonded JPCP overlay were kept constant except for the distressed elastic modulus of the existing JPCP. The design inputs used in the simulation are as shown in Table 2.

The simulations were conducted with these parameters for two climatic regions i.e. moderate climate of Albuquerque (New Mexico) and cold climate of Alamosa (Colorado) to observe the effects of climatic factors on the overlay performance along with the impact of distressed modulus with a constant CTE of overlay concrete.

Table 2. Design inputs for simulation

Parameter	Value
Design life	30 years
JPCP overlay thickness	203 mm (8 in)
Dowel diameter	25 mm (1 in)
Traffic (AADTT)	4000
Initial IRI	0.99 m/km (63 in/mile)
Terminal IRI	2.71 m/km (172 in/mile)
Transverse cracking limit (% Slabs)	15%
Mean joint faulting limit	3 mm (0.12 in)
Reliability	90%
Modulus of rupture of overlay concrete	4.75 MPa (690 psi)
Elastic modulus of overlay concrete	28.9 GPa (4.2×10^6 psi)
Water to cement ratio in overlay concrete	0.42
CTE of overlay Concrete	8.8 $\mu\varepsilon/^{\circ}$C (4.9 $\mu\varepsilon/^{\circ}$F)
HMA (Interlayer) Thickness	50.8 mm (2 in)
HMA Binder Grade	PG 64-22
Existing JPCP Thickness	203 mm (8 in)
Base Course Thickness	254 mm (10 in)
Base Course Resilient Modulus	275 MPa (40000 psi)

6 Simulation Results

The performance parameters of JPCP overlay were evaluated to contrast the effects of distressed elastic modulus on the performance of JPCP overlay for both the climatic regions. It became evident from the simulation results that there is a huge variation in the performance of JPCP overlay with varying distressed modulus and between the two regions. The main reason is the temperature differential of the two regions. The number of freeze thaw cycles in cold region is much greater than the moderate region so along with other factors, the amount of curling also counts for the disparity in the performance results in addition to the existing pavement condition.

6.1 Effects of Distressed Modulus on Transverse Cracking of UBCO

The simulation results of distressed elastic modulus and the transverse cracking of the JPCP overlay over the design life for both climatic regions are shown in Fig. 1(a and b). It is evident from the results that the distressed elastic modulus of existing pavement has a significant impact on the transverse cracking of JPCP overlay and as the distressed modulus decreases, the transverse cracking percentage in the overlay increases, which means that the overlay performance is adversely affected. The distressed modulus is directly related to level of damage to the existing pavement, the higher the level of damage of the existing JPCP, the lower the performance of JPCP overlay. The lower distressed modulus is an indicator of more damaged/cracked existing pavement, which may not be able to provide strong foundation for the overlay resulting in more distresses in overlay pavement. When the distressed modulus falls below 13 GPa (1.89×10^6 Psi) in the moderate climatic region, the transverse cracking percentage increases from 15%, which is the acceptable limit fixed for this overlay design. While in the cold region, all the simulated overlays failed earlier than the designed service life according to the performance threshold. The main reason of this failure is the increase in magnitude of curling which increases the bending stresses in the overlay pavement slab.

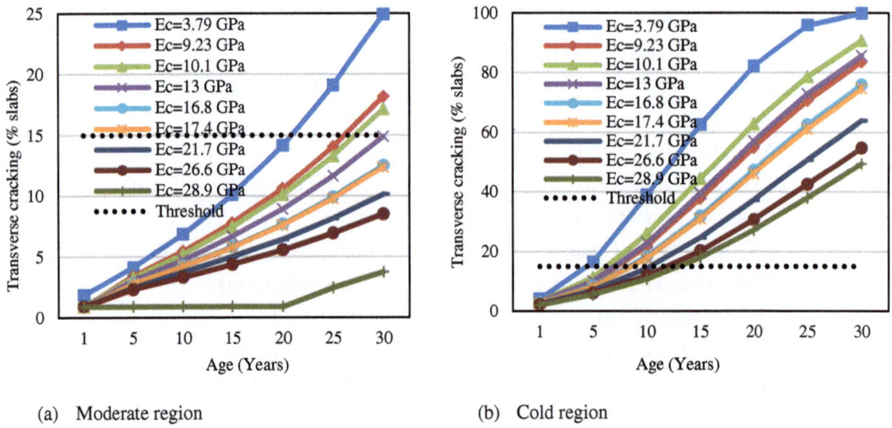

Fig. 1. Effects of distressed modulus on transverse cracking of UBCO

6.2 Impact of Distressed Modulus on IRI of UBCO

The impact of distressed modulus on the IRI of JPCP overlay across the design pavement life for both climatic regions is shown in Fig. 2(a and b). A deliberate analysis of the results makes it clear that distressed elastic modulus of existing JPCP has a direct effect on the terminal IRI of the designed overlay. A lower value of the

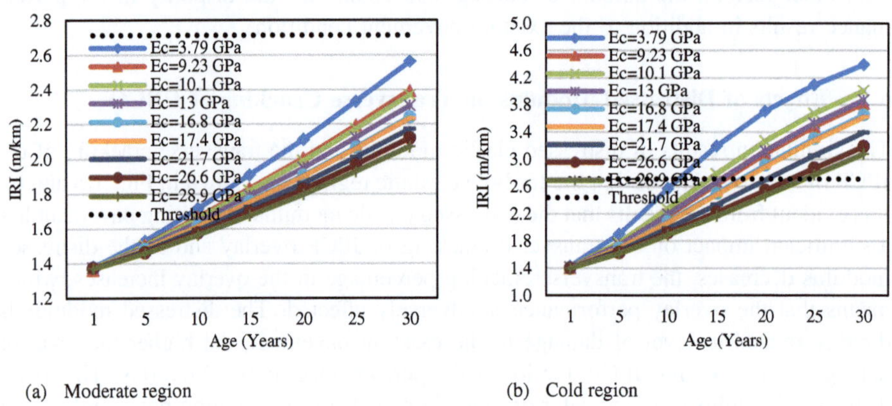

Fig. 2. Impact of distressed modulus on pavement roughness.

distressed modulus results in a higher value of terminal IRI, which means decreased performance of JPCP overlay. This adverse performance of overlay is a function of decreased concrete strength due to damage to existing concrete pavement. According to the results of the moderate climatic region, the distressed modulus impacts the terminal IRI but all the simulated overlays are within the threshold limit of 2.71 m/km (172 in/mile). The range of this impact is 0.55 m/km (35 in/mile). The results of cold region also indicate the adverse impact of distressed modulus on the IRI over the design life. The range for this region is 1.32 m/km (84 in/mile), which means that the impact in cold region is more severe as compared to the moderate region.

6.3 Effects of Distressed Elastic Modulus on Joint Faulting of UBCO

The simulation results for effect of distressed elastic modulus on mean joint faulting of JPCP overlay for both climatic regions have been displayed in Fig. 3(a and b). It is obvious from the results that with decrease in distressed modulus value the terminal joint faulting increases but the increase in faulting is not significant. The range of the impact is 0.76 mm (0.03 in) for the moderate region and 1.52 mm (0.06 in) for cold region. All the simulated cases for moderate climatic region stays within limit of the threshold limit, while all the cold region cases fail according to the prescribed joint faulting limit which means that these should be revised with different pavement properties. This phenomenon can be related to presence of dowel bars as joint faulting is controlled by the action of dowel bars. Hence, it is evident that effect of distressed modulus on joint faulting is minimal.

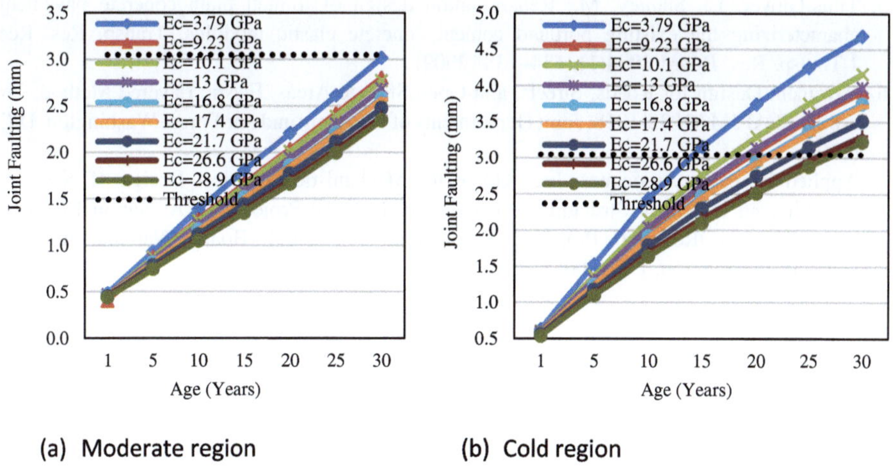

Fig. 3. Effect of distressed elastic modulus on joint faulting.

7 Conclusion

The distressed elastic modulus which is a measure of level of damage of existing concrete pavement has a significant impact on overlay performance. Transverse cracking and IRI are more sensitive to distressed modulus as compared to joint faulting and the impact of existing pavement condition is more severe in the cold climatic region as compared to moderate region. The performance of UBCO decreases significantly with a decrease in distressed modulus. The methods to determine the distressed elastic modulus of existing concrete pavements are mostly empirical so there is a need to develop mechanistic procedures for improved unbonded overlay design. The performance of JPCP overlay can be improved by increasing the overlay thickness and dowel diameter or by reducing the joint spacing.

References

1. Selezneva, O., Rao, C., Darter, M., Zollinger, D., Khazanovich, L.: Development of a mechanistic-empirical structural design procedure for continuously reinforced concrete pavements. Transp. Res. Rec. **1896**, 46–56 (2004)
2. Sabih, G., Tarefder, R.A.: Impact of variability of mechanical and thermal properties of concrete on predicted performance of jointed plain concrete pavements. Int. J. Pavement Res. Technol. **9**, 436–444 (2016)
3. AASHTO, Mechanistic – Empirical Pavement Design Guide, A Manual of Practice, American Association of State Highway and Transportation Officials (2008)
4. Abd El-Hakim, R., El-Badawy, S.: International roughness index prediction for rigid pavements: an artificial neural network application. Adv. Mater. Res. **723**, 854–860 (2013)
5. Titus-Glover, L., Stanley, M.: Rehabilitation design of jointed plain concrete pavement: characterizing the existing portland cement concrete elastic modulus. Transp. Res. Rec. J Transp. Res. Board **2084**(1), 134–138 (2009)
6. Pavement Design for Roads, Streets, and Open Storage Areas, Elastic Layered Method. TM 5-822-13/AFJMAN 32-1018. Joint Departments of the Army and Air Force, Washington, D.C. (1994)
7. Applied Research Associates, Inc., Guide for Mechanistic-Empirical Design of New and Rehabilitated Pavement Structures, Final Report, NCHRP Project 1-37A, National Cooperative Highway Research Program, Transportation Research Board, National Research Council, Washington, DC (2004)

Effects of Modulus of Rupture of Concrete on Performance of Unbonded Jointed Plain Concrete Overlay

Gauhar Sabih(✉) and Rafiqul A. Tarefder

Department of Civil Engineering, University of New Mexico, Albuquerque, NM, USA
gsabih@unm.edu

Abstract. Rehabilitation of existing jointed plain concrete pavements (JPCP) by using unbonded concrete overlay (UBCO) is increasingly being used all over the world. The design and performance of un-bonded overlays is affected by various factors including climate, traffic volume, concrete material properties, existing pavement condition and future design life. Modulus of rupture (MOR)/flexural strength of concrete is an important material property that has significant impact on the performance of concrete pavements. A true estimation of modulus of rupture would improve the accuracy of cracking prediction. This study examines the impact of variability of flexural strength of concrete on the mechanistic-empirical performance predictions of UBCOs. Simulations were conducted using pavement ME design software with varying flexural strength values and keeping all other design variables as constant. The results indicated that MOR has a significant impact on overlay performance. With decrease in concrete's MOR, the performance of UBCOs is adversely affected with increased transverse cracking. The adverse effects of lower flexural strength can be compensated by increasing the thickness of overlay concrete slab or by decreasing the transverse joint spacing.

Keywords: Unbonded · Overlay · Modulus of rupture · Transverse cracking Concrete

1 Background

The increasing highway traffic volume has led to the early deterioration of highway pavements and a greater dependence on the rehabilitation techniques of existing distressed pavements. Jointed plain concrete pavement (JPCP) is the most widely used rigid pavement type in United States and unbonded concrete overlay (UBCO) is a cost-effective rehabilitation technique to improve the performance of the existing deteriorated JPCP. UBCO is a sustainable solution for improved management of pavement assets and the main purpose of unbonded overlay is to restore the structural capacity of a deteriorated/damaged pavement. Factors that affect the performance of UBCO include traffic volume, climate, concrete material properties, existing pavement condition and design life.

X. Shi et al. (Eds.): GSIC 2018, *Proceedings of GeoShanghai 2018 International Conference: Transportation Geotechnics and Pavement Engineering*, pp. 327–334, 2018.
https://doi.org/10.1007/978-981-13-0011-0_35

The latest pavement design and performance prediction tools such as the "pavement mechanistic empirical (ME) design software" has provided the opportunity for analyzing the performance of concrete pavements rehab designs. MOR is the basis for estimating flexural fatigue in concrete. In pavement ME design, the damage calculated for the estimation of transverse cracking is a function of the MOR of the concrete. A true estimation of modulus of rupture would improve the accuracy of cracking prediction. Stoffels et al. (2014) conducted experimental study on correlating concrete flexural strength and fracture parameters and its impact on structural performance and found that concrete MOR is closely related to elastic modulus values. Although MOR is an important parameter in evaluating the design of rigid pavement, it was not given due importance in the past. With the advent of pavement ME design, a lot of emphasis has been given to accurate determination of modulus of rupture and its usage in design of rigid pavement systems.

Sabih and Tarefder (2016) investigated the effects of variability of mechanical properties of concrete on performance of concrete pavements and found that concrete properties such as elastic modulus and MOR significantly affect the performance of JPCP over the pavement design life. Guclu et al. (2009) conducted sensitivity analysis of various inputs/factors on the performance of rigid pavements and found that transverse cracking of concrete pavements is significantly affected by concrete strength properties including flexural strength. Selezneva et al. (2004) identified the material characteristics of concrete including flexural strength, coefficient of thermal expansion (CTE) and ultimate shrinkage as the key design factors that affect the structural performance of concrete pavements. Vandenbossche et al. (2011) evaluated the effects of concrete material properties, pavement structural parameters and the pavement ME design standard fatigue damage and cracking performance curve on slab cracking predictions. The sensitivity analysis suggested that small changes in input values of concrete strength properties including MOR can lead to significant changes in predicted performance. Schwartz et al. (2011) evaluated the effects of concrete stiffness and strength on performance of rigid pavements and found that accurate determination of stiffness values of concrete will result in improved and accurate design of concrete pavements. The available literature is focused on the single layer concrete pavements and there was a need to evaluate the impact of concrete MOR on performance of UBCO. This study focuses on the effects of variability of MOR of concrete on the performance of UBCO and the measures to mitigate the effects of lower MOR.

2 Objectives

- Quantifying the impact of MOR of concrete on the performance of unbonded JPCP overlays including joint faulting, transverse cracking and pavement roughness.
- Evaluating the factors which can minimize the adverse effects of lower MOR of paving concrete to achieve the required pavement performance.

3 Pavement ME Design Procedure

The pavement ME design procedure was originally developed under the National Cooperative Highway Research Program (NCHRP) project 1-37A and further revised under continuing research projects. The procedure was eventually adopted by AASHTO as the standard for pavement design in 2008 (AASHTO 2008). Pavement analysis and design can currently be performed using the software program commonly referred to as Pavement ME design (Rao 2015). Pavement ME design is based on mechanistic-empirical design concepts meaning that the design procedure calculates pavement responses such as stresses, strains, and deflections under axle loads and climatic conditions and then accumulates the damage over the design analysis period. The procedure then empirically relates calculated damage over time to pavement distresses and smoothness based on performance of actual projects throughout the United States. ME design uses a mix of algorithms and models to characterize pavement foundation, structure, layer materials, traffic, and climate and simulate stress/strains/deflection due to the interactions between applied traffic load and climate. The resulting damage manifested as different performance parameters over the design life of a pavement is then calculated (Mallela 2014).

3.1 Performance Prediction Models for Unbonded JPCP Overlays

The performance indicators for unbonded JPCP overlays are joint faulting, transverse cracking and pavement roughness. The threshold and reliability limits for each of the performance indicator is required to be set by the designer (NCHRP 2014).

4 Simulation Methodology

To evaluate the effects of varying MOR of concrete on performance of UBCO, simulations were conducted in pavement ME design software version 2.3 with different values of modulus of rupture ranging from 3.4 to 5.7 MPa while keeping all other design variables as constant. The design parameters are tabulated in Table 1. The effects on the performance parameters were analyzed to contrast the effects of modulus of rupture on overlay performance.

5 Simulation Results and Analysis

5.1 Impact of MOR Variation on Transverse Cracking

The analysis of MOR variation on the transverse cracking parameter of UBCO was conducted and the comparison of results of transverse cracking is presented in Fig. 1. The results show a significant increase in cracking distress with decrease in MOR of overlay concrete with all other design variables being constant. The overlay fails to meet the design criteria of transverse cracking at the age of 26 years when MOR value reaches 4.59 MPa and with further decrease in MOR the overlay fails even earlier. The

Table 1. Design inputs for simulation.

Parameter	Value
Design life	30 years
JPCP overlay thickness	203 mm
Transverse joint spacing	4.57 m
Dowel diameter	25 mm
Traffic (average annual daily truck traffic)	4000
Initial IRI	0.99 m/km
Terminal IRI	2.71 m/km
Transverse cracking limit (% slabs)	15%
Joint faulting limit	3 mm
Reliability	90%
Existing JPCP slab thickness	203 mm
Distressed elastic modulus of existing pavement	16.8 GPa
Coefficient of thermal expansion of overlay concrete	8.8 $\mu\varepsilon/^{\circ}C$
HMA (interlayer) thickness	50.8 mm
Base course	Non-stabilized
Base course thickness	254 mm
Base course resilient modulus	275 MPa

general trend of overlay performance confirms that higher MOR is associated with higher concrete strength to withstand the fatigue loading resulting in lower transverse cracking. Also, according to the pavement ME design prediction model, cracking is a function of MOR which is compatible with the results of the simulations conducted in this study. To minimize the performance issues, either higher MOR concrete should be used or the pavement structure needs to be revised/modified such that overlay can perform within the prescribed performance threshold

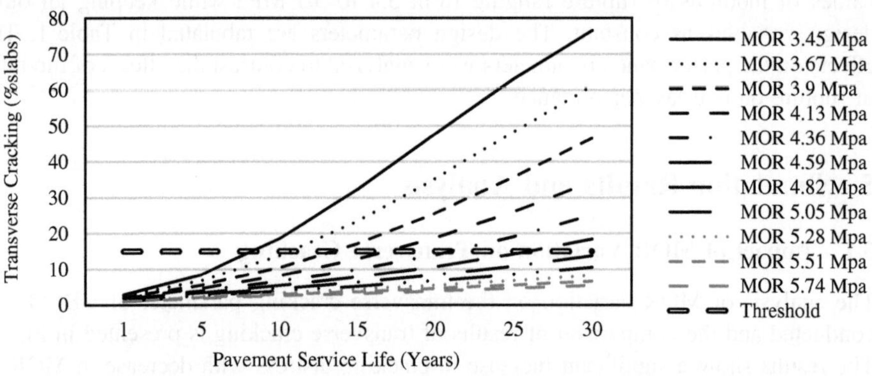

Fig. 1. Effects of MOR on transverse cracking over design life

5.2 Effects of MOR on Pavement Roughness

The effects of MOR variation on the IRI were evaluated and the results are shown in Fig. 2. The overlay fails to meet the pavement roughness criteria at the age of 28 years when MOR reaches the value of 3.45 MPa. The decline in the performance of UBCO with regards to pavement roughness with decrease in MOR of overlay concrete is principally attributed to higher transverse cracking while other design variables being constant. According to the pavement ME design prediction model, IRI is dependent on cracking distress along with other factors. To mitigate the effects of higher CTE and to improve the overlay performance the overlay design needs to be modified to achieve the ultimate design life.

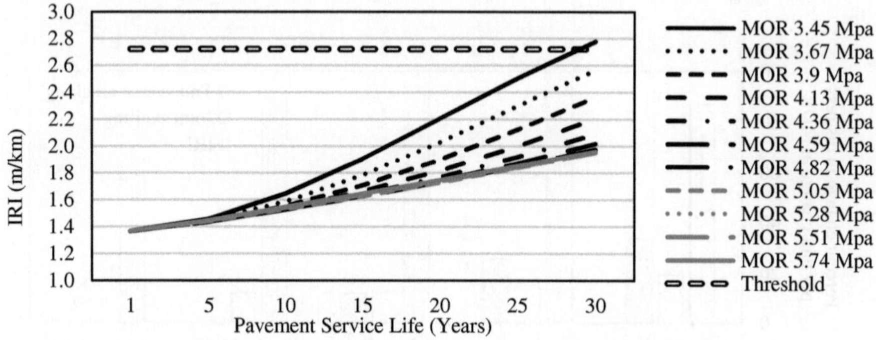

Fig. 2. Effects of MOR on pavement roughness over design life

5.3 Effects of MOR on Joint Faulting

The impact of MOR variation on the performance of overlay with regards to joint faulting was analyzed and the results are presented in Fig. 3. It was observed that there was a slight increase in joint faulting with increasing MOR. The total variation between the minimum and maximum values of faulting is in the range of 0.5 mm which is not very significant as compared to other performance measures.

5.4 Sensitivity Analysis of Performance Parameters

As variation in performance predictions was observed with varying MOR of overlay concrete thus, a sensitivity analysis was performed to quantify the impact of MOR on the three performance parameters and to determine the most significantly affected performance measure. The normalized percent change for cracking, faulting and IRI were obtained for a unit change in MOR for different ranges of MOR values and the results are shown in Fig. 4. According to the results, the transverse cracking is the most affected parameter with percent change of up to 102.3%. Joint faulting is the least affected performance parameter with percent change of up to 19.3%. Pavement roughness is a little more affected then the joint faulting.

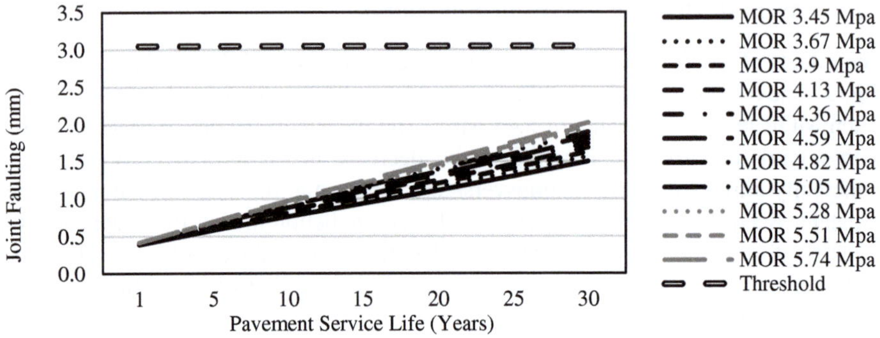

Fig. 3. Effects of MOR on joint faulting over pavement service life

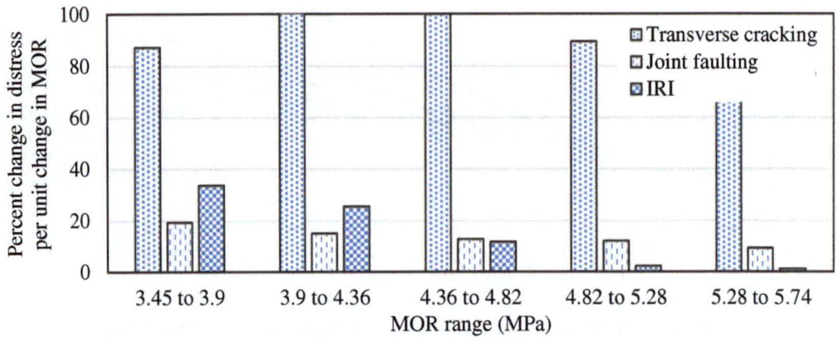

Fig. 4. Sensitivity analysis of performance predictions

6 Minimizing the Effects on Overlay Performance

According to the performance criteria set forth for this study, the JPCP overlay fails if it does not comply to the prescribed performance limits until the end of design life. As evident from the results shown earlier that there are number of design simulations that do not meet the performance threshold of cracking and roughness. There are many inputs which can be revised to improve the performance of the overlays. Out of these, increased overlay slab thickness and decreased transverse joint spacing were evaluated to mitigate the adverse effects of lower concrete flexural strength.

6.1 Effects of Overlay Slab Thickness

Figure 5 presents the effects of increase in slab thickness on pavement performance for different values of MOR. It is evident with the results that with increase in overlay slab thickness transverse cracking and pavement roughness decreases which means improved overlay performance and this procedure can be used to mitigate the performance issues.

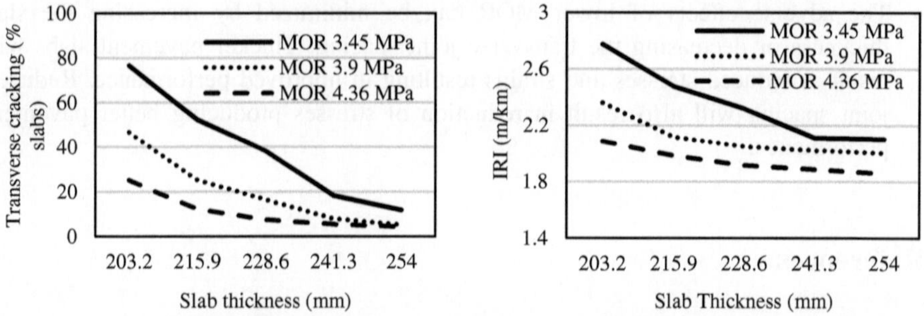

(a) Transverse cracking (b) Pavement roughness

Fig. 5. Effects of slab thickness on performance

6.2 Effects of Transverse Joint Spacing

The simulations were revised with decreased transverse joint spacing to curtail the adverse effects of lower concrete MOR and it was observed that lower joint spacing results into improved overlay performance as it reduces the fatigue loading stresses in the concrete slab. Figure 6 presents the results of varied joint spacing and its impact on the performance parameters. It is evident that as the transverse joint spacing decreases the overlay performance improves with regards to transverse cracking and pavement roughness. So, the adverse effects of lower MOR of concrete can be mitigated by decreasing the transverse joint spacing.

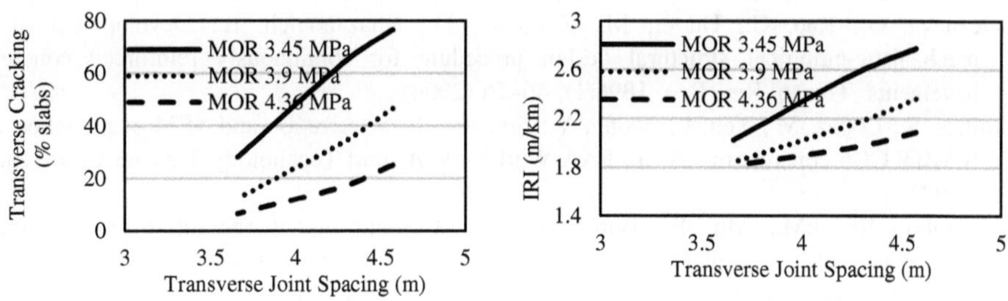

(a) Transverse cracking (b) Pavement roughness

Fig. 6. Effects of joint spacing on performance

7 Conclusions

- Concrete MOR has a significant impact on the performance of unbonded JPCP overlay. Transverse cracking is the most affected parameter with a percent increase of 75.4 to 102.3% with a unit decrease in MOR of overlay concrete. Whereas, there is a slight effect on pavement roughness and joint faulting. As concrete strength decreases the resulting overlay distresses increases and performance is adversely affected.

- The adverse effects of lower MOR can be minimized by increasing the slab thickness or decreasing the transverse joint spacing. Thicker pavement slabs will result in reduced stresses and strains resulting in improved performance. Reduced joint spacing will also result in reduction of stresses producing better pavement performance.

References

AASHTO: Mechanistic – Empirical Pavement Design Guide, A Manual of Practice. American Association of State Highway and Transportation Officials (2008)

Guclu, A., Ceylan, H., Gopalakrishnan, K., Kim, S.: Sensitivity analysis of rigid pavement systems using the mechanistic-empirical design guide software. J. Transp. Eng. **135**(8), 555–562 (2009)

Mallela, J., Titus-glover, L., Bhattacharya, B., Darter, M., Von-Quintus, H.: Idaho AASHTO-Ware Pavement ME Design User's Guide (2014)

NCHRP: Implementation of the AASHTO Mechanistic-Empirical Pavement Design Guide and Software (2014)

Rao, C., Varner, R., Barstis, W.: PCC Material Characterization in Mississippi for AASHTOWare Pavement ME Design (2015)

Sabih, G., Tarefder, R.A.: Impact of variability of mechanical and thermal properties of concrete on predicted performance of jointed plain concrete pavements. Int. J. Pavement Res. Technol. **9**, 436–444 (2016)

Schwartz, C., Li, R., Kim, S., Ceylan, H., Gopalakrishnan, K.: Effect of concrete strength and stiffness characterization on predictions of mechanistic-empirical performance for rigid pavements. Transp. Res. Rec. J. Transp. Res. Board **2226**, 41–50 (2011)

Selezneva, O., Rao, C., Darter, M., Zollinger, D., Khazanovich, L.: Development of a mechanistic-empirical structural design procedure for continuously reinforced concrete pavements. Transp. Res. Rec. **1896**(1), 46–56 (2004)

Stoffels, S., Lopez, M., Yeh, L., Jeong, Y.: Fracture characterization and SEM examination of NAPTF CC6 concrete mixes. In: FAA Worldwide Airport Technology Transfer Conference (2014)

Vandenbossche, J.M., Mu, F., Burnham, T.R.: Comparison of measured vs. predicted performance of jointed plain concrete pavements using the mechanistic–empirical pavement design guideline. Int. J. Pavement Eng. **12**(3), 239–251 (2011)

Error Correction of Ground-Coupled Antenna in Thickness Measurement Based on Superposition Principle

Xing-yu Gu, Tian-jie Zhang, Qiao Dong(✉), and Shan-Shan Tu

School of Transportation, Southeast University, Nanjing 210096, Jiangsu, China
qiaodong@seu.edu.cn

Abstract. As a non-destructive method in pavement detection, ground penetrating radar (GPR) has been attached more importance. Because of different coupling modes of radar antenna, the error, however, detecting pavement thickness is larger in the utilization of ground coupling. Based on the superposition principle of electromagnetic wave, the superposition problem of the antenna coupled wave and the wave reflected by ground is analyzed in the detection of pavement thickness by GPR. In this article, a method for error correction in thickness measurement using ground-coupled system is proposed and the proposed method can effectively reduce the error of thickness measurement.

Keywords: Ground penetrating radar · Ground-coupled antenna
Superposition principle · Error correction

1 Introduction

Ground penetrating radar (GPR) is a non-destructive detecting technique based on the principle of electromagnetic wave propagation. In pavement detection, GPR has rapid and nondestructive characteristics comparing with other traditional detection techniques such as drilling core or nuclear density gauge [1–4]. As a key component of GPR, the antenna will directly affect the detecting depth and resolution ratio. GPR antenna can be classified as ground-coupled antenna and air-coupled antenna according to coupling media. Impossibility for high speed detection and superposition problem of coupled wave are major flaws for ground-coupled antenna. The method for effective error correcting, however, has not been proposed so far [5]. Some discuss the suitable antenna for GPR based on modeling in Finite Difference Time-Domain (FDTD) [6]. GPR system usually uses ultra-wideband short pulse signal, the antenna also has certain bandwidth and directivity due to the existence of ultra-wideband short pulse signal in GPR system. Because of the impedance discontinuity, the antenna often forms a certain reflection at the end resulting in distortion of waveform [7].

X. Shi et al. (Eds.): GSIC 2018, *Proceedings of GeoShanghai 2018 International Conference: Transportation Geotechnics and Pavement Engineering*, pp. 335–343, 2018.
https://doi.org/10.1007/978-981-13-0011-0_36

2 Utilization of Ground Penetrating Radar on Thickness Detection

GPR is designed based on reflection of electromagnetic wave. The radar wave produced by emitter is a wideband and high-frequency pulse electromagnetic wave, the frequency of antenna, for common use, ranges from 1 GHz to 2 GHz for detecting the thickness of asphalt concrete pavement. The frequency guarantees the accuracy of radar and detection depth as required, significant reduction of dispersion of asphalt concrete and insurance of the accurate dielectric constant [8].

Wave reflection occurs at the interface of mediums with different dielectric constant, which is the basic principle in pavement detection. Pavement is a layered structure, so the dielectric constant between the layers varies. When radar waves transfer from antenna to pavement, they firstly reflect at the interface between air and pavement surface. Since the relative permittivity of air is 1, most energy is absorbed by pavement surface. When passing through the layers, the electromagnetic wave will reflect and refract at the interfaces as shown in Fig. 1. According to parameters such as amplitude and time received by GPR, thickness, compaction and moisture content of pavement structure can be analyzed and calculated.

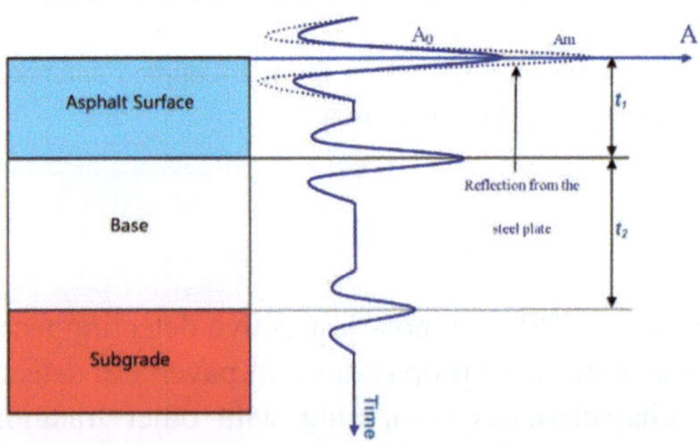

Fig. 1. Principle of GPR pavement detection

The most important step to evaluate pavement performance based on GPR is to calculate the dielectric constant of the pavement structure by analyzing the reflected signal. Measure the amplitude of the wave crest of radar echo waveform, and substitute it into formula (1).

$$\sqrt{\varepsilon} = \frac{1 + \frac{A_0}{A_m}}{1 - \frac{A_0}{A_m}} \tag{1}$$

Where A_0 is the amplitude of an electromagnetic wave reflected between air and pavement; Am is the amplitude of total reflection.

Calculate the thickness of pavement according to formula (2), (3).

$$H = v * \frac{t}{2} \tag{2}$$

$$V = \frac{C}{\sqrt{\varepsilon}} \tag{3}$$

Where H is the thickness of pavement layer; C is the speed of light when traveling in a vacuum; t is the time for electromagnetic waves propagating in the pavement layer.

2.1 Coupling Mode of Antenna

The difference between ground-coupled antenna and air-coupled antenna is the height from the ground. As shown in Fig. 2, the height from the ground of ground-coupled antenna is 0 (H = 0) while height from the ground of air-coupled antenna is more than 0 (H > 0). Due to unevenness in pavement surface, ground-coupled system can only detect pavement surface in a low speed while air-coupled system can conduct high-speed detection.

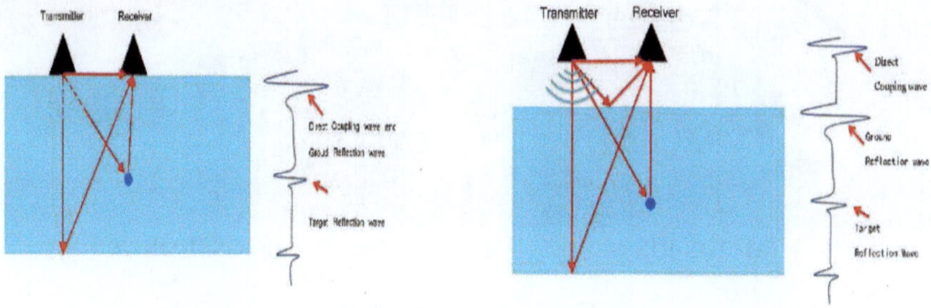

(a) Ground-coupled antenna (b) Air-coupled antenna

Fig. 2. Schematic diagram of antennas with different coupling types

Figure 2(b) shows that the directly coupled wave of antenna and reflected wave by the ground occur separately. In this way, the calculation of dielectric constant can be drawn by measuring the amplitude of the ground reflecting wave and the total reflection amplitude. Because of the weak electromagnetic wave emitted by ground-coupled antenna, it is difficult to penetrate the air, so the antenna emitter and receiver must be close to the ground, which leads to the superposition of directly coupled wave and reflection wave. It makes it hard to calculate the real amplitude of ground reflection wave, which results in error in thickness measurement.

3 Testing Program

The experiment is carried out using IDS-RIS radar (TR-HF: 1600 MHz ground-coupled antenna) and GSSI-SIR radar (2 GHz air-coupled antenna). The objectives of the experiment are to recognize the existence of the coupling wave of antenna and to analyze the error conducting thickness measurement. Pavement specimens with different materials and layer thickness were measured by the radar at different heights.

30 cm * 30 cm asphalt concrete slabs with thickness of 3 cm, 5 cm and 7 cm for rutting test were fabricated as different layers. A total of 9 kinds of layered structure were prepared including 3 cm, 5 cm, 7 cm, 5 cm + 7 cm, 3 cm + 7 cm, 3 cm + 5 cm, 5 cm + 5 cm, and 5 cm + 3 cm + 7 cm. A 40 cm * 40 cm * 5 cm steel plate was placed at the bottom to enhance wave reflection.

The radar antenna was firstly placed at the height of 4.7 cm and 9.4 cm above the specimens to show the superposition of radar waves and then placed on the top surface of asphalt concrete slabs to measure the thickness (Fig. 3).

(a) IDS-RIS (b) GSSI SIR-20

Fig. 3. Radar test process

4 Results and Discussion

4.1 TR-HF Ground-Coupled Antenna

Firstly, the ground-coupled antenna “TR-HF” with a frequency of 1600 MHz was used to test the metal plate and asphalt concrete slab at different heights. The results as shown in Fig. 4.

It can be seen from Fig. 4 that amplitude of the first wave crest is regardless whether no matter what height the antenna is placed above the plate 9.4 cm or 4.7 cm (the wavelength of electromagnetic wave in the 1600 MHz is 18.8 cm), which means that the size of the first wave and the height of radar from the pavement surface are uncorrelated.

Generally broadband antennas will reflect at its end due to impedance discontinuity. Therefore, the first similar waveform is formed by the antenna itself.

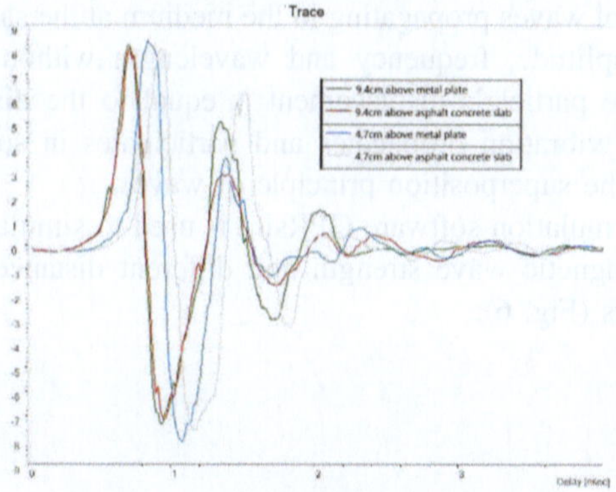

Fig. 4. TR-HF antenna test results

TR-HF antenna is ground-coupled antenna, so when the electromagnetic wave propagate in the air, the strength declines sharply Fig. 4 show that except from the strong reflection (second peaks) at the interface between air and metal plate (or the asphalt concrete slab), those latter ones become very weak. The error in measuring the amplitude of reflection wave will be very large, and the error of pavement thickness detection correspondingly, will become larger.

4.2 GSSI Air-Coupled Antenna

Figure 5 indicates that the first reflection of the radar antenna itself still exists, but its reflection strength has been sharply reduced compared with the result of the IDS-RIS. Meanwhile, the reflection strength of electromagnetic wave at the interface of air and asphalt concrete slabs (or between asphalt concrete slabs and the steel plate) is greatly increased, which is important for calculation of dielectric constant and thickness.

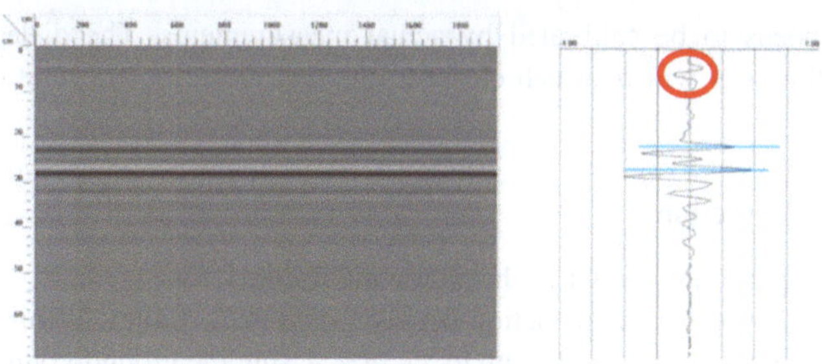

Fig. 5. GSSI SIR-20 radar test results

4.3 Superposition Principle of Waves

Waves are generally described as variations in amplitude in space and time. The wave is a function of the amplitude of each point.

If there are several waves propagating in the medium at the same time, they will be in their original amplitude, frequency and wavelength without disturbance. When waves encounter, the particle's displacement is equal to the displacement of waves alone. The fact that vibration propagates and participates in superposition independently is known as the superposition principle of waves.

A simple GPR simulation software GPRsim is used to simulate the reflected wave with same electromagnetic wave strength and different distances (10 cm, 5 cm and 3 cm) between layers (Fig. 6).

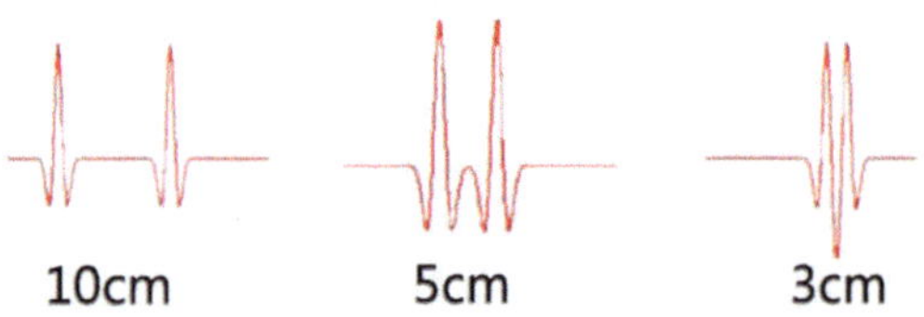

Fig. 6. Electromagnetic reflection of different distance between layers

The directly coupled wave in antenna itself is assumed to be constant size. Because the ground-coupled antenna is close to the ground, the distance between the directly coupled wave and the reflection wave is fixed. It means that when the directly coupled wave and the reflection wave are superposed, the directly coupled wave will superpose a fixed value A_1 on the reflection wave. It is same in the measurement of total reflection wave.

So the formula for calculating the dielectric constant transforms into the following formula (4) as follows:

$$\sqrt{\varepsilon'} = \frac{1 + \frac{A_0 + A_1}{A_m + A_1}}{1 - \frac{A_0 + A_1}{A_m + A_1}} = \frac{A_m + A_0 + 2A_1}{A_m - A_0} \tag{4}$$

The A_1 needs to be calibrated by actual measurements. The dielectric constant obtained in this way will be much closer to the real dielectric constant of the asphalt concrete slab.

4.4 Error Correction

The results of part3 are shown in the following Table 1:

The table shows that the deviation between measured result and true value is very large, where the maximum reaches up to 27.14%. Using the ground-coupled antenna to measure the thickness of pavement, the following formula (5) is needed for error correction.

$$\mathrm{H} = \mathrm{v} \times \frac{\mathrm{t}}{2} = \frac{C}{\sqrt{\varepsilon}} \times \frac{\mathrm{t}}{2} = \frac{C}{\frac{A_m + A_0 + 2A_1}{A_m - A_0}} \times \frac{\mathrm{t}}{2} \tag{5}$$

Table 1. IDS-RIS radar test results

Items	Amplitude A_0	Time Δt	Dielectric constant ε	Thickness calculated (cm)	Deviation (%)
steel plate	13.2676	–	–	–	–
3 cm AC + steel plate	8.0231	0.62	16.48	2.29	−23.64
5 cm AC + steel plate	7.7856	1.08	14.75	4.22	−15.63
7 cm AC + steel plate	7.4639	1.43	12.76	6.00	−14.22
5 cm + 7 cm AC + steel plate	7.9279	2.53	15.76	9.56	−20.33
3 cm + 7 cm AC + steel plate	8.1238	2.02	17.29	7.29	−27.14
5 cm + 3 cm + 7 cm AC + steel plate	7.7629	3.09	14.60	12.13	−19.12
3 cm + 5 cm AC + steel plate	7.9588	1.67	15.99	6.27	−21.69
5 + 5 cm AC + steel plate	7.7556	3.01	14.55	8.22	−17.80

Make H equal to the actual thickness of the asphalt concrete plate, then value of $A_1$is obtained. After calculating all the H in 9 combinations, a prediction range of A_1 from −1.45 to −2.4 can be obtained. Then A_1 was substituted into formula 5 to calculate H. Then the calculation of average and standard deviation between H and the actual thickness of asphalt concrete plate was conducted. The results are shown in Table 2. Figure 7 shows that when $A_1 = -2$, the average error is the smallest. After correcting A_1, the average error dropped from 19.95% to 4.21%, greatly improving the detection accuracy. Therefore, the method is efficient and practical.

Table 2. Error analysis

A_1	Average error	Error standard deviation	A_1	Average error	Error standard deviation
1.45	0.071663	0.050356	1.95	0.043251	0.033518
1.50	0.067349	0.049446	2.00	0.042089	0.032906
1.55	0.063588	0.047794	2.05	0.042155	0.031588
1.60	0.059785	0.046386	2.10	0.042459	0.031108
1.65	0.056135	0.044979	2.15	0.042767	0.031901
1.70	0.053664	0.042187	2.20	0.044516	0.031732
1.75	0.051166	0.03983	2.25	0.04634	0.032681
1.80	0.048639	0.038014	2.30	0.048214	0.034724
1.85	0.046084	0.036843	2.35	0.051595	0.035402
1.90	0.0444	0.035141	2.40	0.055019	0.036958

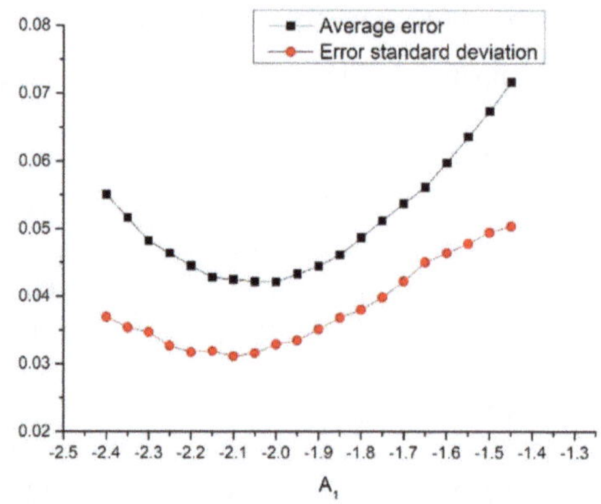

Fig. 7. Average error and error standard deviation

5 Conclusion

Due to the huge error in pavement thickness detecting using ground-coupled antenna, this work carried out a laboratory study to decrease the error.

The laboratory study was conducted to distinguish ground-coupled antenna from air-coupled antenna on account of the fixed amplitude of the first wave crest and figure out the interference from the directly coupled wave in antenna itself. And error correction of ground-coupled antenna in thickness measurement based on superposition principle was proposed, which make the average error dropped from 19.95% to 4.21%, dramatically improving the detection accuracy.

Acknowledgement. This study was sponsored by the project titled "Effectiveness and Applicability of Asphalt Pavement Maintenance Techniques in Zhejiang" (Project no. 8521002166) founded by Zhejiang Transportation Engineering Construction Group Co. and a research project titled "Data mining of Highway Pavement Performance and Evaluation of Maintenance Treatment Effectiveness" (Project no. 3221007409) founded by Southeast University.

References

1. Feng, D.S., Dai, Q.W.: Application of ground penetrating radar in the survey of the pavement thickness in highway. Prog. Geophys. **23**(1), 289–294 (2008)
2. Liu, T.: Research on Dielectric Property and Construction Quality Assessment of Asphalt Pavement Using Nondestructive Testing. South China University of Technology (2016)
3. Guo, S.: Research on GPR Wave Theory in Random Medium for Detecting Highway. China University of Geosciences (2013)
4. Ning, L.: Evaluation of Reclaimed Soil Compaction and Project Quality Based on Ground Penetrating Radar. College of Resources and Environment (2016)
5. Leng, Z., Al-Qadi, I.L.: Prediction of In-situ Asphalt Mixture Density Using Ground Penetrating Radar: Theoretical Development and Field Verification. University of Illinois at Urbana-Champaign (2011)

6. Liu, L., Su, Y., Liu, K.: FDTD analysis of a novel GPR antenna. J. Electronics Inf. Technol. **28**(4), 654–657 (2006)
7. Liu, L., Su, Y., Liu, K.: Radiation characteristics of GPR antenna through FDTD method. J. Microwaves **21**(supplement) (2005)
8. Li, C.: Application of Ground Penetrating Radar in Detecting Water Content and Compactness of Pavement. Zhengzhou University (2003)

Experimental Investigation of the Mechanical Behaviour of Interfaces Between Pavement Layers

Thomas Attia[1(✉)], Hervé Di Benedetto[1], Cédric Sauzéat[1], and Simon Pouget[2]

[1] University of Lyon/Ecole Nationale des Travaux Publics de l'Etat (ENTPE), LTDS (CNRS 5513), Vaulx-en-Velin CEDEX, Auvergne Rhône-Alpe, France
thomas.attia@entpe.fr

[2] Eiffage Infrastructures, Research and Innovation Department, Vaulx-en-Velin CEDEX, Auvergne Rhône-Alpe, France

Abstract. Pavement structure lifetime can be dramatically reduced in case of poor bonding between layers. Study of interfaces between pavement layers is essential to improve pavement durability. An innovative and new experiment named 2T3C Hollow Cylinder Apparatus (2T3C HCA) has been implemented to characterize interface performances. In this test, combined compressive (or tensile) stress and shear stress can be independently applied to the interface. The tested samples are hollow cylinder samples (12 cm height, 17.5 cm external diameter and 12.5 cm internal diameter) constituted of two bituminous pavement layers separated by an interface made with a tack coat. Samples representing a pavement structure are made in laboratory using the French wheel compactor. Digital Image Correlation (DIC) is used to obtain displacements in the three dimensions on the surface of samples. It makes it possible to locate the interface and observe its deformation. In this paper the 2T3C HCA device is presented and first results on bituminous mixtures behaviour and on interface behaviour are presented.

Keywords: Pavement · Interface behaviour · Bituminous mixture
Shear · Digital Image Correlation (DIC) · Hollow Cylinder Apparatus (HCA)
Experimental results

1 Introduction

Interfaces between pavement layers are important to the whole pavement durability. A tack coat is laid at the interface between the two bituminous mixtures layers at the top of pavements subjected to high traffic to ensure that they are bonded. This bonding reduces the maximal strain and thus the damage in the pavement. Yet the condition of bonding at the interface is not properly taken into account in most design methods where adjacent layers made in bituminous mixtures are considered perfectly bonded. Such an assumption can lead to improper design implying shortened pavement lifetime and increased maintenance costs. To improve the design methods, it is necessary to study adequately the behaviour of interfaces between pavement layers.

X. Shi et al. (Eds.): GSIC 2018, *Proceedings of GeoShanghai 2018 International Conference: Transportation Geotechnics and Pavement Engineering*, pp. 344–352, 2018.
https://doi.org/10.1007/978-981-13-0011-0_37

Many tests have been developed to assess interfaces strength, especially their shear strength. Some tests focused on the bonding shear failure like guillotine tests (Leutner 1979; Zofka et al. 2015) and shear box tests (Canestrari et al. 2005). Other tests can apply cyclic loadings (Diakhate et al. 2006; D'Andrea et al. 2016; Cho et al. 2017) and can be used to assess the linear viscoelastic properties or the fatigue properties of interfaces. Nevertheless, all these previous tests are inhomogeneous, the interface shear stress being nil at the edges of the samples. It makes them difficult to analyse because additional hypotheses need to be considered.

The 2T3C Hollow Cylinder Apparatus (2T3C HCA) has been developed at the ENTPE/University of Lyon to create a homogeneous test for the characterization of interfaces between pavement layers. It can apply shear stresses and tensile/compressive stresses to the interface. In this paper the 2T3C HCA is used to obtain linear viscoelastic properties of bituminous mixtures and of the interface of one bilayered material. These properties are very important to characterize the mechanical behaviour of bituminous materials.

2 2T3C Hollow Cylinder Apparatus (2T3C HCA)

The 2T3C HCA can apply independently torsion and tension/compression to hollow cylinder samples made of two bituminous mixtures layers with a tack coat at the interface. The two layers represent the two upper layers of a pavement: the wearing course and the base course. The sample dimensions are 125 mm in height, 172 mm in external diameter and 122 mm in internal diameter for a thickness of 25 mm (Fig. 1). Ratio between internal and external diameters is 0.71 which is between 0.65 and 0.82, the range recommended by (Sayao and Vaid 1991) for a reasonable assumption of homogeneity of shear stresses in the sample when torsion is applied. The samples are loaded with a servohydraulic press and they are placed inside a thermal chamber to control the temperature during a test. Loadings can be monotonic or cyclic.

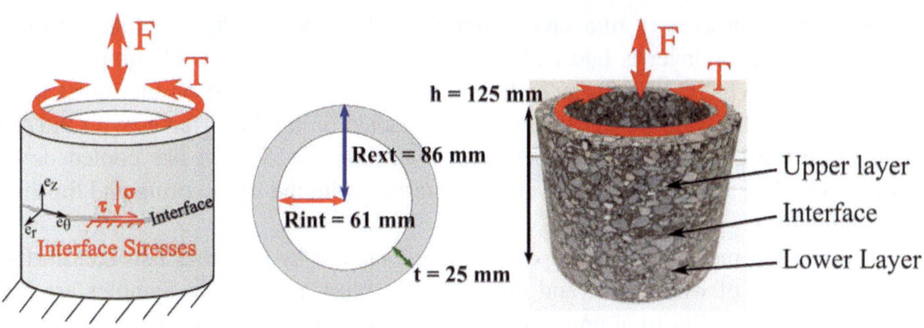

Fig. 1. Hollow cylinder sample of the 2T3C HCA

Using the press load cells (Fig. 2), one can measure the axial force and the torque applied to the sample. Two pairs of non-contact sensors measure the global displacement between the top and the bottom of the sample. In each pair, sensors are diametrically opposed and their measurements are averaged. One pair gives the vertical displacement and the other one gives the displacement related to the rotation.

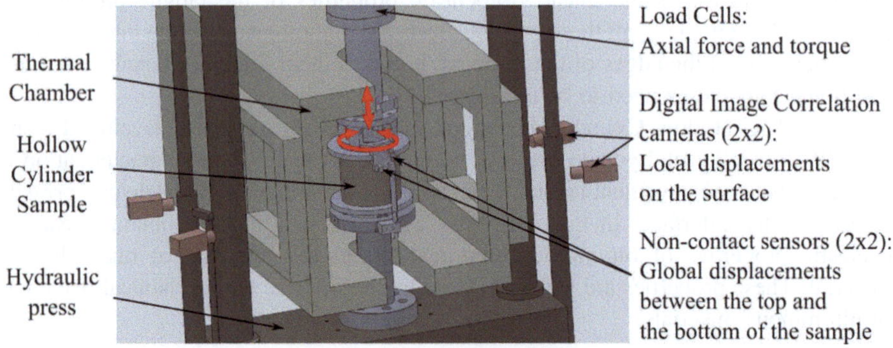

Fig. 2. 3D scheme of the 2T3C HCA with measuring systems

In addition, Digital Image Correlation (DIC) is used to obtain local displacements on the surface of the sample. During a test, pictures of the sample are taken at different states of deformation. At each state of deformation two cameras take two pictures of the same side of the sample and a correlation algorithm computes the displacements on the surface in the three dimensions. The displacements are expressed in the cylindrical coordinates related to the hollow cylinder (u_r, u_θ, u_z). Two pairs of cameras are used, one on each side of the sample. A speckle pattern is applied on samples before each test to improve the accuracy of the DIC.

3 Materials and Sample Preparation

The fabrication process of bilayered materials is described in Fig. 3. First, the lower bituminous mixture layer is laid and compacted with a French wheel compactor (NF EN 12697-33+A1 2007). After a 24 h rest, a tack coat is applied on the lower layer. After another 24 h rest, the upper layer is compacted on the lower layer with the tack coat. Once the bilayered slab of 600 mm × 400 mm × 150 mm has cooled down, hollow cylinder samples are cored vertically starting with the inner coring and finishing with the outer coring.

The speckle pattern is applied on the cored sample by covering its outer surface with a thin coat of white paint and then spraying black paint on it. Samples are then glued with epoxy resin to aluminium plates and mounted on the press.

In this paper, one type of material is studied. The upper layer is made of a so-called BBSG3 0/14 and the lower layer in a EME2 0/14 as specified in the European standards for classification of bituminous mixtures (NF EN 13108-1 2007). Both bituminous mixtures have a maximal aggregate size of 14 mm. The tack coat is a bituminous emulsion of pure bitumen with a residual dosage of 350 g/m^2.

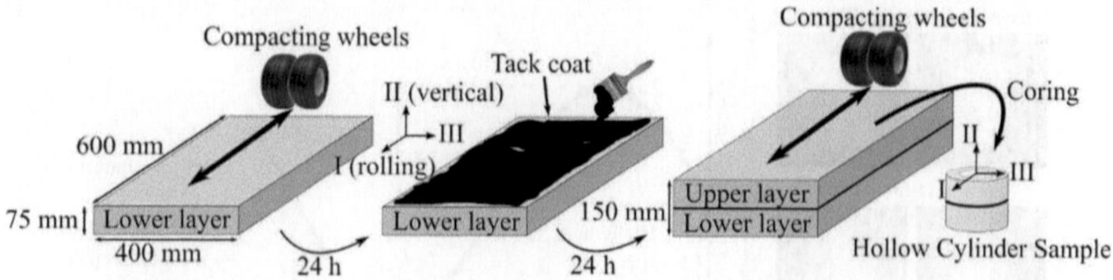

Fig. 3. Fabrication of the slab having two layers and a tack coat at the interface with French wheel compactor

4 Experimental Procedure

To characterize the linear viscoelastic behaviour of bituminous materials, sinusoidal strain loadings at small amplitude are applied to the sample. Two types of tests were performed: axial tests and rotation tests (Table 1). During axial tests, displacement is controlled with the vertical non-contact sensors and torque is maintained nil. During rotation tests, displacement is controlled with the horizontal non-contact sensors and axial force is maintained nil. For each type of test, two amplitudes of displacement and four frequencies were tested with five cycles for each frequency. Temperature was maintained at 20 °C for all the tests.

Table 1. Experimental program at 20 °C

Type of test	Axial loading	Torsion loading	Frequency (Hz)	Amplitude Non-contact sensors*	Name
Axial	Sinusoidal displacement	Zero torque	0.01; 0.03; 0.1; 0.3	ax = 12.5 μm	A100
				ax = 31.3 μm	A250
Rotation	Zero force	Sinusoidal displacement		rot = 45 μm	R100
				rot = 110 μm	R250

* 1 μm of axial displacement with the non-contact sensors corresponds to a global vertical strain ε_{zz_g} of 8 μm/m and 1 μm of rotation displacement corresponds to a global shear strain $\varepsilon_{\theta z\_g}$ of 2.2 μm/m, assuming the deformation is homogeneous in the sample (for instance, if there was no interface and only one bituminous mixture)

For the use of DIC, 50 pictures of the sample were taken during each cycle (Fig. 4) except for 0.3 Hz cycles where only 30 pictures per cycle were taken with the cameras. The capture time of a picture is 6 ms. Cameras are located at 600 mm from the sample surface, the angle between the optical axes of the two cameras being 30°. One pixel of a picture represents 151 μm of material. Only one pair of cameras was used in these tests, taking pictures on one side of the sample.

From DIC, displacements in the three dimensions are obtained on an area of the surface of the sample. To compute vertical strain ε_{zz} on each picture and in both bituminous mixture, horizontal rectangles are drawn on the area (Fig. 5). The average

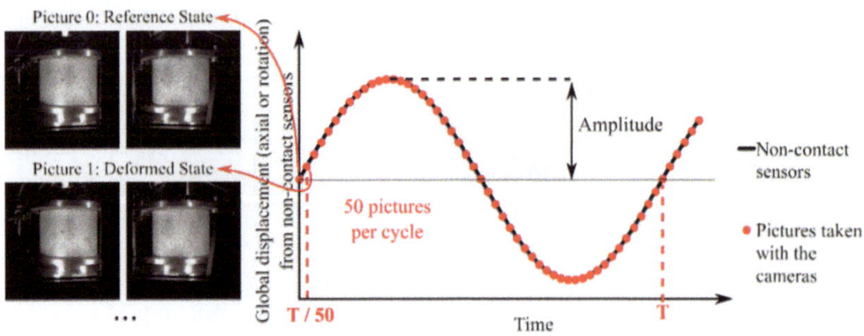

Fig. 4. Acquisition of pictures during the tests (T is the period)

of the vertical displacement u_z of the points located in one rectangle is affected to the center of this rectangle. One can then plot the variation of displacement with the height z in the sample. Using two linear regressions, strain ε_{zz} in both the upper layer and the lower layer is extracted from the diagram z–u_z. The vertical displacement gap at the interface Δu_z is defined as the difference in vertical displacement between the intersections of the two lines obtained using the linear regressions with the interface.

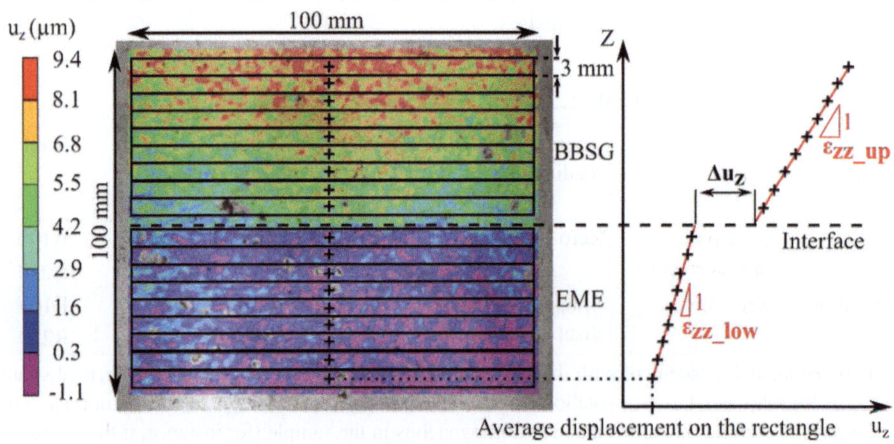

Fig. 5. Computation of ε_{zz} in both bituminous mixtures and of the displacement gap Δu_z at the interface for a picture at 12.5 µm in extension of the test A100 at 0.01 Hz using the Digital Image Correlation

In the Fig. 5, the analysis method is presented for an axial test. The same analysis is conducted for rotation tests in order to obtain the shear strain $\varepsilon_{\theta z}$ in both the upper layer and the lower layer with the difference that the vertical displacement u_z is replaced with the orthoradial displacement u_θ. The rotation displacement gap at the interface Δu_θ is also obtained.

Finally, for each picture the strain $\varepsilon_{\theta\theta}$ of bituminous mixtures in the upper layer and in the lower layer can be found using averages on vertical rectangles drawn in each layer to compute relative variations of perimeters (Attia et al. 2017).

5 Results

The norm of the complex modulus of linear viscoelastic materials is defined as the ratio between the amplitude of the stress sinusoidal signal and the amplitude of the strain sinusoidal signal.

The norm of the axial complex modulus $|E_{zz}^*|$ which is the ratio between axial stress σ amplitude and vertical strain ε_{zz} amplitude is shown in the Fig. 6 for axial tests. The norm of shear complex modulus $|G_{\theta z}^*|$, ratio between the shear stress τ amplitude and distorsion $\gamma = 2\varepsilon_{\theta z}$ amplitude, is represented in Fig. 7 for rotation tests.

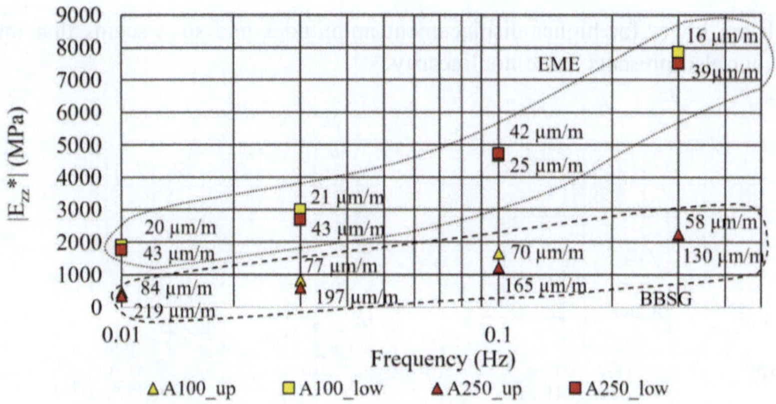

Fig. 6. Norm of complex modulus $|E_{zz}^*|$ for axial tests, in the upper and lower layers (next to each point is indicated the vertical strain ε_{zz} amplitude in the layer)

As expected the norms of the moduli increase with the frequency. It is also possible to notice that the moduli are higher in the lower layer made of EME than in the upper layer made of BBSG. For all the tests except one, the modulus decreased with an increase in strain amplitude, showing nonlinearity in the behaviour of bituminous mixtures in the same way it has already been reported (Mangiafico et al. 2017).

The norm of the interface modulus is defined here as a ratio between a stress and a displacement gap. For axial tests the norm of the vertical interface modulus $|K_{zz}^*|$ is calculated as the ratio between normal stress σ amplitude and vertical displacement gap at the interface Δu_z amplitude. For rotation tests the norm of the shear interface modulus $|K_{\theta z}^*|$ is the ratio between shear stress τ amplitude and rotation displacement gap Δu_θ amplitude.

Again, the norms of the interface moduli increase with the frequency for axial and rotation tests (Fig. 8). It can also be noticed that the norm of vertical interface modulus is higher than the norm of shear interface modulus. For all the tests, the interface

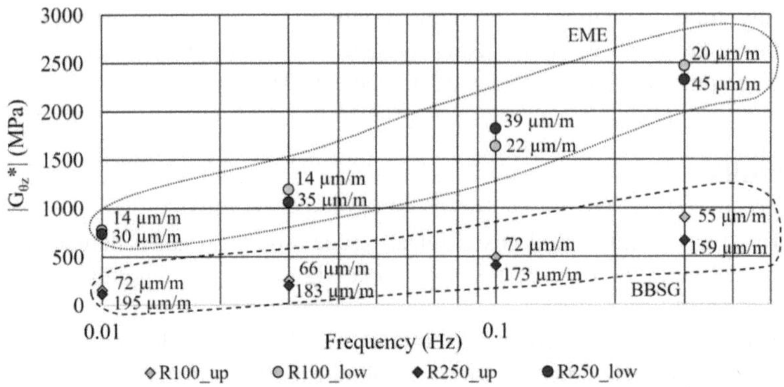

Fig. 7. Norm of complex shear modulus $|G_{\theta z}^*|$ for rotation tests, in the upper and lower layers (next to each point is indicated the shear strain $\varepsilon_{\theta z}$ amplitude in the layer)

moduli are lower for higher displacement amplitudes and so it seems that interface behaviour also presents some nonlinearity.

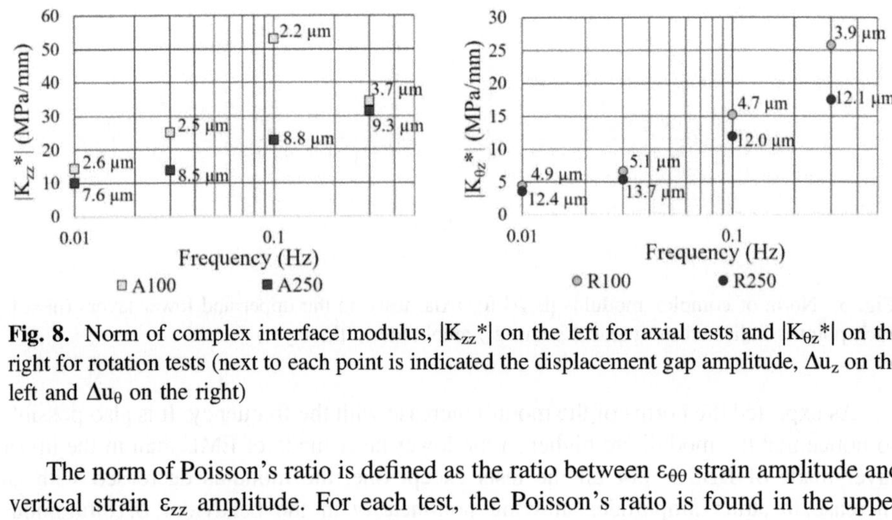

Fig. 8. Norm of complex interface modulus, $|K_{zz}^*|$ on the left for axial tests and $|K_{\theta z}^*|$ on the right for rotation tests (next to each point is indicated the displacement gap amplitude, Δu_z on the left and Δu_θ on the right)

The norm of Poisson's ratio is defined as the ratio between $\varepsilon_{\theta\theta}$ strain amplitude and vertical strain ε_{zz} amplitude. For each test, the Poisson's ratio is found in the upper layer and in the lower layer.

As shown in Fig. 9 for axial tests, values of Poisson's ratio range from 0.3 to 0.4 which is the right order of magnitude for bituminous mixtures at this temperature and these frequencies (Graziani et al. 2017) except for tests at 0.3 Hz where Poisson's ratio can be higher than 0.5, which is possible because bituminous mixtures present an anisotropic behaviour and can be as high as 0.7 (Di Benedetto et al. 2016). But experimental inaccuracies are not to be excluded, the amplitude of $\varepsilon_{\theta\theta}$ being lower than 10 µm/m at this frequency.

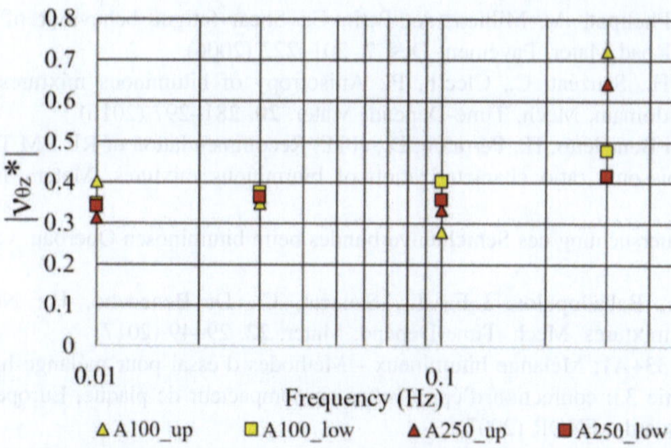

Fig. 9. Norm of complex Poisson's ratio $|\nu_{\theta z}^*|$ for the axial tests in the upper and lower layers

6 Conclusions

An innovative apparatus to characterize the thermomechanical behaviour of interfaces, the 2T3C HCA, has been developed and used to study the linear viscoelastic properties of bituminous mixtures and of an interface. A proposed procedure allowed to measure strain values with a resolution of some μm/m with Digital Image Correlation.

The complex modulus of the two layers have been obtained and nonlinearity observed. Interface moduli could also be obtained for displacement gaps of some micrometers. Interfaces seem more rigid in tension/compression than in shear. From three-dimensional displacements, Poisson's ratio in both layers was also evaluated. All data was frequency dependent.

The DIC applied to hollow cylinders seems to be accurate even at small deformation and has proved to be a relevant tool to study linear viscoelastic behaviour of bituminous mixtures and interfaces.

References

Attia, T., Di Benedetto, H., Sauzéat, C., Olard, F., Pouget, S.: Hollow cylinder apparatus to characterize interfaces between pavement layers. In: International Conference on Advances in Construction Materials and Systems Proceedings, vol. 4, pp. 401–410 (2017)

Canestrari, F., Ferrotti, G., Partl, M., Santagata, E.: Advanced testing and characterization of interlayer shear resistance. Transp. Res. Rec. J. Transp. Res. Board **1929**, 69–78 (2005). http://www.trrjournalonline.trb.org/doi/abs/10.3141/1929-09?journalCode=trr

Cho, S.H., Safavizadeh, S.A., Kim, Y.R.: Verification of the applicability of the time–temperature superposition principle to interface shear stiffness and strength of GlasGrid-reinforced asphalt mixtures. Road Mater. Pavement Des. **18**, 766–784 (2017)

D'Andrea, A., Tozzo, C.: Dynamic tests on bituminous layers interface. Mater. Struct. **49**, 917–928 (2016)

Diakhate, M., Phelipot, A., Millien, A., Petit, C.: Shear fatigue behaviour of tack coats in pavements. Road Mater. Pavement Des. **7**, 201–222 (2006)

Di Benedetto, H., Sauzéat, C., Clec'h, P.: Anisotropy of bituminous mixtures in the linear viscoelastic domain. Mech. Time-Depend. Mater. **20**, 281–297 (2016)

Graziani, A., Di Benedetto, H., Perraton, D., et al.: Recommendation of RILEM TC 237-SIB on complex Poisson's ratio characterization of bituminous mixtures. Mater. Struct. **50**, 142 (2017)

Leutner, R.: Untersuchung des Schichtenverbundes beim bituminosen Oberbau. vol. 41, pp. 84–91 (1979)

Mangiafico, S., Babadopulos, L.F.A.L., Sauzéat, C., Di Benedetto, H.: Nonlinearity of bituminous mixtures. Mech. Time-Depend. Mater. **22**, 29–49 (2017)

NF EN 12697-33+A1: Mélange bitumineux - Méthodes d'essai pour mélange hydrocarboné à chaud - Partie 33: confection d'éprouvettes au compacteur de plaque, European Standard - French Standard AFNOR (2007)

NF EN 13108-1 Mélange bitumineux - Spécification des matériaux - Partie 1 : Enrobés bitumineux, European Standard - French Standard AFNOR (2007)

Sayao, A., Vaid, Y.P.: A critical assessment of stress nonuniformities in hollow cylinder test specimens. Soils Found. **31**, 60–72 (1991)

Zofka, A., Maciej, M., Alexander, B., Ramandeep, J., Audrius, V., Rita, K.: Advanced shear tester for evaluation of asphalt concrete under constant normal stiffness conditions. Road Mater. Pavement Des. **16**, 187–210 (2015)

Analysis on Mechanical Response for Three Kinds of Asphalt Pavement Under Discontinuous Interlayer Condition

Yan-hong Zhang[1,2,3](✉) and Ai-qin Shen[1]

[1] Highway College, Chang'an University, Xi'an 710064, Shanxi, China
240009199@qq.com
[2] China Highway Engineering Consulting Group Limited Company, Beijing 100097, China
[3] Research and Development Center on Pavement Maintenance, CCCC, Beijing, China

Abstract. Semi-rigid asphalt pavement, compound base asphalt pavement and flexible base asphalt pavement are three main structures in China. Each asphalt pavement differs at the structure of base. In order to further analyze the stress characteristic, the paper took base as the starting point of study, set interlayer binding condition as not fully continuous and considered the horizontal load and vertical load under heavy load combining with our national, then made calculation in the full thickness range of pavement with the use of shell designing software Bisar3.0, so the worst distribution area for three kinds of asphalt pavement were obtained. The results showed that discontinuous interlayer binding condition made flexible strain and flexible stress show discontinuous distribution during structure height; the maximum flexible strain all appeared on the bottom of surface layer for three kinds of asphalt pavement, while the maximum flexible stress appeared in different structure layer; impose of double-double made shear stress on the surface of pavement stand greatly high degree; asphalt stabilized crushed stone of compound pavement must be designed to both anti-fatigue and anti-reflection crack; it's suggested that designed indexes matching mechanical destroy mode should be chosen to guide pavement design.

Keywords: Highway engineering · Mechanical response
Not fully continuous · Horizontal load · Destroy mode

1 Introductions

In the process of transport infrastructure construction in China, semi-rigid asphalt pavement plays an important role. However, with the rapid development of transportation, the road traffic composition is becoming more and more heavy-duty and integrated. The traffic volume is soaring and the overloading of vehicles is serious, which causes great traffic pressure. Single semi-rigid base asphalt pavement has been far from meeting the needs of the existing road. On the other hand, through long-term practice shows that due to the high density of semi-rigid material, water which once

X. Shi et al. (Eds.): GSIC 2018, *Proceedings of GeoShanghai 2018 International Conference: Transportation Geotechnics and Pavement Engineering*, pp. 353–363, 2018.
https://doi.org/10.1007/978-981-13-0011-0_38

penetrates the road surface into the base cannot be excluded. After then traffic loads will produce considerable hydrodynamic pressure, resulting in pavement structural damage. Therefore, referring to the foreign experience, China has introduced the compound base asphalt pavement and flexible base asphalt pavement and has tried to solve the problems of serious pavement damage and short life span from the road structure.

From the view of structural characteristics, the main difference between flexible base asphalt pavement, semi-rigid base asphalt pavement and compound base asphalt pavement is the form of base layer. Therefore, in order to reveal the difference between three kinds of pavement in theory, under the discontinuous interlayer binding condition and considering the double impose of the horizontal load and vertical load under heavy load, this paper comprehensively analyzed mechanical response of each kind of pavement, revealed the reason for pavement destruction caused by base and provided a theoretical basis for promotion and application of new pavement structure in China.

2 Calculation Analysis

2.1 Pavement Structure Scheme

Referring to [1], three typical structural schemes of flexible pavement, compound pavement and semi-rigid pavement are selected respectively as shown in Tables 1, 2 and 3.

Table 1. Flexible pavement scheme and material parameters

Material name	Thickness/cm	15 °C compression resilience modulus/MPa	Poisson's ratio
SMA-13	4	1800	0.30
AC-20	6	1600	0.30
AC-25	8	1500	0.30
ATB-25	17	1200	0.30
Graded crushed stone	30	350	0.35
Subgrade		40	0.35

2.2 Pavement Structure Scheme

2.2.1 Wheel Load

Considering the serious overload problem in China, the calculation load is axial load 140 kN, tire pressure 0.81 MPa, wheel pressure radius R is 11.73 cm and double circle center distance is 17.6 cm [2]. In addition, in view of the most unfavorable influence of horizontal load on the structure, the value of horizontal load factor is 0.5 and the one of horizontal load is 70 kN.

Table 2. Compound pavement scheme and material parameters

Material name	Thickness/cm	15 °C compression resilience modulus/MPa	Poisson's ratio
SMA-13	4	1800	0.30
AC-20	6	1600	0.30
AC-25	8	1500	0.30
ATB-25	8	1200	0.30
Cement stabilized crushed stone	20	3200	0.20
Lime soil	35	500	0.20
Subgrade		40	0.35

Table 3. Semi-rigid pavement scheme and material parameters

Material name	Thickness/cm	15 °C compression resilience modulus/MPa	Poisson's ratio
SMA-13	4	1800	0.30
AC-20	6	1600	0.30
AC-25	8	1500	0.30
Cement stabilized crushed stone	38	3200	0.20
Cement stabilized crushed stone	20	3200	0.20
Subgrade		40	0.35

2.2.2 Interlayer Condition

Referring to [3], the analysis was under the condition that interlayer binding is not fully continuous which is more consistent with actual conditions:

(1) The asphalt layers are completely continuous between each other;
(2) The asphalt layer and graded crushed stone layer are calculated by slide;
(3) It is generally believed that asphalt layer or graded gravel and inorganic structural material the stable aggregate can be continuously calculated before the damage of semi-rigid material and calculated by slide after the damage of semi-rigid material. This study was calculated on a less favorable slip basis.

2.2.3 Calculation Mode

In the paper shell designing software Bisar3.0 was used to make calculation. In order to improve calculation accuracy, in the full thickness range of pavement, from the road surface, depth calculation points are increased every 1 cm; in the transverse direction of the road, considering the diffusion effect of vehicle load, force unfavorable points in each layer gradually shifted from the single circle load center to the center of wheel gap [4]. Therefore, in calculation four calculation points are increased between the center of single round load and the center of wheel gap, such as Ci, Di, Fi, Hi, Ii and Ji (i = 1, 2, 3) in Fig. 1,…).

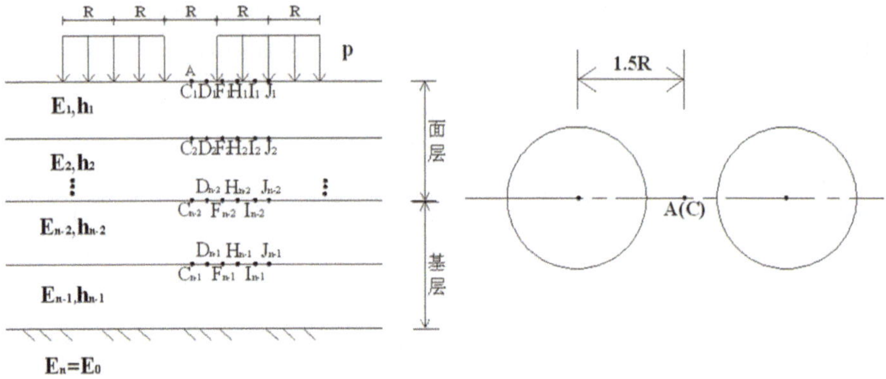

Fig. 1. Mechanical calculation diagram

3 Calculation and Analysis on Mechanical Response for Three Kinds of Asphalt Pavement

The distribution of main mechanical responses of three kinds of asphalt pavement under the condition that interlayer binding is not fully continuous was obtained by Bisar3.0 software. Different mechanical responses will be described below.

3.1 Maximum Flexible Strain of Asphalt Layer

The three-dimensional distribution of flexible strain of three kinds of asphalt pavement in the cross section and the direction of traffic is shown in Fig. 2.

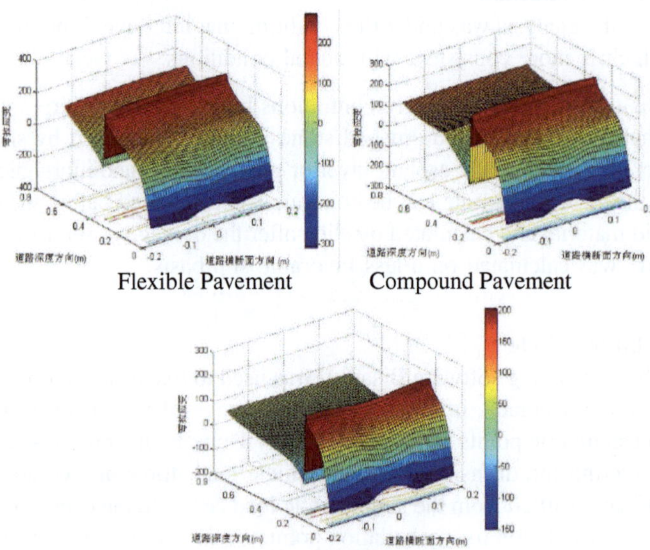

Flexible Pavement Compound Pavement

Semi-rigid Pavement

Fig. 2. Three-dimensional distribution of flexible strain of three kinds of asphalt pavement

It can be seen from Table 4 that when the interlayer is not fully continuous, three kinds of asphalt pavement have the maximum flexible strain of the asphalt layer on the bottom of the surface layer (the surface of compound pavement refers to the ATB layer). The values of the maximum flexible strain are sorted by the numerical size and the rank are flexible pavement > composite pavement > semi- rigid pavement. It also proved from the examples that the possibility of fatigue cracking appearing on the surface of flexible pavement is greater and the force of compound pavement lies between the other two kinds of pavement.

Table 4. Calculation results of the maximum flexible strain of asphalt layer of three kinds of asphalt pavement under discontinuous interlayer condition

Pavement type	Maximum flexible strain of asphalt layer	
	Value/ με	Position
Flexible pavement	256	On the bottom of surface layer in the center of wheel gap
Compound pavement	219	On the bottom of ATB layer in the center of wheel gap
Semi-rigid pavement	196	1.2R from the center of wheel gap on the bottom of surface layer

It can be seen from the distribution diagram of Fig. 2 that the strain distributions of three kinds of asphalt pavement are similar in the upper part of the structure, except that the distribution of the lower half of the basement is different. In general, from the road surface to the table down, the strain on the structure were gradually transferred from compressive strain to flexible strain and suddenly changed into compressive strain on the bottom of base. Specifically speaking:

(1) For flexible pavement, compression range is within 7 cm from road surface, and then compressive strain in the surface layer increases gradually along road depth and reaches the peak on the bottom of surface layer; in the small upper depth of flexible base, the pavement is subjected to compressive strain, and then the tensile strain on flexible pavement increases along the road depth, and reaches the secondary peak value on the bottom of base.

(2) For semi-rigid pavement, when interlayer binding condition is not fully continuous, the width range from the center of wheel gap to the load center of single circular and the 3 cm depth range are dominated by compressive strain. Flexible strain of surface layer increases gradually along road depth and reaches the maximum value on the bottom of surface layer. Since the contact between the asphalt surface and the semi-rigid material layer is completely smooth, the strain mutated to compressive strain once at the depth of base and was converted into flexible strain in the middle part of upper base and the strain value increases with the depth. Therefore, it can be seen that large flexible strain appears on the bottom of asphalt layer due to the deterioration of the interface between the surface layer and the base layer.

(3) The flexible strain distribution of compound pavement is generally similar to that of the first two kinds of asphalt pavement, except that the range of flexible and compressive strain is different. Specifically, 5 cm down the road surface is the range of compressive strain; the maximum flexible strain of the structure appears on the bottom of asphalt stabilized crushed stone; the base is still subjected to flexible and compressive stress alternately, the upper part bears compressive stress and the lower part bears flexible stress; the secondary peak value of flexible appears on the bottom of base

In general, when interlayer binding condition is not fully continuous, the change of flexible strain during the structural height is discontinuous mutation, and the mutant horizons of three kinds of asphalt pavement are all located at the combination of surface layer and base. Generally speaking, three kinds of asphalt pavement have the maximum flexible strain on the bottom of surface layer, so the asphalt mixture corresponding to the layer must have good fatigue resistance. For composite pavement, on one hand, it is required that the asphalt stabilized crushed stone, as the transition layer, can depend on its coarse gradation and large gap to play the role of blocking the upward spread of the cracks at the base and on the other hand, the mixture have a certain fatigue resistance, so as to prevent fatigue cracking on the bottom of surface (Table 5).

Table 5. Calculation results of the maximum flexible stress of three kinds of asphalt pavement under discontinuous interlayer condition

Pavement type	Maximum flexible strain	
	Value/MPa	Position
Flexible pavement	0.377	On the bottom of surface layer in the center of wheel gap
Compound pavement	0.261	On the bottom of upper base in the center of wheel gap
Semi-rigid pavement	0.228	On the bottom of base in the center of wheel gap

3.2 Maximum Flexible Stress of Structure

The three-dimensional distribution of flexible stress of three kinds of asphalt pavement in the cross section and the direction of traffic is shown in Fig. 3.

It can be seen that the maximum flexible stress appears at the center of wheel gap for three kinds of pavement, but the depth and the horizon are different. Flexible Pavement has the maximum flexible stress on the bottom of surface layer; the base bottom of semi-rigid pavement is subjected to the peak level of flexible stress; due to structural characteristics of compound pavement, the position under concentrated flexible stress is transferred to the bottom of upper base.

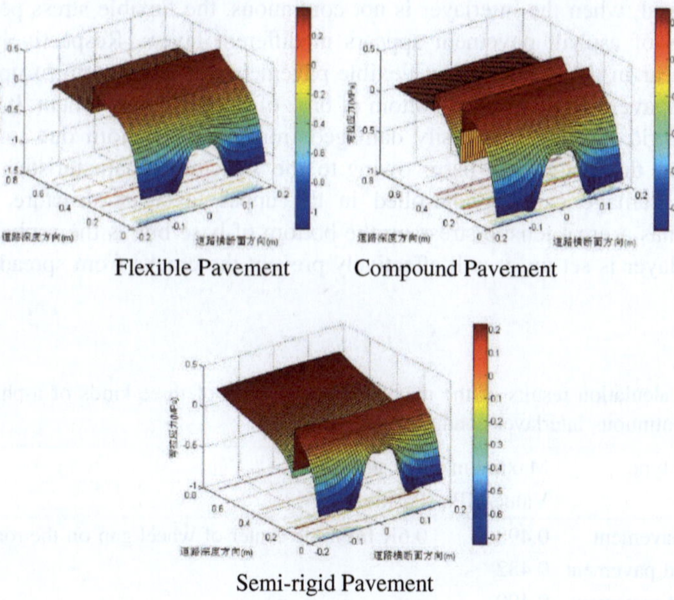

Fig. 3. Three-dimensional distribution of flexible stress of three kinds of asphalt pavement

In general, the flexible stress distribution of three kinds of pavement is similar to the flexible strain distribution. Specifically speaking:

(1) For flexible pavement, the compression range is within 14 cm from road surface, and then changed to flexible strain which increases gradually along road depth and reaches the peak on the bottom of surface layer; in the small upper depth of flexible base, the upper part of flexible base is subjected to compressive stress and the lower part bears flexible stress that increases gradually along road depth.

(2) For the sake of convenience, the asphalt stabilized crushed stone layer is called the transition layer and its lower two semi-rigid bases are used as the upper and lower layers respectively. The surface layer (including asphalt stabilized crushed stone layer) of compound pavement is still in the status that the upper part is under compressive stress and the lower is under flexible stress. And then flexible strain reaches the extreme level on the bottom of asphalt stabilized crushed stone, becomes compressive strain at the combination of the surface layer and the base layer and stays in an alternative statues of compressive stress and flexible stress. On the bottom of upper base, the flexible stress of structure reaches the highest level and flexible stress of subbase is greatly weakened. But its value increases with the depth. As can be seen from the figure, the flexible distribution of compound pavement is more tortuous along the depth direction.

(3) Compared with tensile strain, the compression region of semi-rigid pavement in surface layer and base layer has expanded and the extreme value of flexible stress is on the bottom of base. The discontinuity between layers also leads to great flexible stress on the bottom of asphalt layer, but its value is less than the maximum value of tensile stress.

In general, when the interlayer is not continuous, the flexible stress peak value of three kinds of asphalt pavement appears in different layers. Respectively, the peak values appear in the upper part of flexible pavement, on the bottom of upper base of compound pavement and on the bottom of base of semi-rigid pavement. It can be seen that semi-rigid pavement is easily damaged from top to bottom due to the fatigue cracking on the bottom of base; owing to the structural characteristics of flexible pavement, damage can be controlled in the upper layer of structure. Composite pavement has a great tensile stress on the bottom of base but as the asphalt stabilized macadam layer is set up, it will effectively prevent the cracks from spreading upward (Table 6).

Table 6. Calculation results of the maximum shear stress of three kinds of asphalt pavement under discontinuous interlayer condition

Pavement type	Maximum shear strain	
	Value/MPa	Position
Flexible pavement	0.498	0.6R from the center of wheel gap on the road surface
Compound pavement	0.482	
Semi-rigid pavement	0.490	

3.3 Maximum Shear Strain of Structure

The three-dimensional distribution of shear stress of three kinds of asphalt pavement in the cross section and the direction of traffic is shown in Fig. 4.

When interlayer is not fully continuous and horizontal load is considered, the maximum shear stress appears on the road surface of three kinds of asphalt pavement, and the most unfavorable position is basically the same. It can be seen from the shear stress distribution diagram in Fig. 4 that, unlike the traditional shear stress distribution pattern, the shear stress does not soar to the peak value from the road surface while it directly decreases from the road surface. Specifically speaking:

(1) The shear stress level on the surface of flexible pavement is very high. It declines along the road depth down deep and increases from the depth of 25 cm to the bottom of surface layer; the shear stress level on the base has greatly reduced. Since the top surface of base, the shear stress increases along the road depth, but the shear stress plunges only after reaching the bottom of flexible base.
(2) Composite pavement also takes a higher level of shear stress on the road surface, and then the shear stress decreases along the depth of the road, but shear stress slightly increased in the transitional layer and decreases sharply after reaching the base. Specifically, within the upper base range, shear stress decreases slightly along the road depth and then increases gradually. The shear stress decreases sharply after reaching the bottom of subbase, and is basically on the decline within the depth range of subbase.

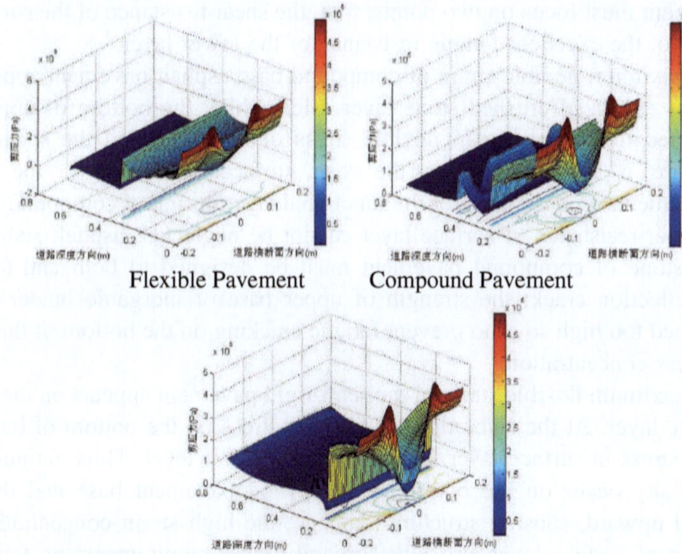

Flexible Pavement Compound Pavement

Semi-rigid Pavement

Fig. 4. Three-dimensional distribution of shear stress of three kinds of asphalt pavement

(3) The shear stress of semi-rigid pavement decreases gradually from the road surface, decreases a little until the depth of 15 cm, and gradually increases along the depth. In the base range, the shear stress firstly increases at the depth of 10 cm, and then decreases gradually along. After reaching the bottom of subbase, it increases slightly, but the overall level is still on the low side.

In general, when the horizontal load and vertical load act simultaneously, most shear stress of structure is concentrated on the road surface, especially at the distance of about 7 cm from the center of wheel gap. Therefore, from the point of view of shear resistance, no matter what kind of asphalt pavement, under the most unfavorable heavy load (including horizontal load and vertical load), the surface must have excellent shear resistance.

4 Comprehensive Evaluation of Three Kinds of Asphalt Pavement

(1) The maximum flexible strain and flexible stress of flexible pavement appears on the bottom of surface layer; at the same time, the level of shear stress on the road surface is the highest and the most unfavorable anti-fatigue layer is concentrated on the bottom of surface layer in the upper part of structure of; relatively speaking, the stress condition on base is more favorable. It shows that the destruction of flexible pavement is mainly from top to bottom rather than structural damage. Under horizontal load and vertical load, the pavement mixture design of flexible

pavement must focus on two points: first, the shear resistance of the surface layer; and two, the excellent fatigue resistance of the lower layer.

(2) The maximum flexible strain of compound base asphalt pavement appears in the asphalt stabilized crushed stone layer. Meanwhile, the bottom of upper base is subjected to the maximum flexible stress of structure, and the shear stress of structure is concentrated on the road surface. Accordingly, the following requirements are proposed for the functional layer design of compound pavement: the shear resistance of surface layer cannot be neglected; asphalt stabilized crushed stone of compound pavement must be designed to both anti-fatigue and anti-reflection crack; the strength of upper base of inorganic binder cannot be designed too high so as to prevent fatigue cracking on the bottom of the layer due to stress concentration.

(3) The maximum flexible strain of the semi-rigid pavement appears on the bottom of surface layer. At the same time, the tensile stress on the bottom of base and the shear stress in surface layer are both at the highest level. Thus, fatigue cracking can easily occur on the bottom of semi-rigid pavement base and then cracks spread upward, causing structural damage; the high strain concentration on the bottom of surface layer also puts forward higher requirements on fatigue resistance of lower layer.

5 Conclusions

The paper set interlayer binding condition as not fully continuous and considered the horizontal load and vertical load under heavy load combining with our national, which is recommended by Professor Shen Jin-an, then made calculation in the full thickness range of pavement with the use of shell designing software Bisar3.0, comprehensively analyzed the mechanical response and distribution law of three kinds of asphalt pavement and came to the conclusions as follows:

(1) Discontinuous interlayer binding condition makes flexible strain and flexible stress show discontinuous distribution during structure height and mutant layers of three kinds of asphalt pavement are all located at the combination of surface layer and base.

(2) There is a certain similarity between flexible strain distribution and flexible stress distribution, but the specific flexible and compressive ranges are different. Both surface layer and base are in the state of tension and pressure alternation. Generally speaking, to a certain extent, surface layer and the top part of base are in a compressive state and surface layer and the lower part of base are mainly subjected to flexible stress.

(3) Under discontinuous interlayer binding condition, the maximum flexible strain all appears on the bottom of surface layer for three kinds of asphalt pavement, while the maximum flexible stress appears in different structure layer: on the bottom of surface layer of flexible pavement, on the bottom of upper base of compound pavement and on the bottom of base of semi-rigid pavement.

(4) When horizontal load and vertical load act simultaneously, most shear stress of structure is concentrated on the road surface, especially at the distance of about 7 cm from the center of wheel gap.
(5) The distribution areas of the flexible pavement are concentrated in the upper part of structure, including the bottom of the layer with high stress-strain level and the road surface with high shear level and the failure mode is from top to bottom.
(6) The unfavorable distribution areas of compound pavement are the bottom of asphalt stabilized crushed stone layer and the bottom of upper base layer. The asphalt stabilized gravel layer must t crushed stone layer to prevent road surface from structural damage.
(7) Fatigue cracking can easily occur on the bottom of semi-rigid pavement base and due to its own structural characteristics, the cracks are mainly bottom-up structural damage.
(8) Designed indexes matching mechanical destroy mode should be chosen to guide pavement design. Shear stress of the surface and flexible strain on the bottom of surface layer are suggested as design controlling index for flexible pavement and composite pavement.

References

1. JTG D50-2006: Specifications for Design of Highway Asphalt Pavement. China Communications Press, Beijing (2006)
2. Zeng, F., Huang, X.: Effect of overload on asphalt pavement. J. Transp. Eng. **4**(3), 8–10 (2004)
3. Shen, J., Li, F., Chen, J.: Research on Early Damage of Expressway Asphalt Pavement and Its Countermeasures. People's Communications Press, Beijing (2004)
4. CJJ37-90: Design Code for Urban Road. People's Communications Press, Beijing (1991)

Evaluation of Semi-rigid Base Performance Through Numerical Simulation and Data Mining of Pavement Deflection Basin

Xuqiu Cui, Qiao Dong(✉), Fujian Ni, and Xiaolong Liang

School of Transportation, Southeast University, Nanjing 210096, China
cxq_seu@foxmail.com, qiaodong@seu.edu.cn

Abstract. More than 90% of the highway pavements in China use cement treated base or semi-rigid base. Due to the relatively high stiffness of this type of base layer, its structural capacity plays a vital role for the stability, integrity, performance and durability of the pavement structure. However, there are little studies on the long-term structural capacity of in-service semi-rigid base and the influence of other pavement or non-pavement related factors. This paper evaluated the long-term structural capacity of the base layer by investigating the deflection basin data of a typical section of highways. The deflection basin parameters (DBPs) measured by falling weight deflectometer (FWD), such as the D_{20}–D_{60}, were investigated by simulation and used to characterize the structural capacity of the semi-rigid base. Data mining methods including decision tree and multiple linear regression were adopted to identify and quantify the influence of factors including base material properties, traffic level, upper layer thickness and materials. Findings from this study can be used to evaluate the structural performance of in-service semi-rigid base and to help design pavement structural preservation treatments.

Keywords: Semi-rigid base · Structural capacity · Deflection basin
Data mining · Numerical simulation

1 Introduction

Pavement base layer, which bears the whole surface structure and transfers traffic loads to the subgrade, plays a key role in the pavement structure. Bearing capacity of the base has an important impact on road performance in the long term. Cement treated base, or Semi-rigid material, has been widely used in the base and subbase layers of highways due to its higher stiffness and cost-effectiveness compared with flexible material. Accordingly, more than 90% of the highway pavements in China use semi-rigid base.

According to the theory of pavement mechanics, the key point for pavement structural capacity is the determination of elastic modulus. The highway is affected by traffic load and environmental factors during its service period, which may lead to the changes in modulus of each layer. To detect the condition of pavement structure, traditional methods include core drilling and testing although it is inefficiency and causes damage to the road. Falling Weight Deflectometer (FWD) is a non-destructive testing devices that receives extensive application all over the world for its advantages

X. Shi et al. (Eds.): GSIC 2018, *Proceedings of GeoShanghai 2018 International Conference: Transportation Geotechnics and Pavement Engineering*, pp. 364–371, 2018.
https://doi.org/10.1007/978-981-13-0011-0_39

of high efficiency, economy and safety. The deflection basin is a comprehensive response of modulus and thickness of the pavement structure under the dynamic load. Based on FWD deflection basin data, the inversion of pavement structure modulus has been studied in recent years. However, some researchers found it difficult to actually reflect the performance of the pavement with inversion method as its discreteness, complex calculation and limited precision [1]. On the other hand, there is a strong correlation between the geometric characteristics of the deflection basin and the strength of the pavement structure. Different FWD parameters have been proposed to evaluate pavement structural capacity. Widely used deflection basin parameters (DBPs) for flexible base asphalt pavement are listed in Table 1 [2–5].

Table 1. Deflection basin parameters for flexible base asphalt pavement.

Deflection basin parameter	Equation	Parameter's objective
Center point value	D_0 (EU)	Condition of whole structure
Surface curvature index	SCI = D_0–D_r ($r \in [20, 60]$)	Condition of surface layer
Base damage index	BDI = D_{30}–D_{60} (EST); D_{12}–D_{24} (USA)	Condition of base layer
Base curvature index	BCI = D_{48}–D_{60} (USA); D_{90}–D_{120} (FIN)	Condition of subbase/subgrade layer
Area	$AREA = \frac{6(D_0 + 2D_{12} + 2D_{24} + D_{36})}{D_0}$	Shape of defection basin
Deflection ratio	DR = D_0/D_r (EU)	Condition of the layer at the equivalent depth

Some researchers studied the deflection basin of semi-rigid base pavement in China and found that there is an inflection point at a distance of 20 cm from the origin on the deflection basin curve for semi-rigid based pavement, which is rather different from that of flexible base pavement [1, 6]. The sensitivity analysis of deflection and moduli of some highways in China was carried out by Ding and Wang, and they proposed an index, D_{20}–D_{60}, to evaluate the bearing capacity of semi-rigid base. The applicability of this index need to be further verified as the research they made was only for some specific structures.

The objective of this study is firstly to verify the correlation between D_{20}–D_{60} (or other parameters) and modulus of semi-rigid base. The finite element model of typical pavement structures is established using COMSOL software to calculate deflection basin with multiple combinations of moduli. Based on identified indicator, the historical FWD data were analyzed to investigate the change of semi-rigid base capacity over time and the effects of influencing factors including traffic level, materials, time etc. The relationship between DBP and the factors considered is established, and further the long-term structural capacity variation law of semi-rigid base is explored.

2 Numerical Simulation

2.1 Finite Element Model

In this study, a 2D model of pavement structure is developed using COMSOL software for its batch processing function. The moduli of structure layers are changed to calculate deflection basin in different combinations. The model is 6 m along the direction of traffic and 3 m depth in longitudinal direction. The uniformly distributed loads with radius of 0.15 m are applied at the middle axle of the surface. Based on the static equivalent principle (Eq. 1), standard load 0.7 MPa in 3D is converted to 164,928 Pa m in 2D. The thickness (h), modulus (E) and density (ρ) of each layer are shown in Table 2.

$$A * p_{3d}/2R = p_{2d} \tag{1}$$

Where,

R = the radius of the uniformly distributed loads, 0.15 m;
A = the area of the uniformly distributed loads in 3D, m^2;
p_{3d} = the standard load in 3D, MPa;
p_{2d} = the equivalent load in 2D, MPa·m.

Table 2. Values of parameters of model structure layers.

Layers		Thickness (m)	Modulus (MPa)	Density (kg/m^3)
Surface	Top	0.04	range(800, 100, 1200)	2300
	Middle	0.06	range(800, 100, 1200)	2300
	Lower	0.08	range(800, 100, 1200)	2300
Base		0.4	range(1300, 100, 1700)	2200
Subbase		0.2	range(1200, 100, 1600)	2100
Subgrade		2.22	200	1800

Note: Range(x, y, z)—range of values from x to z every other y.

2.2 Analysis of Calculation Results

The moduli of three surface layers (top, middle & lower) are 1000 MPa while the moduli of subbase and subgrade are 1400 MPa and 200 MPa. Under this combination, the modulus of base is changed from 1300 MPa to 1700 MPa every 100 MPa, respectively. Figure 1 shows the DBP D_{20}–D_{60} (mm) calculated correspondingly, which demonstrates a power function relation between D_{20}–D_{60} and modulus of base (E_3) with other parameters unchanged. Then, the modulus of each layer is changed according to Table 2 to calculate all the deflection basins of multiple modulus combinations. Quantitative relationship between D_{20}–D_{60} and the moduli is established by multiple non-linear fitting. In consideration of the above-mentioned power function

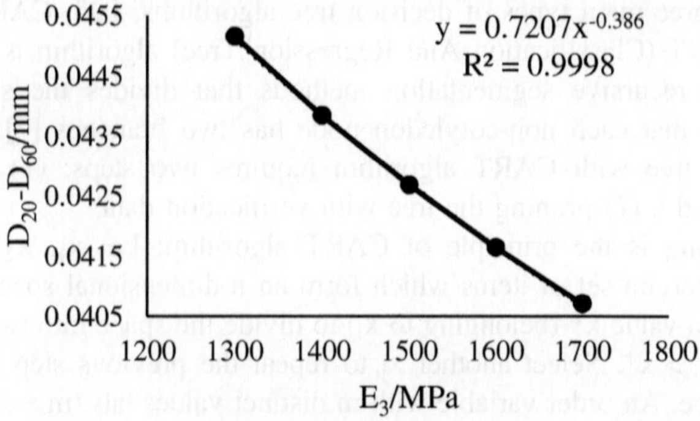

Fig. 1. D_{20}–D_{60}—E_3 curve.

relation, the result is shown as follow. To sum up, the DBP D_{20}–D_{60} is a good indicator to evaluate the bearing capacity of semi-rigid base.

$$lg(D_{20} - D_{60}) = -0.38584 lg(E_3) - 0.1128 1g(E_4) + 0.44969 \\ R^2 = 0.85992 \tag{2}$$

Where, E3, E4 = the moduli of base and lower surface layers.

3 Data Analysis

3.1 Data Collection

The construction of the expressways in Jiangsu Province started relatively early with developed road net and varied types of pavement structure. Based on the Pavement Management System (PMS), historical FWD data as well as traffic load and initial road properties data of Feng Guan Expressway in Jiangsu are selected and merged as the analysis dataset. Feng Guan Expressway was opened to traffic in 2002 and the maintenance proportion is relatively small up to now. The FWD data collected between 2009 and 2013 were used as the targets in the data analysis model. Statistical methods including classification tree and multiple linear regression are then used to analyze the correlation degree of the DBP and factors above to identify the key factors influencing semi-rigid base modulus.

3.2 Classification and Regression Tree

Decision tree is a frequently used data mining tool to solve classification problems in practical applications. The essence of decision tree is the inductive learning algorithm based on the training set. It reaches the classifying rule that infers the decision tree manifestation through a group of disorderly and unsystematic cases. The program of the tree would judge the branch down from the internal node according to different attribute values. And then it comes to the conclusion at the leaf node of the tree.

There are three main types of decision tree algorithms, ID3, CART and C4.5. In this paper, CART (Classification And Regression Tree) algorithm is adopted. It is a type of binary recursive segmentation methods that divides the sample into two sub-samples so that each non-cotyledon node has two branches [7]. Constructing a concise binary tree with CART algorithm requires two steps: (1) partitioning the sample recursively; (2) pruning the tree with verification data.

The following is the principle of CART algorithm: Let x_1, x_2 … x_n be the n attributes of a certain set of items which form an n-dimensional space. Then select a variable x_i and a value x* (belonging to x_i) to divide the space into two parts, namely, $x_i \leq x^*$ and $x_i > x^*$. Select another x_j to repeat the previous step several times to complete the tree. An order variable with m distinct values has (m − 1) splits, and the selection of the split point is based on Gini impurity. Gini impurity is an index of how often a randomly chosen element from the set would be incorrectly labeled if it was randomly labeled according to the distribution of labels in the subset. Gini impurity can be computed by the formula as shown below. Gini impurity reaches its minimum (Gini = 0) when all the items fall into a single target category, while it reaches maximum (Gini = (N − 1) * N/2) when all the classes have the same fraction.

$$Gini(A) = 1 - \sum\nolimits_{i=1}^{N} p_i^2 \tag{3}$$

Where,

A = a certain node;
N = number of classes;
P_i = the fraction of items labeled with class i in the set.

3.3 Discussion

There has been no special treatment for the semi-rigid base of Feng Guan Expressway since it was opened to traffic, indicating that the base layer of this section is still in good condition. The distribution of historical D_{20}–D_{60} data of the pavement is shown in Fig. 2 from 2009 to 2013. It indicates that the values of D_{20}–D_{60} centralized distributed around 10 μm in the first three years. The modulus of semi-rigid base was at a stable stage along with traffic operation before 2012. While the values of D_{20}–D_{60} increased dramatically in 2013, and the number of extreme values (>40 μm) was significantly greater than ever before, which demonstrated it turned into the attenuation stage.

The deterioration of modulus of semi-rigid base in service is mainly influenced by two types of factors. One is the factors of traffic load such as operating time (Age) and average equivalent standard axle load (AESAL). The other is initial property parameters of the layers such as compressive strength, compaction and thickness and so on. For the latter, the decision tree of D_{20}–D_{60} is constructed as shown in Fig. 3 with the initial property parameters of bases (upper and lower base layers) and under surface layer as the factors. According to the nodes of the tree, the compressive strength, compaction and thickness have significant influence on D_{20}–D_{60}.

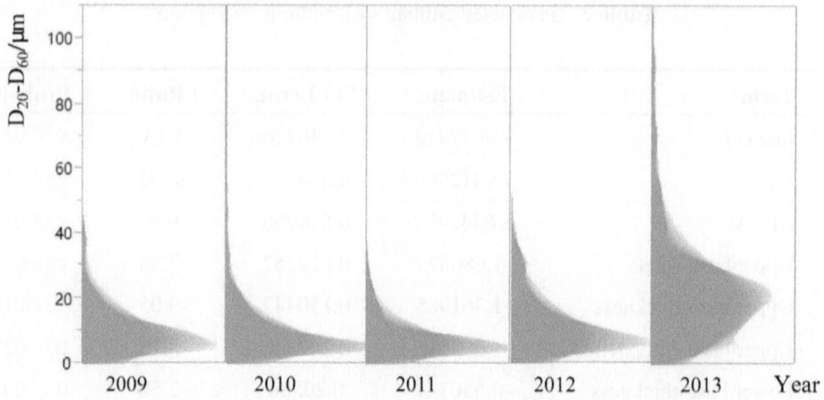

Fig. 2. Historical distribution of D_{20}–D_{60} of Feng Guan Expressway

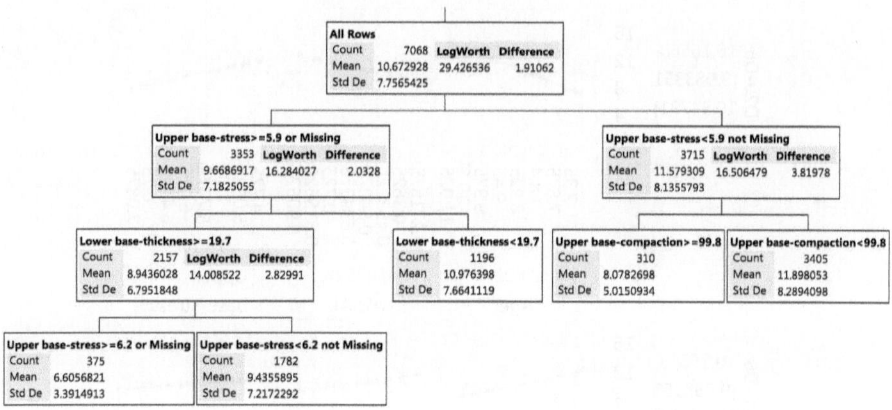

Fig. 3. Decision tree of D_{20}–D_{60}

Based on the factors identified by the CART tree, multiple regression method, shown below was utilized to quantify the influence of different factors (Xi).

$$Y = \beta_0 + \beta_1 X_1 + \cdots + \beta_i X_i + \cdots + \beta_k X_k + \varepsilon \tag{4}$$

Where,

$\beta_0, \beta_1, \ldots, \beta_i, \ldots, \beta_k$ = Estimates of the regression parameters, β_i is the magnitude and direction change in response with each one-unit increase in predictor i, when other predictors are held constant.

ε = random error term.

Table 3 and Fig. 4 present the multiple regression results of all the factors with D_{20}–D_{60}, including the significance test (p-value) of each factor and the prediction profiler. As is shown in Table 3, the p-value of each factor is significant (less than 0.05),

Table 3. Parameter estimates of multiple regression

Term	Estimate	Std Error	t Ratio	Prob>|t|
Intercept	136.17712	26.30426	5.18	<.0001*
Age	1.9412969	0.064265	30.21	<.0001*
AESAL	5.6342479	0.574966	9.8	<.0001*
Upper base-stress	-1.680879	0.214257	-7.85	<.0001*
Upper base-thickness	-1.361985	0.150447	-9.05	<.0001*
Upper base-compaction	-1.003316	0.267348	-3.75	0.0002*
Lower base-thickness	-0.530727	0.20586	-2.58	0.0100*

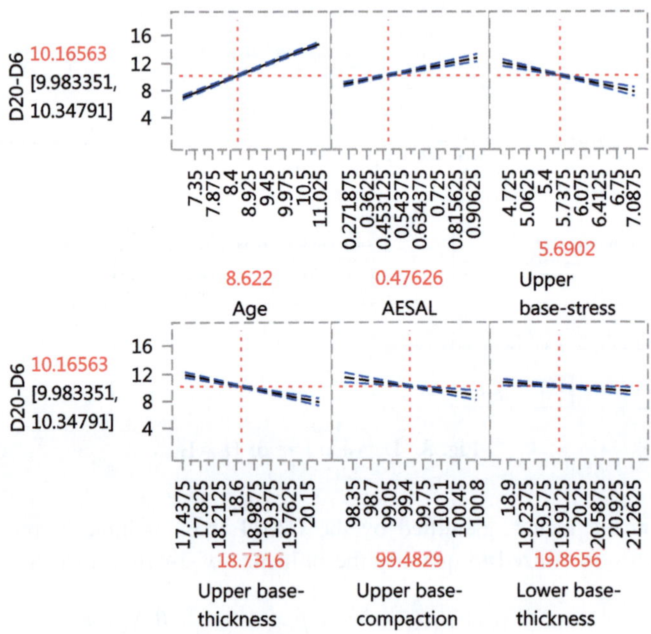

Fig. 4. Prediction profiler of multiple regression

indicating the traffic factors as well as initial property factors significantly affect the modulus of base. Figure 4 indicates that D_{20}–D_{60} increases markedly with the increase of Age and AESAL. In addition, road sections with higher initial compressive strength, compaction and thickness of the base tend to have higher moduli under traffic loads of years.

4 Conclusion

In this study, the FEM of the pavement structure is firstly established to verify the correlation between DBP and moduli of layers. Besides, the factors associated with deflection basin are discussed by investigating historical FWD data. The following conclusion can be drawn:

1. D_{20}–D_{60} is a good indicator to evaluate the relative strength condition of semi-rigid base for its good correlation with the moduli of base and under layer.
2. The development of the modulus of semi-rigid base includes two stages. In the first stage, the modulus increases slightly due to the continued hydration of cement. In the second stage, the modulus showed clear decrease and the bearing capacity of the layer deteriorates slightly.
3. Both Classification tree and multiple regression showed that traffic volume and initial condition of structure layer have significant influence on the deterioration of modulus. The increasing of AESAL lead to the reduction of modulus. The compressive strength, compaction and thickness of base are negatively correlated to D_{20}–D_{60}, which indicates the better of the initial condition of base, the lower of the modulus reduction.

Acknowledgement. This study was supported by the project titled "Thirteen fifth research project, Part I: Highway pavement and subgrade structure and material performance evaluation" (Project no. 7621000131) founded by Jiangsu Department of Transportation and a re-search project titled "Data mining of Highway Pavement Performance and Evaluation of Maintenance Treatment Effectiveness" (Project no. 3221007409) founded by Southeast University.

References

1. Ding, J.F.: Evaluation of the asphalt concrete pavement structural strength with semi-rigid base based on deflection basin. Dissertation of Southeast University (2009)
2. Talvik, O., Aavik, A.: Use of FWD deflection basin parameters (SCI, BDI, BCI) for pavement condition assessment. Baltic J. Road Bridge Eng. **4**(4), 196–202 (2009)
3. Transport Research Cost 336 Falling Weight Deflectometer. European Commission (1996)
4. Bing, X.U.: Assessing pavement layer condition using FWD deflection data. Ph.D. dissertation of North Carolina State University (2000)
5. Kim, Y.R., Lee, Y.C., Ranjithan, S.R.: Flexible pavement condition evaluation using deflection basin parameters and dynamic finite element analysis implemented by artificial neural networks. In: Tayabji, S.D., Lukanen, E.O. (ed.) ASTM STP 1375 Non-destructive Testing of Pavements and Backcalculation of Moduli, vol. 3, pp. 514–530. American Society for Testing and Materials (ASTM International), West Conshohocken (2000). ISBN 0803128584
6. Wang, H.Y.: Evaluation system of structural bearing capacity of the asphalt concrete road with semi-rigid base. Dissertation of Southeast University (2013)
7. Loh, W.-Y.: Classification and regression trees. Wiley Interdiscip. Rev.: Data Min. Knowl. Discov. **1**(1), 14–23 (2011)

Application Analysis of Osculating Value Method in Pavement Performance Evaluation

Zhaorong Wu(✉)

Department of Transportation Engineering, Tongji Zhejiang College, Jiaxing 314051, China
wuzhaorong110@126.com

Abstract. A new evaluation method is introduced in the comprehensive evaluation model of pavement performance. This method shows it is simple, convenient and operational through the analysis of the practical engineering application. The results show that evaluation results from the osculating value method have good consistency with that from the specifications. Meanwhile, each selected indices are endowed with weights in the process of evaluation by means of this method and its principle is very simple, taking a full consideration to the validity of the evaluation results so that the accuracy of the results are improved as well.

Keywords: Pavement performance · Evaluation index
Osculating value method · Engineering example

1 Introduction

Pavement performance synthetically estimation based on recommended approaches from *Highway technical condition evaluation criteria* for the reference in most of china currently. Therefore, there are lack of diversity and comparability.

The first choice of effectiveness appraising, quality management, project decision-making, and other fields in system engineering is osculating value method, which can solve a finite alternative multiple objects decision-making problem effectively. And from the whole decision-making processes, multi-objective decision and evaluation is the same concept, therefore, the decision is the process of analysis and evaluation virtually [1]. The method in recent years has been widely used in economics, society, medicine, environmental protection and other fields, such as water quality evaluation, investment project optimization evaluation, safety assessment of mineral, medical work quality evaluation and so on, is a kind of effective method in comprehensive evaluation.

As a consequence, osculating value method is introduced to be theoretical basis of pavement performance evaluation model in this paper, in order to solve problems about influence of artificial factors, complex calculation, etc. exist in previous comprehensive evaluation for pavement surface condition and to avoid disadvantages from other evaluation methods which give equal weight to each index. A new method is hope to put forward for pavement performance evaluation and maintenance decisions.

X. Shi et al. (Eds.): GSIC 2018, *Proceedings of GeoShanghai 2018 International Conference: Transportation Geotechnics and Pavement Engineering*, pp. 372–378, 2018.
https://doi.org/10.1007/978-981-13-0011-0_40

2 General Principle of Osculation Value Method

Osculating value method is a multiple attribute synthetical evaluation methods, dividing the evaluation index into forward index and backward index and finding the optimal point and the worst point in each evaluation index by normalization processing, and then figuring out the distance between an evaluation unit and these two points respectively, that is, calculating their close degree. In the end, according to the size of osculation-value confirm preferential order of evaluation unit. When osculating value method is brought in pavement performance evaluation models, every road section remains to be evaluated is an evaluation unit. Optimal point and worst point correspond respectively to optimal road section and worst road section, then calculating Euclidean distance between each road section and optimal road section or worst road section so that figure out their osculation-value for rank, as a result, the quality of the whole section will be obtained.

2.1 Establish the Indicator Matrix

Suppose there are m trial road sections, denoted by $M_1, M_2, \ldots, M_m$ and there are n evaluation indexes, denoted by $N_1, N_2, \ldots, N_n$ correspond to f–level evaluation standard, denoted by $F_1, F_2, \ldots, F_f$. The real samples' standard initial matrix $\mathbf{C}$ is make up of these three parts and composed of $m+f$ pending ordered samples and n evaluation indexes, containing $(f+m) \times n$ elements, that is $\mathbf{C} = \left(\mathbf{C}_{ij}\right)_{(m+f)\times n}$.

2.2 Establish the Dimensionless Index Matrix

Every index's dimension is different in matrix, therefore, a deviation will be caused and the evaluation results won't be accurate if the initial matrix used directly. So that index of initial matrix should be standardized according to the formula (1), and then get the dimensionless index matrix $\left(\mathbf{r}_{ij}\right)_{(m+f)\times n}$.

$$r_{ij}\begin{cases} \frac{c_{ij}-E_j}{E_j}, (forward\ index) \\ \frac{E_j-c_{ij}}{E_j}, (backward\ index) \end{cases} \tag{1}$$

Of which, E_j is the target value of index and takes the maximum when it is forward index and vice versa.

2.3 Establish the Virtual Sets of Optimal and Worst Road Sections

Choosing the maximal and minimal value of each evaluation standard from the dimensionless matrix form the virtual sets of optimal and worst road sections:

Optimal road section set indicates every index of that is optimal value, that is $G = \left(r_1^+, r_2^+, \ldots, r_n^+\right)$. Worst road section set indicates every index of that is worst value, that is $B = \left(r_1^-, r_2^-, \ldots, r_n^-\right)$. Of which, $r_n^+ = \max\left\{r_{1n}, r_{2n}, \ldots, r_{(m+f)n}\right\}$, $r_n^- = \min\left\{r_{1n}, r_{2n}, \ldots, r_{(m+f)n}\right\}$.

2.4 Calculate Osculation Value

Using euclidean distance calculates the distance $d_{i,G}$ between target section r_{ij} and virtual optimal section $\left((r_{ij})_G\right)$ as well as the distance $d_{i,B}$ between target section r_{ij} and virtual worst section $\left((r_{ij})_B\right)$. Optimal section and worst section don't have to be a line, so the greater $d_{i,G}$ is doesn't mean the greater $d_{i,B}$ is. Therefore, it is necessary to use osculation value to comprehensively reflect the distance between the target section and optimal section or worst section. The greater optimal osculation value is indicates the section is far from optimal section and the worse pavement performance is and vice versa.

According to the proportion of each index in pavement performance evaluation and the recommended weight values in *Highway technical condition evaluation criteria* for calculation, the formula is as follows:

$$d_{i,G} = \sqrt{\sum_{j=1}^{n} W_j\left[r_{ij} - (r_{ij})_G\right]^2} \tag{2}$$

$$d_{i,B} = \sqrt{\sum_{j=1}^{n} W_j\left[r_{ij} - (r_{ij})_B\right]^2} \tag{3}$$

Of which, W_j represents weight value of the jth index.

Optimal osculation value:

$$E_{i,G} = \frac{d_{i,G}}{\min\{d_{1,G}, d_{2,G}, \ldots, d_{(m+j),G}\}} - \frac{d_{i,B}}{\max\{d_{1,B}, d_{2,B}, \ldots, d_{(m+j),B}\}} \tag{4}$$

Worst osculation value:

$$E_{i,B} = \frac{d_{i,B}}{\min\{d_{1,B}, d_{2,B}, \ldots, d_{(m+j),B}\}} - \frac{d_{i,G}}{\max\{d_{1,G}, d_{2,G}, \ldots, d_{(m+j),G}\}} \tag{5}$$

3 Engineering Example

This paper is based on the five-year maintenance planning project of wenzhou section of shenhai G15 highway, five sections' traffic data are chosen randomly. It is taken pavement surface condition index (PCI), riding quality index (RQI), rutting depth index (RDI), skidding resistance index (SRI) and pavement structure strength index (PSSI) as pavement performance evaluation index. Firstly, on the basis of recommend calculation formula of pavement quality index (PQI) in the standard for evaluation, of which the values of the parameters determined by *Highway technical condition evaluation criteria* (see Table 1). The citation of Table 2 is also from *Highway technical condition evaluation criteria*. All the sample sections' pavement performance indexes and comprehensive evaluation results summarized by Table 3.

Table 1. Weight proposed value of each index in PQI

Index	W_{PCI}	W_{RQI}	W_{SRI}	W_{RDI}	W_{PSSI}
High-grade highways	0.35	0.40	0.10	0.15	0.20

Table 2. Highway pavement performance evaluation standard (hundred mark system)

Index	Evaluation grade				
	Excellent	Good	Middle	Second-rate	Bad
RQI	90–100	80–90	62–80	40–62	0–40
PSSI	97–100	92–97	80–92	59–80	0–59
PCI	85–100	70–85	55–70	40–55	0–40
SRI	85–100	62–85	40–62	30–40	0–30
PQI	85–100	70–85	50–70	40–50	0–40

Table 3. Selected sections' pavement performance comprehensive evaluation results

Section stake (K)			Mile	Evaluation grade	PQI	PCI	RQI	RDI	SRI	PSSI
Start	End	No.								
1703 + 000	1704 + 000	Section 1	1000	Good	82.6	80.1	90.3	89.9	94.4	99.5
1700 + 000	1701 + 000	Section 2	1000	Excellent	86.5	90.9	90.0	90.9	94.9	99.7
1678 + 000	1679 + 000	Section 3	1000	Excellent	90.5	98.9	92.8	91.1	97.7	98.9
1768 + 000	1769 + 000	Section 4	1000	Middle	69.7	56.0	82.2	81.7	90.9	99.9
1769 + 000	1770 + 000	Section 5	1000	Middle	69.9	48.5	88.2	84.00	95.3	95.7

Comprehensive evaluation model of pavement performance:

$$PQI = w_{PCI}PCI + w_{RQI}RQI + w_{SRI}SRI + w_{RDI}RDI \quad (6)$$

Of which, w_{PCI}, w_{RQI}, w_{SRI}, w_{RDI} respectively represents weight value in PQI.

Then, start calculating by means of osculation value method after collecting and organizing data of the sections, from which the results are compared and analyzed with that obtained from recommend evaluation method in the standard. How to verify the feasibility of this method in the comprehensive evaluation of pavement performance? As described below:

3.1 Establish the Initial Matrix

Five selected sections' index value and lower limiting value of the standard's recommend values form initial matrix **C** of which 1–5 represent five selected sections, E, G, M, S represent excellent, good, middle and second-rate, four evaluation grade respectively. There is no bad section because all the sections' performance is above second-rate. Each column of matrix represents one index, including PCI, RQI, SRI and PSSI, which can make the evaluation results more accurately and comprehensively.

$$\mathbf{C} = \begin{array}{c} 1 \\ 2 \\ 3 \\ 4 \\ 5 \\ E \\ G \\ M \\ S \end{array} \begin{array}{c} \begin{array}{cccc} PCI & RQI & SRI & PSSI \end{array} \\ \begin{bmatrix} 80.11 & 90.30 & 94.40 & 99.95 \\ 90.99 & 90.00 & 94.90 & 99.57 \\ 98.94 & 92.80 & 97.70 & 98.92 \\ 56.05 & 82.20 & 90.90 & 99.91 \\ 48.49 & 88.20 & 95.30 & 95.97 \\ 85.00 & 90.00 & 85.00 & 97.00 \\ 70.00 & 80.00 & 62.00 & 92.00 \\ 55.00 & 70.00 & 40.00 & 80.00 \\ 40.00 & 60.00 & 30.00 & 59.00 \end{bmatrix} \end{array}$$

3.2 Establish the Dimensionless Index Matrix

It's known by journal articles [2], all the pavement performance evaluation indexes are forward indexes.

According to initial matrix **C**, all indexes' target value respectively are: $E_1 = 98.9$, $E_2 = 92.8$, $E_3 = 97.7$, $E_4 = 99.9$, and then the initial matrix **C** is normatively processed through formula (1), the dimensionless matrix **r** is obtained at the end.

$$\mathbf{r} = \begin{bmatrix} -0.190 & -0.040 & -0.040 & 0 \\ -0.080 & -0.030 & -0.035 & -0.004 \\ 0 & 0 & -0.006 & -0.010 \\ -0.433 & -0.114 & -0.075 & 0 \\ -0.510 & -0.050 & -0.031 & -0.040 \\ -0.141 & -0.030 & -0.135 & -0.029 \\ -0.293 & -0.138 & -0.369 & -0.079 \\ -0.444 & -0.246 & -0.593 & -0.200 \\ -0.596 & -0.353 & -0.695 & -0.410 \end{bmatrix}$$

3.3 Assume Virtual Sets of Optimal and Worst Road Sections

Virtual sets of optimal road sections is make up of maximum value per column in dimensionless matrix **r**, that is $G(0, 0, -0.006, 0)$. Virtual sets of worst road sections is make up of minimum value per column in dimensionless matrix **r**, that is $B(-0.596, -0.353, -0.695, -0.410)$.

3.4 Calculate Osculation Value

The distance between each evaluation road section and virtual optimal worst) road sections is $d_{i,G}$ ($d_{i,B}$), which can be figured out by formulas (2) and (3), then the optimal osculation value $E_{i,G}$ is obtained by formulas (4) and (5). All sections' evaluation are forward comparison, so the worst osculation value doesn't need be calculated. Therefore, all road sections' pavement performance evaluation grade can be ordered by distance between each road section and optimal road section. The calculation results as shown in Table 4.

All the optimal osculation values are shown on the number line (see Fig. 1) so that all road sections' comprehensive performance status can be observed more clearly.

Table 4. Osculation value of evaluation sections

Evaluation sections & grade	$d_{i,G}$	$d_{i,B}$	$E_{i,G}$
Section1	0.1160	0.4098	22.385
Section2	0.0520	0.4541	9.481
Section3	0.0050	0.4974	0.009
Section4	0.2663	0.3181	52.616
Section5	0.3038	0.3262	60.096
Excellent	0.0963	0.4114	18.429
Good	0.2269	0.2844	44.817
Middle	0.3651	0.1483	72.721
Second-rate	0.4998	0.0004	99.953

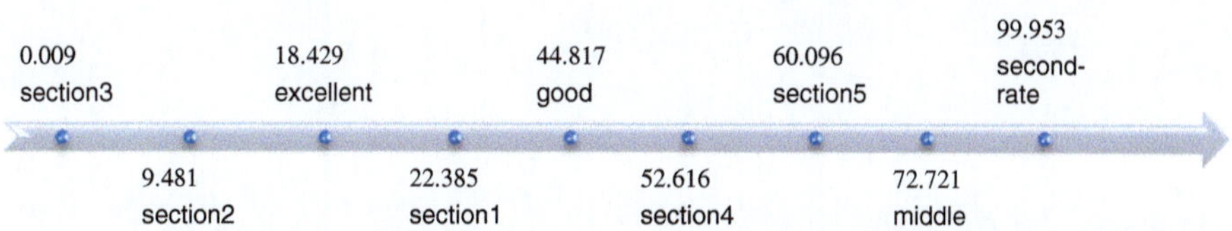

Fig. 1. A number line diagram about road sections' osculation values rank

4 Results and Discussion

Figure 1 shows that the optimal osculation value of section2 and section3 is greater than that of optimal road section, so these two sections' comprehensive performance grade is excellent. The optimal osculation value of section1 is in the range of optimal road section and worst road section, so its comprehensive performance grade is good. The optimal osculation value of section4 and section5 is in the range of better road section and worse road section, so its comprehensive performance grade is middle, compared with the results based on evaluation method recommended by the standard. It finds that the results of two kinds of methods are identical, so that osculation value method is introduced in the pavement performance comprehensive evaluation system, which is reasonable and applicable.

5 Conclusion

This paper analyzed several typical evaluation models of pavement performance of Chinese highway management system and introduced a new evaluation method for studying, exploring the building of an evaluation model that fits damage mechanics and maintenance characteristic of pavement in the jiaxing Zhejiang area. According to the practical application, this new method is proved to be feasible and its accuracy of calculation is improved as well. Therefore, this new method provides a valuable reference for the maintenance decision work of China's highway management department.

References

Rahman, M.M., Uddin, M.M., Gassman, S.L.: Pavement performance evaluation models for South Carolina. Ksce J. Civ. Eng. **21**(7), 1–12 (2016)

Wang, L., Liu, R., Wu, Q., et al.: The application of osculating value method in the performance evaluation of technological innovation. Value Eng. **26**(7), 43–44 (2007)

Effectiveness Evaluation of Asphalt Pavement Maintenance Treatments Based on Equivalent Area Method

Linyi Yao, Qiao Dong(✉), Fujian Ni, and Jiwang Jiang

Department of Highway and Railway Engineering, College of Transportation, Southeast University, 2# Sipailou, Nanjing 210096, Jiangsu, China
qiaodong@seu.edu.cn

Abstract. The effectiveness and cost-effectiveness of different pavement maintenance treatments in Zhejiang, China were evaluated based on the pavement performance data and costs of collected maintenance projects. Equivalent area method and 3.5-year unit cost were proposed to calculate the treatment effectiveness and cost-effectiveness. Three types of asphalt overlays including AC-13, AC-16 and SMA-13 and three preventive treatments including ultra-thin friction overlay, hot-in-place rehabilitation and micro surfacing were included in this research. PCI (pavement surface condition index), RQI (riding quality index) and RDI (rutting depth index) were adopted to evaluate the treatment performance. Multiple linear regression analysis method was then applied to analyze the influence of factors on the effectiveness and cost-effectiveness. Analysis results showed that both effectiveness and cost-effectiveness of asphalt overlays were higher than those of preventive treatments. Among them, SMA-13 showed relatively higher effectiveness and ultra-thin friction showed higher cost-effectiveness while micro surfacing seemed to be least effective and cost-effective. The cost-effectiveness decreased clearly with the increase of AADT (annual average daily traffic). It was found that maintenance effectiveness tended to be higher after crack sealing treatment. The geometric characteristics of road sections also affected the effectiveness and cost-effectiveness. It is worth noting that effectiveness and cost-effectiveness of sections with big curvature and long and steep slope were relative lower.

Keywords: Pavement maintenance · Pavement condition index
Roughness · Rutting · Cost-effectiveness

1 Introduction

1.1 Research Background

In order to maintain pavement performance and provide a comfortable, safe driving condition, a large amount of budget was invested on pavement maintenance every year. However, pavement riding quality and service level decrease with the increase of pavement age. Maintenance task became increasingly serious. Optimizing the

X. Shi et al. (Eds.): GSIC 2018, *Proceedings of GeoShanghai 2018 International Conference: Transportation Geotechnics and Pavement Engineering*, pp. 379–388, 2018.
https://doi.org/10.1007/978-981-13-0011-0_41

application of various pavement maintenance treatments becomes an important task for pavement maintenance engineers. To assist engineers selecting appropriate pavement treatments, effectiveness or cost-effectiveness analysis of maintenance treatments were indispensable.

The effectiveness of maintenance treatments includes improving pavement performance, prolonging pavement service life, delaying the decline of pavement performance as well as saving travel time of road users and improving travel safety and riding comfort. Various effectiveness analysis methods were proposed in previous studies. Treatment service life and improved pavement performance were two main measurements adopted by researchers [1]. Several existing effectiveness measures include performance jump (immediate improvement of pavement performance index after treatment) [2], improved long-term average pavement condition [3], treatment service life [4, 5], the extended pavement life [6], the reduction of performance deterioration rate [7], the increased area bounded by the performance curve and the threshold line [8, 9], benefit-cost ratio [10], etc.

1.2 Objectives

Since the pre-treatment pavement condition data in Zhejiang, China has not been preserved, this paper aims to establish a more practical and simple treatment effectiveness evaluation method and then compare the effectiveness and cost-effectiveness of different maintenance treatments widely used in Zhejiang. Afterwards, the influences of factors including traffic level, crack sealing treatment and geometric characteristics of road sections on treatment effectiveness were analyzed utilizing multiple linear regression analysis method.

2 Scope and Data Collection

This study evaluated the effectiveness and cost-effectiveness of three types of asphalt overlay (AC-13, AC-16 and SMA-13) and three preventive treatments (ultra-thin friction overlay, hot-in-place rehabilitation and micro surfacing) widely used in Zhejiang. A total of 45 maintenance projects were investigated. Table 1 shows the sample size of each maintenance treatment. One third of these projects have received a secondary treatment. Of all the SMA-13 overlay projects, only one had failed which indicated a relatively better performance of SMA-13 overlay. The average service life of those failure samples are shown in Table 2. Hot-in-place rehabilitation did not have a specific service life owing to the fact that all the hot-in-place rehabilitation projects are still in service. It can be found that the service life of asphalt overlays was higher than that of preventive treatments. Among them, the only failed project of SMA-13 overlay had served up to 6.5 years while micro surfacing had the shortest service life.

Table 1. Sample size of each maintenance treatment

Treatments	Overlay			Preventive treatment			Total
	AC-13	AC-16	SMA-13	Ultra-thin friction overlay	Hot-in-place rehabilitation	Micro surfacing	
In service	10	2	11	2	3	2	30
Proportion (%)	22.2	4.4	24.4	4.4	6.7	4.4	66.7
Out of service	5	7	1	1	0	1	15
Proportion (%)	11.1	15.6	2.2	2.2	0	2.2	33.3

Table 2. The average service life of each maintenance treatment

Treatments	Overlay			Preventive treatment		
	AC-13	AC-16	SMA-13	Ultra-thin friction overlay	Hot-in-place rehabilitation	Micro surfacing
Average service life (years)	3.9	3.5	6.5	3.5	>2.5	2.5

3 Equivalent Area Method

Usually both pre- and post- treatment pavement condition are considered to evaluate the long-term treatment effectiveness. The typical "effectiveness area" is defined as the area bounded by the upper or lower threshold, the pre-treatment and post-treatment performance curves. However, since treatment service life and observation time differ, it was problematic to compare the effectiveness of various treatments through this method. Therefore, "equivalent area method" was proposed in this paper to compare the effectiveness of different treatments by using a unified standard service time. This method supposed the same pre-treatment pavement conditions for different road sections. Since the pavement structure and weather condition in Zhejiang is similar, this assumption was considered to be appropriate when the pre-treatment performance data was deficient. Moreover, in order to simplify the establishment of performance models, this method presumed that performance indices reach the ultimate value through a linear process.

To calculate the equivalent effectiveness area, step1 is to calculate the sum of each trapezoid area bounded by the post-treatment performance polyline and the threshold. Step 2 is to draw the equivalent trapezoid with the same total area. Step 3 is to identify a unified standard service time and figure out the equivalent area. Figure 1 shows the specific calculating illustration. Among them, Fig. 1(a) and (b) are for non-decreasing performance indices (e.g., IRI, RD, DR) while Fig. 1(c) and (d) was for non-increasing indices (e.g., PCI, RQI, RDI). Equations (1) and (3) is used to calculate the equivalent effectiveness when adopting non-decreasing pavement performance indicators and Eqs. (2) and (4) is used for non-increasing performance indicators.

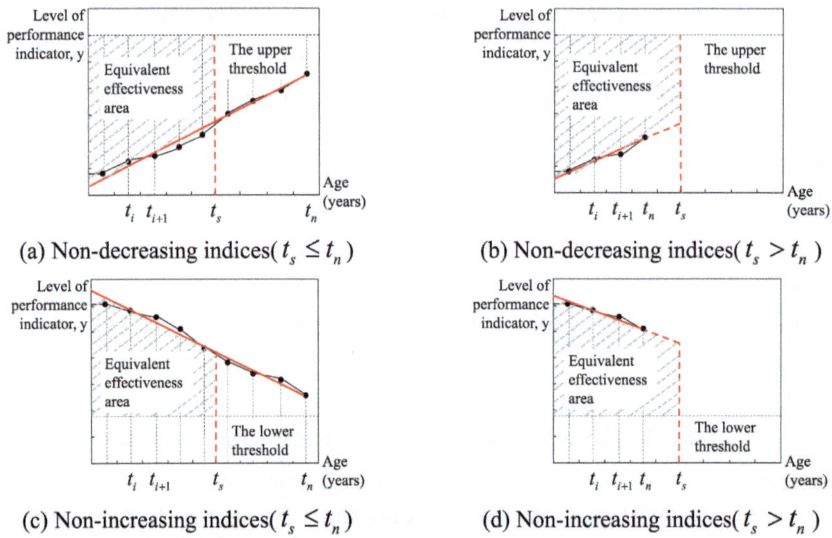

Fig. 1. The specific calculating illustration

$$Eff_1 = \sum_{i=0}^{n-1} \frac{1}{2}(2y_u - y_i - y_{i+1})(t_{i+1} - t_i) \tag{1}$$

$$Eff_2 = \sum_{i=0}^{n-1} \frac{1}{2}(y_i + y_{i+1} - 2y_l)(t_{i+1} - t_i) \tag{2}$$

$$Eff_{1e} = Eff_1 - (y_u - y_n)(t_n - t_s) - (1 - \frac{t_s}{t_n})^2 [Eff_1 - (y_u - y_n)t_n] \tag{3}$$

$$Eff_{2e} = Eff_2 - (y_n - y_l)(t_n - t_s) - (1 - \frac{t_s}{t_n})^2 [Eff_2 - (y_n - y_l)t_n] \tag{4}$$

where Eff_1, Eff_2 = the original treatment effectiveness (Eff_1 is for non-decreasing indices and Eff_2 is for non-increasing indices); Eff_1, Eff_2 = the equivalent treatment effectiveness;y_i = the value of performance index when it's the i-th time to detect (y_0 represents the instant value after treatment, if it's unknown, then $y_0 = y_1$); t_i = the treatment age when it's the i-th time to detect (years); t_s = the standard service time (years); y_l = the lower threshold (used for non-increasing indices); y_u = the upper threshold (used for non-decreasing indices); n = the total detection times.

4 Treatment Effectiveness

4.1 Performance Indicators and the Threshold

The existing asphalt pavement performance evaluation system in China uses five pavement condition indices, including pavement surface condition index (PCI), riding quality index (RQI), rutting depth index (RDI), skidding resistance index (SRI) and pavement structural strength index (PSSI) as well as a comprehensive index named pavement quality of performance index (PQI). All these indices are between 0 and 100, with 100 indicating the best pavement condition. In this research, PCI, RQI and RDI were employed to calculate the equivalent effectiveness as a consequence of data availability.

The lower threshold value of a pavement condition index, also called the maintenance trigger value, usually indicates that a pavement is in very bad condition and needs immediate maintenance treatment. Highway maintenance quality evaluation method stipulated that pavement condition index should be kept higher than 75. Otherwise, maintenance treatment must be conducted to improve pavement condition and service level [11]. Therefore, the lower threshold value in this research was set to 80.

4.2 Effectiveness Evaluation

In order to evaluate the effectiveness of various treatments and investigate the influence of different factors, multiple linear regression model incorporating four factors including maintenance treatments, AADT, crack treatment and the geometric characteristics of road sections was established.

Figure 2 presents the analysis results. As can be seen from Fig. 2, asphalt overlays have higher effectiveness compared to preventive treatments, particularly in RDI effectiveness which indicates that preventive treatments showed poor performance in repairing deep rutting.

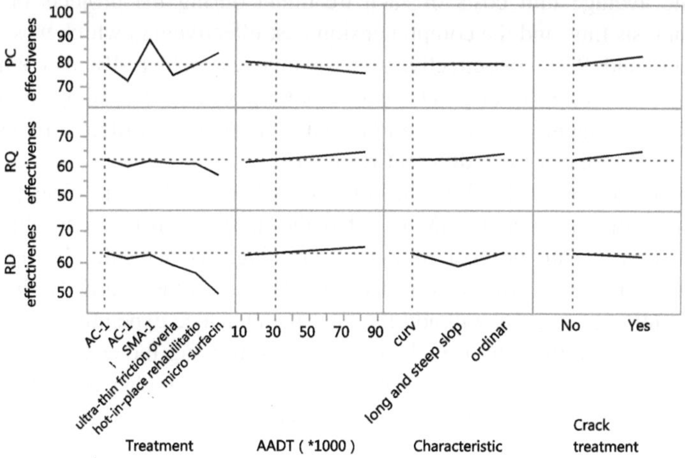

Fig. 2. Influence of factors on treatment effectiveness

SMA-13 overlay shows strong capacity in repairing pavement distresses. AC-16 has the surprisingly lower effectiveness since nearly half of the road sections using AC-16 asphalt overlay are curves or long and steep slopes. Micro surfacing has the lowest RQI and RDI effectiveness but the comparatively high PCI effectiveness. Since most of the road sections had high PCI values, it indicates that micro surfacing can repair surface distress only when the pavement was in relatively better condition. However, for those with high IRI and deep rutting depth, the effectiveness of micro surfacing was fairly poor.

It can also be observed that the PCI effectiveness decreased with the increase of AADT as heavy traffic volume accelerated the deterioration of pavement performance. However, RQI and RDI effectiveness showed different trends which might be attributed to the more uniform compaction effect of higher traffic volumes.

Moreover, the treatment effectiveness was also influenced by the geometric characteristics of road sections. As is shown in Fig. 2, road sections with big curvature and long and steep slope tend to have the lower effectiveness, especially for RDI effectiveness which is caused by the low speed and frequently braking on the curves and long and steep slopes.

Figure 2 also exhibits the impact of crack treatment prior to maintenance treatment. It can be found that the effectiveness was higher with the treatment of cracks. PCI effectiveness was affected mostly, followed by RQI effectiveness as crack pre-treatment can effectively prevent the cracks from being reflected to the new layer so that delay the appearance of distress on the new layer. Further, it seems that crack pre-treatment had little effect on RDI effectiveness.

5 Treatment Cost-Effectiveness

5.1 Treatment Cost

In addition to the effectiveness, cost-effectiveness was also investigated. Figure 3 presents the average unit costs of each treatment during the projects or within the 3.5-year analysis time and the comprehension cost-effectiveness which was the average of PCI, RQI and RDI cost-effectiveness. In this research, a 4% discount rate was considered, converting the original costs to present values. The high variation in the treatment costs is mainly due to variations in asphalt prices at different times as well as variations in aggregates from different resources.

As can be seen from Fig. 3(a), the costs of asphalt overlays are clearly higher than those of preventive treatments. SMA-13 has the most expensive cost which is nearly four times higher than the cost of micro surfacing. As a result, the least effective treatment turned to the most cost-effective one. However, the relatively shorter service life of preventive treatments will add the later maintenance expenditures which appear to be uneconomic in the long-run. Thus, simply comparing the cost of each treatment during a single project was unreasonable. The seemingly higher cost-effectiveness of preventive treatments will mislead the maintenance decision-making process for highway agencies.

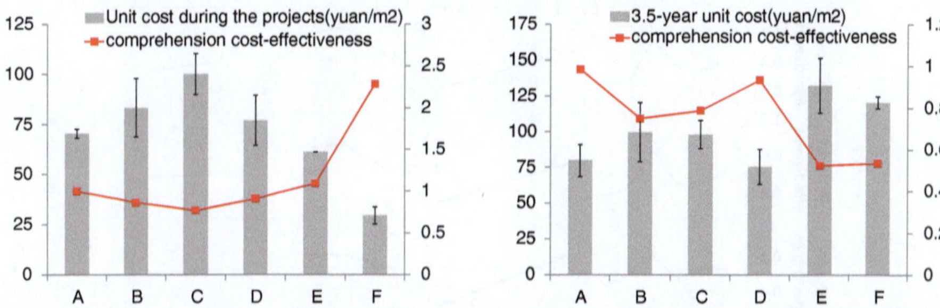

(a) The average unit costs during the projects

(b) The average unit costs within the 3.5-year analysis time

A=AC-13 overlay, B=AC-16 overlay, C= SMA-13 overlay, D= ultra-thin friction overlay, E=hot-in-place rehabilitation, F= micro surfacing

Fig. 3. The unit cost of each treatment

Therefore, a 3.5-year analysis time was put forward to measure the costs of distinct treatments for long-term consideration as 3.5 years was the boundary of service life between asphalt overlays and preventive treatments. In order to determine the 3.5-year unit costs, three cases were taken into account:

(1) If the treatments served more than 3.5 years, the first 3.5 years' total costs calculated were defined as the 3.5-year unit costs.
(2) If the treatments failed within 3.5 years, the costs of second treatments before the end of 3.5 years were added in the 3.5-year unit costs.
(3) For those projects with observation time less than 3.5 years, different assumptions were made for asphalt overlays and preventive treatments. As mentioned above, the average service life of asphalt overlays was more than 3.5 years while that of preventive treatments was less than 3.5 years. Thus, for asphalt overlays, their service life was expected to be more than 3.5 years which was back to the first case. Whereas preventive treatments were assumed to fail at 3.5 years and the costs of milling and resurfacing should be added to calculate their 3.5-year unit costs.

The 3.5-year unit costs were displayed in Fig. 3(b) and the comprehension cost-effectiveness were presented for comparison. As can be seen, the costs of hot-in-place rehabilitation and micro surfacing were higher than others when considering their extra expenditure afterwards. Hence, the cost-effectiveness of asphalt overlays became higher than that of preventive treatments which was more acceptable.

5.2 The Influencing Factors Analysis for Cost-Effectiveness

As for cost-effectiveness which was the ratio of effectiveness to 3.5-year unit cost, multiple linear regression analysis was also performed to explore the influence of different factors. It can be seen from Fig. 4, asphalt overlays tend to be more cost-effective than preventive treatments. Ultra-thin friction overlay had the

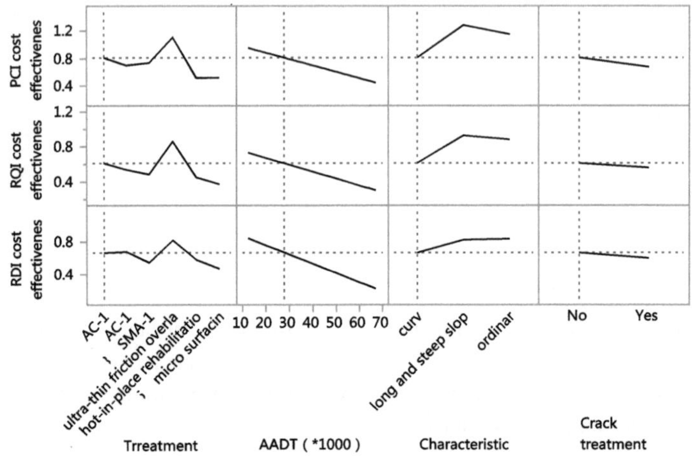

Fig. 4. Influence of factors on treatment cost-effectiveness

unexpectedly high cost-effectiveness which might be attributed to the low cost but relatively longer service life as well as better pavement condition after ultra-thin friction overlay in this research.

Regarding the impact of traffic volume, Fig. 4 indicates that the cost-effectiveness decrease markedly with the increase of AADT. Combining the relationship of AADT and effectiveness which is shown in Fig. 2, it can be implied that higher traffic volume lead to higher maintenance treatment cost.

The geometric characteristics of road sections also affect the cost-effectiveness of maintenance treatments. Those with big curvature tend to be less cost-effective. In addition, as illustrated in Fig. 4, crack pre-treatment hardly influence the cost-effectiveness since crack pre-treatment also need some cost which weaken the advantages of crack pre-treatment.

6 Conclusion

The effectiveness and cost-effectiveness of six different maintenance treatments frequently used in Zhejiang were evaluated based on the investigation of pavement conditions and cost of maintenance projects. Several conclusions can be summarized as follows:

1. The proposed "equivalent area method" can effectively overcome the defect of typical effectiveness area method due to the use of unified standard service time. And it is quite applicable when the pre-treatment performance data is deficient and the original pavement conditions are similar.
2. Asphalt overlays had higher effectiveness compared to the preventive treatments, particularly in RDI effectiveness which indicates that preventive treatments showed poor performance in repairing deep rutting. SMA-13 showed the highest effectiveness while micro surfacing seemed to be least effective.

3. With the increase of traffic level, PCI effectiveness decreased due to the acceleration of traffic volume on performance deterioration. However, RQI and RDI effectiveness got improved as the growth of traffic volume. Road sections with big curvature and long and steep slope showed relatively lower effectiveness. Crack treatment improves the treatment effectiveness.
4. The concept of 3.5-year unit cost considered the extra expenditures for those with short service life. As a result, asphalt overlays tend to be more cost-effective than preventive treatments. Ultra-thin friction overlay had the highest cost-effectiveness due to its low cost but relatively longer service life as well as better pavement condition in this research.
5. The cost-effectiveness decreased significantly with the increase of AADT. Road sections with big curvature tend to be less cost-effective. Moreover, crack treatment almost had no influence on cost-effectiveness of maintenance treatments.

Acknowledgement. This study was supported by the project titled "Effectiveness and Applicability of Asphalt Pavement Maintenance Techniques in Zhejiang" (Project no. 8521002166) funded by Zhejiang Transportation Engineering Construction Group Co., a research project titled "Data mining of Highway Pavement Performance and Evaluation of Maintenance Treatment Effectiveness" (Project no. 3221007409) funded by Southeast University, the thirteen fifth research project titled "Highway asphalt pavement structure long term preservation" (Project no. 7621000132) funded by Jiangsu Department of Transportation and the thirteen fifth research project titled "Highway asphalt pavement intelligent maintenance (Project no. 7621000133), funded by Jiangsu Department of Transportation. This study used the data collected by Zhejiang Transportation Engineering Construction Group Co. The engineers are acknowledged for their assistance on data inquiry. The contents of this paper reflect the views of the authors, who are responsible for the facts and the accuracy of the data presented herein, and do not necessarily reflect the official views or policies of any organization, nor do the contents constitute a standard, specification or regulation.

References

1. Irfan, M., Khurshid, M.B., Labi, S., et al.: Evaluating the cost effectiveness of flexible rehabilitation treatments using different performance criteria. J. Transp. Eng. **135**(10), 753–763 (2009)
2. Labi, S., Sinha, K.C.: Measures of short-term effectiveness of highway pavement maintenance. J. Transp. Eng. **129**(6), 673–683 (2003)
3. Labi, S., Sinha, K.C.: Effectiveness of highway pavement seal coating treatments. J. Transp. Eng. **130**(1), 14–23 (2004)
4. Eltahan, A., Daleiden, J., Simpson, A.: Effectiveness of maintenance treatments of flexible pavements. Transp. Res. Rec.: J. Transp. Res. Board **1680**, 18–25 (1999)
5. Irfan, M., Khurshid, M., Labi, S.: Determining the service life of thin hot-mix asphalt overlay by means of different performance indicators. Transp. Res. Rec.: J. Transp. Res. Board **2108**, 37–45 (2009)
6. Bardaka, E., Zhang, Z., Labi, S., Sinha, K.C., Mannering, F.: Statistical assessment of the cost effectiveness of highway pavement warranty contracts. J. Infrastruct. Syst. **22**(3), 04016017-1–04016017-10 (2016)

7. Chen, X., Zhu, H., Dong, Q., Huang, B.: Optimal thresholds for pavement preventive maintenance treatments using LTPP data. J. Transp. Eng. Part A: Syst. **143**(6), 04017018-1–04017018-9 (2017)
8. Dong, Q., Huang, B.: Evaluation of effectiveness and cost-effectiveness of asphalt pavement rehabilitations utilizing LTPP data. J. Transp. Eng. **138**(6), 681–689 (2011)
9. Chen, X., Dong, Q., Huang, B., Zhu, H.: Case study: performance effectiveness and cost-benefit analyses of open-graded friction course pavements in Tennessee. Int. J. Pavement Eng. 1–14 (2016). https://doi.org/10.1080/10298436.2016.1138112
10. Dong, Q., Huang, B., Richards, S.H., et al.: Cost-effectiveness analyses of maintenance treatments for low-and moderate-traffic asphalt pavements in Tennessee. J. Transp. Eng. **139** (8), 797–803 (2013)
11. Highway maintenance quality evaluation method. China Communications Press. Beijing, China. Department of Transportation (2002)

Short-Term Cost-Effectiveness Evaluation of Maintenance Treatments for Asphalt Pavement in Jiangsu, China

Hui Du, Fujian Ni(✉), Song Li, and Guoqiang Liu

School of Transportation, Southeast University, Nanjing 210096, Jiangsu, China
duhui1105@163.com, nifujian@gmail.com

Abstract. The short-term cost-effectiveness of maintenance treatments in asphalt pavement were evaluated based on the pavement historical maintenance records and pavement performance data from the pavement management system (PMS) in Jiangsu province, China. This paper took a deep investigation into the short-term cost-effectiveness and created a short-term cost-effectiveness analyses scheme. Applying the principles of engineering economy, this study analyzed the performance vary quantity, performance vary slope and performance vary rate of both pretreatment and posttreatment pavement performance trends and selected cost-effectiveness ratio as the short-term treatment cost-effectiveness evaluation index. Some statistical data analysis methods were used to eliminate the abnormality data, then formed a short-term cost-effectiveness scheme. Finally the short-term cost-effectiveness of four kinds of maintenance treatments (thin ECA overlay, mill and fill, micro surfacing and hot-in-place recycling) were evaluated and verify the feasibility of the scheme. The analyses indicates that the effectiveness of maintenance treatments on moderate-traffic and AK pavement structure are better than on low-traffic and SMA asphalt pavements, respectively. Analyses results that thin ECA overlay has more short-term cost-effective than other three treatments.

Keywords: Pavement management · Maintenance
Short-term treatment cost-effectiveness · Cost-effectiveness ratio

1 Introduction

1.1 Research Background

Because of longer service life of pavement, high riding comfort, durability and lower accident rate, pavement maintenance of highway is necessary and critical. The main purpose of maintenance is to retard the deterioration and rehabilitation without reconstruction [1]. Especially, According to SHRP, the cost of maintenance will be saved 45%–50% through 3–4 times maintenance across the pavement life cycle [2]. Cost-effectiveness of different maintenance treatments can be confirmed by the cost of pavement maintenance treatment in a given period and the improvement of pavement performance, which lead to different maintenance effectiveness [3–5]. Not only for good short-term effectiveness, but also obtaining long-term effectiveness through

X. Shi et al. (Eds.): GSIC 2018, *Proceedings of GeoShanghai 2018 International Conference: Transportation Geotechnics and Pavement Engineering*, pp. 389–399, 2018.
https://doi.org/10.1007/978-981-13-0011-0_42

maintaining pavement. Many studies have been done, including by Dong [6, 7], Morian [8] and Freeman [9], to investigate different methods to evaluate the cost-effectiveness of maintenance treatments. However, according to the result of FHWA, most studies were focused on the long-term effectiveness and ignore the data change of pavement performance shortly after maintaining, which underestimate the availability and effectiveness of maintenance treatments even misjudgement. Therefore, it is critical to study the short-term cost-effectiveness to make economic maintenance strategies.

How to select the effective maintenance treatments is vitally important considering pavement condition and traffic level. Because of lacking of related data of pavement condition and traffic, limited finding about the pavement performance in China were obtained in previous studies. But, highway weight charge system has been extensively used and the pavement condition data are recorded in the developed pavement management system (PMS) in Jiangsu Province in China. Detailed information of pavement historical, traffic level, axle loads, performance data and maintenance historical treatments have been collected in PMS, which are very important to cost-effectiveness of maintenance treatments.

1.2 Research Objectives

The objectives of this study were as follows:

- Investigating the pavement performance changes of pre-treatment and post-treatment over times.
- Creating a short-term cost-effectiveness analysis scheme, including performance data investigation and process, short-term cost-effectiveness evaluation indicators and evaluating procedure.
- Estimating this short-term cost-effectiveness scheme through detailed practical example, comparing four different maintenance treatments in Jiangsu, China (thin ECA overlay, mill and fill, micro surfacing and hot-in-place recycling).

2 Evaluation Index

Because of the different application time of maintenance treatments and the length of analysis period, it is difficult to evaluating the economic and social benefits in short-term. As a result, the performance changes of pre-treatment and post-treatment and treatment cost are used to evaluate short-term cost-effectiveness of different maintenance treatments.

Two kinds of performance change condition were existed among all pavement performance data (see Fig. 1), there is a certain change characteristic looked similar to the letter "Z". Then some evaluation indicators are decided to describe the short-term maintenance treatment cost-effectiveness.

Notes:

A is at a time t_1 before treatment
C is just before treatment
D is just after treatment
E is at a time t_2 after treatment
B is the treatment application time, just as an reference point
K_1 is the slope of performance jump before treatment
K_2 is the slope of performance jump after treatment
K_3 is the slope of performance jump between pre-treatment and post-treatment.

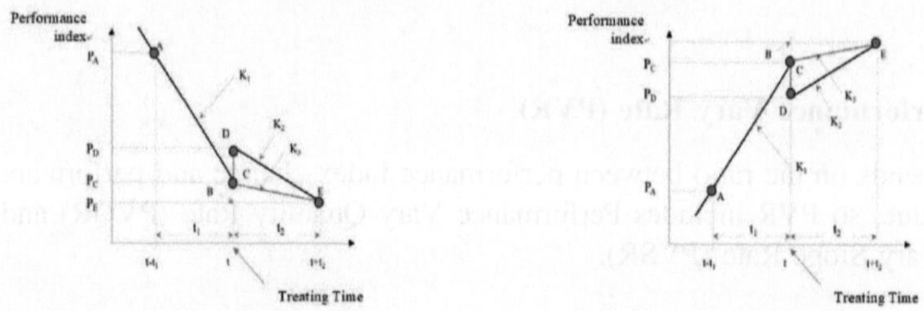

(a) Incremental Performance (b) Descending Performance

Fig. 1. Pavement performance indicators jump of pre-treatment and post-treatment

2.1 Performance Vary Quantity (PVQ)

As shown in Fig. 1,

$$PVQ = |P_D - P_C| = |P_D - (P_A - K_1 \times t_1)| \tag{1}$$

Assume $K_1 = K_2$ and $t_1 = t_2 = 1$,

$$PVQ = |P_D - P_C| = |P_D - (P_A - K_1 \times t_1)| = |P_D - P_A + K_1 \times t_1| = |P_D - P_A + K_2|$$

$$= |P_D - P_A + P_D - P_E| = |2P_D - P_A - P_E| \tag{2}$$

$$PVQ = |P_D - P_C| = |P_E + K_2 \times t_2 - (P_E + K_3 \times t_2)| \tag{3}$$

2.2 Performance Vary Slope (PVS)

See Fig. 1(a),

$$PVS = K_1 - K_2 = (P_A - P_C)/t_1 - (P_D - P_E)/t_2 \quad (K_1, K_2 < 0) \tag{4}$$

If $t_1 = t_2 = 1$,

$$PVS = K_1 - K_2 = P_A - P_C - P_D + P_E \tag{5}$$

See Fig. 1(b),

$$PVS = K_1 - K_2 = (P_C - P_A)/t_1 - (P_E - P_D)/t_2 \quad (K_1, K_2 > 0) \tag{6}$$

If $t_1 = t_2 = 1$,

$$PVS = K_1 - K_2 = P_C - P_A - P_E + P_D \tag{7}$$

2.3 Performance Vary Rate (PVR)

PVR depends on the ratio between performance index change and performance index initial value, so PVR includes Performance Vary Quantity Rate (PVQR) and Performance Vary Slope Rate (PVSR).

$$PVQR = |P_D - P_C|/P_C = |P_D - (P_A - K_1 * \mathrm{t}_1)|/P_C \tag{8}$$

When $K_1 = K_2$ and $t_1 = t_2 = 1$,

$$PVQR == |2P_D - P_A - P_E|/P_C \tag{9}$$

In addition,

$$PVSR = (K_1 - K_2)/K_1 \tag{10}$$

When $K_1, K_2 < 0$ and $t_1 = t_2 = 1$,

$$PVSR = (P_A - P_C - P_D + P_E)/(P_A - P_C) \tag{11}$$

When $K_1, K_2 > 0$ and $t_1 = t_2 = 1$,

$$PVSR = (P_C - P_A - P_E + P_D)/(P_C - P_A) \tag{12}$$

As can be seen from Eqs. (8) and (9), PVQR is always non-negative. The bigger the PVQR value, the better the treatment effectiveness. Similarly, only $K_1 > 0, \quad K_1 > K_2$ or $K_1 < 0, \quad K_1 < K_2$, PVSR value is a positive number. Then $PVSR > 0$ presents that the performance changes slower after treatment, the greater the PVSR value, the slower the performance varying, the more pronounced the treatment effectiveness-and vice versa. In conclusion, either PVQR or PVSR is used to evaluate treatment benefit, the result value is larger, the treatment effectiveness will be more remarkable.

2.4 Cost-Effectiveness Ratio (ECR)

According to PVR, PVSR and PVQR, the variation characteristics of pavement performance indexes after treatments can be explored immediately, which will effects treatment short-term cost-effectiveness indirectly. Furthermore, the treatments cost must be taken into account to determine short-term cost-effectiveness of different treatments.

$$ECR = PVR/Cost \tag{13}$$

Where *PVR* contains *PVQR* and *PVSR*, *Cost* presents the unit price of different treatments. With great *ECR* comes great maintenance treatment cost-effectiveness.

3 Short-Term Cost-Effectiveness Analysis Scheme

Based on the historical performance detection data and cost-effectiveness indexes, the short-term cost-effectiveness analysis procedure was designed as follow:

(1) Collecting highway pavement performance data, rutting depth, IRI, SFC and damage condition.
(2) Filtering and eliminating anomalous data and pavement segment through Python language.
(3) Exploring the maintenance treatment effectiveness of different treatments based on the SPSS and Auto2fit software.
(4) Evaluating the short-term cost-effectiveness of different treatments according to short-term evaluation indicators.

Table 1. Detection program and indicator

Program	Deflection	IRI	Rut	Skid resistance	Damage	Transverse crack
Test equipment	FWD	Inspection car	Inspection car	Inspection Car	Inspection Car/PDA	Inspection Car/PDA
Test frequency	Once a year	Once a year	Twice a year	Once a year	Twice a year	Twice a year
Test time	November	October–December	March–May/November	October–December	March–May/November	March–May/November
Test lane	Outside lane	All lanes	All lanes	All lanes	All lanes	All lanes

3.1 Performance Data Detection

Five kinds of pavement performance were been tested in Jiangsu Province. Table 1 shows the test information in detail.

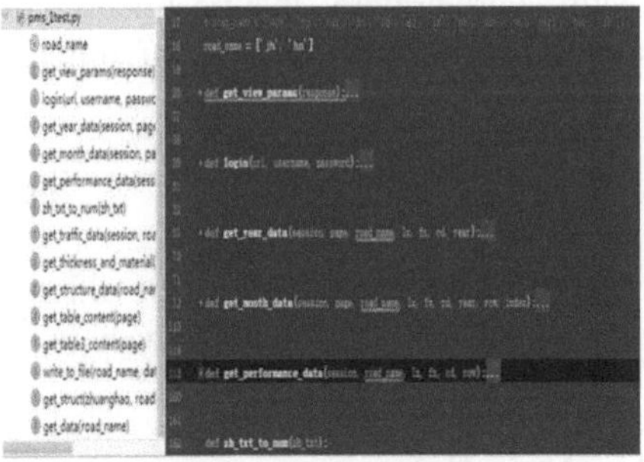

Fig. 2. Evaluating treatment cost-effectiveness and effectiveness database

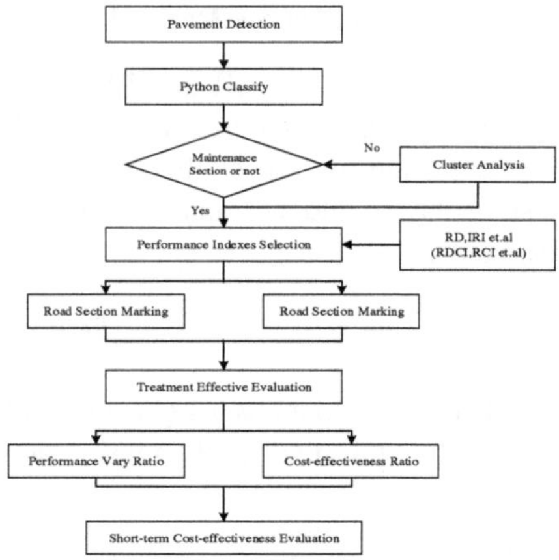

Fig. 3. Short-term cost-effectiveness evaluating scheme

3.2 Performance Data Analysis

Based on mathematical statistics and probability theory, anomalous data among detection results can be removed making use of computer programing languages (Python, C# or Vba).

Because of the random error of test equipment, some objectives factors and inspector operation, some test results had a huge difference with pavement actual condition, which must be eliminated from performance test results. Many methods can be used to finish this process, t test criterion was used in this study.

According to PMS system and Python language, all performance data were classified depended on road basic information (lane and direction), maintenance history situation (maintenance section, treatment, cost, unit cost etc.), pavement performance data (pre-treatment data and post-treatment data) and road material information, at last, the database for evaluating treatment effectiveness and cost-effectiveness was created (see Fig. 2, only an example). Then applying SPSS and Auto2fit software to carry out variance analysis and regression analysis.

3.3 Evaluation Scheme Design

A short- term cost-effectiveness evaluation scheme in this paper was designed as shown in Fig. 3.

4 Engineering Application Practice

Owing to many achievements about performance test, collecting and processing, so these courses will be not described in detail in this paper. Four widely used asphalt pavement treatments in Jiangsu Province, namely thin ECA overlay, mill and fill, micro

Table 2. The summary of treatments relative information

Section number	Treatment	Road name	Stake number range	Direction	Lane number	Surface layer style	Traffic level
LD-1	Mill and fill	Yan Jing	23.733–23.834	YJ	1	SMA-13	Light
LD-2			98.2–99.2	YJ	1	AK-13	Moderate
LD-3		Ning Tai	0.466–0.516	NT	1	AK-13	Light
LD-4			78.52–78.632	NT	1	SMA-13	Moderate
LD-5	Thin ECA overlay	Xin Yang	134.315–134.424	YX	1	AK-13	Moderate
LD-6		Huai Xu	197.324–201.768	XH	2	SMA-13	Moderate
LD-7		Lian Xu	104.669–105.024	LX	2	AK-13	Moderate
LD-8			107.426–111.111	XL	2	AK-13	Moderate
LD-9	Hot-in-place recycling	Xi Yi	113.77–114.402	XY	2	AK-13	Moderate
LD-10			145.39–146.24	YX	2	AK-13	Moderate
LD-11		Yan Jing	62.38–62.694	YJ	2	SMA-13	Light
LD-12			72.232–72.53	JY	2	AK-13	Light
LD-13	Micro surfacing	Yan Jing	145.669–146.51	YJ	2	AK-13	Light
LD-14		Lian Xu	226.134–227.238	XL	1	SMA-13	Extra heavy
LD-15			84.397–84.992	LX	2	SMA-13	Moderate
LD-16			156.815–157.996	LX	2	AK-13	Moderate

surfacing and hot-in-place recycling (HIPR) as evaluated objects were studied in this paper. Table 2 summarizes the relative main information of these four treatments.

Rut Depth (RD) as performance index was used to show the evaluation result in this study, for example. Table 3 describes the RD change characteristic between pre-treatment and post-treatment of all studied road sections. In addition, the short-term cost-effectiveness indicators results of all studied road sections were seen in Table 4.

Figures 4 and 5 indicate the short-term cost-effectiveness condition of different road sections and different maintenance treatments, respectively.

Table 3. RD performance change characteristics of all treatments

Section number	Treatment	Road name	Stake number range	P_A	P_C	P_D	P_E
LD-1	Mill and fill	Yan Jing	23.733–23.834	3.61	4.26	1.14	3.62
LD-2			98.2–99.2	9.54	10.84	3.42	3.49
LD-3		Ning Tai	0.466–0.516	9	9.48	3.75	4.84
LD-4			78.52–78.632	4.88	6.59	3.35	3.88
LD-5	Thin ECA overlay	Xin Yang	134.315–134.424	11.18	11.84	3.2	3.61
LD-6		Huai Xu	197.324–201.768	4.86	6.15	4.82	5.48
LD-7		Lian Xu	104.669–105.024	11.62	15.6	3.83	5.59
LD-8			107.426–111.111	13.58	16.71	3.28	4.93
LD-9	Hot-in-place recycling (HIPR)	Xi Yi	113.77–114.402	8.35	8.68	7.08	7.94
LD-10			145.39–146.24	11.74	13.08	0.77	1.59
LD-11		Yan Jing	62.38–62.694	10.1	10.49	5.89	11.59
LD-12			72.232–72.53	9.07	10.54	5.8	7.13
LD-13	Micro surfacing	Yan Jing	145.669–146.51	9.38	9.61	4.49	5.19
LD-14		Lian Xu	226.134–227.238	16.48	17.68	5.41	6.24
LD-15			84.397–84.992	5.02	5.81	1.5	2.39
LD-16			156.815–157.996	9.35	9.77	6.52	10.45

Table 4. The short-term cost-effectiveness indicators results

Section number	Treatment	*PVQ*	*PVS*	*PVR*		*ECR*	
				PVQR	*PVSR*	*PVQR/Cost*	*PVSR/Cost*
LD-1	Mill and fill	3.12	−1.83	0.73	−2.82	0.77%	−2.96%
LD-2		7.42	1.23	0.68	0.95	0.72%	1.00%
LD-3		5.73	−0.61	0.60	−1.27	0.64%	−1.34%
LD-4		3.24	1.18	0.49	0.69	0.52%	0.73%
LD-5	Thin ECA overlay	8.64	0.25	0.73	0.38	2.08%	1.08%
LD-6		1.33	0.63	0.22	0.49	0.62%	1.40%
LD-7		11.77	2.22	0.75	0.56	2.16%	1.59%
LD-8		13.43	1.48	0.80	0.47	2.30%	1.35%
LD-9	Hot-in-place recycling (HIPR)	1.6	−0.53	0.18	−1.61	0.28%	−2.47%
LD-10		12.31	0.52	0.94	0.39	1.45%	0.60%
LD-11		4.6	−5.31	0.44	−13.62	0.67%	−20.95%
LD-12		4.74	0.14	0.45	0.10	0.69%	0.15%
LD-13	Micro surfacing	5.12	−0.47	0.53	−2.04	2.13%	−8.17%
LD-14		12.27	0.37	0.69	0.31	2.78%	1.23%
LD-15		4.31	−0.1	0.74	−0.13	2.97%	−0.51%
LD-16		3.25	−3.51	0.33	−8.36	1.33%	−33.43%

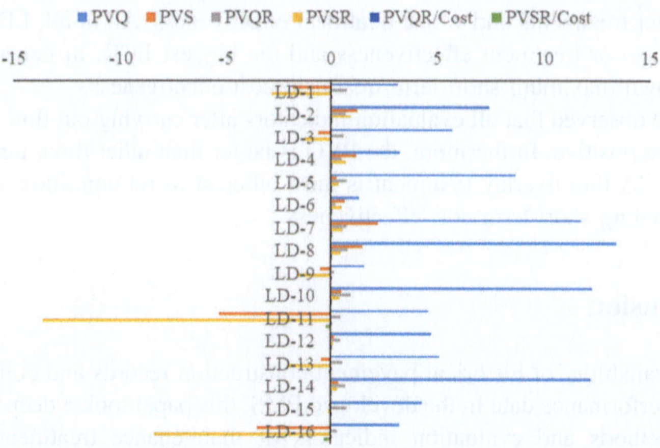

Fig. 4. Evaluation short-term cost-effectiveness condition of different road section

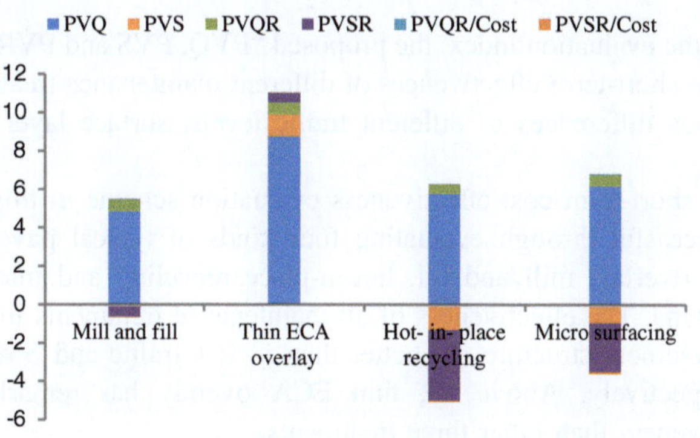

Fig. 5. Evaluation short-term cost-effectiveness condition of different maintenance treatments

As shown in Tables 3, 4 and Fig. 4, regarding to the impact of traffic level and surface layer style, the maintenance effectiveness and cost-effectiveness of moderate traffic sections are better than light traffic level, AK sections have more remarkable short-term cost-effectiveness compared to SMA. Besides, the original RD is higher in pre-treatment period lead to the higher RD variation at that time. Moreover, the PVQ values of LD-7, LD-8, LD-10 and LD-14 surpass 10 mm, which indicates RD reduces more than 10 mm after treatments, so the relative PVQR of these four sections are bigger than other sections and their short-term cost-effectiveness are better than others. It can be seen that the PVQ and PVQR values of LD-6 and LD-9 are the minimum, having small change ranges after treatments, then their treatment effectiveness are poorer.

LD-7 and LD-8 perform good effectiveness because of their PVS and PVSR results are maximum, but the post-treatment RD vary slope is slower than pre-treatment vary slope. LD-11 and LD-16 show the smallest PVS and PVSR, causing the larger RD variation after treatments and worse treatment effectiveness. All in all, LD-7 and LD-8 show the superior treatment effectiveness and the biggest ECR, in general, the LD-7 and LD-8 own maximum short-term treatment cost-effectiveness.

It can be observed that all evaluation indicators after carrying out thin ECA overlay treatment are positive, furthermore, the PVQ is larger than other three treatments. In a word, the ECA thin overlay treatment is most efficient to rut condition and it has the most outstanding short-term cost-effectiveness.

5 Conclusion

With the availability of historical pavement construction records and collected related pavement performance data in the developed PMS, this paper took a deep investigation into the methods and evaluation indicators of maintenance treatments short-term cost-effectiveness. Then a reasonable short-term cost-effectiveness analysis scheme was established and estimating this scheme through practice cases in Jiangsu Province. Based on the study above, several conclusions can be summarized as follows:

1. Regarding the evaluation index, the proposed "PVQ, PVS and PVR" can effectively describe the short-term effectiveness of different maintenance treatments. And they have obvious differences to different traffic levels, surface layer styles and road sections.
2. Creating a short-term cost-effectiveness evaluation scheme in this paper and it is proven successful through evaluating four kinds of typical pavement treatments (thin ECA overlay, mill and fill, hot-in-place recycling and micro surfacing) in Jiangsu, China. The effectiveness of all maintenance treatments in moderate-traffic and AK pavement structure are better than in low-traffic and SMA asphalt pavements, respectively. Above all, thin ECA overlay has remarkable short-term cost-effectiveness than other three treatments.

References

1. Huang, B., Dong, Q.: Optimizing pavement preventive maintenance treatment applications in Tennessee (Phase I). Report from Department of Civil and Environmental Engineering (2009)
2. Collura, J., Spring, G.: Service Lives and Costs of Local Highway Maintenance and Rehabilitation Treatments, Transportation Research Record. J. Transp. Res. Board, No. 1397 (2014). https://trid.trb.org/view/383228
3. Peshkin, D.G., Hoerner, T.E., Zimmerman, K.A.: Optimal Timing of Pavement Preventive Maintenance Treatment Applications. Transportation Research Board, NCHRP Report 523 (2004)
4. Peterson, D.E.: NCHRP Synthesis of Highway Practice 135: Pavement Management Practices. Transportation Research Board, National Research Council, Nov. 1987-139 (1987)
5. Geoffroy, D.W.: NCHRP Synthesis of Highway Practice 223: Cost Effective Preventive Pavement Maintenance. Transportation Research Board, National Research Council, 1996-103 (1996)

6. Dong, Q., Huang, B.: Evaluation of effectiveness and cost-effectiveness of asphalt pavement rehabilitations utilizing LTPP data. J. Transp. Eng. **138**(6), 681–689 (2012)
7. Dong, Q., Huang, B., Yan, X.: Evaluation of effectiveness and cost-effectiveness of different pavement maintenance treatments in Tennessee. J. Perform. Constr. Facil. **138** (2012)
8. Morian, D., Gibson, S., Epps, J.: Maintaining flexible pavements–the long term pavement performance experiment. SPS-3 5-Year Data Analysis. Report No. FHWA-RD-9-102, Federal Highway Administration, U.S. Department of Transportation, Washington, D.C. (1998)
9. Freeman, T.J.: Supplemental maintenance effectiveness research program (SMERP). In: Presentation at the AGC Pavement Preservation Seminar, 4–5 October 1999

Numerical Simulation of Long Term Pavement Temperature Field

Xingyu Gu, Xiaolong Liang, and Qiao Dong(✉)

School of Transportation, Southeast University, Nanjing, Jiangsu, China
qiaodong@seu.edu.cn

Abstract. Asphalt pavements subject to solar radiation and air temperature can reach high temperatures causing not only environmental problems such as the heat island effect on cities but also structural damage due to rutting or hardening as a result of thermal cycles. This paper established a finite element model to analyze the temperature distribution in asphalt pavement due to thermal environmental conditions. By considering the continuous solar radiation, air temperature and wind velocity, pavement temperature was calculated in the heat transfer module in Comsol Multiphysics. It is found that the amplitude of the pavement temperature is mainly dependent on air temperature and solar radiation. Another 3-D model was established to test the high temperature distribution of asphalt pavement in Nanjing, China. Then the simulation result was compared to SHRP and LTPP prediction, and pavement temperature could agree well with SHRP model in certain depth. The model could predict the maximum, minimum, and amplitude of the pavement temperature with good accuracy, which could provide guide for asphalt pavement design and to develop cool pavements.

Keywords: Finite element method · Thermal environmental conditions
Pavement temperature

1 Introduction

Thermal environmental conditions, to which pavements are exposed, significantly influence pavement mechanical responses and long-term performance. Accurate prediction of the temperature profile in pavements greatly aids pavement engineers in the design process. In 1957, barber [1] built a theoretical model to predict the maximum temperature in pavement. A one-dimensional, finite-difference heat transfer model is developed by Schenk and improved by Williamson [2] using FORTRAN program. In 1974, Berg [3] developed a series of analytical/statistical and numerical approaches to assess the thermal behavior of asphalt pavements as a function of thermal environmental conditions. Rumney and Jimenez [4] did a field test to measure the maximum temperature of asphalt pavement in desert climate, and built a norm curve to predict temperature at 2 inches and 4 inches deep in pavement structure. Based on the energy balance principle, Solaimanian and Kennedy [5] developed a prediction model to determine the extreme temperature with maximum air temperature and solar radiation, which was proved to be satisfying and adopted by SHRP. Mohseni [6] proposed

X. Shi et al. (Eds.): GSIC 2018, *Proceedings of GeoShanghai 2018 International Conference: Transportation Geotechnics and Pavement Engineering*, pp. 400–407, 2018.
https://doi.org/10.1007/978-981-13-0011-0_43

revisions to the SHRP performance grading system for asphalt binder selection, specifically for low temperature applications. Superpave [7] took the minimum air temperature as the minimum design temperature of asphalt pavement, but C. SHRP in Canada proved that the temperature of Superpave model is lower than the actual temperature in pavement. In 1990s, the LTPP [8] in America developed prediction models for the maximum and minimum temperature in pavement.

It is important to note that, although currently available methods with some modifications have been implemented with varying success for use in the practice of pavement engineering, the accuracy and reliability of the currently available methods have been questioned since proposed correlations stipulate constant values for a number of asphalt mix properties, surface convective conditions, and solar irradiation.

2 Energy Exchange on Pavement Surface

The temperature distribution in an asphalt pavement is directly affected by the outside thermal environmental conditions. The heat transfer processes include solar radiation, thermal and long-wave radiation between the pavement surface and the sky, free and forced convection, and conduction inside the pavement.

The intensity of solar radiation is dependent on diurnal cycles, the latitude, and the angle between the surface and sun's rays. Thermal or long-wave radiation accounts for radiative heat transfer between the pavement surface and the sky, which usually has a cooling effect, especially during nighttime hours. The energy balance on pavement surface means that the sum of all heat transfer through the surface of the pavement must be equal to the heat conducted in the pavement.

Solar radiation and air temperature are the two main factors affecting the temperature field in pavement. Previous study usually adopted the result of Barber's [1] research to represent the variation of radiant intensity, which is expressed as

$$q(t) = \begin{cases} 0 & 0 \leq t < 12 - c/2 \\ q_0 \cos m\omega(t-12) & 12 - c/2 \leq t \leq 12 + c/2 \\ 0 & 12 + c/2 < t \leq 24 \end{cases} \tag{1}$$

Where q_0 = maximum radiant intensity (W/m^2), $q_0 = 0.1313mQ$, $m = 12/c$; Q = total amount of solar radiation; c = actual effective time of solar radiation(h); ω = angular frequency, $\omega = 2\pi/24$(rad). Eq. (1) is commonly used in the simulation of pavement temperature in ABAQUS, and usually expressed in Fourier series.

And previous study usually adopted the combination of two sine functions to represent the variation of air temperature in a day, which is expressed as

$$T_a = \bar{T}_a + T_m[0.96\sin\omega(t-t_0) + 0.14\sin 2\omega(t-t_0)] \tag{2}$$

Where $\bar{T}_a$ = average daily temperature (°C), $\bar{T}_a = (T_a^{\max} + T_a^{\min})/2$; T_m = amplitude of daily temperature (°C), $T_m = (T_a^{\max} - T_a^{\min})/2$, $T_a^{\max}$ and $T_a^{\min}$ represent the maximum and minimum temperature in a day; t_0 = initial phase, usually $t_0 = 9$ h.

It adopted the actual thermal environment data in every hour from South Carolina [9] in America, to calculate the simulation of pavement temperature. The intensity of solar radiation input was by every hour, variation of solar radiation in a summer day is shown in Fig. 1. As we can see from Fig. 1, the sun rises at about 5am and goes down at 5 pm, the radiant intensity reaches the peak at about 12 o'clock. The actual solar radiation agrees well with the Barber's formulation.

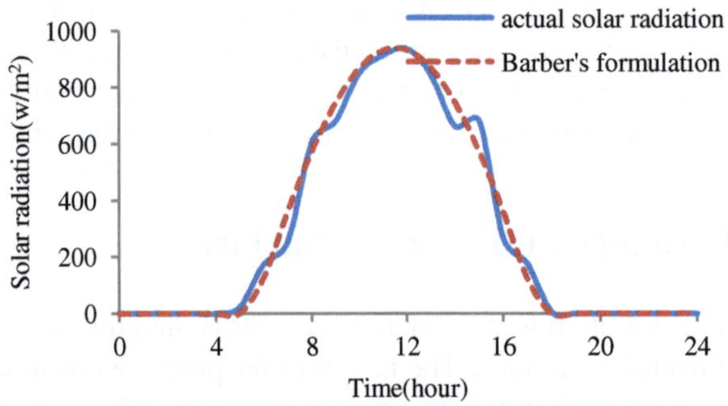

Fig. 1. Variation of solar radiation in a summer day.

The air temperature in every hour in a day is shown in Fig. 2. As can be seen from the graph, air temperature rise rapidly as the sun rises and goes down slowly after it reach the maximum at about 2 pm. Compared to the actual air temperature, though formulation (2) agree well with air temperature around the maximum and minimum temperature, it could not agree well in other times. Therefore, it is more accurate to directly use the actual measured air temperature than using the formula.

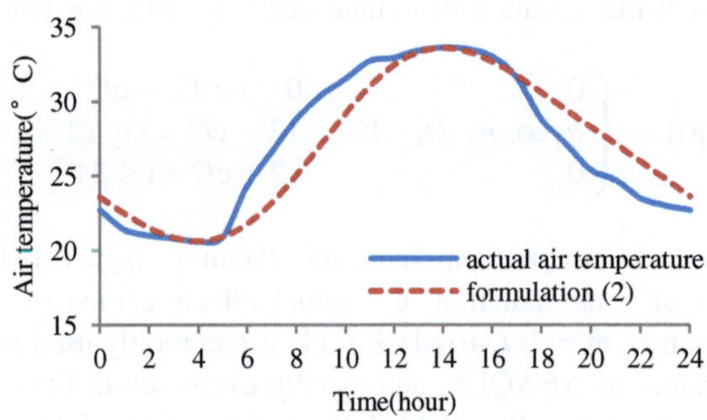

Fig. 2. Variation of air temperature in a day.

3 Pavement Structure and Material Parameters

A typical pavement structure is selected, including three AC layers (the total thickness is 18 cm), cement treated base course (40 cm) and lime stabilized subbase layer (20 cm), as shown in Fig. 3.

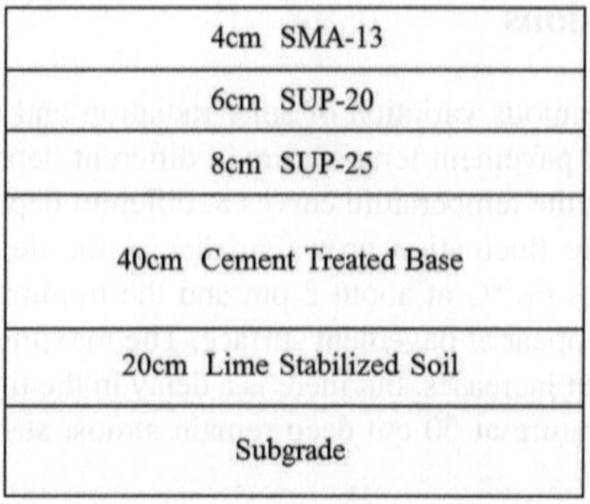

Fig. 3. Asphalt pavement structure.

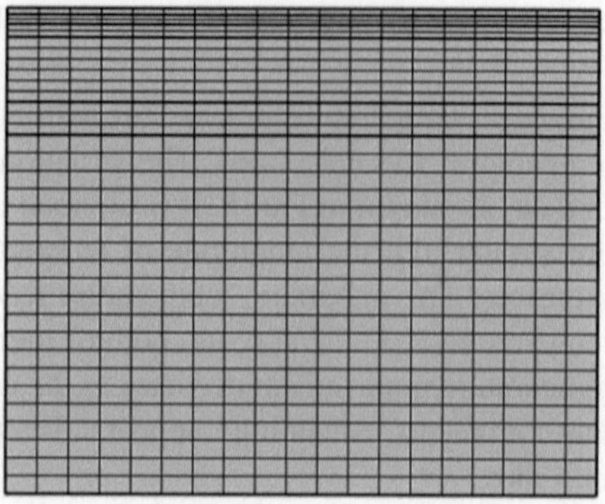

Fig. 4. Meshes of pavement structure.

This study adopted a 2D model to analyze the temperature distribution of the pavement structure, which has a dimension of 3.75 m wide and 3 m high. Dimension of meshes is set as extremely fine in pavement structural layers, and meshes of subgrade is relatively fine, as shown in Fig. 4.

As for the boundary conditions, the bottom of model is set as outflow; the left and right edges of pavement structure are considered to be adiabatic for sufficiently large horizontal expansions since spatial temperature changes in the vertical direction will be much greater than horizontal changes at pavement edges, and any heat transfer through pavement edge surfaces can be neglected. The primary temperature of the entire model is set as 20 °C, and wind velocity is a constant at 3 m/s.

Asphalt mixture is a typical temperature-sensitive material, so its mechanical property and performance vary significantly with temperatures. However, based on previous research, the thermal property of asphalt mixture remain stable as the variation of temperature. Therefore, the thermal property of the pavement structure is set to be constant in this simulation. The thermal and physical parameters of pavement materials are shown in Table 1.

Table 1. Thermal parameters of pavement materials for temperature analysis.

Parameters	Asphalt			Base	Subbase	Subgrade
Thermal conductivity k (J/m•h•°C)	SMA-13	Sup-20	Sup-25	5616	5148	5616
	7380	4500	4176			
Density ρ (kg/m^3)	2100	2400	2600	2200	2100	1800
Specific heat capacity c (J/kg•°C)	1168.0	894.0	921.0	911.7	942.9	1040.0
Solar absorptance α_s			0.9			
Pavement surface emissivity ε			0.81			
Absolute zero temperature T_Z (°C)			−273			
Stefan–Boltzmann constant σ (J/(h•m^2•K^4))			2.041092×10^{-4}			

4 Simulation Results and Conclusions

Pavement temperature in a month with continuous variation of solar radiation and air temperature was calculated. The variation of pavement temperature at different depths in a day is shown in Fig. 5. It is obvious that the temperature curves at different depths are similar, but the amplitude of temperature fluctuation grows smaller as the depth increases. The maximum temperature reaches 68 °C at about 2 pm and the minimum temperature is 32 °C at 6 am, both of them appear at pavement surface. The maximum temperature in pavement decrease as the depth increases, but there is a delay in the time that maximum temperature appears. Temperature at 50 cm deep remain almost stable in a day.

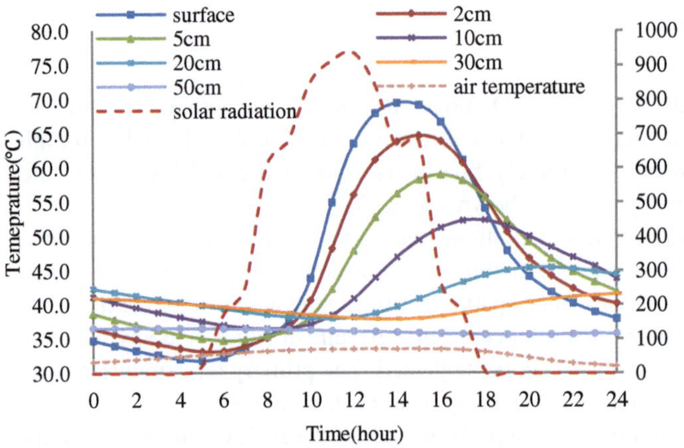

Fig. 5. Variation of pavement temperature at different depths in a day.

Figure 6 shows the Variation of pavement temperature at different time along pavement depth. As solar radiation intensify, pavement temperature increases, and temperature of upper layers is greater than lower layers; after solar radiation reaches its peak, temperature of surface layers start to decrease, but temperature inside pavement structure continue to increase. The reason is that the heat energy continues to transfer in pavement structure. After the sun goes down, without solar radiation, the heat exchange between pavement and cold air is dominant, causing the temperature of surface layers decrease rapidly; during this period, temperature in middle layers is greater than surface layers and layers beneath it. It is notable that temperature under 50 cm to pavement surface remain almost stable, it proves that pavement structure could reduce the effect of temperature variation on subgrade.

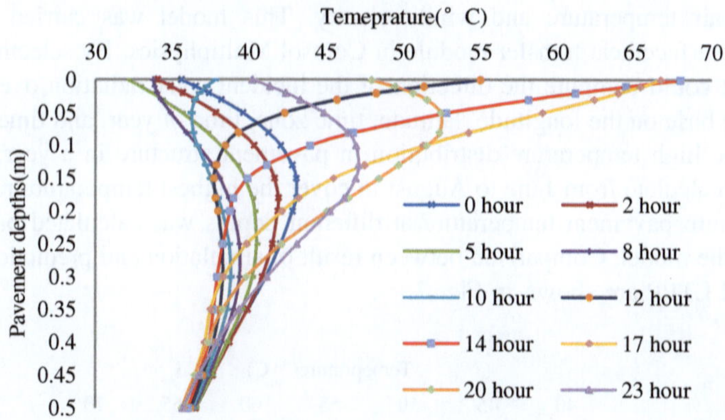

Fig. 6. Variation of pavement temperature at different time along pavement depth.

5 Comparison with SHRP and LTPP Equations

In 1998, LTPP developed the high temperature and low temperature prediction model in pavement, which is expressed as

$$\mathrm{T}_{d(\max)} = 54.32 + 0.78T_{a(\max)} - 0.0025lat^2 + 15.14\lg(d + 25) \tag{3}$$

Where $\mathrm{T}_{d(\max)}$ = the maximum pavement temperature (°C), $T_{a(\max)}$ = the maximum air temperature (°C), *lat* = latitude (°),d = depth.

A report from SHRP pointed out that temperature in pavement surface is determined by the hot air flow in surface, but the hot air flow is subjected to many factors. In order to simplify calculation, SHRP did a regression analysis based on air temperature and heat balance in pavement surface, and developed a regression formula as followed

$$T_{s(\max)} = T_{a(\max)} + 0.00618lat^2 + 0.2289lat + 24.4 \tag{4}$$

Where $T_{s(\max)}$ = the maximum temperature in pavement surface(°C), $T_{a(\max)}$ = the maximum air temperature(°C), *lat* = latitude(°).

And the maximum temperature in certain depths could be calculated by

$$T_{d(\max)} = \left[T_{s(\max)} + 17.78\right](1 - 2.48 \times 10^{-3}d + 1.085 \times 10^{-5}d^2 - 2.441 \times 10^{-8}d^3) - 17.78 \tag{5}$$

Where $T_{d(\max)}$ = the maximum temperature in certain depths, d = depth to pavement surface(mm).

To compare with the SHRP and LTPP data, a 3-D pavement model was established for the simulation of pavement temperature. The environmental conditions of Nanjing in Jiangsu, China was investigated in this model, with the continuous variation of solar radiation, air temperature and wind velocity. This model was carried out in the surface-to-surface heat transfer module in Comsol Multiphysics. By selecting a city or location, it could compute the direction of the incident solar radiation over the simulation time base on the longitude, latitude, time zone, time of year, and time of day. To identify the high temperature distribution in pavement structure in a year, the model was set to calculate from June to August to cover the highest temperature range. Then, the maximum pavement temperature at different depths was calculated based on the results of the model. Comparison between result of simulation and prediction model of SHRP and LTPP are shown in Fig. 7.

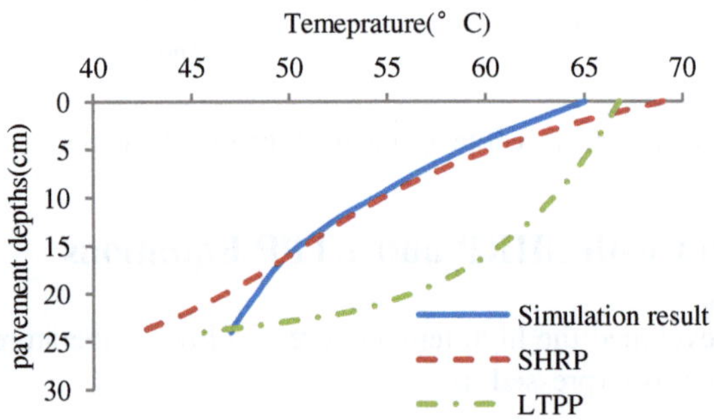

Fig. 7. Comparison of simulation result and SHRP and LTPP prediction model.

As can be seen from Fig. 7, the result of simulation agrees well with SHRP prediction from about 5 cm to 15 cm, and the temperature curve is similar to the SHRP model from pavement surface to 15 cm deep. The maximum temperature in pavement surface is close to each other. But the simulation result agrees poorly with the LTPP prediction.

In addition, there is still some difference between the simulation result and SHRP prediction. In the model, pavement temperature is derived from the heat balance equation in pavement surface. However, the only parameter in SHRP model is latitude, which is simplified and may not be applicable to all locations and weather conditions.

Although this model could predict pavement temperature accurately, there is some limitation. The amount of cloud in the sky has effect on pavement temperature as well, it could reflect the long-wave radiation of air and pavement surface, which will heat up air and pavement structure. And variation of weather condition, such as rainfall, usually has a cooling effect on pavement temperature. However, this model could not reveal the influence of cloud and rainfall process.

6 Conclusions

This paper built a finite element model to predict the temperature distribution in pavement structure. The model was carried out in the surface-to-surface heat transfer module in Comsol Multiphysics, using continuously observed weather data from weather station. Simulation results could agree well with previous study, comparing with SHRP and LTPP predictions, it is most closed to SHRP equations. This model could predict pavement temperature with good accuracy, which would assist pavement engineers in the design process.

Acknowledgement. This study was supported by the project titled "Effectiveness and Applicability of Asphalt Pavement Maintenance Techniques in Zhejiang" (Project no. 8521002166) founded by Zhejiang Transportation Engineering Construction Group Co., a research project titled "Data mining of Highway Pavement Performance and Evaluation of Maintenance Treatment Effectiveness" (Project no. 3221007409) founded by Southeast University, the Thirteen fifth research project, Part IV: Highway asphalt pavement structure long term preservation (7621000132), funded by Jiangsu Department of Transportation; and the Thirteen fifth research project, Part V: Highway asphalt pavement intelligent maintenance (7621000133), funded by Jiangsu Department of Transportation.

References

1. Barber, E.S.: Calculation of maximum pavement temperatures from weather reports. Highw. Res. Board Bull. **168**, 1–8 (1957)
2. Williamson, R.H.: Effects of environment on pavement temperatures. In: International Conference on Structural Design Proceedings, pp. 144–158 (1972)
3. Berg, R.L.: Energy balance on a paved surface. Technical Report Number 26, sub-Project 42, U.S. Army Cold Regions Research and Engineering Laboratory. Hanover, N.H (1974)
4. Rumney, T.N., Jimenez, R.A.: Pavement temperatures in the southwest. Highw. Res. Rec. **361**, 1–19 (1971)
5. Solaimanian, M., Kennedy, T.W.: Predicting maximum pavement surface temperature using maximum air temperature and hourly solar radiation. Transportation Research Record 1417, Transportation Research Board, Washington, D.C., pp. 1–11 (1993)
6. Mohseni, A.: LTPP seasonal asphalt concrete (AC) pavement temperature models. FHWA-RD-97-103 (1998)
7. Kennedy, T.W., et al.: Superior performing asphalt pavements (Superpave): The product of the SHRP Asphalt Research Program. SHRP-A-410 (1994)
8. Mohseni, A., Symons, M.: Improved AC pavement temperature models from LTPP seasonal data. In: Transportation Research Board 77th Annual Meeting, Washington D.C. (1998)
9. National Centers for Environmental Information. https://www.ncdc.noaa.gov. Accessed 06 July 2017

Harvesting Energy from Pavements

Asha M. Nair[1(✉)], Varchasvi Surya[2], R. Apoorva[1], Kumar Gaurav Singh[1], K. Chandan[1], and Bakhar Hassan Nidha[1]

[1] Department of Civil Engineering, CMR Institute of Technology, Bengaluru, Karnataka, India
asha.n@cmrit.ac.in
[2] Department of Electronics and Communication Engineering, CMR Institute of Technology, Bengaluru, Karnataka, India

Abstract. Rapid Industrialization and Urbanization has amplified the dependency of Indian economy on import of fossil fuels such as Oil, petroleum, and Natural Gas for energy production. The excessive consumption of these fossil fuels is always associated with high greenhouse gas emissions which can affect the Climatic changes and ecological balance of the country. Hence, the need for innovative technologies to harvest energy has increased. *Energy Harvesting* or *Scavenging* is the process of capturing the wasted energy from naturally occurring energy sources, accumulating and storing for later use. India has the second largest road network in the world. Hence, connecting road network can prove to be beneficial if used as a source of energy. From pavements two sources of energy can be identified viz., solar energy and mechanical energy. Technologies dealing with solar energy are – Asphalt Solar Collector combined with piping system, photovoltaic cells and thermoelectric Generators. Technology dealing with Mechanical energy is by using embedded piezoelectric crystals. The primary focus of this paper is to quantify the electric voltage generated by a bituminous specimen for different positions of piezoelectric crystals for same impact loading. The harvested energy can be used for small scale road applications like road furniture, lighting, roadside advertising or railway and airport signage where the installation and maintenance cost are low.

Keywords: Energy harvesting · Piezoelectric crystals · Bituminous mix

1 Introduction

Conservation of energy is of paramount importance as far as sustainability is concerned. The different natural resources that accomplish the global energy demand are fossil fuels, water currents, wind energy, tidal energy, solar energy, geothermal energy, biomass energy and nuclear energy. According to the statistics presented by UN Division of Sustainable Development, around 2.8 billion people do not have access to modern energy services and over 1.1 billion do not have electricity which demonstrates the severity of the energy crisis. By 2035, Global energy demand is expected to increase by more than one-third with Asian countries China, India and the Middle East accounting for about 60% of the increase. This warrants the development of alternate, viable and sustainable sources for energy.

X. Shi et al. (Eds.): GSIC 2018, *Proceedings of GeoShanghai 2018 International Conference: Transportation Geotechnics and Pavement Engineering*, pp. 408–415, 2018.
https://doi.org/10.1007/978-981-13-0011-0_44

The electricity consumption in India is expected to rise by around 2280 BkWh by 2021–22 and around 4500 BkWh by 2031–32 [1]. The energy needs include basic needs of human beings (like heating, lighting, cooking and refrigeration,), indoor lighting and outdoor lighting during night times [2] has quoted that for the year 2010–2011, 1% of the total electricity consumption in India (equal to 6.7 TWh) is utilised for street lighting system costing \$500 M. Similarly railway lines also claim electrical energy not only at night times for lighting but also for safe design of crossings. Henceforth, it is critical to develop alternate resources for energy harnessing or economically utilise the existing energy resource so as to enhance safety and efficiency in commutation. The statistical figures given by Ministry of New and Renewable Energy, 2011 highlight the meagre hike in renewable energy consumption in India (from 2% in 2002 to 11% in 2010) and this enunciate befitting integration of Science and Technology to develop novel energy resources that are renewable and sustainable.

2 Literature Review

In the recent years with the advent of electronic and electrical circuits, attempts have been made to harness energy from small scale and large scale mechanical motions and vibrations that are common in day life. This energy exploitation sources include, small scale and common daily movements such as footsteps, knee movements, running and walking, vibrations and motions from high rise buildings, towers, bridges, and framed structures under wind and earthquake loads. A comprehensive literature review of the alternate sources of energy and the harnessing mechanisms that can be sought for different sources of motions and vibrations has been presented by few researchers such as [3] and [4, 5] has simulated movements on a railway track and it was observed that, on harnessing horizontal and vertical vibrations for a simulated speed of 55 mph, the average power production was 215.2 W for a loaded train. A rack and pinion gear system was used to harness the vertical motions induced in rails and convert to translational motion. [6] has explored the potential of harnessing energy from rail track vibrations to develop an efficient rail transport management and thereby enhance safety in rail mode.

In the recent years, research studies have been carried out to study the potential of harnessing mechanical stresses imposed by wheel loads on pavements to serve as a perennial source of energy. In the studies carried out by [7] it is reported that piezoelectric based energy harvesting is more promising than solar panels. In the experimental studies carried out by [8] attempts have been made to harness the energy expelled during impact loading of vehicle tyres on pavements using piezoelectric transducers. From the experimental studies it was observed that the height of the top layer has minimal effect on the output voltage and stacking of PZT increased the output voltage. The voltage generated in the experimental studies ranged from 20 V–80 V.

Though this technology is used abroad, this technology is a far distant dream for developing country like India. This paper details the attempt made to understand the feasibility of placing piezoelectric crystals in bituminous specimen. Since, bitumen is heated before placing, the difficulties involved in preparing bituminous specimen and sandwiching the piezoelectric crystals is also discussed.

3 Principle of Piezoelectric Crystal

Piezoelectric crystals, as the name implies produce electricity from pressure. Though there can be a number of methods that can be used to convert mechanical energy to electrical energy, majority of the literature supports the use of piezoelectric crystals for the same. The piezoelectric crystals accumulate electric charges in response to applied mechanical stresses and can be stored in a rechargeable battery for later use. A system can be developed consisting of piezoelectric elements embedded in the pavement layers and subjecting it to varying loads. These systems generate an electric voltage when subject to alterations in their dimensions caused by mechanical stresses (vibrations). On application of mechanical stresses, the piezoelectric materials get polarized. The dipole moment generates electricity and the generation of electrical energy from mechanical energy is called as direct piezoelectric effect. The principle of piezoelectric transducer as explained by [9] is shown in Fig. 1.

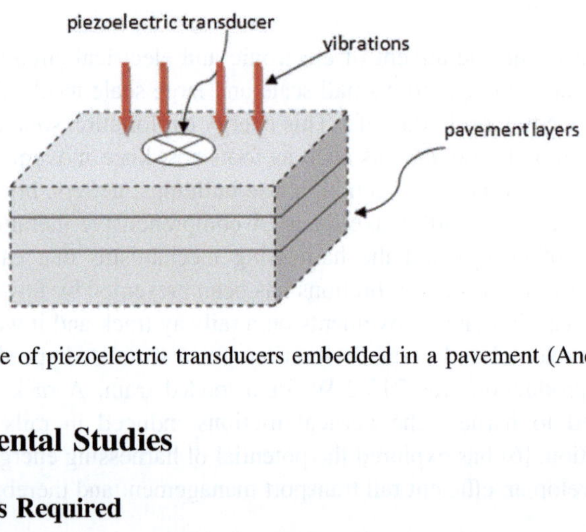

Fig. 1. Principle of piezoelectric transducers embedded in a pavement (Andriopoulou 2012)

4 Experimental Studies

4.1 Materials Required

The different materials required for the experimental studies are Bitumen, Aggregates, Piezoelectric elements, Aluminum sheet with circuit connections and mould.

Preparation of Bituminous Specimens. For the experimental studies bituminous specimens were prepared by mixing hot aggregates with bitumen of flowing consistency. Before the bituminous specimens were prepared, the properties of the different constituents were estimated. The properties of bitumen were determined as per Bureau of Indian Standards and are summarized in Table 1. Three different materials viz., Coarse Aggregates, Sand and Quarry dust were proportioned to prepare bituminous specimen. The grain size distribution curve of the different materials used for the test is presented in Fig. 2.

To prepare the bituminous specimen, the coarse aggregates, sand and quarry dust are heated to a temperature of 160 °C. The bitumen was also heated to a flowing consistency (to a temperature of 130–140 °C) and 4% by weight of bitumen was added

Table 1. Properties of bitumen

Property	BIS Code	Value obtained
Specific gravity of bitumen	IS 1202-1978	1.04
Ductility value, cm	IS 1208-1978	80.7
Softening point of bitumen, °C	IS 1205-1978	62.5
Flash point of bitumen, °C	IS 1209-1978	190
Fire point of bitumen, °C	IS 1209-1978	220
Viscosity of bitumen, seconds	IS 1206-1978 (Part 2)	23.25 s at 80 °C for 10 mm orifice
Penetration grade	IS 1203-1978	30/40

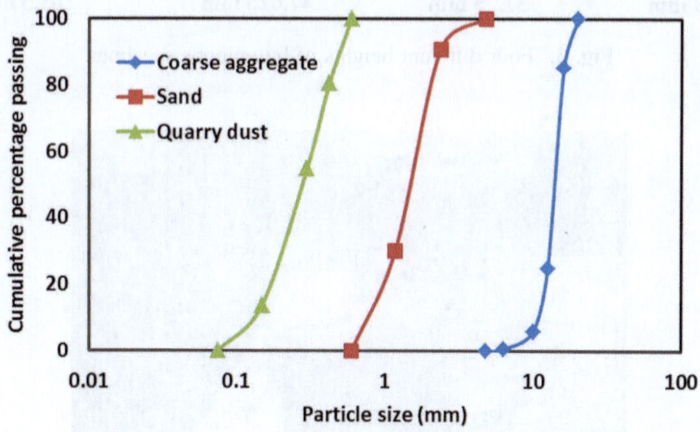

Fig. 2. Grain size distribution curve of the different constituents used for making bituminous specimen

to the aggregates and the mixture was sorted to ensure uniformity of the mix. The mix was heated to a temperature of 190–200 °C. The hot mix was placed into a mould of 63.5 mm height and 100 mm diameter. The hot mix was compacted using a hammer of 4.5 kg mass falling from a height of 450 mm. The quantities of aggregates were so proportioned so as to produce four different heights of sample of same density. A uniform density of 2.4 g/cc was maintained for all the specimens and the number of blows applied using rammer was varied such that uniform compaction energy was maintained in all specimens. The different heights of the specimen prepared were 63.5 mm, 47.625 mm, 31.75 mm and 15.878 mm. The photograph of the different specimens prepared is shown in Fig. 3.

4.2 Electronic Circuit

To measure the electrical voltage, four piezoelectric crystals were collected in series using an aluminum sheet and sandwiched between two bituminous specimens. In the experiments, the bottom height of the specimen was maintained constant (63.5 mm), whereas the top height of the specimen was varied. A photograph of the experimental study is shown in Fig. 4.

15.878 mm **31.75 mm** **47.625 mm** **63.5 mm**

Fig. 3. Four different heights of bituminous specimen

Fig. 4. Photograph of the experimental study

Piezoelectric circuit was made by drilling holes in the aluminum plate using a radial drilling machine and the piezoelectric crystals were connected to it in series by soldering as shown in Fig. 5. It was then connected to diode, resistor and capacitor on the bread board and recording positive variation of average voltage was observed in Digital Storage Oscilloscope (DSO). The circuit diagram is shown in Fig. 6.

4.3 Application of Impact Load

After the circuit is prepared and sandwiched between the bitumen specimens, an impact load was applied using a rammer of 4.5 kg falling from a height of 450 mm and the maximum voltage was recorded. The experiment was repeated for different heights of top specimen and the variation in voltage recorded with respect to height of specimen was studied. Figure 7 shows application of wheel load using a rammer and Fig. 8

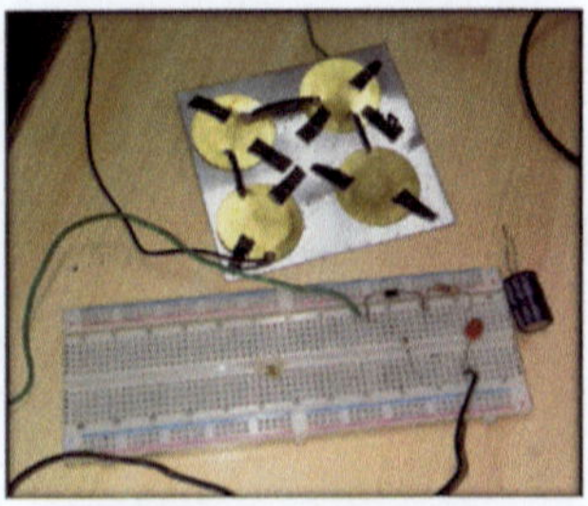

Fig. 5. Circuit connections

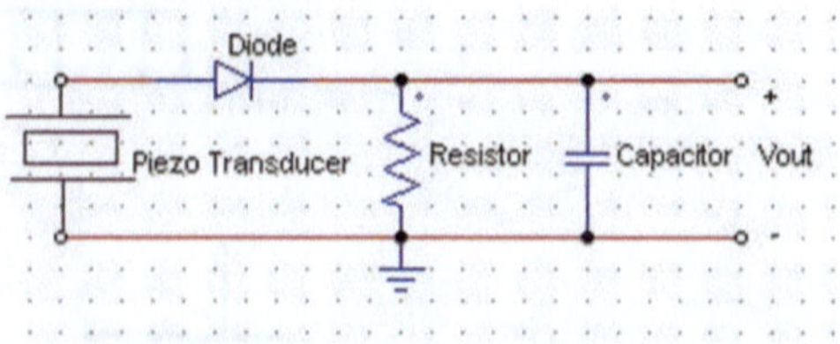

Fig. 6. Schematic of piezoelectric circuit

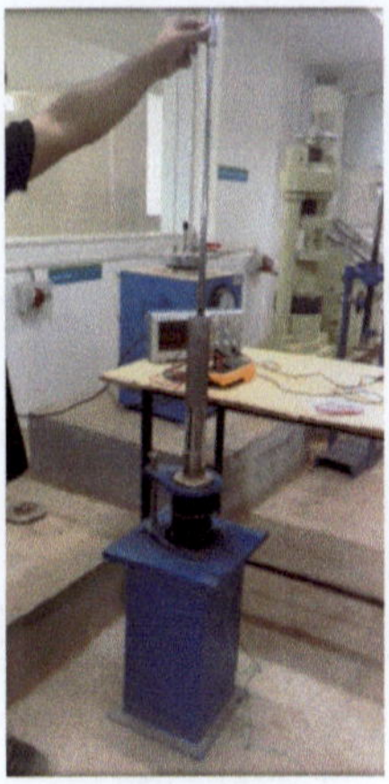

Fig. 7. Application of impact load bitumen specimens

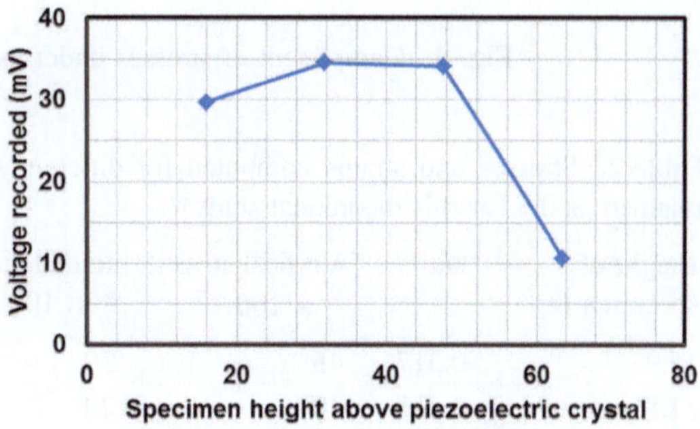

Fig. 8. Voltage recorded for on the different specimen heights

shows the maximum voltage recorded. Though the voltage recorded is very less, mechanisms can be developed to magnify the same.

5 Discussion on Experimental Results

Pavements can be considered as elastic half-space and on application of wheel loads they are subjected to four stresses namely, vertical stresses, radial stresses, tangential stresses and shear stresses (as shown in Fig. 9) and the corresponding strains namely, vertical strains, radial strains, tangential strains and shear strains. Piezoelectric crystals generate voltage impulse when normal strains are induced on it. For the pilot experiments conducted, the three normal stresses are computed using charts developed by Foster and Ahlvin on the basis of Boussinesq's principle. Using the three normal stresses, the vertical/normal strains are computed using Eq. (1) and presented in Table 2. For computations r/a was taken as one (since load was uniformly applied).

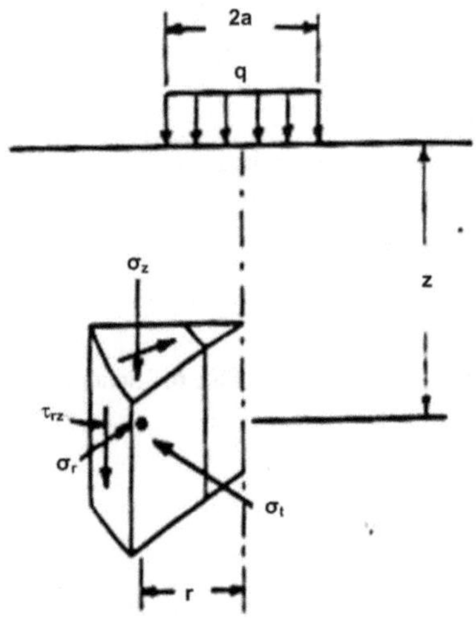

Fig. 9. Component of stresses under axisymmetric loading [10]

Table 2. Stresses and strains computed for different specimen heights (z) (poisson's ratio is assumed as 0.5 for this theoretical study)

Height of specimen (z)	z/a	Vertical stess $\frac{\sigma_y}{q} \times 100$	Radial stess $\frac{\sigma_r}{q} \times 100$	Tangential stess $\frac{\sigma_t}{q} \times 100$	Vertical strain $\frac{qE\epsilon_y}{100}$
15.9	0.317	45	30	20	20
31.8	0.635	40	20	10	25
47.6	0.953	34	18	6	22
63.5	1.270	28	12	4	20

$$\epsilon_y = \frac{1}{E}\left[\sigma_y - \vartheta(\sigma_r + \sigma_t)\right] \tag{1}$$

For a constant (qE/100), it is observed that the vertical strains induced, increase slightly with depth and then decreases. The experimental results are in agreement to theoretical results which justifies the experimentation. Though the experiments conducted do not represent a half-space continuum, the theoretical results give an insight on the mechanical behavior of bitumen specimens.

6 Summary and Scope for Future Work

The experimental work presented here is a preliminary study undertaken to investigate the difficulties involved in placing piezoelectric crystals within hot specimens. However, flexible pavement can be considered as a system, wherein performance of such sections is dependent upon the performance of individual layers and also the energy storage mechanism. This is not considered in the present study. Majority of the literature reported also emphasizes on the studies carried out on bitumen specimens and also on harvesting energy from pavements based on simulation studies. This opens up the scope for further research, wherein the energy harnessing potential of pavement as a system can be studied. Also there is a need to develop a technology by which harvested energy can be stored and made available for public usage.

Acknowledgements. The authors express their sincere thanks to Dr. I. C. Lekshmi, Dr. D. P. Giridhar and Prof. H. C. Guruprasad of CMR Institute of Technology for the useful discussions and suggestions provided by them in carrying out the experiments.

References

1. Garg, P.: Energy scenario and vision 2020 in India. J. Sustain. Energy Environ. **3**(1), 7–17 (2012)
2. Johnson, A., Phadke, A., de la Rue du Cann, S.: Energy Savings Potential for Street Lighting in India. Report Submitted to Ernest Orlando Lawrence Berkeley National Laboratory (2014)
3. Zuo, L., Tang, X.: Large-scale vibration energy harvesting. J. Intell. Mater. Syst. Struct. **24** (11), 1405–1430 (2013)
4. Mitcheson, P.D., Yeatman, E.M., Rao, G.K., Holmes, A.S., Green, T.C.: Energy harvesting from human and machine motion for wireless electronic devices. Proc. IEEE **96**(9), 1457–1486 (2008)
5. Phillips, K.J.: Simulation and control system of a railroad track power harvesting device. M. Sc thesis submitted to Graduate College of at the University of Nebraska (2011)
6. Penamalli, G.R.: Structural health monitoring and energy harvesting for railroad. M.Sc thesis submitted to Graduate College of at the Stony Brook University (2011)
7. Kim, S., Shen, J., Ahad, M.: Piezoelectric-based energy harvesting technology for roadway sustainability. Int. J. Appl. Sci. Technol. **5**(1), 20–25 (2015)
8. Roshani, H., Dessouky, S.: Feasibility study to harvest electric power from highway pavements using laboratory investigation. In: Proceedings of the 2015 ASEE Gulf–Southwest Annual Conference (2015)
9. Andriopoulou, S.: A review on energy harvesting from roads, Report No [TSC-MT 12-017] (2012)
10. Huang, Y.H.: Pavement Analysis and Design, 2nd edn. Pearson Education, Upper Saddle River (2008)

Structure Optimization of Nano Electromechanical Energy Harvester Using Isogeometric Analysis

Qimin Wang[1] and Xiaoying Zhuang[1,2,3](✉)

[1] Department of Geotechnical Engineering, Tongji University, Shanghai 200092, China
xiaoyingzhuang@tongji.edu.cn
[2] State Key Laboratory of Disaster Reduction in Civil Engineering, Tongji University, Shanghai 200092, China
[3] State Key Laboratory of Structural Analysis for Industrial Equipment, Dalian University of Technology, Dalian 116024, People's Republic of China

Abstract. This paper utilized Isogeometric Analysis with the trimmed surface analysis to simulate the piezoelectric effect and the flexoelectric effect of micro-electromechanical systems (MEMS). Based on the methods, structural optimization considering circular defects was conducted in nano flexoelectric beam. Studies were made first on the theory and formulae of electromechanical mechanisms and the isogeometric analysis. Then the topology optimization based on the trimmed surface analysis was utilized in terms of defects location and defects size. The authors used Matlab to realize the above content and used the program to calculate a classical benchmark example, which comprised the basic verification of the program. Finally, the paper put up some key results in structural optimization and anticipated the further research based on this study.

Keywords: Isogeometric analysis · Flexoelectricity · Topology optimization

1 Introduction

Energy harvesters have been an increasingly attractive alternative to conventional batteries for remote Micro/Nano Systems (MNSs) in smart structures and smart infrastructures, e.g., Environmental Monitoring (EM) and Structural Health Monitoring (SHM) sensors, which are deployed throughout the whole tunneling project including geotechnical survey, construction, operation and maintenance. Embedded with effective energy harvesters, the remote cracks detectors and deformation sensors in railway tunnels, can capture minute amounts of vibration energy excited by passing high-speed metros, and then convert the mechanical energy into electricity for storage. Moreover, whether it is for cave excavating or slope/pit monitoring, the reliable performance of electronic equipment strongly depends on reliable functions of the powering unit, especially for applications with limited human accessibility like sensors set at the excavating face of a highly unstable rock tunnel.

Several review articles [1–3] have emphasized on power harvesting using piezoelectric materials from mechanical vibrations. The main advantages of piezoelectric

X. Shi et al. (Eds.): GSIC 2018, *Proceedings of GeoShanghai 2018 International Conference: Transportation Geotechnics and Pavement Engineering*, pp. 416–424, 2018.
https://doi.org/10.1007/978-981-13-0011-0_45

materials are the large power densities and ease of application. But they usually have low efficiency [4]. Therefore, researches are meaningful on increasing energy conversion rate of electromechanical transduction mechanism for energy harvesting.

NURBS (Non-Uniform Rational B-Splines) has been one of the most thoroughly developed CAD (Computer Aided Design) splines and thus be used most widespread. With NURBS in CAD software, the authors can build freely whatever they think is a good structure. We proposed a prototype of the complex flexoelectric energy harvester in Fig. 1, which is composed by functional flexoelectric layers and supplementary substrates. The geometry model is easy to build but the flexoelectric effect is hard to simulate for it is governed by high-order partial differential equations in Eq. 1.

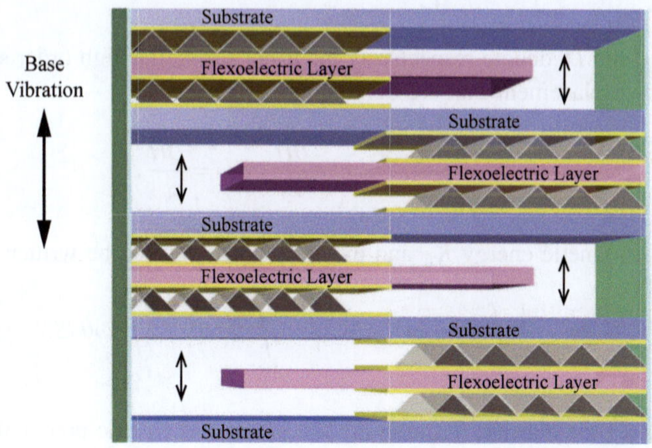

Fig. 1. Illustration of a sandwich-like electromechanical energy harvester

Unlike the common practices where great efforts are made in applying traditional FEM (Finite Element Method) to solving that problem, this study implements a spline-based FEM called isogeometric analysis (IGA) [5] in the analysis framework, so as to optimize the energy conversion rate of a basic flexoelectric layer in Fig. 1. The high-order continuity of NURBS also simplifies finite element analysis of high-order problems. In the following, IGA refers to IGA based on NURBS without notification.

2 Flexoelectric Isogeometric Analysis Formulation

2.1 Control Function for Piezoelectric and Flexoelectric Problems

For the linear dielectric continuum possessed with the piezoelectric and flexoelectric effects, the electric enthalpy density h is determined by the linear strain tensor ε_{ij}, the electric field E_i, i.e., $h(\varepsilon_{ij}, E_i)$ and the strain gradient $\varepsilon_{jk,l}$ [6]. Thus we have

$$h\left(\varepsilon_{ij}, E_i, \varepsilon_{jk,l}\right) = \frac{1}{2}C_{ijkl}\varepsilon_{ij}\varepsilon_{kl} - e_{ikl}E_i\varepsilon_{kl} - \mu_{ijkl}E_i\varepsilon_{jk,l} - \frac{1}{2}\kappa_{ij}E_iE_j, \tag{1}$$

where E_i is the electric field intensity; e_{ikl}, κ_{ij} and C_{ijkl} denote the piezoelectricity tensor, the dielectric tensor and the elasticity tensor, respectively. μ_{ijkl} is a material tensor defined as $\mu_{ijkl} = d_{ijkl} - f_{ijkl}$. f_{ijkl} and d_{ijkl} are the direct and the reverse flexo-electricity tensors, respectively.

Integrating h on the domain, we have the total electric enthalpy

$$H = \frac{1}{2}\int_{\Omega} (\hat{\sigma}_{ij}\varepsilon_{ij} + \tilde{\sigma}_{ijk}\varepsilon_{ij,k} - \hat{D}_iE_i)d\Omega, \tag{2}$$

where $\hat{\sigma}_{ij}$, $\tilde{\sigma}_{ijk}$ and $\hat{D}_i$ denote respectively the usual stress, the high order stress and the usual electric displacement that are defined as

$$\hat{\sigma}_{ij} = \frac{\partial H}{\partial \varepsilon_{ij}},\ \hat{D}_i = -\frac{\partial H}{E_i},\ \tilde{\sigma}_{ijk} = \frac{\partial H}{\varepsilon_{ij,k}}. \tag{3}$$

The system kinetic energy K_E and external work W_{ext} can be written as

$$K_E = \frac{1}{2}\int_{\Omega} \rho\dot{u}_i\dot{u}_i d\Omega, \quad W_{ext} = \int_{\Gamma_t} \bar{t}_i u_i dS - \int_{\Gamma_D} \omega\theta dS, \tag{4}$$

where ρ denotes the density; θ and ω are the prescribed electric potential and surface charge density while u_i and $\bar{t}_i$ are the prescribed mechanical displacements and tractions. Using Hamilton principle without damping, we obtain

$$\delta\int_{t_1}^{t_2} (K_E - H + W_{ext})dt = 0, \tag{5}$$

and

$$\delta\int_{t_1}^{t_2}\left(\frac{1}{2}\int_{\Omega} \rho\dot{u}_i\dot{u}_i d\Omega - \frac{1}{2}\int_{\Omega} (\hat{\sigma}_{ij}\varepsilon_{ij} + \tilde{\sigma}_{ijk}\varepsilon_{ij,k} - \hat{D}_iE_i)d\Omega + \int_{\Gamma_t} \bar{t}_i u_i dS - \int_{\Gamma_D} \omega\theta dS\right)dt = 0. \tag{6}$$

Changing the sequences of operations and ignoring the inertia contribution yields

$$\int_{\Omega} \left(C_{ijkl}\delta\varepsilon_{ij}\varepsilon_{kl} - e_{kij}E_k\delta\varepsilon_{ij} - \mu_{ijkl}E_l\delta\varepsilon_{ij,k} - \kappa_{ij}\delta E_iE_j\right)d\Omega - \int_{\Gamma_t} \bar{t}_i\delta u_i dS + \int_{\Gamma_D} \omega\delta\theta dS = 0. \tag{7}$$

2.2 IGA-FEM Formulation

In our study, the strain gradients field has been involved in the analysis as shown in Eq. 1. The classical FEM cannot handle the problem well without increasing degrees of freedom for the strain gradients and thus complexing the analysis process. IGA differs classical FEM in using NURBS as the base function, which meets right the continuity requirement of high-order partial differential equations.

NURBS are originated from B-splines. The base function $N_{i,p}(u)$ of a p-order B-splines with $(n + 1)$ control points can be recursively written as

$$\begin{cases} N_{i,p}(\xi) = \frac{\xi-\xi_i}{\xi_{i+p}-\xi_i} N_{i,p-1}(\xi) + \frac{\xi_{i+p+1}-\xi}{\xi_{i+p+1}-\xi_{i+1}} N_{i+1,p-1}(\xi) \\ N_{i,0}(\xi) = \begin{cases} 1, & \xi_i \le \xi < \xi_{i+1} \\ 0, & \text{others} \end{cases} \end{cases}, \tag{8}$$

where ξ is knot vector in the parametric space composed by non-decreasing coordinates set $\Xi = \{\underbrace{a, \cdots, a}_{p+1}, \xi_{p+1}, \cdots, \xi_{m-p-1}, \underbrace{b, \cdots, b}_{p+1}\}$. By introducing weight ω_i to B-splines base functions, a p-order NURBS surface with $(m + 1)(n + 1)$ control points is defined as

$$S(\xi,\eta) = \sum_{i=0}^{n}\sum_{j=0}^{m} R_{i,j}(\xi,\eta)P_{i,j} = \sum_{i=0}^{n}\sum_{j=0}^{m} \frac{N_{i,p}(\xi)N_{j,q}(\eta)\omega_{i,j}}{\sum_{k=0}^{n}\sum_{l=0}^{m} N_{k,p}(\xi)N_{l,q}(\eta)\omega_{k,l}} P_{i,j}. \tag{9}$$

The mechanical displacements $\mathbf{u}$ and electric potential ϕ can be written as

$$\mathbf{u}(s,t) = \sum_{N} R_i(s,t)\mathbf{u}_i,\ \phi(s,t) = \sum_{N} R_i(s,t)\phi_i. \tag{10}$$

According to the standard FEM process, the discretized control equation for the static piezoelectric and flexoelectric systems in Eq. 7 can be rewritten as

$$\begin{aligned} 0 = \tfrac{1}{2}\sum_{N}\mathbf{u}^T\left(\int_{\Omega}\mathbf{B}_{\mathbf{u}}^T C_{ijkl}\mathbf{B}_{\mathbf{u}}d\Omega\right)\mathbf{u} + \sum_{N}\mathbf{u}^T\left(\int_{\Omega}\mathbf{B}_{\mathbf{u}}^T e_{kij}\mathrm{B}_{\phi}d\Omega\right)\phi \\ + \sum_{N}\mathbf{u}^T\left(\int_{\Omega}\mathbf{H}_{\mathbf{u}}^T\mu_{ijkl}\mathbf{B}_{\phi}d\Omega\right)\phi - \tfrac{1}{2}\sum_{N}\phi^T\left(\int_{\Omega}\mathbf{B}_{\phi}^T\kappa_{ij}\mathbf{B}_{\phi}d\Omega\right)\phi, \\ - \sum_{N}\mathbf{u}^T\left(\int_{\Gamma_t}\bar{t}R_{i,j}dS\right) + \sum_{N}\phi^T\left(\int_{\Gamma_D}\omega R_{i,j}dS\right) \end{aligned} \tag{11}$$

where $\mathbf{B}_{\mathbf{u}}$ and $\mathbf{B}_{\phi}$ are the corresponding gradient operators for displacements $\mathbf{u}$ and electric potential ϕ; $\mathbf{H}_{\mathbf{u}}$ is Hessian operator for displacements $\mathbf{u}$. Those items can all be derived from the explicit expressions of NURBS base functions in Eqs. 8–10.

One more thing to note about IGA, is that the discretization is finished in parametric space and then the discretized geometries are mapped into the physical space using a unified mapping relationship in one NURBS patch, unlike the classical FEM.

This property helps IGA to eliminate conversion errors from inaccurate model description and to save time for mapping computations.

3 Verification

To verify the numerical results, we took the same simulation conditions of the Galerkin methods by Abdollahi et al. [6]. The material parameters of $BaTiO_3$ and the benchmark case have been shown in Table 1 and Fig. 2, respectively. In this case, a cantilever beam is clumped at one end and applied with a point load at the other end. Two main types of electric boundaries as shown in Fig. 2, i.e., (a) the open circuit condition connecting only one surface of the beam to the ground, and (b) the equipotential closed circuit condition where both the top and bottom surfaces are connected to the ground.

Table 1. Material parameters and loading conditions

Parameter	Value	Unit	Parameter	Value	Unit
E	100	GPa	κ_{11}	11	$\mathrm{nC}/(\mathrm{V}\cdot m)$
ν	0.37		κ_{13}	12.48	$\mathrm{nC}/(\mathrm{V}\cdot m)$
e_{13}	-4.4	$\mathrm{C/m^2}$	χ_{33}	1408	
μ_{12}	1	$\mu\mathrm{C/m}$	F	100	μN

The metric used for comparison is the electromechanical coupling coefficient (EMCC) representing the energy conversion ability. EMCC of 1-D beam [7]

$$k_{eff} = \frac{\chi_{33}}{1+\chi_{33}}\sqrt{\frac{1}{E\cdot\kappa_{33}}\left(e_{31}^2 + 12\left(\frac{\mu_{12}}{h}\right)^2\right)}, \tag{12}$$

where h denotes the nano beam thickness. And the normalized EMCC is defined as

$$\bar{k}_{eff} = \frac{k_{eff}}{k_{piezo}}, \tag{13}$$

where k_{piezo} denotes the pure piezoelectric effect by assigning $\mu_{12} = 0$ in Eq. 12.

Under the same circumstances, the numerical results are shown in Fig. 3 with the normalized beam thickness ($h' = -e_{31}h/\mu_{12}$). From the normalized EMCC changing with the beam thickness, we could see that no matter for 1-D simulation or 2-D simulation, the outcomes of both methods are quite similar. In addition, Fig. 3 demonstrates that flexoelectricity playing a leading role in micro scale structure. By comparing simulation results with and without the flexoelectric effect, the rapid increasing of energy conversion rate could be mainly attributed to flexoelectricity.

Note that in Fig. 3(a), the horizontal axis should not be interpreted as starting from zero thickness, which is neither feasible in the present theoretical analysis nor in the numerical simulation. The actual horizontal tick label at the start point is missing in the

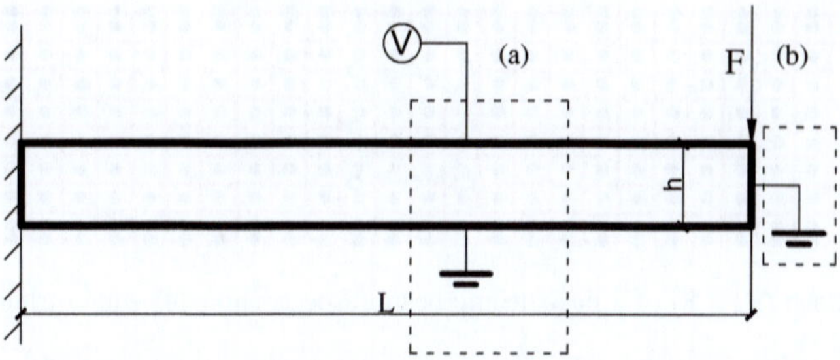

Fig. 2. The electromechanical coupling benchmark: (a) Open circuit; (b) Closed circuit

referred paper [6] and can be inferred to be about 0.5 of the normalized thickness by comparing the tick length of nearby intervals and by comparison with our results.

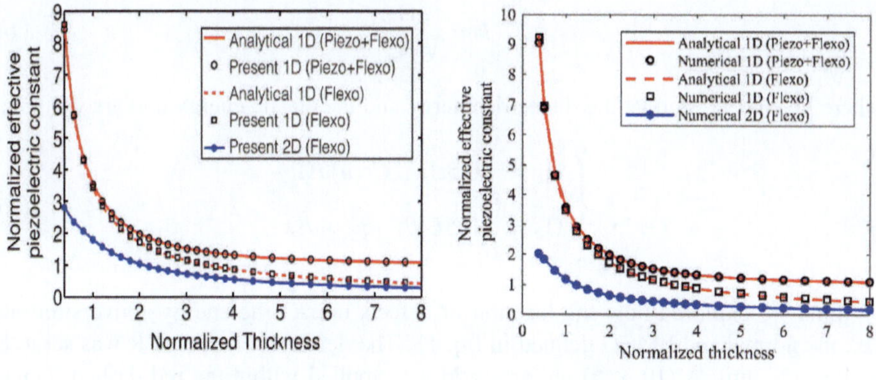

(a) Abdollahi et al. [6]; (b) Our method.

Fig. 3. Numerical results of the benchmark test

4 Topology Optimization

For the isogeometric topology optimization, we realized NURBS trimming technique firstly proposed by Kim et al. [8] into our IGA analysis framework to simulate high-order flexoelectric effect, which we believe is in the first attempts in nano structure design and optimization with trimmed NURBS. Our method has enables single NURBS patch to represent geometries with complicated configurations like holes and interfaces while avoiding the stress discontinuity emerging in multi-NURBS-patch models.

To simplify the optimization process while remaining representative, we used circle to indicate the defect's size and location. Figure 4 shows the IGA mesh of a basic flexoelectric layer in Fig. 1 along with one circular trimming curve. The length of the beam L is 0.909 μm and the thickness of the beam h is 0.227 μm. During the optimizing

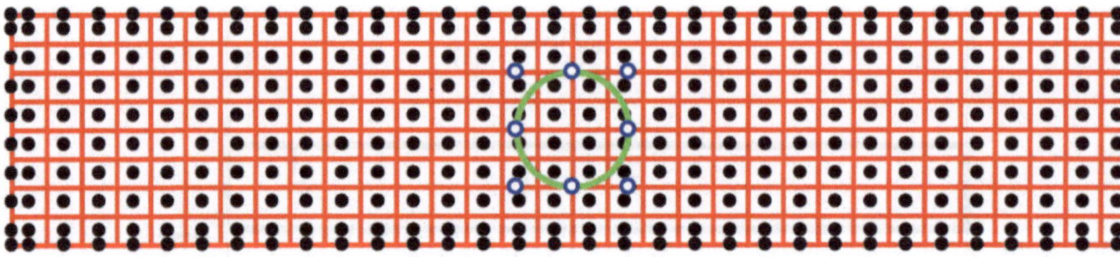

Fig. 4. Mesh (32 × 8) of a flexoelectric beam (aspect ratio = 4) with a trimming curve

process, the trimming curve moved freely but the mesh of the beam body remained fixed. The work in each iteration was to determine whether an element was trimmed or not and then to take suitable numerical integration methods as presented in [8].

The loading condition was the same as that of Fig. 2, where a point load was applied at the top-right corner of the free surface. And the optimization objective was to maximize the electromechanical coupling coefficient (EMCC) [9]

$$k = \sqrt{\frac{\Pi_d}{\Pi_e}}, \tag{14}$$

where Π_d and Π_e denote the dielectric energy and the elastic energy and are written as

$$\begin{cases} \Pi_e = \int\limits_{\Omega} 0.5\varepsilon(u)^T C\varepsilon(u) d\Omega \\ \Pi_d = \int\limits_{\Omega} 0.5E(\theta)^T \kappa E(\theta) d\Omega \end{cases}. \tag{15}$$

First we explored how the location of defects impact the energy conversion rate, i.e., the normalized EMCC defined in Eq. 13. The defect circle radius R was set to be 0.1h = 22.7 nm. A (10 × 5) uniform grid was applied within the red dashed domain such that the curve would not intersect with the beam surfaces and boundary conditions remained valid. The results are shown in Fig. 5. When the circular defect locates at (0.307 μm, 0.193 μm), i.e., (0.34*L*, 0.85*h*), EMCC reaches the maximum value of 0.1410.

Then we studied the influence of the defect's size on the normalized EMCC by fixing the center of the circular defect at the geometry center of the nano beam, i.e., (0.5*L*, 0.5*h*). Varying the radius of the circle from 0.05h to 0.4h, the changing of EMCC is demonstrated in Fig. 6. When the radius of the circular defect is about 0.17h, EMCC reaches its Maximum. This is because when radius is 0.17h, the dielectric energy reaches its maximum and the elastic energy arrives at a local minimum as well.

Actually, if both the size and location of defects are the design variables at the same optimizing problem, Grid Search is still a simple but effective way in multi-parameter optimizations.

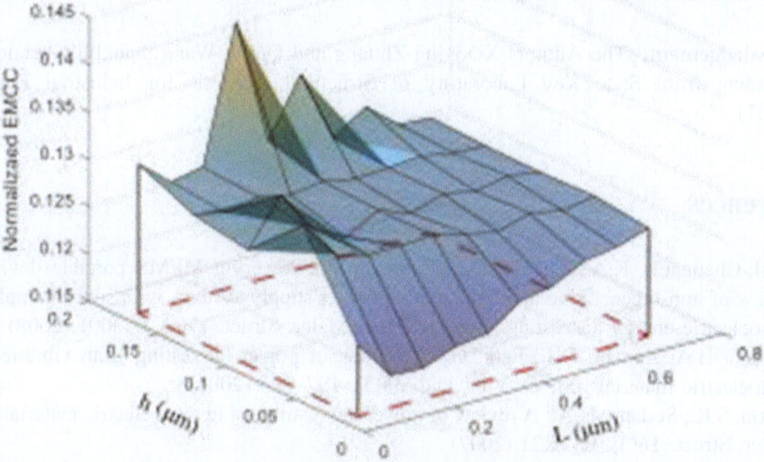

Fig. 5. Grid search (10 × 5) of EMCC within the allowable region of defects

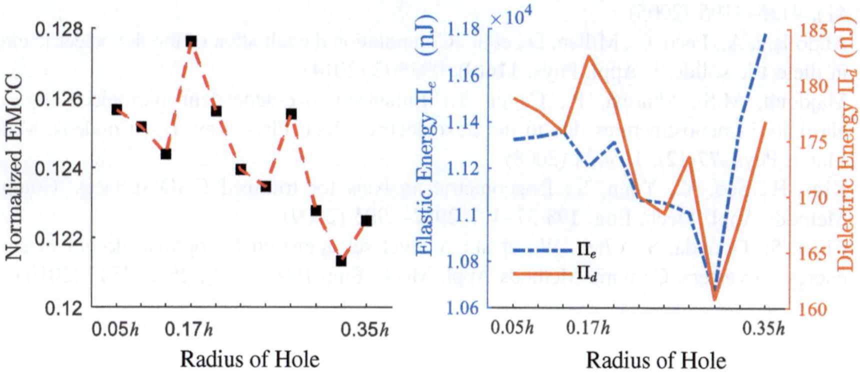

Fig. 6. EMCC, elastic energy, dielectric energy changing with the defect size increasing

5 Conclusions

This study realized the IGA-FEM analysis framework of optimizing the defects' size and location in the nano harvesting structure, by using IGA with trimming NURBS techniques. The IGA-based methods could be instructive for researches on relevant fields, especially for high-order physic problems of complex structures since IGA with NURBS basis is inherently good at simulating high-order physical fields.

In nano structure design and optimization, objectives like stress concentration and manufacturing constraints should also be considered. The obvious stress concentration is observed along the defect boundary of an optimized energy harvester, and can be more hazardous when coupled with electric fields. Meanwhile, the holing operation is rather costly at such a micro scale. The tradeoff between structural behaviors and the overall manufactory costs requires further studies.

Acknowledgement. The Authors Xiaoying Zhuang and Qimin Wang thankfully acknowledge the funding from State Key Laboratory of Structural Analysis for Industrial Equipment (GZ1607).

References

1. Cook-Chennault, K.A., Thambi, N., Sastry, A.M.: Powering MEMS portable devices - A review of non-regenerative and regenerative power supply systems with special emphasis on piezoelectric energy harvesting systems. Smart Mater. Struct. **17**(4), 043001 (2008)
2. Sodano, H.A., Inman, D.J., Park, G.: A review of power harvesting from vibration using piezoelectric materials. Shock Vibr. Dig. **36**(3), 197–205 (2004)
3. Anton, S.R., Sodano, H.A.: A review of power harvesting using piezoelectric materials. Smart Mater. Struct. **16**(3), R1–R21 (2007)
4. Wang, Z., Wu, W.: Nanotechnology-enabled energy harvesting for self-powered micro-/ nanosystems. Angew. Chem. Int. Ed. **51**(47), 11641–11903 (2012)
5. Hughes, T.J.R., Cottrell, J.A., Bazilevs, Y.: Isogeometric analysis: CAD, finite elements, NURBS, exact geometry and mesh refinement. Comput. Methods Appl. Mech. Eng. **194**(39–41), 4135–4195 (2005)
6. Abdollahi, A., Peco, C., Millan, D., et al.: Computational evaluation of the flexoelectric effect in dielectric solids. J. Appl. Phys. **116**(9), 093502 (2014)
7. Majdoub, M.S., Sharma, P., Cagin, T.: Enhanced size-dependent piezoelectricity and elasticity in nanostructures due to the flexoelectric effect. Phys. Rev. B – Condens. Matter Mater. Phys. **77**(12), 125424 (2008)
8. Kim, H., Seo, Y., Youn, S.: Isogeometric analysis for trimmed CAD surfaces. Comput. Methods Appl. Mech. Eng. **198**(37–40), 2982–2995 (2009)
9. Chen, S., Gonella, S., Chen, W., et al.: A level set approach for optimal design of smart energy harvesters. Comput. Methods Appl. Mech. Eng. **199**(37–40), 2532–2543 (2010)

The Compaction Properties of Construction Waste Slag-Clay Mixtures

Chao Hu[1], Jiru Zhang[2], Xiaoqiang Gu[1(✉)], and Kai Xu[3]

[1] Department of Geotechnical Engineering, Tongji University, Shanghai, China
guxiaoqiang@tongji.edu.cn
[2] School of Civil Engineering and Architecture, Wuhan University of Technology, Wuhan, China
[3] State Key Laboratory of Hydrology-Water Resource and Hydraulic Engineering, Nanjing Hydraulic Research Institute, Nanjing, China

Abstract. The construction waste slag mixed with certain proportion of clay has been used as filling material, which has significant social economy and environmental protection benefits. In order to reveal the compaction properties of the slag-clay mixture, a series of compaction tests are conducted to investigate the influence of the slag content and its initial gradation on the behavior and particle breakage. The results show that the optimum water content for slag-clay mixture is smaller than that for clay, and the maximum dry density for slag-clay mixture is larger than that for clay. As the slag content increases, the optimum water content for the slag-clay mixture increases, and the maximum dry density increases firstly and then decreases. It was found that the optimum slag content is 30% in the slag-clay mixture despite the different initial gradation of slag. The maximum dry density is greater for slag with larger particle size, which leads to well compaction of wide-graded slag-clay mixture. The compaction properties of slag-clay mixture are explained by linking the structural response of slag-clay mixtures with the slag content and behavior of particle breakage.

Keywords: Construction slag-clay mixture · Compaction test
Optimum water content · Maximum dry density · Particle breakage

1 Introduction

The rapid development of urban construction has given rise to a large number of construction waste (mainly composed of waste concrete, cement, brick, lime ash etc.). It has eroded the living environment, and intensified the tension of shortage of urban land resources. Therefore, it is of great significance to recycle and make full use of the construction waste. However, the composition of the construction waste residue is complex, and has diverse properties. Besides, in the residue, the coarse particles are relatively concentrated, and the fine particles are far from enough to fill in the pores formed by the coarse particles. Therefore, the waste residue has poor compaction performance and cannot be directly used as the roadbed filling material [1, 2]. Fortunately, a mobile crushing and screening equipment has been put into use in recent years and the construction waste can be broken and formed into slag. But the slag is still a

X. Shi et al. (Eds.): GSIC 2018, *Proceedings of GeoShanghai 2018 International Conference: Transportation Geotechnics and Pavement Engineering*, pp. 425–433, 2018.
https://doi.org/10.1007/978-981-13-0011-0_46

poor graded coarse-grained mixture and it is lack of fine particles to fill in the voids. Therefore, the waste slag cannot be largely applied as filling material, especially the anti-seepage material.

It is well known that the construction slag features the coarse particles while the clay soil features the fine particles. If the two are mixed by a certain proportion to form a slag-clay mixture with a broad size gradation, the mixture itself is expected to become a filling material with good compaction performance. In fact, a wide-gradation gravelly soil, formed by mixing a certain proportion of gravel with natural soil, has already existed. This gravelly soil shows great compaction property and is widely used as the impermeable material in the dam core wall [3]. Many researchers have carried out in-depth investigations on the compaction properties of the wide-gradation gravelly soil, and has achieved fruitful results [4–7]. However, although the wide-gradation gravelly soil and the slag-clay mixture have similar size distribution, the slag-clay mixture has larger discreteness, and thus has more complex engineering and mechanical properties. The main components in a slag-clay mixture, such as concrete and brick, have a much lower critical crushing stress than the gravel material, and thus are more easily to be broken during the compaction process. Consequently, the particle components, arrangement and the contact methods will be changed, and the engineering characteristics will be finally affected. In addition, the current research about the particle crushing is mainly focused on the effect of coarse grain content on the particle crushing, but the influence of the initial size gradation and particle size on the particle crushing is rare.

Based on the above discussion, this study prepares construction waste slags of different initial size gradations and particle sizes, and then it is mixed with the clay by different proportions to form different types of slag-clay mixtures. On the basis of this, the effects of the slag content and the initial gradation of the slag on the compaction performance of the slag-clay mixture are investigated. Meanwhile, the particle breaking characteristics of the slag-clay mixtures are also investigated.

2 Material and Method

2.1 Tested Clay

The clay was taken from a foundation pit in Wuhan. The soil was dried first and then ground. The gradation of the clay is listed in Table 1. Its specific gravity G_s is 2.75. The plastic limit w_p and the liquid limit w_L are 21% and 41.3%, respectively.

2.2 Tested Slag

The construction waste was taken from a waste treatment station in the city of Zhengzhou. Table 2 lists the properties of main components of the waste. It is found that 96.36% of the waste is concrete, stone and brick. The natural water content of the waste is 1.5–2.0%. The brick is light and has high water absorption, while the concrete and the stone are heavy and have low water absorption.

Table 1. Initial gradation of the clay and the slag used in the tests

Material	Name	Size/mm								
		20–10	10–5	5–2	2–1	1–0.5	0.5–0.25	0.25–0.1	0.1–0.075	<0.075
Clay	C	0	0	0	0	0	1.10	2.49	12.71	83.70
Slag	S1	100	0	0	0	0	0	0	0	0
	S2	0	21.70	19.56	4.92	6.99	11.8	7.80	8.32	18.91
	S3	0	100	0	0	0	0	0	0	0
	S4	0	0	100	0	0	0	0	0	0
	S5	50	10.85	9.78	2.46	3.50	5.90	3.90	4.16	9.45
	S6	80	10	10	0	0	0	0	0	0
	S7	33.33	33.33	33.33	0	0	0	0	0	0

Table 2. Physical properties of various components of the construction waste slag

Components	Density/g·cm^{-3}	Water absorption/%
Concrete	2.2	6.5
Brick	1.6	22.5
Stone	2.5	2.0

The construction waste was first crushed by the crushing equipment, and then the iron pieces were removed by the iron removal system. Meanwhile, the light materials such as plastic, wood and foam in the slag were also separated. Finally, the slag was sieved into two kinds of slags by the screening system, including the slag S1 with particle size ranging from 20 mm to 10 mm and the slag S2 with particle size less than 10 mm. Then the S2 slag was artificially sieved into slag S3 (10–5 mm) and slag S4 (5–2 mm). Figure 1 shows the four slags.

In order to investigate the effects of the initial particle size of the slag and the uniformity of the initial gradation on the compaction properties and the particle crushing properties, S1 and S2 were mixed at a ratio of 1:1 to prepare the slag S5. Meanwhile, S1 (80%) was mixed with S3 and S4 (10% each) to prepare the slag S6. S1, S3 and S4 (1:1:1) were mixed to prepare S7. The size gradation of each slag is listed in Table 1.

2.3 Test Method

The compaction test was carried out according to the Geotechnical Test Method Standard (GBT50123-1999). The weight of hammer is 4.5 kg and the falling altitude is 457 mm, corresponding to a compaction energy of 2684.9 kJ/m^3. The sample was divided into 3 layers and each layer was compacted by 94 blows. In order to study the effect of the slag content and the uniformity of the initial gradation on the compaction properties of the slag-clay mixture, the slag S5 was mixed with clay to form CS5, and the slag content R ranged from 0 to 90% with an interval of 10%. The S1 and S6 were also mixed with clay to form CS1 and CS6 respectively, and the slag content R ranged

(a) S1 (10 mm < particle size<20 mm)

(b) S2 (particle size<10 mm)

(c) S3 (5 mm<particle size<10 mm)

(d) S4 (2 mm<particle size<5 mm)

Fig. 1. The slag used in the tests

from 10% to 90% with an interval of 20%. The compaction test was conducted on each slag–clay mixture.

The CS1 mixtures with different slag contents were compacted first at their optimum water content, and then the slag in the mixture was separated from the mixture. In order to study the effect of slag content on the particle crushing, the size gradation of the separated slag was analyzed and the crushing of the slag particles was quantified. Besides, in order to investigate the effect of the initial particle size and the uniformity of the size gradation on the particle crushing, the CS1, CS3, CS4, CS7 mixture with the same slag content were compacted at their optimum water content.

In this paper, the relative breaking rate B_r suggested by Hardin [8] was used to quantify the crushing of the slag particles under compaction. As shown in Fig. 2, the area compassed by the size gradation curve of the slag and the 0.075 mm cut-off line, is defined as the breaking potential. The difference between the crushing potentials before and after the compaction is defined as the crushing amount B_t. B_r is the ratio of B_t to the crushing potential B_p before crushing. It is obvious that the concept of B_r is clear and can reflect the overall change in the gradation curve before and after compaction.

3 Results and Discussions

3.1 The Compaction Curve of the Slag-Clay Mixture

Figure 3 shows the relationship between the water content w and compacted dry density ρ_d of the slag-clay mixture with different slag contents. It is found that the compaction curve of the slag-clay mixture is similar to that of the pure clay.

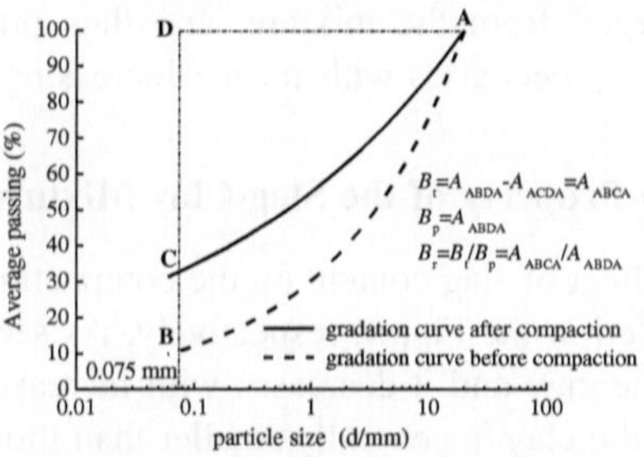

Fig. 2. Definition of Hardin's relative breaking rate

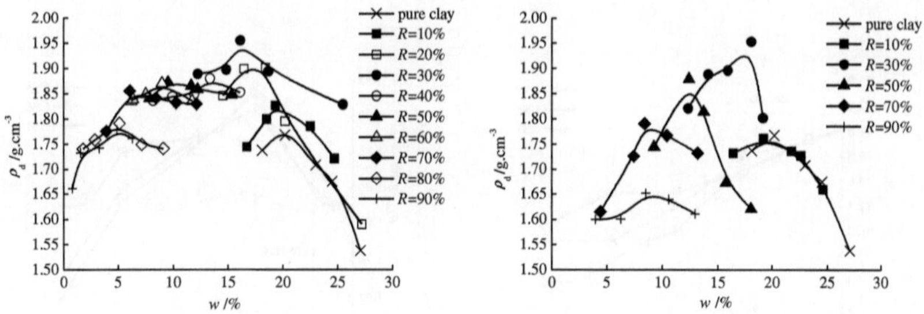

(a) CS1 (b) CS6

Fig. 3. The compaction curves of slag-clay mixtures with different slag contents *R*

The compaction curve is convex as expected. When w increases to the optimum water content w_{op}, ρ_d increases to the maximum dry density ρ_{dmax}. The law of ρ_d-w curve can be explained by the combined water film lubrication theory and electrochemical properties theory [9]:

(1) When w of the mixture is low, the combined water film on the soil surface is thin, resulting in small grain spacing and large attraction force. Under compaction, it is difficult for the mixture particle to move due to the strong cohesive force formed by the inter-granular attraction force and the friction between particles. As a consequence, the mixture shows poor compaction property and smaller ρ_d.

(2) With increasing w, the combined water film begins to thicken, resulting in increasing grain spacing and decreasing attraction force. Meanwhile, the thickening water film lubricates the particles, reducing the friction between the coarse particles. Therefore under compaction, the particles are more easily to move and become denser. Consequently the mixture shows good compaction property and ρ_d increases with increasing w. Finally, ρ_d increases to ρ_{dmax} when the optimum water content w_{op} is reached.

(3) With further increasing w after w_{op}, the free water begins to appear and occupies the inter-granular pores, although the thickening combined water film further weakens the cohesive force. Under compaction, the free water in the pores is not

easy to be discharged from the mixture, and thus prevents the particles from moving. Therefore ρ_d decreases with further increasing w.

3.2 The Compaction Property of the Slag-Clay Mixture

To better illustrate the effect of slag content on the compaction property, the w_{op}-R and ρ_{dmax}-R curves are plotted in the Fig. 4, respectively. As seen in Fig. 4(a), w_{op} of the mixture is lower than the clay and it decreases with increasing R. On the other hand, ρ_{dmax} (=1.77 g/cm^3) of the clay is generally smaller than those of the mixtures, except for CS1 and CS6 with $R > 70\%$. It is interesting that ρ_{dmax} first increases with increasing R and it reaches the maximum value when R becomes an optimum value R_{op}. With further increasing R, ρ_{dmax} begins to decrease.

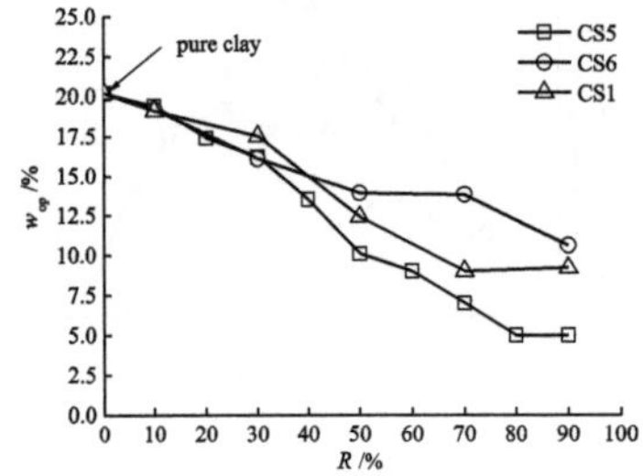

(a) w_{op}—R curve

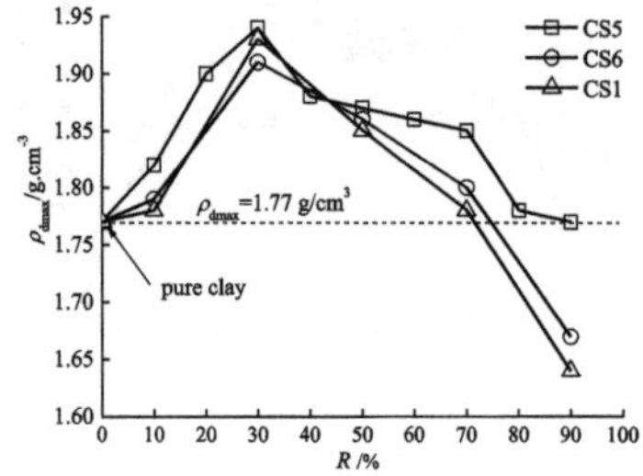

(b) ρ_{dmax}—R curve

Fig. 4. The effect of slag content on the compaction properties of mixtures

It also can be seen in Fig. 4 that the ρ_{dmax}-R curve of CS5 is generally located above the curves of CS1 and CS6, while the w_{op}-R curve of CS5 is generally located below the curves of CS1 and CS6. It indicates that at the same slag content, the mixture with a less uniform size gradation tends to have a larger ρ_{dmax} and smaller w_{op}. The ρ_{dmax} of CS5 is 1.94 g/cm^3, which is slightly larger than 1.93 g/cm^3 of CS1 and 1.91 g/cm^3 of CS6, and the corresponding w_{op} is 15%, slightly lower than 16% of CS1 and 15.8% of CS6. Moreover, it is interesting that R_{op} of each mixture is 30%, indicating that the initial distribution of slag has little effect on R_{op}.

The structure of the mixture varies due to different aggregate gradation and is usually described in three forms: suspension-dense, skeleton-void, skeleton–dense, as shown in Fig. 5. These three forms have been widely used to describe the characteristics of the structure of gravelly soil with different slag contents and analyze its engineering properties [2, 4, 5, 7, 10]. Therefore, the effect of slag content on the compaction properties of the slag-clay mixture can be analyzed as follows:

(1) When R is small, the coarse slag particles are separated by the fine particles and cannot directly contact each other to form a skeleton. The coarse grains are suspending between the fine particles, showing a suspension-dense structure, as

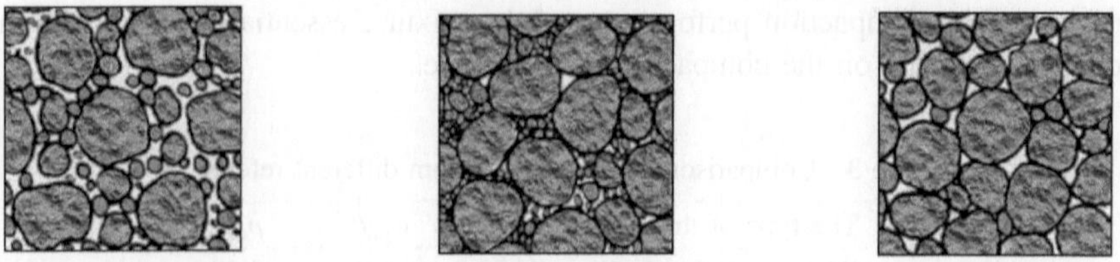

(a) suspension-dense (b) skeleton-dense (c) skeleton-void

Fig. 5. The structure of slag-clay mixture

shown in Fig. 5(a). Because the densities of the concrete and stone are higher than the density of clay, the increase of R causes the increase of ρ_{dmax}.

(2) As R increases, the growing amount of coarse particles begin to move closer and form a skeleton. At the same time, a sufficient amount of fine particles can fill in the skeleton pores. Thus a skeleton-dense structure is formed, as shown in Fig. 5 (b). When R increases to R_{op} and ρ_{dmax} reaches a maximum value.

(3) With further increasing R, the coarse slag particles increase and the fine clay particles decrease, resulting in the lack of sufficient fine particles to fill in the skeleton pores. Then a skeleton-void structure is formed and ρ_{dmax} decreases.

It is clear that the relationship between the structure and the slag content is useful and can explain the compaction property of the slag-clay mixture. Other experimental phenomena can also be explained using the structure of the mixture. As shown in Fig. 4(b), when R increases to 90%, the ρ_{dmax} of CS5 decreases to 1.77 g/cm^3, while the ρ_{dmax} of CS1 and CS6 decrease to 1.64 g/cm^3 and 1.67 g/cm^3 respectively. The reason is that when R reaches 90%, the slag-clay mixture has already formed a skeleton-void structure. The clay content in CS1 and CS6 is very low and it is far from enough to fill in the skeleton pores, causing a significant decrease of ρ_{dmax}. However, although the clay content in CS5 is also very low, the good gradation of S5 slag itself contains a large number of small particles (see Table 1), and they can fill the skeleton pores. Hence, the decrease of ρ_{dmax} of CS5 is much less than CS1 and CS6.

3.3 Discussions

Some researchers have carried out compaction tests on the construction filling materials, and obtained the related compaction property indicators, as listed in Table 3. Although the composition of each material, particle size and distribution, test methods, the impact of energy are different, we can still find some clues after careful analysis. It shows that R_{op} of the slag-clay mixture is generally less than 50%, while R_{op} of other types of coarse-grained soils such as gravelly soils is generally higher than 70%. The reason for such significant difference of R_{op} may be that the slag particles are much easier to be broken. In fact, R reflects the initial relative content of the coarse particles in the slag and the fine particles in the mixture. The crushing of the slag particles during compaction promotes the change in relative content between the coarse particles and the fine particles. Therefore, the influence of the slag content and the crushing of particles on the compaction performance of the mixture essentially reflects the influence of the grading on the compaction performance.

Table 3. Comparison among results from different references

References	The type of the soil	R_{op}/%	w_{op}/%	ρ_{dmax}/g·cm^{-3}
[4]	Wide-graded gravelly soil	70.0	6.0	2.20
[6]	Soil-stone mixture	75.0	8.0	2.15
[1]	Construction waste	45.6	5.0	1.82
[2]	Construction waste	34.7	18.0	1.74
[10]	Construction waste	50.0	15.5	1.81
[11]	Construction waste	35.9	16.4	1.73
This study	Slag-clay mixture	30.0	15.0–16.0	1.91–1.94

Table 3 also shows that at the optimum slag content, the w_{op} of the slag-clay mixture is larger than that of the gravelly soil, but the ρ_{dmax} is smaller than that of the gravelly soil. It is also related to the composition and particle breaking characteristics of the slag-clay mixture. Since the density of concrete, cement, brick and other main components in the slag is generally smaller than the density of gravel, ρ_{dmax} of the slag-clay mixture is smaller than the gravel soil. Moreover, due to the presence of brick, the w_{op} of slag-clay mixture is generally much higher. Brick is a lightweight, porous structure with strong water absorption capacity. In addition, the crushing of the bricks in the compaction process further increases the specific surface area of the bricks, leading to further enhancement of the water absorption capacity of the bricks.

4 Conclusions

In this paper, the compaction characteristics and particle crushing effect of the slag-clay mixture with different gradations and slag contents were studied by compaction tests. The preliminary conclusions are as follows:

(1) The compaction properties of the slag-clay mixture is similar to that of clay, but the optimum water content of the slag-clay mixture is lower than that of clay, and the maximum dry density is higher than that of clay.

(2) The optimum water content of the slag-clay mixture decreases with increasing slag content, and the maximum dry density first increases and then decrease with increasing slag content. There is an optimum slag content of about 30%. Compared with the slag-clay mixture with smaller particles size, the maximum dry density of the slag-clay mixture with larger particles is higher. Therefore the mixture with large slag particles has better compaction performance.

Acknowledgement. The work presented in this paper is supported by National Key Research and Development Program (Grant no. 2016YFC0800204), National Natural Science Foundation of China (Grant no. 41772283) and the Fundamental Research Funds for the Central Universities. These supports are gratefully acknowledged.

References

1. Yao, Z.X.: Experimental study on pavement performance of building debris as subgrade backfill. Ind. Constr. **40**(S1), 773–777 (2010)
2. Zhang, Q.F., Wangd, Q., Jiang, C.G.: A study on road performance of construction waste as municipal road filling. Highway **11**, 157–160 (2006)
3. Liu, J., Zhang, X.: Study on filter design of broadly-graded soil. Chin. J. Geotech. Eng. **18** (6), 1–9 (1996)
4. Chen, Z.B., Zhu, J.G., Wang, Q.: Compaction properties of wide grading gravelly soil. Chin. J. Geotechn. Eng. **30**(3), 446–449 (2008)
5. Wang, T., Chen, Z.B., Li, F.Z.: Experimental study on improving for compact performance of wide-grading gravelly soil. J. Water Resour. Architect. Eng. **11**(5), 52–55 (2013)
6. Liu, L.P., She, X.S.: Study on compaction property of earth-rock mixture. Chin. J. Rock Mech. Eng. **25**(1), 206–210 (2006)
7. Li, Z.H.: The test research on wide grading gravelly soil compaction and shear resistance characteristics. Shanxi Architect. **40**(33), 221–222 (2014)
8. Hardin, B.O.: Crushing of soil particles. J. Geotechn. Eng. ASCE **111**(10), 1177–1192 (1985)
9. Zhao, M.H., Yu, X.: Soil Mechanics and Foundation Engineering. Wuhan University of Technology Press, Wuhan (2003)
10. Li, Y.Y., Li, Z.: A study on construction waste as municipal road filling. Highway **7**, 235–239 (2013)
11. Wang, D.Q., Chen, P., Liu, C.R., et al.: Study on the application of constructional waste to municipal roadbed works. Architect. Technol. **36**(2), 145–146 (2005)

Mechanical Property Tests for Waste Tires Steel Fiber Reinforced Concrete

Zhang Yancong[1(✉)], Gao Lingling[2], and Liu Shaowen[1]

[1] Key Laboratory of Highway Construction and Maintenance in Loess Region, Shanxi Transportation Research Institute, Taiyuan 030006, Shanxi People's Republic of China
568223103@qq.com

[2] Department of Management Engineering, Shanxi Conservancy Technical Institute, Taiyuan 030027, Shanxi, People's Republic of China

Abstract. In order to research the application possible of waste fiber steel fiber reinforced concrete in road engineering, the effect of fiber types and strength grade of base concrete on cube compressive strength, splitting strength, flexural-tensile strength and other mechanical properties of waste tires steel fiber and conventional steel fiber reinforced concrete were determined. The results show that: the waste tires steel fiber has a significant enhancement effect on the strength of concrete, and it can be used as green road material as the waste tire powder. The strength of waste steel fiber reinforced concrete is lower than that of conventional steel fiber. To achieve the same strength, the amount of waste tires steel fiber required to be about 40% higher than conventional steel fiber. In addition, the effect of strength grade of base concrete on the growth ratio of flexural tensile strength of waste tires steel fiber reinforced concrete is weak, and the influence on growth ratio of the compressive strength and splitting strength can be neglected.

Keywords: Cement concrete · Steel fiber · Waste tires · Mechanical property

1 Introduction

With the development of the automobile industry, the recycling of waste tires has received more and more attention. Waste tires are mainly composed of rubber, carbon black and metal, etc., with a high resource recycling value, but inappropriate storage, regeneration, disposal and other processes will cause environmental and safety issues, affecting people's health [1, 2]. Therefore, how to rationally and effectively use and disposal of waste tires, to prevent environmental pollution has become a problem people must face.

At present, the old tires are mainly used for refurbishing new tires. Waste tires are mainly used for recycled rubber and the production of rubber powder, while the metal wire part is no better use of channels [3, 4]. As the tire wire is mostly cold drawn or tempered steel wire, tensile strength up to 1800 MPa, with excellent tensile, bending performance [5]. With the development of steel fiber reinforced concrete [6–10] and the development of tire recycling process, the use of waste tires to produce steel fiber, and

X. Shi et al. (Eds.): GSIC 2018, *Proceedings of GeoShanghai 2018 International Conference: Transportation Geotechnics and Pavement Engineering*, pp. 434–441, 2018.
https://doi.org/10.1007/978-981-13-0011-0_47

then the production of waste tire steel fiber reinforced concrete is possible. In this paper, the mechanical properties such as compressive strength, splitting strength and flexural tensile strength of waste tire steel fiber reinforced concrete were studied in order to provide reference to recycle waste tire.

2 Raw Material

2.1 Steel Fiber

In order to compare the difference between the steel fiber and the conventional steel fiber, the paper uses two kinds of steel fiber raw materials, namely Zhitai waste tire steel fiber and Zhitai conventional milling steel fiber with end hook. The performance of the two fiber's indicators is shown in Table 1. Among them, the waste tire steel fiber is mainly obtained by the following process: pull-cut off-shaping.

Table 1. Parameters of steel fiber

Type	Tensile strength/MPa	Bending performance	Equivalent diameter/mm	Length/mm	Process	Shape	Remarks
Waste tires steel fibe	1260	Qualified	1.0	35.34	Cutting	With indentation	Uniform, neatly cut, little surface with a small amount of rubber adhesion
Conventional steel fiber	990	Qualified	1.0	36.02	Milling	With end hook	Uniform, neatly cut, the surface glossy

2.2 Cement

P.O 42.5 cement used in this test was produced in Shanxi Weidun. The physical mechanical performance indicators are shown in Table 2 and meet Chinese specification "General Portland cement" GB175-2007 requirements.

Table 2. Performance indicators of Cement

Finess/(m^2/kg)	Setting time/min		Compression strength/MPa		Flexural tensile strength/MPa		Stability (boiling method)
	Initial	Final	3d	28d	3d	28d	
350	180	275	35.6	49.8	6.1	8.9	Confirmed

2.3 Coarse Aggregate

The coarse aggregate is made of limestone and has a particle size of 5 mm to 20 mm and the gradation is shown in Table 3.

Table 3. Coarse aggregate gradation

Mesh diameter/mm	Cieve quality/g	Cieve percentage/%	Cumulative sieve percentage/%
19	125	0.8	0.8
16	930	6.2	7
13.2	2870	19.1	26.1
9.5	5320	35.5	61.6
4.75	5360	35.7	97.3
0	400	2.7	100

2.4 Fine Aggregate

The source of the fine aggregate is Duling River. Its gradation is shown in Table 4.

Table 4. Fine aggregate gradation

Mesh diameter/mm	Sieve quality/g	Sieve percentage/%	Cumulative sieve percentage/%
4.75	9	1.8	1.8
2.36	47	9.3	11.1
1.18	57	11.3	22.5
0.6	141	28	50.5
0.3	151	30	80.5
0.15	57	11.3	91.7
0	41	8.2	100

2.5 Water

Water used in this test is Taiyuan municipal tap water.

3 Experimental and Analysis on Mechanical Properties of Steel Fiber Reinforced Concrete

In order to characterize the mechanical properties of waste fiber steel fiber reinforced concrete, C30 and C40 base concrete were used to prepare conventional steel fiber reinforced concrete and waste fiber steel fiber reinforced concrete with different volume ratio. Then, the cube compressive strength, splitting strength and flexural tensile strength was compared.

3.1 Test Methods

Concrete mix design reference to Chinese specification "highway cement concrete pavement construction technical rules" JTG /T F30-2015. The mix of C30, C40 concrete is shown in Table 5. Among them, the amount of water-reducing agent is meet the requirement of 30–55 mm slump.

Table 5. Concrete mix

Strength grade	Water/kg	Cement/kg	Sand/kg	Stone/kg	Water reducing agent/kg
C30	195	380	833	981	5.3
C40	195	460	783	998	5.8

After several tests, we found that to achieve the equal strength, the volume ratio of waste tire steel fiber was higher than conventional steel fiber. According to the test results, the volume ratio of the two kinds of steel fiber is shown in Table 6.

Table 6. Steel fiber volume ratio

Type	Volume ratio/%				
conventional steel fiber	0.5	1.0	1.5	2.0	2.5
waste tires steel fiber	1.2	1.8	2.4	3.0	3.6

The test of cube compressive strength, splitting strength and flexural tensile strength of steel fiber concrete is referenced to Chinese specification "highway cement and cement concrete test procedures" JTG E50-2005.

3.2 Test Results and Analysis

Cube Compressive Strength. When the concrete strength grade is C30 and C40, the effect of fiber volume ratio on the cube compressive strength of two kinds of steel fiber reinforced concrete is shown in Fig. 1.

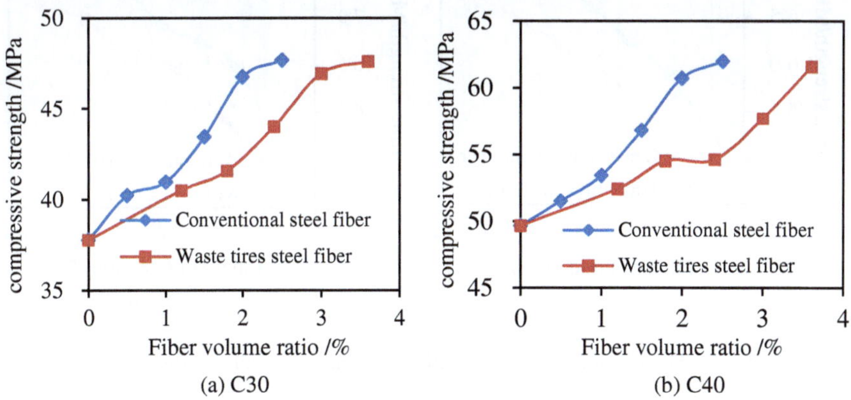

Fig. 1. The relationship between fiber volume ratio and cube compressive strength

When the fiber volume ratio is the same, the compressive strength of the conventional steel fiber reinforced concrete is obviously higher than that of the waste tire steel fiber. With the increase of the volume ratio of steel fiber, the compressive strength of the two kinds of steel fiber reinforced concrete is increasing, the growth ratio shown in Fig. 2.

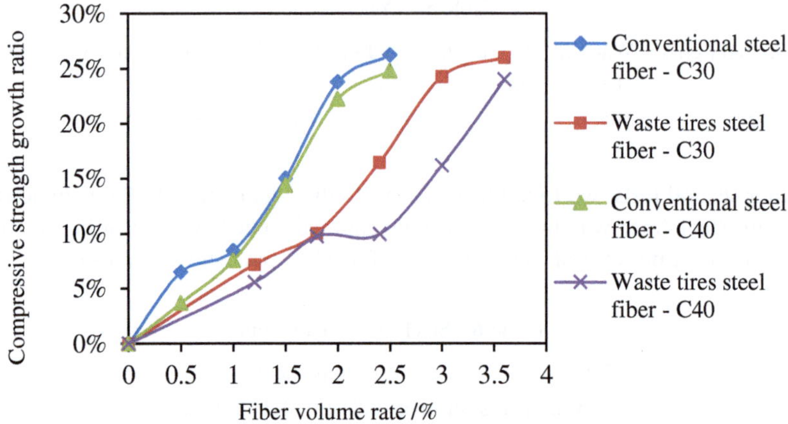

Fig. 2. Growth curve of cube compressive strength

When the steel fiber type is same, the strength growth ratio of C30 and C40 concrete is similar, that is, the strength grade of the concrete has little effect on the strength growth ratio. When the type is different, the compressive strength growth ratio of conventional steel fiber reinforced concrete is higher than that of waste tire steel fiber reinforced concrete. As the waste tire steel fiber processing technology, shape characteristics and other factors, to achieve the same growth ratio, the volume ratio of conventional steel fiber is about 38.7% higher.

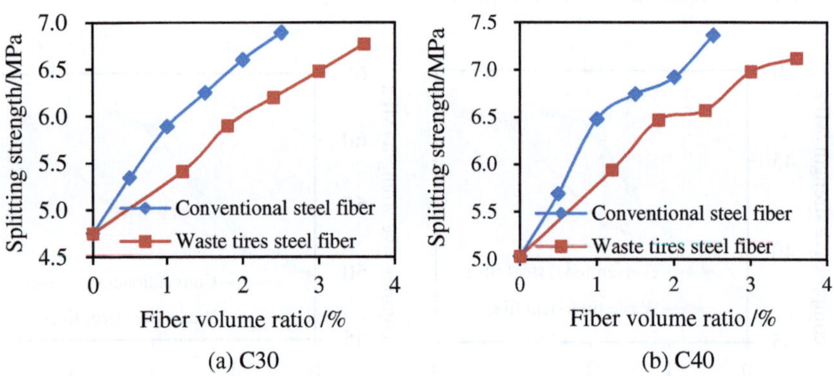

Fig. 3. The relationship between fiber volume ratio and splitting strength

Splitting Strength. When the strength grade of the base concrete is C30 and C40, the effect of the fiber volume ratio on the splitting strength of the two kinds of steel fiber concrete is shown in Fig. 3. When the fiber volume ratio is the same, the splitting strength of conventional steel fiber concrete is higher than that of waste tire steel fiber.

With the increase of the fiber volume ratio, the splitting strength of the two kinds of steel fiber reinforced concrete is increasing, as shown in Fig. 4. To the C30 concrete, the splitting strength of the two kinds of steel fiber concrete increases uniformly with the volume ratio. Compared with C30 concrete, when the strength grade is 40 MPa, the splitting strength growth ratio decreased when the steel fiber volume ratio is greater than 2%. The strength grade of concrete has little effect on the growth ratio of splitting strength, and the growth ratio of conventional steel fiber reinforced concrete is higher than that of waste tire steel fiber concrete. To achieve the same growth ratio, the amount of waste tire steel fiber is about 42.9% higher than that of conventional steel fiber.

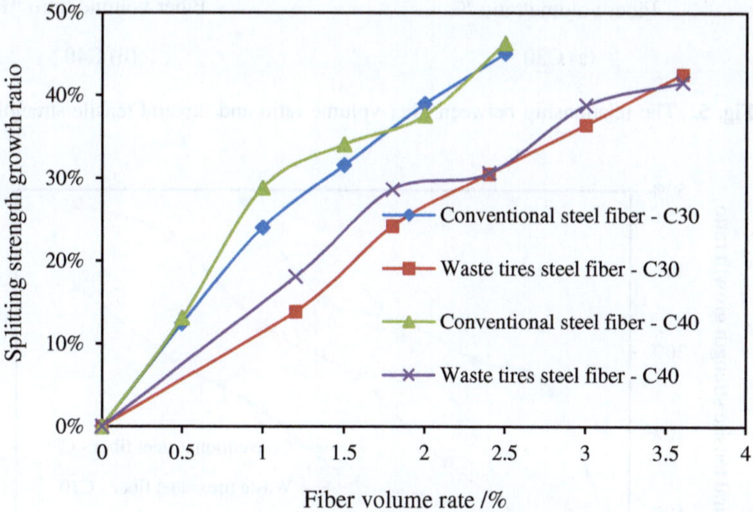

Fig. 4. The growth curve of splitting strength

Flexural Tensile Strength. The flexural tensile strength of conventional steel fiber reinforced concrete is higher than that of waste tire steel fiber, as shown in Fig. 5, when the fiber volume is the same. With the increase of the steel fiber volume ratio, the flexural tensile strength of the two kinds of steel fiber reinforced concrete is increasing, as shown in Fig. 6. The fiber volume ratio of waste tire steel fiber is about 44.6% higher than that of conventional steel fiber to achieve the same strength. The strength grade of the concrete and the steel fiber type has a great influence on the growth ratio of the flexural tensile strength. The difference is that the effect of the steel fiber type is more significant than the strength grade.

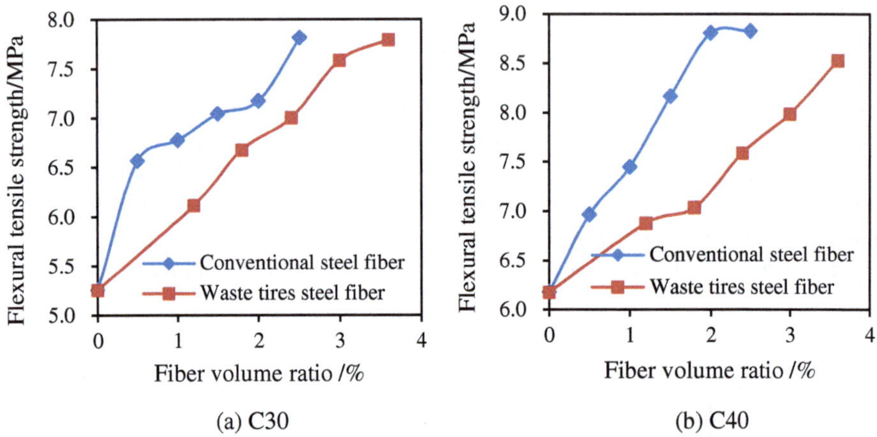

Fig. 5. The relationship between fiber volume ratio and flexural tensile strength

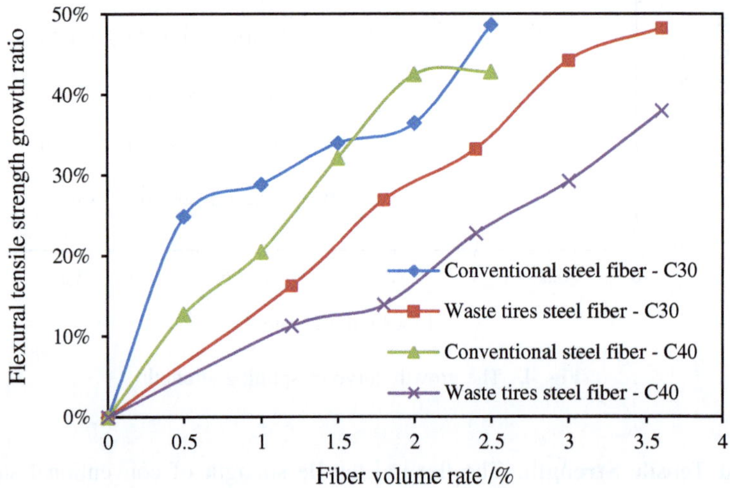

Fig. 6. The growth curve of flexural tensile strength

4 Conclusions

In order to characterize the mechanical properties of waste fiber steel fiber reinforced concrete, C30 and C40 concrete with two kinds of steel fiber (conventional steel fiber and waste fiber steel fiber), five kinds of volume ratio were prepared, and the cube compressive strength, splitting strength and flexural tensile strength was measured.

The waste tires steel fiber has a significant enhancement effect on the strength of concrete, and it can be used as green road material as the waste tire powder.

The strength of waste steel fiber reinforced concrete is lower than that of conventional steel fiber. To achieve the same strength, the amount of waste tires steel fiber required to be about 40% higher than conventional steel fiber.

The effect of strength grade of base concrete on the growth ratio of flexural tensile strength of waste tires steel fiber reinforced concrete is weak, and the influence on growth ratio of the compressive strength and splitting strength can be neglected.

Acknowledgements. This research was partially supported by the Shanxi natural science foundation project (No. 2015021115, No. 201701D221147), Shanxi province communications department science and technology project (No. 2015-1-26, No. 2017-1-17), and Shanxi Province Scientific and Technological Achievements Transformation project (No. 201604D132036).

References

1. Yu, Q.: Recycling and reuse of waste tires in the world. World Rubber Ind. (02), 50–54 (2010)
2. Cao, W., Wang, C., Han, H.: Application of waste tires in road engineering. J. Transp. Stand. (06), 78–82 (2005)
3. Yu, J.: Waste tire recycling status and utilization of ways. Renew. Resour. Res. (05), 7–13 (2003)
4. Li, R.: China's waste tire recycling industry status and development countermeasures. China Resour. Compr. Util. (03), 3–6 (2003)
5. Application of waste steel wire instead of steel fiber in highway pavement. Coal Eng. (10) 27 (2004)
6. Wang, Z., Meng, X.: Study on basic mechanical properties of steel fiber reinforced concrete. J. Concr. (04), 78–81+86 (2014).
7. Yang, R., Yin, J., Xiao, H., Liu, Q.: Experimental study on mechanical properties of steel fiber reinforced concrete. J. Concr. (01), 27–29+42(2006).
8. Xi'an, Liu, X., Yao, J.: Regression analysis of splitting strength and flexural tensile strength of steel fiber reinforced concrete. Highw. Eng. (05), 112–114 (2013)
9. Tang, W., Tang, G., Meng, H.: Application and experimental study of steel fiber reinforced concrete in bridge engineering. J. Highw. Eng. (04), 79–81+86 (2007).
10. Hao, J.: Application of steel fiber reinforced concrete in expressway tunnel pavement structure (06), 158–162 (2006)

Flexural Modulus of Cement-Stabilized Materials for the Mechanistic Pavement Design Approach

Peerapong Jitsangiam[1], Korakod Nusit[2(✉)], and Hamid Nikraz[3]

[1] Faculty of Engineering, Chiang Mai University, Chiang Mai 50200, Thailand
[2] Faculty of Engineering, Naresuan University, Mueang 65000, Phitsanulok, Thailand
korakodn@nu.ac.th
[3] Faculty of Science and Engineering, Curtin University, Bentley, WA 6102, Australia

Abstract. The cement-stabilized road base is a mixture of a conventional road base material, a prescribed amount of cement, and water with an aim to improve the performance of road pavement structure. To obtain the appropriate thickness of pavement structure layers, the elastic moduli of pavement materials are usually required in the process of the mechanistic pavement design approach. In Australia, the flexural modulus obtained from the four-point bending test is recommended by the design guideline. However, this recommended flexural modulus can be determined by the static or cyclic four-point bending tests. Therefore, the flexural modulus derived from the both loading conditions were characterized in this research. Based on the research testing programme, test results reveal that the static flexural moduli measured from monotonic four-point bending tests are generally higher than those obtained from cyclic flexural modulus of cement-stabilized specimens. Furthermore, the cyclic four-point bending tests in this research were performed under the stress-controlled and strain-controlled testing conditions. According to the four-point bending tests setup used in this research, the stress-controlled test seems to provide more consistent results than the strain-controlled test.

Keywords: Cement-stabilized base · Cyclic flexural modulus
Mechanistic pavement design

1 Background

The cement-stabilization technique is popularly used in road construction, particularly for improving engineering properties of poorly-graded or sub-standard materials for being as a proper pavement base, due to well-graded materials become more expensive or even unavailable. Cement-stabilized road base, based on the cement-stabilization technique, is essentially a mixture of a standard road base material, a prescribed amount of cement, and water. A main objective in adding cement to a pavement base is to increase the strength of the road structure. However, the performance of cement-stabilized road base is greatly influenced by various parameters. These

X. Shi et al. (Eds.): GSIC 2018, *Proceedings of GeoShanghai 2018 International Conference: Transportation Geotechnics and Pavement Engineering*, pp. 442–451, 2018.
https://doi.org/10.1007/978-981-13-0011-0_48

parameters are: cement content, water-cement ratio, porosity, curing duration, and method of casting [1].

According to the mechanistic pavement design method, a thickness of a cement-stabilized layer is generally determined based on critical strains produced by traffic loads on a pavement surface. Therefore, the moduli, the main component in deriving strains, of all pavement materials are necessary in the design [2, 3]. Austroads [3] recommended the flexural modulus obtained by the four-point bending test is one of the design inputs for a cement-stabilized layer. Accordingly, the critical tensile strain at a bottom of cement-stabilized layer can be estimated based on the design load. The flexural modulus recommended would be because of its potential to describe the response of a cement-stabilized layer when subjected to a loading regime in the field [3]. Powell et al. [4] conducted an intensive program of field investigations to characterize traffic-induced cracking of cement-treated road bases. It is evident from their study that longitudinal cracking is generally observed when induced stresses are greater than the 28-day flexural strength values. Accordingly, the flexural test, or in this research, the four-point bending test may be used to simulated the responses of cement-stabilized layer in the field.

According to the design guideline [3], the flexural modulus of cement-stabilized materials recommended for mechanistic pavement design can be determined by either monotonic four-point bending test or cyclic four-point bending test [5]. However, the behaviors of flexural modulus measured from monotonic and cyclic loading tests are considerably different. The recent literatures also indicate that, the material differently responses to the stress-controlled and strain-controlled loading tests. Consequently, the aim of this research is characterizing the flexural modulus measured from monotonic (static) and cyclic four-point bending tests. In addition, the cyclic four-point bending test performed under the stress-controlled (constant stress) and strain-controlled (constant strain) loading conditions were also investigated in this research.

2 Cement-Stabilized Specimens

2.1 Parent Materials

Crushed Rock Base (CRB) from a local quarry in Western Australia was employed as the parent material for cement-stabilized crushed-rock specimens. The selected CRB was classified as granite/diorite. The geotechnical properties of CRB used in this research was determined by [6, 7]. This CRB can be categorized as a suitable material for roadway construction and it is used to produce cement-stabilized road base in Australia [8].

2.2 Cement-Stabilized Mixtures

According to Austroads [3], cement-stabilized materials for road pavement can be categorized into two types: modified and stabilized materials. Austroads [3] recommended that the minimum cement content required for stabilized material is 3%. A cement content of 10% is considered as the maximum value for economic use in road construction [9]. However, Nusit et al. [7] discovered that adding cement greater than 7% to this CRB would result in minor improvements to its material strength. Accordingly, the cement content of cement-stabilized specimens in this research was limited to a minimum value of 3% and a maximum value of 6%.

Nusit et al. [7] determined the optimum water content (OMC) and maximum dry density (MDD) of the cement-stabilized materials prepared from the same CRB used in this research. The compaction test reveals that the OMC and MDD of cement-stabilized materials are 6.2% and 2.29 g/cm^3, respectively. The test results also demonstrated that the cement content varied from 3% to 6% contributed a little effect to the OMC and MDD values [7]. Therefore, the OMC and MDD values determined by Nusit et al. [7] were employed for the specimen preparation process in this research.

2.3 Specimen Preparation

The standard-size beam specimens - 400 mm in length, 100 mm wide, and 100 mm thick – were prepared in the rectangular steel mold shown in Fig. 1. The cement-stabilized mixtures were prepared from dry CRB, cement, and water with an amount equivalent to OMC, using a concrete drum mixer. Following this, two equal thicknesses of cement-stabilized material layers were immediately compacted in the steel mold to achieve at least 95% of the MDD. The compacted specimens were left in the steel mold for a minimum of 48 h to gain sufficient strength prior to removal from the mold. The specimens were then cured for 28 days after the compaction was completed.

Fig. 1. Preparation of large beam specimens.

3 Laboratory Testing and Results

3.1 Flexural Strength Test

The flexural strength tests of the cement-stabilized specimens were conducted according to Austroads [5]. The test results are summarized in Table 1.

Table 1. Flexural strengths of cement-stabilized specimens.

Specimen	Cement content (%)	Curing time (Days)	*FS* (MPa)	E_s (MPa)	ε_b (με)
F-1	4	28	1.99	22,800	103
F-2	5	28	2.53	25,000	127
F-3	6	28	2.87	24,600	148

FS = Flexural strength
E_s = Secant flexural modulus calculated at 30% of *FS*
ε_b = Breaking strain at 95% of *FS* [5]

3.2 Cyclic Flexural Modulus Test

Methodology. The four-point bending test configuration illustrated in Fig. 2 was employed to conduct the cyclic flexural modulus tests in this research. It should be noted that the cyclic flexural modulus test in this research was an initial part of the fatigue test. A beam specimen, 100 mm in height and width, and 400 mm in length, was tested in accordance with Austroads's testing protocol [5]. According to the testing protocol, a haversine loading waveform with a frequency of 1 Hz (250 milli-second loading and a 750 milli-second rest period) was applied to beam specimens for one hundred cycles. The cyclic flexural modulus of cement-stabilized specimens was then assigned as the average value of the flexural modulus calculated from the 51st cycle to the 100th cycle. The flexural modulus of the individual loading cycle was determined according to Eq. (1). Based on the testing protocol, a suitable applied stress level for cyclic flexural modulus determination is between 20% and 40% of flexural strength (0.20*FS* and 0.40*FS*) [5]. A maximum applied stress level of 0.40*FS* was recommended to ensure the elastic response of test specimen.

$$E_{flex} = \frac{23PL^2}{108bd^3\delta} \times 1,000 \tag{1}$$

in which, E_{flex} is the flexural modulus (MPa), P is the maximum applied load (N), L is the span length (mm), b is the width of the specimen (mm), d is the depth of the specimen, and δ is the peak mid-span displacement (mm).

Figure 2 shows the hydraulic universal testing machine employed for the four-point bending test. The machine was equipped with a four-point bending jig for the flexural strength tests, cyclic flexural modulus measurements and fatigue tests for cement-stabilized materials in this research. A linear variable differential transformer (LVDT) was attached to the aluminium frame and located at the centre-top of the beam specimen. Accordingly, the maximum vertical displacement of beam specimens induced by cyclic load were continually monitored by the LVDT and recorded throughout the test.

In this research, the cyclic flexural modulus of every cement-stabilized specimen was measured prior to the commencement of fatigue tests. An applied stress level of 0.30*FS* was employed for the cyclic flexural modulus measured by a stress-controlled test.

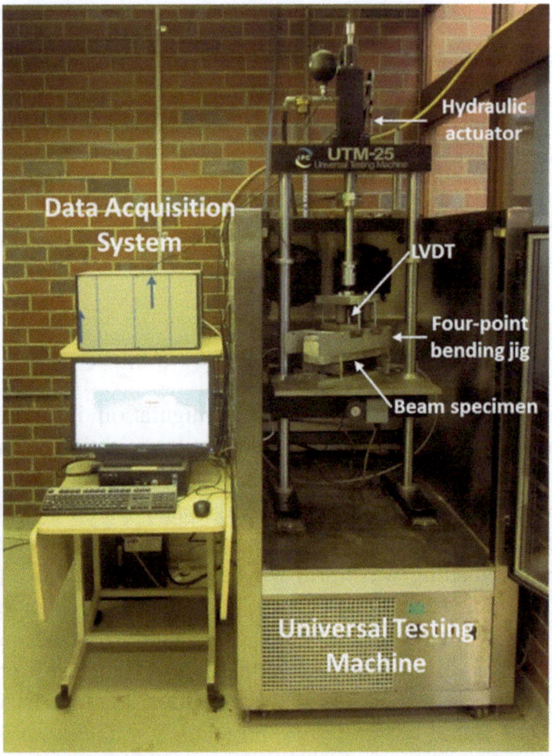

Fig. 2. Universal testing machine equipped with four-point bending jig.

The specified stress level generally generated maximum tensile strain at the bottom of the beam fibre equivalent to 30–40 με. This induced strain range accords with the values recommended by Austroads [5] if the specimen is to be subsequently tested for fatigue behavior. During the measurement of the cyclic flexural modulus, the maximum applied loads and maximum vertical displacements were continuously recorded until one hundred loading cycles were completed. The maximum applied tensile stress and maximum tensile strain could then be calculated, using Eqs. (2) and (3), respectively [5].

$$\sigma_{t,max} = \frac{3(L-a)P}{2bd^2} \tag{2}$$

$$\varepsilon_{t,max} = \frac{108d\delta}{23L^2} \tag{3}$$

The flexural modulus of an individual loading cycle was then calculated by dividing the maximum tensile stress by the maximum tensile strain ($\sigma_{t,max}/\varepsilon_{t,max}$), or computed using Eq. 1.

Testing. In this research, the beam specimens were cured in a moist chamber for 25 to 29 days prior to the cyclic flexural modulus measurements. During the cyclic flexural modulus measurement, the beam specimens were sealed in plastic film to prevent moisture loss. The cyclic modulus measurements in this research was conducted under strain-controlled (constant strain) and stress-controlled (constant stress) testing conditions. The test details for each specific beam are summarised in Table 2. According to Table 2, the total number of cement-stabilized specimens tested in this research was 29. The number of specimens tested under stress-controlled testing conditions alone was 23, while 6 specimens were tested using strain-controlled testing conditions. For the stress-controlled test, the maximum values of applied stress for 4%, 5%, and 6% specimens were 0.60 MPa, 0.76 MPa, and 0.86 MPa, respectively. Therefore, the maximum level of the applied stress ratio (σ_t/FS) was kept at 30% in this research. The values of flexural strength (*FS*) for 4%, 5%, and 6% specimens are summarised in Table 1.

To maintain the stress ratio at 30% during the strain-controlled test, a target applied strain level of 35 με was estimated and assigned for the cyclic flexural modulus measurements. However, the real stress ratios achieved during the strain-controlled tests fluctuated between 24% and 29%.

Test Results. The cyclic flexural moduli measured from the stress-controlled and strain-controlled tests are summarised in Table 2. The variations in cyclic flexural modulus, measured from the stress-controlled test, with respect to changes in curing time are plotted in Fig. 3. It can be seen from Fig. 3 that the curing time, which varied from 25 days to 29 days, seems to have a negligible effect on the cyclic flexural modulus of cement-stabilized specimens.

Table 3 summarizes the average cyclic flexural moduli determined from the four-point bending test results. The cyclic flexural modulus of cement-stabilized specimen increased with respect to the rise in the specimen's cement content. The cyclic flexural modulus of the 5% specimen, measured using strain-controlled testing conditions, was slightly smaller than the value obtained from the stress-controlled test. The difference may due to two main reasons; (1) that there were differences between the magnitude of applied stress and applied strain achieved during the stress-controlled and strain-controlled tests, and (2) that the fluctuated tensile strain was unavoidably occurred during the strain-controlled test. It has been proven in previous literature that the cyclic flexural modulus is influenced by the magnitude of applied tensile strain (or stress) during testing [5, 10]. Austroads [5] reported that an increase in strain of 1 (με) then led to a reduction in cyclic flexural modulus which was approximately equal to 40 MPa. Apart from the effects of applied tensile strain (or stress) seen during the test, high fluctuations in targeted tensile strain observed during the strain-controlled test may be another rationale for explaining the variation in the cyclic flexural modulus. The variations in cyclic flexural modulus and tensile strain measured during stress-controlled and strain-controlled tests are illustrated in Fig. 4a and b, respectively. It can be clearly seen from Fig. 4 that the cyclic flexural modulus measured from the stress-controlled test was more consistent than the values obtained from the strain-controlled test.

Table 2. Cement-stabilized specimens and cyclic flexural modulus measurement results.

Specimen	Cement content (%)	Curing duration (Days)	Tensile stress (MPa)	Tensile strain (με)	E_{flex} (MPa)
Stress-Controlled Test					
M-1	5	27	0.76	30	25,100
M-2	5	28	0.76	40	18,800
M-3	5	26	0.76	37	20,300
M-4	5	28	0.76	39	19,500
M-5	5	25	0.76	37	20,400
M-9	5	29	0.76	36	21,100
M-10	5	27	0.76	40	18,700
M-14	5	29	0.76	36	20,900
M-15	5	29	0.76	38	20,200
M-16	5	27	0.76	41	18,500
M-17	5	27	0.76	36	21,100
M-18	5	28	0.76	37	20,400
M-19	5	28	0.76	41	18,500
M-28	5	28	0.76	36	21,300
M-29	5	25	0.76	41	18,500
M-20	4	28	0.60	35	17,300
M-21	4	27	0.60	36	16,600
M-22	4	26	0.60	35	16,900
M-23	4	28	0.60	36	16,600
M-24	6	27	0.86	41	20,800
M-25	6	27	0.86	41	20,800
M-26	6	27	0.86	40	21,400
M-27	6	28	0.86	36	23,900
Strain-Controlled Test					
M-6	5	26	0.61	33	17,900
M-7	5	29	0.72	34	21,000
M-8	5	26	0.74	34	21,300
M-11	5	29	0.66	34	19,400
M-12	5	27	0.60	34	17,300
M-13	5	29	0.65	34	19,000

Besides presenting the cyclic flexural modulus of cement-stabilized material, Table 3 also summarises the secant flexural moduli (E_s) determined from the flexural strength test in this research (see Table 1). The values of the secant modulus of cement-stabilized materials were generally higher than the cyclic flexural modulus. However, it should be highlighted that the secant moduli of cement-stabilized specimens were computed from only one specimen for each percentage of cement content.

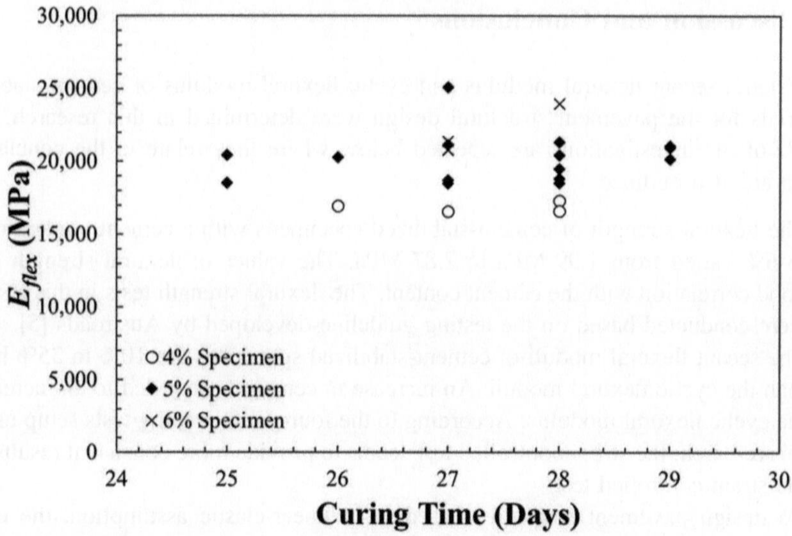

Fig. 3. Relationship between cyclic flexural modulus and curing time.

Table 3. Average cyclic flexural modulus of cement-stabilized specimens.

Cement content (%)	No. of specimens	Average E_{flex} (MPa)	E_s (MPa)	SD (MPa)	COV (%)
Stress-Controlled Test					
4	4	16,900	22,800	362	2.14
5	15	20,200	25,000	1,710	8.45
6	4	21,700	24,600	1,531	7.04
Strain-Controlled Test					
5	6	19,300	25,000	1,603	8.30

E_s = Secant flexural modulus from flexural strength test (see Table 1)
SD = Standard deviation
COV = Coefficient of variation

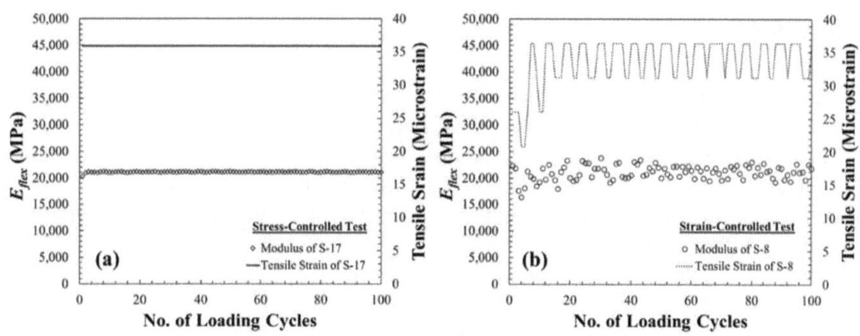

Fig. 4. The variations in cyclic flexural modulus and tensile strain during (a) stress-controlled test, and (b) strain-controlled test.

4 Discussion and Conclusions

The (static) secant flexural modulus and cyclic flexural modulus of cement-stabilized materials for the pavement structural design were determined in this research. Brief details of the investigations are repeated below where they relate to the conclusions which are also outlined.

- The flexural strength of cement-stabilized specimens with a cement content of 4% to 6% varied from 1.99 MPa to 2.87 MPa. The values of flexural strength had a good correlation with the cement content. The flexural strength tests in this research were conducted based on the testing guideline developed by Austroads [5].
- The secant flexural moduli of cement-stabilized specimens are 10% to 35% higher than the cyclic flexural moduli. An increase in cement content led to an increase in the cyclic flexural modulus. According to the four-point bending tests setup used in this research, the stress-controlled test seems to provide more consistent results than the strain-controlled test.
- To design pavement structures based on a linear-elastic assumption, the elastic modulus is generally applied as the design input parameter. In addition, this input parameter should adequately explain the material response under dynamic loading conditions. However, the cyclic flexural modulus determined from the four-point bending test is restricted by two limitations: (1) the inevitable factors associated with the four-point bending test generally lead to uncertain results, (2) the precise value of the elastic modulus is rarely accurately determined from the four-point bending test due to the effects of shear stress. Consequently, the dynamic response of cement-stabilized material under cyclic-compressive loads requires further examinations by the other testing platform.

Acknowledgement. The authors would like to thank the Department of Civil Engineering, Curtin University, Australia for laboratory testing contributions to this research. Moreover, the research team of the Civil Engineering Department, Chiang Mai University and Naresuan University, Thailand are also gratefully acknowledged for providing overview guidance and valuable inputs into the testing part of this work.

References

1. Destrebecq, J.F.: Cyclic and dynamic loading fatigue of structural concrete. In: Mechanical Behavior of Concrete, pp. 185–224. ISTE Ltd., London (2010)
2. NCHRP: Guide for mechanistic-empirical design of new and rehabilitated pavement structures. Final Report, Project No. 1-37A, Washington, D.C. (2004)
3. Austroads: Guide to pavement technology part 2: Pavement structural design. AGPT04A-08, Sydney, NSW (2010)
4. Powell, W.D., Potter, J.F., Mayhew, H.C., Nunn, M.E.: The structural design of bituminous roads. TRRL Laboratory Report 1132, Berkshire, UK (1984)
5. Austroads: Cemented materials characterization. Final report (AP-R462-14), Sydney, NSW (2014)

6. Jitsangiam, P., Nikraz, H.: Mechanical behaviours of hydrated cement treated crushed rock base as a road base material in Western Australia. Int. J. Pavement Eng. **10**(1), 39–47 (2009)
7. Nusit, K., Jitsangiam, P., Kodikara, J., Bui, H.H., Leung, G.L.M.: Advanced characteristics of cement-treated materials with respect to strength performance and damage evolution. J. Mater. Civil Eng. **29**(4), 04016255 (2017)
8. MRWA: Specification 501-Pavements, Perth, WA (2012)
9. NAASRA: Guide to stabilization in roadworks, including metric addendum, Sydney, NSW (1970)
10. Austroads: A laboratory study of the influenced of multiple axle loads on the performance of a cement treated material. Interim report (AP-T185-11), Sydney, NSW (2011)

Mechanical Properties of Cement Stabilized Macadam Mixed with Emulsified Asphalt

Qin L. Huang[1(✉)], Chen J. Quan[1], Zhuang Yang[2], and J. Xu[3]

[1] Tongji University, Shanghai 201804, China
hql04@tongji.edu.cn
[2] Changjiang Survey, Planning, Design and Research Co. Ltd., Wuhan 430010, China
[3] Shanxi Highway Bureau, Taiyuan 030006, China

Abstract. At present, the performance evaluation of cement stabilized macadam mixed with emulsified asphalt is mainly based on strength and shrinkage deformation, however, lack of consideration of its fatigue performance, resulting in a relatively less application in practical engineering. This paper aims at evaluating comprehensively the material properties of cement stabilized macadam mixed with emulsified asphalt based on its mechanical strength and fatigue performance. The conventional mechanical strength, compressive resilient modulus and fatigue life were tested by MTS test machine, and the flexural tensile resistance performance of the mixture was analyzed based on the strain work required by flexural tensile failure, finally combining the flexural tensile strength and fatigue test results, a compressive strength modification method was proposed for the reference of engineering acceptance. The test results show that the strength indicators of the mixture with 1.2% pure asphalt content decrease, but its flexural tensile resistance performance improves: the unconfined compressive strength and splitting strength decrease by 20%, the flexural tensile strength decreases by 4%, the compressive resilient modulus decreases by 13%, but the flexural tensile strain work increases by 18.7%. At the same time, the long-term performance of the mixture improves: the fatigue life increases by 31%.

Keywords: Emulsified asphalt · Cement stabilized macadam
Mechanical properties

1 Introduction

Cement stabilized macadam is widely used in highway pavement structure in China with the advantages of high strength, good integrity, but its rigidity and brittleness, the easiness to crack throughout the pavement due to moisture loss and temperature change, seriously affect the service life of the road [1]. The research shows that the cement stabilized macadam mixed with emulsified asphalt can effectively increase the flexibility of the mixture and improve the reflection crack of pavement [2–5]. Oruc [2] used the emulsified asphalt cement stabilized macadam as pavement base in 2006; Zhang [6] constructed the testing road based on 237 highway expansion project in 2014, results show that the addition of emulsified asphalt reduces the reflection cracks of base, but the early strength is not as good as that of cement stabilized macadam.

X. Shi et al. (Eds.): GSIC 2018, *Proceedings of GeoShanghai 2018 International Conference: Transportation Geotechnics and Pavement Engineering*, pp. 452–462, 2018.
https://doi.org/10.1007/978-981-13-0011-0_49

In practical engineering application, the addition of emulsified asphalt results in a decrease in resilient modulus of the mixture, an increase in deflection value, and when the emulsified asphalt content is high, the mixture strength does not meet the requirements of the existing specifications for cement stabilized macadam base [4, 7], increasing the difficulty of engineering acceptance.

At present, there are mainly two types of methods in China and abroad used for stabilization: adding cement in asphalt mixture and adding emulsified asphalt in cement stabilized macadam. As for the cement stabilized macadam mixed with emulsified asphalt, Oruc [2], Wang [3], Li [8] et al., have studied the shrinkage performance, results show that the addition of emulsified asphalt can effectively improve the flexibility of base: the temperature shrinkage coefficient can reduce by 10%–18% with the emulsified asphalt content 2.5%–3%. Jia [4], Fu [9], Jiang [10] have studied the conventional mechanical strength indicators and compressive resilient modulus of the mixture, results show that the mechanical strength indicators of the mixture with emulsified asphalt all decrease: when the emulsified asphalt content is 2%, the compressive strength reduces by 15%–25%, the splitting strength reduces by 21%–23%, the flexural tensile strength reduces by 0–10%; and when the emulsified asphalt content is 1%–3%, resilient modulus decreases by 15%–30%. Fu [9] has analyzed the flexural tensile resistance performance of the material based on the strain work required by flexural tensile failure (the area integral of the load curve about the middle point deflection during the loading process), results show that the addition of emulsified asphalt can effectively improve the flexural tensile resistance performance of the material. Overall, the mechanical properties of cement stabilized macadam with emulsified asphalt is less studied in China and abroad, and single scholar mostly aims at one or more of the mechanical performance indicators to study, comprehensive evaluation of mechanical properties and fatigue performance of the mixture are not common. The addition of emulsified asphalt has changed the original performance of cement stabilized macadam, and the performance of the mixture cannot be evaluated directly by the indicators of cement stabilized macadam, so it is necessary to modify the testing results for the reference of the performance evaluation and engineering acceptance of semi rigid base material with emulsified asphalt.

Based on the problems above, this paper used the indoor static pressure shaping method, and carried out the conventional mechanical strength and four point flexural tensile fatigue test [11] by the MTS810 testing machine. This paper systematically studied the influence of the emulsified asphalt addition on the conventional mechanical strength, flexural tensile strain work and fatigue performance of cement stabilized macadam materials, and comprehensively evaluated the mixture performance, finally combining the flexural tensile strength and fatigue test results, proposed an unconfined compressive strength modification method suitable for the strength evaluation and engineering acceptance of emulsified asphalt cement stabilized macadam.

2 Test Materials and Mix Ratio

2.1 Test Materials

The cement is Conch brand P.O.42.5 ordinary Portland cement, and its indicators meet the requirements of "General Portland cement standard" (GB175-2007) [12]. The technical indicators of cement is shown in Table 1.

Table 1. Technical indicators of cement.

Technical indicators	Compressive strength/MPa		Rupture strength/MPa		Setting time/min		Apparent density/g·cm^{-3}	Soundness
	3d	28d	3d	28d	Initial set	Final set		
Standard value	≥ 16	≥ 42.5	≥ 3.5	≥ 6.5	≥ 45	≤ 600	-	-
Test value	20.7	46.6	4.7	7.9	100	240	3.112	Qualified

The emulsified asphalt is East China Sea brand cationic slow-crack emulsified asphalt (PC-2) produced by Guangdong province, and its indicators meet the requirements of "Technical specification for construction of highway asphalt pavement" (JTG F40-2004) [13]. The technical indicators of emulsified asphalt is shown in Table 2.

Table 2. Technical indicators of emulsified asphalt.

Technical indicators	Particle charge	Demulsification speed	Evaporation residue			Standard viscosity $C_{25,3}/s$
			Asphalt content/ %	Needle penetration/25 °C, 0.1 mm	Softening point/°C	
Standard value	-	-	≥ 50	50–200	-	10–25
Test value	Cationic	Slow-crack	61.9	66.5	48.5	19

The coarse aggregate is the limestone, which is divided into four levels: 15–25 mm, 5–15 mm, 3–5 mm and 0–3 mm, and its indicators meet the requirements of "Technical guidelines for construction of highway road bases" (JTG/T F20-2015) [14]. The technical indicators is shown in Table 3. In addition to the above coarse aggregate, fine aggregate is the machine-made fine sand with a size of 0.15–0.3 mm.

2.2 Mix Ratio

Mix Ratio of Mixture. External-adding method was used to design the admixture content of the mixture. Under the condition of 4% cement content, the mechanical

Table 3. Technical indicators of course aggregates.

Technical indicators	Crushing value/%	Los Angeles wear value/%	Firmness/ %	Polish value/ %	Content of needle-like particles/%	Apparent density/g·cm^{-3}
Standard value	≤ 30	≤ 28	≤ 8	≥ 40	≤ 15	≥ 2.60
Test value	12.9	16.4	3.5	47	8.4	2.87

properties of the mixture with 1.2% pure asphalt content (2% emulsified asphalt) were compared to the mixture without emulsified asphalt. The aggregate gradation was designed according to the middle value of skeleton dense structure gradation range recommended by "Technical guidelines for construction of highway road bases" (JTG/T F20-2015) [14]. The aggregate gradation is shown in Table 4.

Table 4. Aggregate gradation of skeleton dense structure.

Gradation	Mass fraction of passing through the following sieves size (mm)/%											
	26.5	19	16	13.2	9.5	4.75	2.36	1.18	0.6	0.3	0.15	0.075
Upper limit of gradation	100	90	80	74	68	50	38	30	22	17	12	7
Nether limit of gradation	100	67	57	51	45	29	18	13	8	5	3	0
Design gradation	100	83.8	72.5	65.7	59.4	40.6	23.6	17.3	13.4	7.9	5.1	2.7

Optimum Water Addition and Maximum Dry Density. The optimum water content and maximum dry density of the mixture under two types of working conditions can be determined according to the A-type heavy compaction method of "Test methods of materials stabilized with inorganic binders for highway engineering" (JTG E51-2009) [11]. The emulsion water contained in emulsified asphalt can be the free water in the mixing process, and it leads to a great difference between the actual water content and the designed water content. So, the emulsion water amount can be subtracted from the optimum water content to obtain the optimum water addition, which can be a controlling indicator about the water addition in the mixing process. The optimum water addition and maximum dry density of two types of mixture is shown in Table 5.

Table 5. Optimum water addition and maximum dry density.

Cement content/%	Pure asphalt content/%	Optimum water addition/%	Maximum dry density/g·cm^{-3}
4	0	5.1	2.341
	1.2	4.2	2.316

3 Specimen Shaping and Test Method

3.1 Specimen Shaping Method

According to the standard compaction test results, in accordance with the requirements of "Test methods of materials stabilized with inorganic binders for highway engineering" (JTG E51-2009) [11], the indoor static pressure shaping method was used to make the relevant test specimens with the 98% compaction degree. There were two types of cylinder specimens size: Ø 100 mm × 100 mm, Ø 150 mm × 150 mm, and were used to test the unconfined compressive strength and splitting strength, compressive resilient modulus of the mixture, respectively. The size of middle beam specimens is 400 mm × 100 mm × 100 mm. After shaping, the specimens were wrapped in plastic film, and put into the standard curing room for 28 days. The 5 parallel specimens were shaped in every working condition.

3.2 Test Method

The specimens were cured for 28 days. On the last day of curing period, the specimens were put into the water tank with constant temperature 20 °C ± 2 °C to soak for 24 h. Then according to the testing methods of "Test methods of materials stabilized with inorganic binders for highway engineering" (JTG E51-2009) [11], the specimens were tested for the strength, modulus and fatigue performance by using MTS810 testing machine. The testing temperature is 20 °C ± 2 °C.

4 Test Results and Analysis

4.1 Unconfined Compressive Strength, Splitting Strength and Flexural Tensile Strength

The unconfined compressive strength, splitting strength and flexural tensile strength of two types of mixture were tested according to the mechanical strength test procedure of "Test methods of materials stabilized with inorganic binders for highway engineering" (JTG E51-2009) [11]. In the testing process of flexural tensile strength, the middle deflection of the beam corresponding with bearing flexural tensile strength was recorded by the displacement sensor, setting the maximum deflection is 2 mm. In order to coincide with the loading rate of dynamic load during the fatigue test, the rapid loading mode is adopted in the process of flexural tensile test [11]. The 5 parallel specimens were tested in every working condition, and the mean value of the 5 parallel specimens is represented as the strength in every working condition of the mixture, and the results are shown in Table 6.

From Table 6, when adding 1.2% pure asphalt content, the 28d unconfined compressive strength and splitting strength of the mixture, to a certain extent, decrease: the unconfined compressive strength reduces from 5.6 MPa to 4.5 MPa, decreasing 20%; the splitting strength reduces from 0.49 MPa to 0.39 MPa, decreasing 20%. However, the addition of emulsified asphalt has little influence on the flexural tensile strength of the mixture: the flexural tensile strength reduces from 1.28 MPa to 1.23 MPa, decreasing 4%.

Table 6. The mechanical test results of the mixture with curing for 28d.

Cement content/(%)	Pure asphalt content/(%)	Unconfined compressive strength/(MPa)	S	Splitting strength/(MPa)	S	Flexural tensile strength/(MPa)	S
4	0	5.6	0.45	0.49	0.06	1.28	0.11
	1.2	4.5	0.38	0.39	0.04	1.23	0.09

Note: S is the standard deviation.

4.2 Compressive Resilient Modulus

According to the test results of the unconfined compressive strength, the half of the compressive strength value of the corresponding working conditions was used as the maximum load value of this working condition to carry out the static resilience modulus test. During the test, the 0–2.5 MPa loading sequence and the 0.5 MPa interval unit pressure were applied to load step by step. The 5 parallel specimens were tested in every working condition, and the mean value of the 5 parallel specimens is taken as the compressive resilient modulus in every working condition of the mixture, shown in Table 7.

Table 7. The compressive resilient modulus of the mixture with curing for 28d.

Cement content/(%)	Pure asphalt content/(%)	Compressive resilient modulus/(MPa)	S
4	0	1,204	101
	1.2	1,046	80

Note: S is the standard deviation.

From Table 7, after adding 1.2% pure asphalt, the 28d compressive resilient modulus of the mixture decreases from 1,204 MPa to 1,046 MPa, decreasing by 13%. The results show that the addition of emulsified asphalt decreases the compressive resilient modulus of the mixture, making its rigidity reduce.

4.3 Flexural Tensile Resistance Performance

In the process of flexural tensile strength test, the vertical deformation of the beam specimen increases with the increasing load. When the vertical deformation increases to a certain amount, the beam specimen cracks from the bottom of the middle and gradually expands, leading to the ultimate fracture failure.

The flexural tensile resistance performance of the mixture should be reflected by the flexural tensile strain work [9], which has taken the flexural tensile load and the flexural tensile deformation into consideration. The flexural tensile strain work W is the integral of the flexural tensile load about the middle point deflection, and the formula is shown as follows:

$$W = \int_0^{l_{max}} F\mathrm{d}l \tag{1}$$

In the formula, F is flexural tensile load, l is flexural tensile deformation, l_{max} is mid-span ultimate deflection.

The flexural tensile strain work of two types of mixture beam specimen can be calculated by formula (1), and calculation examples of two specimens are shown in Fig. 1: flexural tensile strain work of the specimen (one of the 5 specimens) without emulsified asphalt is the integral of the flexural tensile load about the middle point deflection in the 0–0.58 mm range, and flexural tensile strain work of the specimen (one of the 5 specimens) with 1.2% pure asphalt content is the integral of the flexural tensile load about the middle point deflection in the 0–0.83 mm range.

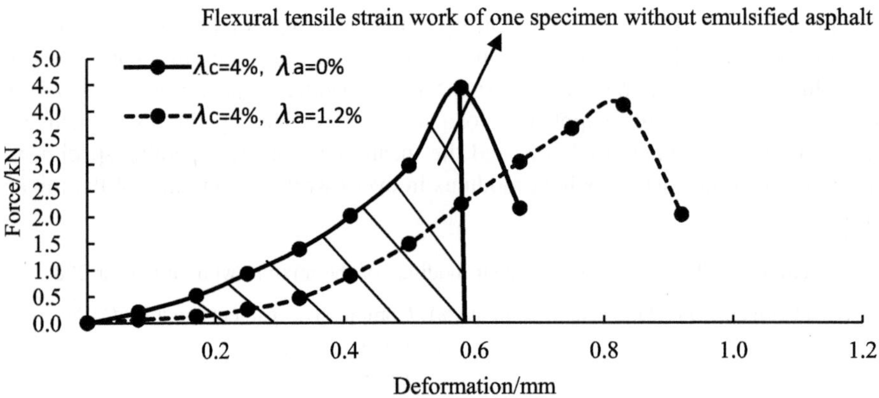

Fig. 1. Calculation example figure of two specimens' flexural tensile strain work.

The 5 parallel specimens were tested in every working condition, and the mean value of ultimate flexural tensile deflection of the 5 parallel specimens is taken as the ultimate flexural tensile deflection in every working condition of the mixture. The flexural tensile strain work is the same. All the calculation results are shown in Table 8.

Table 8. The flexural tensile strain work of the mixture with curing for 28d.

Cement content/(%)	Pure asphalt content/(%)	Ultimate flexural tensile deflection (mm)	S	Flexural tensile strain work (J)	S
4	0	0.53	0.044	0.91	0.099
	1.2	0.85	0.072	1.08	0.107

Note: S is the standard deviation.

From Table 8, with the addition of emulsified asphalt, the ultimate flexural tensile deflection and strain work required by flexural tensile failure of the beam specimen both increase. After adding 1.2% pure asphalt, the ultimate flexural tensile deflection increases from 0.53 mm to 0.85 mm, increasing by 60.4%; the flexural tensile strain work increases from 0.91 J to 1.08 J, increasing 18.7%.

According to Tables 6 and 8, the results show that after the addition of emulsified asphalt, the flexural tensile strength of the mixture slightly reduces, but the ultimate flexural tensile deflection and flexural tensile strain work greatly increase, so adding emulsified asphalt can improve the toughness and flexural tensile resistance ability of semi rigid base material.

4.4 Fatigue Performance

According to the fatigue performance test procedure, the four points flexural tensile fatigue test was carried out for two types of mixture beam specimens with curing for 28d. In the test, the ultimate flexural tensile strength adopts the results in Table 7, and the stress ratio is 0.75. The maximum load P_{max} can be determined based on the ultimate flexural tensile strength and stress ratio, and set 10% maximum load P_{max} as the minimum load P_{min} of Havesine wave, so the load cycle characteristic value $P_0 = P_{max} - P_{min} = 0.9\ P_{max}$. The 5 parallel specimens were tested in every working condition, and the mean value of the 5 parallel specimens is taken as the fatigue life results in every working condition of the mixture, shown in Table 9.

Table 9. The fatigue test results of the mixture with curing for 28d.

Cement content/(%)	Pure asphalt content/(%)	Flexural tensile strength/(MPa)	Stress ratio	Fatigue life $\overline{N}$ (times)	S
4	0	1.28	0.75	19,858	15,664
	1.2	1.23	0.75	26,021	16,945

Note: S is the standard deviation.

From Table 9, after adding 1.2% pure asphalt, the fatigue life of the mixture increases greatly: under the load condition of 0.75 stress ratio, the fatigue life increases from 19,858 times to 26,021 times, increasing by 31%. The main reason is that the toughness and flexural tensile strain work of the mixture increase with the addition of emulsified asphalt, so the mixture has less structural micro cracks under the same stress ratio load. The test result shows that the addition of emulsified asphalt makes semi rigid base materials have better fatigue performance.

5 Overall Evaluation and Strength Modification Method

5.1 Overall Evaluation of the Material

Based on the study on mechanical performance of cement stabilized macadam with emulsified asphalt, the change rules of the related indicators were obtained. Compared to cement stabilized macadam materials, the unconfined compressive strength and splitting strength of the mixture with 1.2% pure asphalt content both decrease by 20%, and the compressive resilient modulus decreases by 13%, the flexural tensile strength decreases by 4%.

From the above analysis, the mechanical strength and compressive resilient modulus of cement stabilized macadam both decrease, while the addition of emulsified asphalt effectively improves the flexibility and anti-shrinkage performance [2–5] of the material. Therefore, the material performance of cement stabilized macadam with emulsified asphalt cannot be fully evaluated by these indicators, such as mechanical strength, compressive resilient modulus and shrinkage performance.

In order to further evaluate the performance of the mixture, this paper, in addition to the above indicators, takes the strain work required by flexural tensile failure of the material to explore its flexural tensile resistance performance, and studies its fatigue performance by simulating the flexural tensile state in the practical use. From the test results, the flexural tensile strain work of the mixture with 1.2% pure asphalt content increases by 18.7%. Therefore, adding the emulsified asphalt effectively improves the toughness of the mixture and the ability of resisting flexural tensile deformation. From the fatigue test results, the fatigue life of the mixture with 1.2% pure asphalt content increases by 31%, which makes the material have better long-term performance.

Combining all these indicators to fully and comprehensively evaluate the cement stabilized macadam with emulsified asphalt, it is found that the overall performance of the material is better than that of cement stabilized macadam.

5.2 Strength Modification Method

According to the regulations of "Technical guidelines for construction of highway road bases" (JTG/T F20-2015) [14], the 7d unconfined compressive strength is one of the most important construction quality control indicators of cement stabilized macadam base, and the control standards of various highway levels and traffic conditions in the specification are different, such as the 7d unconfined compressive strength standard of the base is 4 MPa–6 MPa in the heavy traffic highway. On the other hand, adding emulsified asphalt makes the mechanical strength of cement stabilized macadam decrease, and when the emulsified asphalt content is high, the strength of the mixture may not meet the control requirements of the existing specifications for the 7d compressive strength of cement stabilized macadam base.

Considering the above specification control on the compressive strength mainly aims at the cement stabilized macadam materials, so this paper has proposed a modification method for the compressive strength of emulsified asphalt cement stabilized macadam, and the modified strength values can be used for engineering acceptance and evaluation by the existing specifications. In this paper, the curing period of the specimens used for testing the unconfined compression strength is 28 days, so the mechanical strength and fatigue life of the mixture with curing for 28 days were used to illustrate the modification method.

Combining the flexural tensile strength and fatigue life of the mixture, the modification coefficient F of the unconfined compressive strength was obtained, and its calculation formula is shown as follows:

$$F = \frac{R_{s0}}{R_{s1}} \times \frac{N_1}{N_0} \tag{2}$$

In the formula, R_{s0} and R_{s1} are the 28d flexural tensile strength of the mixture without emulsified asphalt and the mixture with emulsified asphalt, respectively; N_0 and N_1 are the 28d fatigue test results of two types of mixture, respectively.

The modified unconfined compressive strength can be calculated by formula (3):

$$R_c = F \times R_{c1} \tag{3}$$

In the formula, R_c and R_{c1} are the modified and unmodified 28d unconfined compressive strength of cement stabilized macadam with emulsified asphalt, respectively.

Based on the above modification method, the coefficient F can be used to modify the unconfined compressive strength of cement stabilized macadam with emulsified asphalt, and the modified strength can be considered the compressive strength of cement stabilized macadam under the same conditions, such as calculating the modified strength by using testing results of this paper, and the calculation results are shown in Table 10.

Table 10. The modified results of 28d unconfined compressive strength.

$\overline{R_{s0}}$ (MPa)	$\overline{R_{s1}}$ (MPa)	$\overline{N_0}$ (times)	$\overline{N_1}$ (times)	F	$\overline{R_{c1}}$ (MPa)	$\overline{R_c}$ (MPa)
1.28	1.23	19,858	26,021	1.36	4.5	6.12

The modification method proposed in this paper is simplified based on some assumptions: the compressive strength of two types of mixture can be transformed based on the linear relationship by their flexural tensile strength and fatigue life, and this paper does not further explore their specific transformation relationship, therefore the modification method is still imperfect, and the modification idea on the compressive strength can be used for reference for project acceptance engineers.

6 Conclusions

With the indoor static pressure shaping method, this paper has tested the conventional mechanical strength, compressive resilient modulus and fatigue life of cement stabilized macadam mixed with emulsified asphalt by MTS testing machine, and studied their effect rules; and analyzed the flexural tensile resistance performance of the material based on the strain work required by flexural tensile failure, finally fully and comprehensively evaluated the emulsified asphalt cement stabilized macadam material. The following conclusions are obtained:

(1) With 1.2% pure asphalt content, the 28d unconfined compressive strength and splitting strength of cement stabilized macadam both decrease by 20%, compressive resilient modulus decreases by 13%, flexural tensile strength decreases by 4%, flexural tensile strain work increases by 18.7%; and the test results are consistent with the conclusions of existing papers in change rules.

(2) With 1.2% pure asphalt content, the strain work of the mixture required by flexural tensile failure increases by 18.7%, improving its toughness; the flexural tensile fatigue life increases by 31%, improving its durability. The results show that adding emulsified asphalt can effectively improve the ability of resisting the deformation and fatigue performance of cement stabilized macadam.
(3) Based on the flexural tensile strength and fatigue life of the mixture, a modification method about the unconfined compressive strength was proposed, which is suitable for the strength evaluation and engineering acceptance of cement stabilized macadam with emulsified asphalt.
(4) From the overall performance of the material, adding emulsified asphalt reduces the strength and rigidity of cement stabilized macadam, but improves its shortages, such as the easiness to shrinkage cracking, and increases its flexural tensile strain work and fatigue life. On the whole, cement stabilized macadam with emulsified asphalt is better than cement stabilized macadam.

References

1. Sha, Q.L.: Semi-rigid Base Asphalt Pavement of High Grade Highway, 2nd edn. China Communications Press, Beijing (1998)
2. Oruc, S., Celik, F., Vefa Akpinar, M.: Effect of cement on emulsified asphalt mixtures. J. Mater. Eng. Perform. **16**(5), 578–583 (2007)
3. Wang, Y.Q., Tan, Y.Q., Wang, K.S.: Temperature shrinkage characteristics of cement emulsified asphalt stabilized crushed stones. J. Build. Mater. **18**(4), 584–588 (2015)
4. Jia, K.C.: The affect of emulsified asphalt on strength characteristics and mechanical properties of cement stabilized macadam. Highway Eng. **40**(3), 213–217 (2015)
5. Sun, D.G., Wang, B., Zhu, X.S.: Performance and application of cement emulsified asphalt macadam base material. J. Highway Transp. Res. Dev. **24**(8), 27–31 (2007)
6. Zhang, H.H.: Application of emulsified asphalt cement stabilized macadam in highway reconstruction and extension project. Constr. Archit. **11**, 78–80 (2014)
7. Wang, M., Zhang, X., Zhang, Z.C.: Performance effect of cement stabilized macadam base mixed with emulsified asphalt. China Sci. Technol. Inf. (Eng.) **34**, 293–297 (2008)
8. Li, Z.G., Deng, X.Y.: Shrinkage characteristics of cement emulsified asphalt stabilized crushed stones. J. PLA Univ. Sci. Technol. **9**(2), 177–181 (2008)
9. Fu, J., Yang, W.L., Tan, Z.M.: Experimental research on deformation behavior of emulsified asphalt cement-stabilized macadam. Highway **6**, 202–206 (2015)
10. Jiang, L.W.: Emulsified asphalt cement stabilized graded gravel research on performance test. Sichuan Build. Sci. **39**(3), 222–225 (2013)
11. Industry standard of PRC: Test methods of materials stabilized with inorganic binders for highway engineering. JTG E51-2009. China Communication Press, Beijing (2009)
12. Industry standard of PRC: General Portland cement standard. GB 175-2007. Standards Press of China, Beijing (2007)
13. Industry standard of PRC: Technical specification for construction of highway asphalt pavement. JTG F40-2004. China Communication Press, Beijing (2004)
14. Industry standard of PRC: Technical guidelines for construction of highway road bases. JTG/T F20-2015. China Communication Press, Beijing (2015)

Design and Performance Evaluation of Cement Enhanced Cold Patch Asphalt Mixture

Jiawei Yuan, Qiao Dong(✉), Tianjie Zhang, and Xingyu Gu

School of Transportation, Southeast University, Jiangsu 210096, China
qiaodong@seu.edu.cn

Abstract. Cutback asphalt mixture, which is less sensible to temperature, easy to compact, and environmental-friendly, has been widely used as the pothole patch material. Since most pavement potholes are caused by water or winter freeze-thaw, it is vital to improve the water stability and ensure the durability and stability of the pothole patch material. In this study, a new method is proposed to improve the moisture stability of cold patch asphalt mixture by incorporating cement. The performance of cold patch asphalt mixture with different cement content is tested through Marshall test and freeze thaw split test. Test results showed that cutback asphalt mixtures had a relatively weak initial strength and adding cement significantly increased the strength as an effective mineral filler. The TSR of designed cutback cold mix was lower than 80%. Adding cement significantly improved the TSR, indicating an enhanced resistance to freeze-thaw damage. For the 3% and 6% cement content, the TSR were even higher than 100%, clearly indicating the hydration of cement during the freeze-thaw cycle. Further, for longer soaking time, higher cement content showed improved strength, which is also a good indicator of the hydration of cement mixed in the cutback mixes. Findings from this study can be utilized directly to fix the pavement pothole and to improve the water stability of the cold asphalt mixture for practical application.

Keywords: Cold patch asphalt mixture · Cement · Water stability Pavement pothole

1 Introduction

Pothole is one of the main distresses in asphalt pavement, which is a great threaten to traffic safety and pavement durability [1]. It is caused by various factors such as the surface aging, the reflection cracks in the long-term use and the freeze-thaw cycling in winter. As a kind of asphalt mixtures commonly used in practices for pothole repairs [2], cutback asphalt mixtures is less sensible to temperature, which overcome the shortcoming [3] that traditional hot-patch working cannot be carried out in low temperature or raining day. Previous studies found that, compared to the hot mixed patching mixtures, the cold mixes are less easily impacted by environmental temperature, repair locations or other factors. In addition, cutback asphalt mixture have some excellent properties, such as, easy storage and compaction, independent of professional equipment and simple construction. Hence, it is widely used for pothole repair materials.

X. Shi et al. (Eds.): GSIC 2018, *Proceedings of GeoShanghai 2018 International Conference: Transportation Geotechnics and Pavement Engineering*, pp. 463–470, 2018.
https://doi.org/10.1007/978-981-13-0011-0_50

Since the cutback asphalt is configured through a special method, the cold mixes have loose qualities and a good workability. But this also results in the relatively low initial strength, longer maintenance time and lack of moisture stability. Therefore, it is of great importance for cold mix patches to improve its initial strength, moisture stability and freeze-thaw resistance.

There have been practices using lime, cement or other alkaline materials as slag, which are added into the asphalt mixture to improve its water stability. Its mechanism of action is as follows: when the water intrudes into the asphalt-mineral powder-aggregate interface, since the water molecules are polar substances, water is more easily combined with the aggregate, which will make the asphalt be peeled off from the aggregate surface. Asphalt contains a small amount of carboxylic acid. The calcium hydroxide in the alkaline mineral powder contacts with carboxylic acid, which will produce a kind of alkaline earth salt, which can adhere to the aggregate surface firmly. Therefore, polar water molecules cannot make the asphalt mortar be peeled off easily from the aggregate surface. In addition, previous studies found that the initial strength of the cold patching mixes can be greatly improved by adding a certain amount of cement, indicating that this study about adding cement additive into cutback asphalt mixtures has clear merits.

The objective of the study is to investigate the feasibility of improving the initial strength and the resistance to moisture damage of cutback asphalt mixtures by adding Portland cement. The contribution of this study is to investigate if cement hydration could be utilized to help enhance the performance of cutback asphalt mixture.

2 Materials Preparation

2.1 Raw Material

Commercially available Portland cement was used in this study. The aggregate used in this study was basalt. 0# diesel and 70# base asphalt were used as solvent and binder. The properties of the 70# base asphalt are presented in Table 1.

2.2 The Strength Composition of Cold Patch Mixtures

The strength composition of the cold patching mixes is essentially the same as the hot asphalt mixes. The inner friction from the aggregate and the bond or cohesion between the asphalt and the aggregate are the main contributor of strength. The most striking difference between them is that the latter will cool down immediately to form a high strength after paving. While the initial strength of the former is very low due to the role of solvent. When the cutback asphalt mixture is paving on the road, the solvent evaporates, leaving the strength of mixtures increasing gradually.

2.3 Gradation Composition

2.3.1 Mineral Aggregate Gradation

Cold patching mixes is commonly used to repair potholes on asphalt pavement. If the suspend-dense gradation such as traditional dense asphalt concrete (AC) is used, the

Table 1. Properties of the 70# base asphalt

Properties		Unit	Test results	Test methods
Penetration (25 °C 100 g 5 s)		0.1 mm	71	T0604-2011
Ductility (15 °C·5 cm/min)		cm	36	T0605-2011
Ductility (15 °C·5 cm/min)		cm	>100	T0605-2011
Softening point		°C	48.4	T0606-2011
Penetration index (PI)			−0.71	T0604-2011
Flash point		°C	311	T0611-2011
Paraffin content (distillation method)		%	1.4	T0615-2011
Solubility (trichloroethylene)		%	99.98	T0607-2011
Density (15 °C)		g/cm^3	1.035	T0603-2011
The film was heated	Weight loss	%	−0.113	T0609-2011
	Penetration ratio	%	78.1	T0604-2011
	Ductility (15 °C·5 cm/min)	cm	9	T0605-2011
	Ductility (15 °C·5 cm/min)	cm	70	T0605-2011

porosity of the mixture is too low and the performance is dramatically affected by environmental factors such as temperature. In addition, since the particles with large size are separated apart from each other, the strength is not fully exerted. On the other hand, skeleton-gap gradation such as the open graded friction course (OGFC) has too large porosity and relatively poor durability. Hence, a new material for pavement pothole repairs called Easy-Compact Asphalt (ECA) was selected in this study. The curve of ECA is shown as Fig. 1.

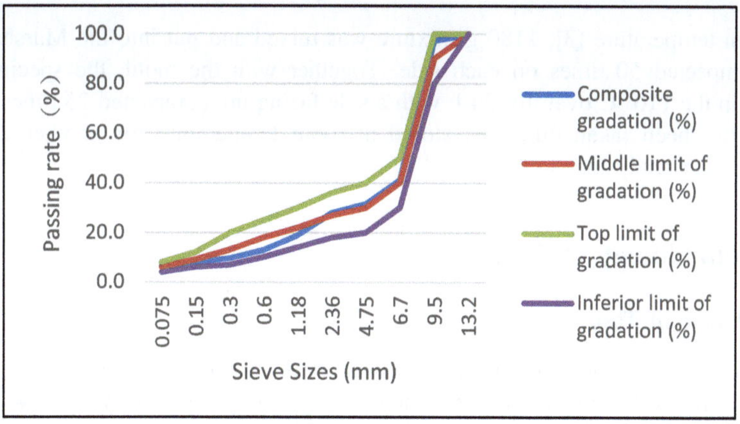

Fig. 1. The curve of ECA

2.3.2 Optimum Content of Cutback Asphalt

Cutback asphalt is composed of asphalt, solvent and other additives, the content of which is usually determined by empirical method. The Eq. (1) was used in this study to calculate the content of asphalt [4].

$$P = 0.021 \times a + 0.056 \times b + 0.099 \times c + 0.12 \times d + 1 \quad (1)$$

Where,

P = the content of cutback asphalt (%);
a = the percentage of particles whose diameter are larger than 2.36 mm (%), which is 68.5% in this study;
b = the percentage of particles whose diameter are between 0.3 mm and 2.36 mm (%), which is 22% in this study;
c = the percentage of particles whose diameter are between 0.075 mm and 0.3 mm (%), which is 3.1% in this study;
d = the percentage of particles whose diameter are between 0 mm and 0.075 mm (%), which is 6.5% in this study.

The content of the asphalt is determined to be 5.0% according to the empirical formula.

2.3.3 Optimum Content of Solvent

The quality of the cold patching mixes is mainly determined by the properties and amount of the solvent. The content of solvent was selected as 20% according to relevant information [1].

2.4 The Making Method and Process of Specimens

At room temperature [3], 1180 g mixture was mixed and put into the Marshall mold and compacted 50 times on each side. Together with the mold, the specimen was placed in the 110 °C oven for 24 h with a side facing up, compacted 25 times on both sides after been taken out, then stored at room temperature for another 24 h and extruded out ready for the testing.

3 Methodology of Tests

3.1 Marshall Test

Marshall test was conducted in this study to evaluate the initial strength of the cold patching mixes. Specimens whose cement content were respectively 0%, 3%, 6% and 9% were prepared and all compacted for a total of 75 times on each side. After 6 h of preserving at room temperature, Marshall test was conducted to measure the Marshall stability.

3.2 Freeze-Thaw Split Test

According to related specifications, freeze-thaw split test was conducted to evaluate the moisture stability of patching materials. Specimens whose cement content were respectively 0%, 3%, 6% and 9% were prepared and divided into two groups. The first group including 4 specimens of each kind of cement content was placed at room temperature for 48 h and then was used for split tensile strength test. The second group was saturated through standard saturation method.

The saturated specimens were wrapped with a layer of plastic bag, which contains 10 mL water, placed in a freezer at −18 °C for 16 h, and then immersed in a 25 °C water bath for 24 h.

The splitting tensile strength and tensile strength ratio (TSR) can be calculated by Eqs. (2) and (3).

$$R_T = 0.006287 \times P_t/h \tag{2}$$

Where,

R_T = splitting tensile strength (MPa);
P_t = maximum load (N);
H = specimen height (mm).

$$TSR = R_{T2}/R_{T1} \times 100 \tag{3}$$

Where,

TSR = tensile strength ratio (%);
R_{T1} = average tensile strength under the freeze-thaw conditions (MPa);
R_{T2} = average tensile strength at 25 °C (MPa).

4 Discussion of Results

4.1 Evaluation of Initial Strength

Figure 2 presents the Marshall test results. It can be seen that adding cement significantly improved the initial strength, while higher cement content reduced the strength. The improvement of the initial strength is mainly because that cement is an effective mineral filler, while excessive filler reduced effective asphalt which reduced the cohesion and thus caused lower strength.

4.2 Evaluation of Moisture Stability

Figure 3 presents the results of freeze-thaw split test. It can be seen that the splitting tensile strength of different cement content underwent different changes after a freezing-thawing cycle. The strength of the specimen with the cement content of 0% was significantly reduced, with TSR less than 80%, due to the damage during the

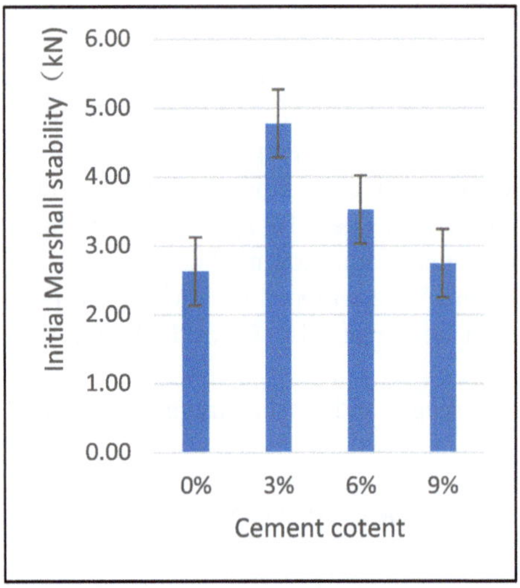

Fig. 2. Marshall test result with different cement content

freeze-thaw cycles. For the 3% and 6% cement content, the strength was even improved after the freeze-thaw with TSR higher than 100%, indicating that the cement hydrated during the freeze-thaw cycle which enhance the strength of the mixture.

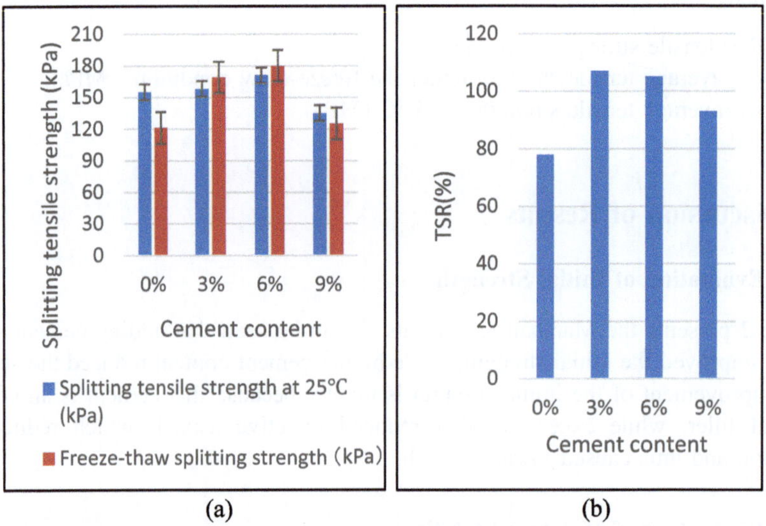

Fig. 3. Result of freeze thaw split test (a) splitting tensile strength with different cement content, (b) tensile strength ratio with different cement content

In addition, a slight decrease appeared for the cement content of 9% with TSR still higher than 90%, indicating the excessive cement is less effective. It was proved that cement plays an important role in improving the resistance to moisture damage.

Figure 4 presents the results of spilt tensile test at 25 °C. It can be seen that the splitting tensile strength changed greatly after adding cement. When the cement content was between 0% and 6%, the strength increased with the increasing of the cement content and reached the maximum at 6%. However, if the cement content exceeded 6%, the strength began to decrease. It is noted that the optimum cement content for initial strength was 3%, which is different from this test result. It is because the split tensile test requires to soak the specimens for 2 h before the splitting test, while the Marshall test only involves 0.5 h soaking. More cement hydrated during the longer soaking time and thus increased the strength. Therefore, test results indicated that cement hydration effectively improved the strength of the cutback asphalt mixes and the optimum cement content is 6%.

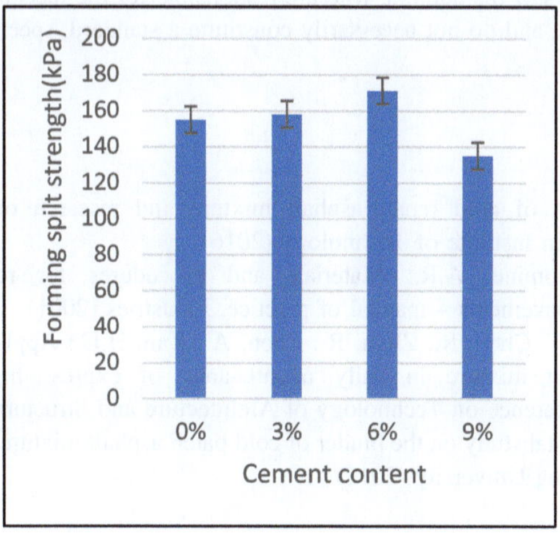

Fig. 4. Splitting test at 25 °C

5 Conclusions

In this study, laboratory tests were carried out to evaluate the initial strength and water stability of cutback asphalt mixtures by adding Portland cement. Based on the test results, the following conclusions can be drawn:

(1) Cutback asphalt mixtures had a relatively weak initial strength and adding cement significantly increased the strength. Cement is an effective mineral filler and the optimum content was 3%.

(2) For the split tensile test, 6% cement content had the highest strength while that for the Marshall test is 3%. It is because the split tensile test requires soaking the specimens for 2 h, while the Marshall test only involves 0.5 h soaking, indicating that additional cement in the cutback mix hydrated during longer soaking time.

Therefore, it is feasible to utilize cement hydration to improve the strength and resistance to moisture damage of cutback mixes.

(3) The TSR of designed cutback cold mix was lower than 80%. However, adding cement significantly improved the TSR, indicating an enhanced resistance to freeze-thaw damage. For the 3% and 6% cement content, the TSR were even higher than 100%, clearly indicating the hydration of cement during the freeze-thaw cycle.

Acknowledgement. This study was supported by the project titled "Effectiveness and Applicability of Asphalt Pavement Maintenance Techniques in Zhejiang" (Project no. 8521002166) founded by Zhejiang Transportation Engineering Construction Group Co., a research project titled "Data mining of Highway Pavement Performance and Evaluation of Maintenance Treatment Effectiveness" (Project no. 3221007409) founded by Southeast University and a research project titled "Highway pavement and subgrade structure and material performance evaluation" (Project no. 7621000131) funded by Jiangsu Department of Transportation. The contents of this paper reflect the views of the authors, who are responsible for the facts and the accuracy of the data presented herein, and do not necessarily constitute a standard, specification or regulation.

References

1. Wang, J.: Design of cold repair asphalt mixture and research on durability. Doctoral dissertation, Harbin Institute of Technology (2016)
2. Wilson, T.P., Romine, A.R.: Materials and procedures for repair of potholes in asphalt-surfaced pavements – manual of practice. Industries (2001)
3. Liu, L., Chang, Q., Chen, R., Zhou, R., Chen, A., Yan, H.U.: Application and analysis of cold-patch asphalt mixture in daily maintenance of express highway. In: Shanghai International Conference on Technology of Architecture and Structure (2009)
4. Li, M.: Experimental study on the binder of cold patch asphalt mixture. Doctoral dissertation, Chongqing Jiaotong University (2012)

Evaluation and Assessment of Moisture Condition of Asphalt Pavement

Lin Cong[1], Tiantong Zhu[1(✉)], Jintang Peng[2], and Nan Ouyang[1,3]

[1] Key Laboratory of Road and Traffic Engineering of the Ministry of Education, Tongji University, Shanghai 201804, China
ztt_2004@163.com

[2] Fujian Communications Planning & Design Institute, Fuzhou 350004, China

[3] Guizhou Transportation Planning Survey & Design Academe Co. Ltd., Guiyang 550001, China

Abstract. This paper studied the hydraulic behavior of asphalt pavements and provides an assessment method for evaluating drainage conditions. The effect of evaporation on the temperature of asphalt pavement surfaces is studied using an infrared thermal method, and the hydrological condition of pavements characterized using optical fiber water level sensors. The test results indicate that with the increase of permeability, the water duration time as in unsaturated status increases. Besides, the evaporation rate is responsible for the differences in temperature of the pavement surfaces. Then, the assessment for evaluating field drainage conditions is established by comparison with the temperature difference of the selected standard model road condition. The assessment also categorizes the drainage conditions into three general zones. This study provides foundations for further research on evaluating the contaminant concentration within pavement structures.

Keywords: Drainage condition · Hydraulic behavior · Infrared thermal test Optical fiber sensor · Permeability

1 Introduction

Rain, especially storm runoff, has been recognized as a source of serious pollution problems in many areas due to direct pollution of receiving water and overloading of treatment facility [1–4]. As a result, contaminants such as chemical waste, automobile exhaust, air-deposited substances contained in urban runoffs were washed into underground drainage systems without enough treatment [4, 5].

Permeable pavements, which are able to hold rain in pavement structures, have been proved as an effective means to reduce the pollutant substances into drainages without treatment [6, 7]. However, this approach also raises other urban water pollution concerns. The rain as well as the contaminants within the pavement structure infiltrates into the subgrade and seeps into underground water systems in the end. Previous studies also found out that more contaminants will dissolve into the water when the duration time of water in the pavement structure increases.

X. Shi et al. (Eds.): GSIC 2018, *Proceedings of GeoShanghai 2018 International Conference: Transportation Geotechnics and Pavement Engineering*, pp. 471–482, 2018.
https://doi.org/10.1007/978-981-13-0011-0_51

In evaluating the impact of pavement structures on the concentration of contaminates within the structure, it is important to investigate the water movements from infiltration to evaporation in pavement structures first.

The objectives of this study mainly focus on two areas.

1. Investigating the hydraulic behavior of pavement structures by characterizing the infiltration and evaporation process of water within the structures.
2. Evaluating field drainage conditions by comparison with the temperature difference of a selected model road condition.

The strategy of this study is illustrated in Fig. 1.

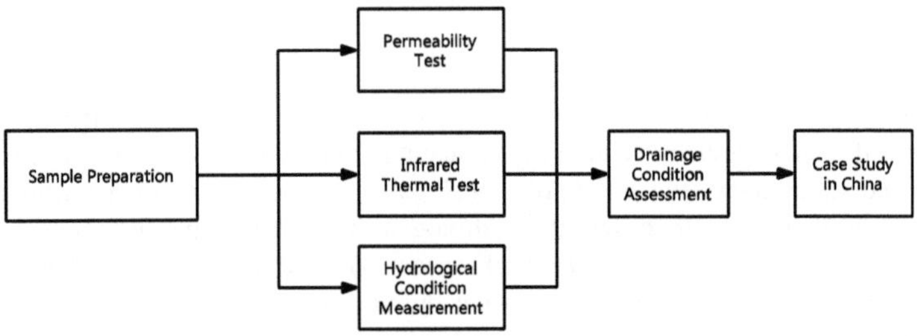

Fig. 1. Flowchart of this study

2 Methodology

The surface water without drainage mainly flows into three directions: infiltration, storage, evaporation [8]. First, at a low speed, some surface water infiltrates into lower layers of asphalt pavement, and the draining speed is dependent on the air voids. Second, some water is stored in the air void of the mixtures. Third, the rest surface water is evaporated.

Evaporation is a phenomenon that liquid gasification merely occurs on its surface. The evaporation rate, K, is used to characterize the velocity of evaporation, measured in kg/m^2s. It is closely related to the liquid temperature, liquid superficial area, liquid types, air velocity, and air humidity. The evaporation rate can be calculated as:

$$K = (k_1 + k_2 v)(P_l - P_a)^{k_3} \tag{1}$$

where P_l is the saturated vapour pressure of water at the surface temperature, P_a is partial pressure of water vapour in air, v is air velocity, and k_1, k_2, k_3 are three coefficients.

When near the surface water molecules have sufficient high kinetic to overcome liquid-phase intermolecular forces, evaporation takes place. Therefore, water evaporates faster under a higher temperature, larger surface or better-ventilated circumstances. The heat of gasification during evaporation process is defined as the evaporation heat rate, expressed as γ_0 and measured in J/kg. It represents the heat absorbed by per unit mass of liquid to convert to steam in the same temperature. As evaporating water escapes, the rest molecules have lower average kinetic energy and the temperature decreases. It was found that, in one standard air pressure, 1 kg water gasification releases 2260 kJ heat, which can increase a temperature of 10 °C from 250 kg asphalt mixture.

The heat transfer formula can be written as follow:

$$Q = \gamma_0 KAt \tag{2}$$

Where, Q is absorbed heat due to evaporation, A is evaporation area, t is time, γ_0 is evaporation heat rate and K is velocity of evaporation.

As described above, the temperature diversity of the samples is mainly caused by the evaporation capacity of water in per unit time. Therefore, the difference in evaporation capacity in each time increment of these samples can be obtained from the inspection of temperatures.

3 Materials and Methods

3.1 Materials

Five different aggregate gradations, Asphalt Concrete (AC), Stone Mastic Asphalt (SMA) and Open Graded Friction Course (OGFC), with air void ranging from 4% to 22%, were produced to investigate the relationship between air void and evaporation behavior.

The used gradations are AC-13, SMA-13, and OGFC-13 according to China Technical Specifications for Construction of Highway Asphalt Pavement JTG F40-2004. The AC-13 was used to produce samples with less than 6% air void. The OGFC-13 were compacted to produce the samples with greater than 14% air void. The air void of SMA-13 is between AC-13 and OGFC-13.

Plates with a dimension of 300 mm × 300 mm × 50 mm were produced according to China T0703 Test Specification for later permeability test. The permeability test was according to China T0971-2008 Test Specification. Three repeated samples of each variant were prepared for obtaining reliable results.

For the evaporation test and the water level measurement, cylinder samples with 100 mm in height, 150 mm in diameter, with a 10 mm-diameter hole in the center was produced. The optical sensors were placed in the hole, and then the hole was glued to prevent water evaporating. Later on, the samples were saturated in the water tank for 48 h before testing.

The aggregates used in this study were limestone. The physical properties of aggregates are shown in Table 1.

3.2 Evaporation Test

This evaporation process reduces the surface temperature and water level of pavement. These differences in surface temperature are initially derived from pavement permeability characteristics, evidenced by using the infrared thermal test method.

Table 1. Physical properties of aggregate

Physical properties	Size (mm)		
	0–3	3–6	5–15
Maximum specific gravity	2.705	2.718	2.719
Dry bulk specific gravity	–	2.558	2.667
Bulk specific gravity	–	2.465	2.636

The infrared thermal test is an effective, fast, comprehensive, and non-destructive method for investigating the surface temperature of an object since it emits infrared radiation across a spectrum of wavelengths [9, 10]. Kordatos et al. used this technology to identify the crack under concrete's surface [9, 11, 12]. In this method, detectors collect infrared radiation to create a thermal image showing the superficial temperature distribution [7, 10]. The sensors of the device collect its radiation and illustrate the temperature changes on the surface of an object [12, 13]. It is also typically used to check the pavement surface temperature and segregation of asphalt mixtures [10, 13].

The bottom sides and round sides of the cylinder samples were sealed with rubber membranes to prevent from evaporating and the top sides of the samples were left open. The sealed sample was directly heated by the iodine tungsten light and surface temperature was measured every 15 min by an infrared thermal device.

At the beginning of the experiments, the room temperature was 18 °C. The iodine tungsten lamp was placed at 500 mm away from the sample surface. For simulating the evaporation characteristics of pavement surface caused by solar radiation heating, the heating power was set at 1 KW. The iodine tungsten lamp was able to heat the surface temperature of the sample to 65 °C in proximately two hours, which was close to pavement temperature in the summertime of Shanghai, China [14]. The surface average temperatures of asphalt samples were recorded every 15 min for a total of two hours. The selected location was the surface center of the samples.

3.3 Hydrological Condition Measurement

The hydrological condition of asphalt pavements was characterized by optical fiber water level sensors. As shown in Fig. 2, the high-precision optical fiber water level sensors were placed to record the water level in asphalt mixtures. The hydraulic head

Fig. 2. Placement of high-precision optical fiber sensor

pressure and the water level were obtained by measuring the wavelength. The influence of the hydraulic head pressure on the wavelength is about 0.1 nm, while the measurement accuracy is 1 pm, which is able to satisfy the experimental precision requirement.

Before the test, the cylinder samples were sealed using rubber membranes to prevent from evaporating and the top side of the sample was left open for evaporation, as shown in Fig. 3. Then the sealed samples were directly irradiated by the iodine tungsten light and water levels were measured during the evaporation process by connecting an optical fiber sensor analyzer.

Fig. 3. Samples before test

4 Test Results and Discussion

4.1 Permeability Test Results

The permeability test results are presented in Fig. 4. It can be noted that when air void is less than 6%, the asphalt pavement is almost impermeable, then with the increase of the air voids, the permeability increases. The relationships between permeability coefficients and air voids of asphalt pavement can be summarized and conducted a regression model as follow.

Typical asphalt pavement structure in China, as shown in Fig. 5, is used to calculate infiltration behavior of water in different gradations and air void pavements. The results are shown in Table 2.

4.2 Evaporation Test Results

The pavement surface temperature during evaporation process is illustrated in Fig. 6. Only negligible changes of sample temperatures were observed in the first 15 min. With the increase of the exposure time, the surface temperatures of samples rise with an increasing rate, and temperature difference begins to occur among samples with various air voids. After being heated for 30 min, the surface temperature reaches about 30 °C, while the temperature difference is approximately 4 °C between samples with 4% and 12% air voids.

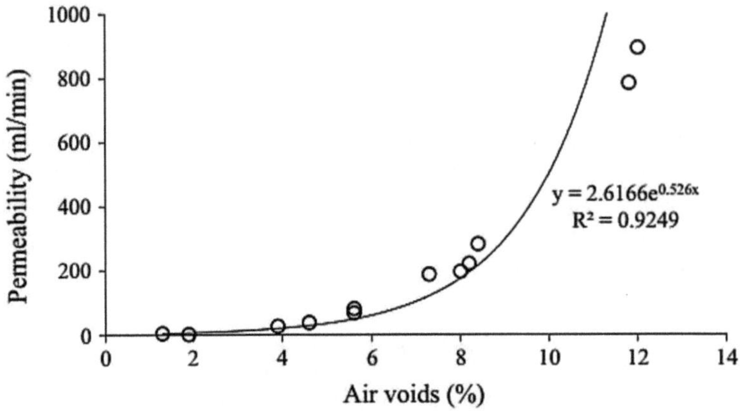

Fig. 4. Relationship between air voids and permeability

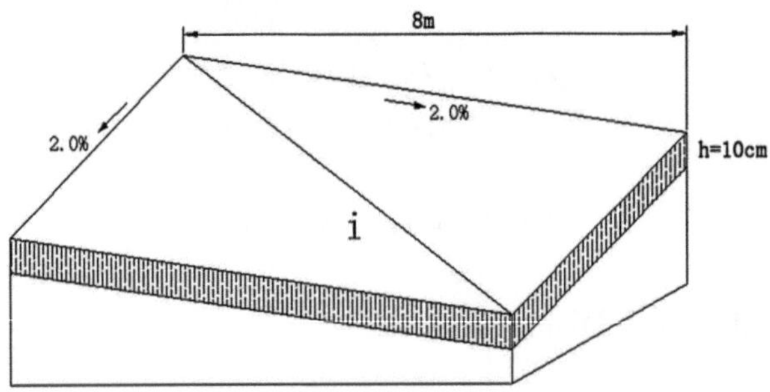

Fig. 5. Typical flexible pavement structure in China

Table 2. Results of infiltration time of water in asphalt pavement

Gradation	Parameters				
	Permeable coefficient (m/h)	Air voids (%)	Length of infiltration path (m)	Infiltration velocity (m/h)	Infiltration time (h)
Dense gradation	0.85	6.5	11.3	0.5	22.78
Semi-open gradation	2.47	8.0	11.3	0.9	12.95
	5.67	10.0	11.3	1.6	7.06
Open gradation	32.56	18.0	11.3	7.9	1.43

When an hour passed, the surface temperatures of all samples were higher than 40 °C. Meanwhile, the evaporation rate of samples increases significantly and cause a larger temperature difference (up to 5 °C). After 1.5 h, the evaporation process of low air void sample finishes, but there is still moisture on the surface of high air void sample.

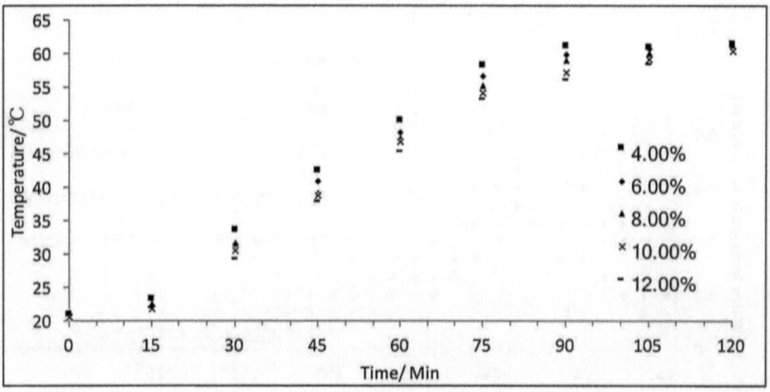

Fig. 6. Pavement surface temperature changes during evaporation process

At the same time, the temperature difference peaks at 6 °C. After 2 h heating, all samples were dry, reached and maintained their maximum surface temperature, around 60 °C.

4.3 Results of Hydrological Condition Measurements

The relationships between water levels and evaporating times are presented in Fig. 7. It can be observed that the water levels of the samples gradually decreased from 100 mm to 0 mm. It took 270 min and 450 min for voids samples respectively with 8% and 22% air to dry out. As the air voids increased, more water was held in the structure. Less water retained in a lower air void content samples and mainly existed as capillary water, compared with a higher air void sample. As a consequence, water evaporated constantly by the capillary action and water level dropped as temperature increases.

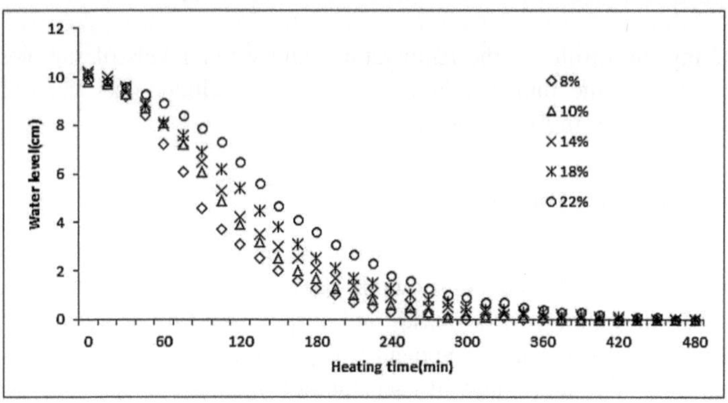

Fig. 7. Water level changes versus heating time

The relationships between water level gradients and evaporating times are presented in Fig. 8, and trend lines for the results of each air voids samples are added. As shown in Fig. 8, the evaporation rates first increased, then decreased gradually.

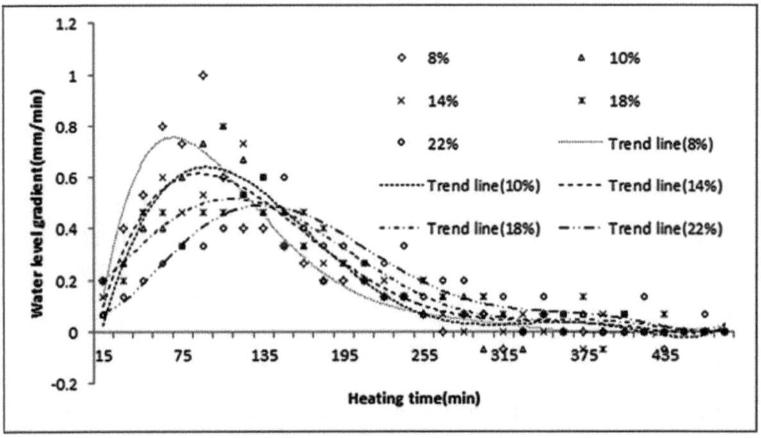

Fig. 8. Water level gradient changes versus heating time

Samples with lower air voids reached the peak evaporation rate in a shorter time and the peak rates were higher than samples with higher air voids.

In the beginning, the decline rate of water level increased with the increase of temperature due to the radiation of the iodine tungsten lamp. As a result of the evaporation process, the water level reduced and more power was needed to keep water evaporating. During evaporation in porous media, the internal suction increases as water contained in the pores decreasing [15], which makes water evaporation increasingly difficult. The water level in low air voids samples decreased quickly and then achieved the threshold below which the pore suction overcame the forces powering the upward flow of water [16]. More water was held in the samples with higher air voids, thus the water level decreased more slowly compared to the lower air voids samples.

Combining the results of the temperatures and water levels of samples under the radiation of the iodine tungsten lamp, it can be concluded that the differences in temperature results from the evaporation rates.

4.4 Retained Water Duration

Based on the previous discussion, infiltration and evaporation time of pavements with different air voids are obtained and shown in Table 3. It can be noted that when air void increases, the infiltration time decreases and evaporation time increases.

The durations of water in a pavement was predicted based on infiltration time and evaporation time. The rainy days of each rainfall intensity in Shanghai are shown in Table 4 [14]. Then, the duration time of surface water in OGFC mixtures without drainage facilities in Shanghai was calculated and shown in Table 5. As Table 5 presents, water retained in the pavement structures for about four months of a year, which is long enough for weakening the mixtures performance and dissolve much contaminant.

Table 3. Infiltration and evaporation time of pavement with different air voids

Gradation	Permeable coefficient (m/h)	Air voids (%)	Infiltration time (h)	Evaporation time (h)
Dense graded	0.85	6.5	22.78	5.50
Semi-open graded	2.47	8.0	12.95	5.75
	5.67	10.0	7.06	6.00
Open graded	32.56	18.0	1.43	7.25

Table 4. Rain days of different rainfall intensity in Shanghai

Rainfall intensity	Drizzle	Moderate	Heavy	Rainstorm	Total
Rain days (d)	94	24	9	3	130

Table 5. Duration of different moisture existing status

Air voids (%)	Permeable coefficient (mm/h)	Infiltration time (h)	Evaporation time (h)	Saturated days (h)	Unsaturated days (h)	Humidity balance days (h)
18.0	32.56	1.43	7.25	38.8	196.9	129.3

5 A Case Study in China

A recently constructed road with semi-rigid base in Shanghai, China was selected as the typical road for standardization of drainage capacity. The standard test was chosen on a sunny day, and the environment temperature was 38 °C and the wind velocity of was 3 m/s. Firstly, to measure the permeability, water was slowly poured on this pavement until it stopped infiltrating. Then, the infrared thermal test was conducted to record the temperature, as shown in Fig. 9.

Fig. 9. Pavement surface temperature inspection used infrared thermal device

The wavelength was recorded during the test. Based on the relationship between hydraulic head pressure and wavelength, the water level was calculated and presented in Fig. 10. The water level increased sharply when first infiltrated into the asphalt pavement and then decreased gradually during the evaporation and drainage process.

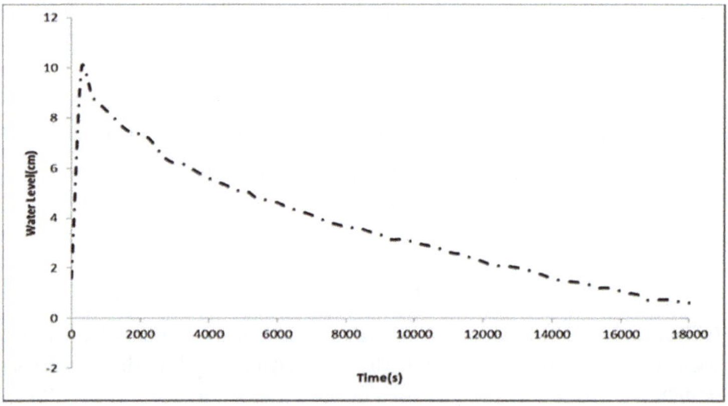

Fig. 10. Water level curve of the field test

In addition, without reasonable drainage facilities in the pavement structure, the evaporation process would last for a long time especially at low temperature without sunlight. As shown in Fig. 11, the temperature of pavement decreased after watering.

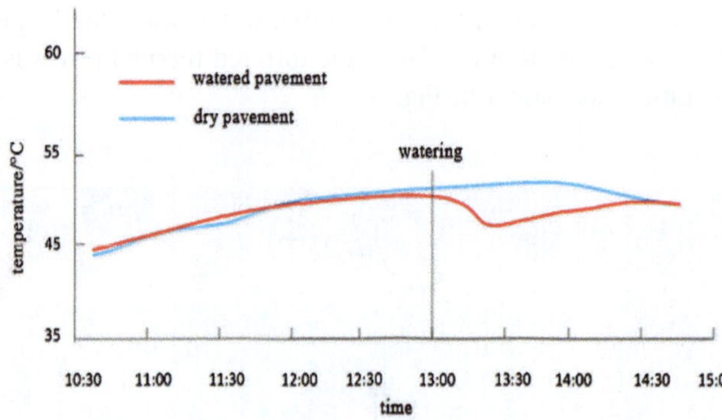

Fig. 11. Temperature of watered and dry pavement

Three sections of an expressway in Shanghai were chosen as the drainage samples. Based on the test results in both laboratory and field, the categorized results of the drainage conditions are presented in Fig. 12. As shown in Fig. 12, it can be observed that, the pavements in Zones I and II have good moisture resistances while other

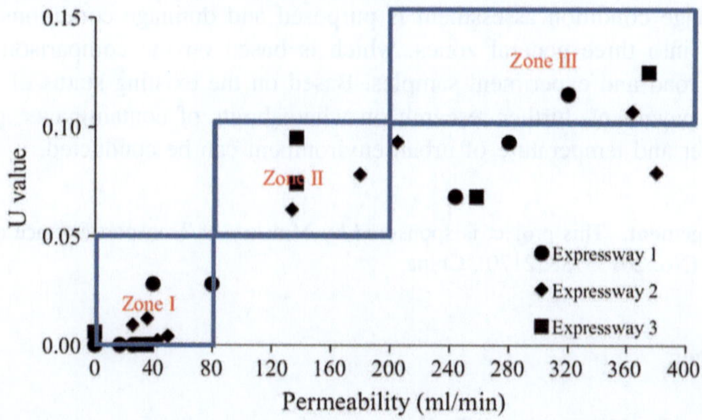

Fig. 12. Temperature differences of three expressways in terms of drainage

pavements in Zone III exhibit high permeability coefficients and air voids. The pavements below Zone III indicate good moisture resistances while keeping high permeability.

Meanwhile, based on the test results in the laboratory and field, a simple evaluation parameter can be summarized in Table 6. When the temperature difference (U value) between inspected pavement and standard pavement is less than 5%, it can be categorized as level one. If U values are in range of 5%–10%, the pavement conditions were categorized as level two. When U values are greater than 10%, it is categorized as level three. This definition is very simple and easy to use for the engineer and technician to identify the pavement condition.

Table 6. Assessment of pavement drainage

Pavement permeability rank	Difference in surface temperature (U)
I	$U < 5\%$
II	$5\% < U < 10\%$
III	$U > 10\%$

6 Conclusion

Based on the results from laboratory tests and field investigations, the following summaries and findings can be reached:

1. The infiltration and evaporation behavior of pavements with different air void are characterized. The water in pavement with high air void infiltrated faster than pavements with low air void. With the increase of permeability, the time for water existing as unsaturated status increases.
2. Based on the results of the temperatures and water levels of samples under the radiation of the iodine tungsten lamp, it can be concluded that the differences of pavement surface temperature were caused by the different evaporation rates.

3. A drainage condition assessment is purposed and drainage conditions were categorized into three general zones, which is based on the comparison between a control road and experiment samples. Based on the existing status of moisture in asphalt pavement, further research on whereabouts of contaminants, pollution of rainwater and temperature of urban environment can be conducted.

Acknowledgement. This project is sponsored by Ministry of Transport Science and Technology Project (No. 2015318822170), China.

References

1. Deletic, A.B., Maksimovic, C.T.: Evaluation of water quality factors in storm runoff from paved areas. J. Environ. Eng. **124**(9), 869–879 (1998)
2. Lee, J.H., et al.: First flush analysis of urban storm runoff. Sci. Total Environ. **293**(1–3), 163 (2002)
3. Maniquiz-Redillas, M.C., Kim, L.H.: Evaluation of the capability of low-impact development practices for the removal of heavy metal from urban stormwater runoff. Environ. Technol. **37**(18), 2265 (2016)
4. Nikolaeva, O., Rozanova, M., Karpukhin, M.: Distribution of traffic-related contaminants in urban topsoils across a highway in Moscow. J. Soils Sed. **17**(4), 1045–1053 (2017)
5. Shorshani, M.F., et al.: Road traffic impact on urban water quality: a step towards integrated traffic, air and stormwater modelling. Environ. Sci. Pollut. Res. **21**(8), 5297–5310 (2014)
6. Jiang, W., et al.: Experimental study on filtration effect and mechanism of pavement runoff in permeable asphalt pavement. Constr. Build. Mater. **100**, 102–110 (2015)
7. Khankhaje, E., et al.: Sustainable clean pervious concrete pavement production incorporating palm oil fuel ash as cement replacement. J. Clean. Prod. **172**, 1476–1485 (2017)
8. Cooley, L.A., Brown, E.R., Maghsoodloo, S.: Developing critical field permeability and pavement density values for coarse-graded superpave pavements. Transp. Res. Rec. J Transp. Res. Board **1761**(1), 41–49 (2001)
9. Barreira, E., de Freitas, V.P.: Evaluation of building materials using infrared thermography. Constr. Build. Mater. **21**(1), 218–224 (2007)
10. Clark, M.R., Mccann, D.M., Forde, M.C.: Application of infrared thermography to the non-destructive testing of concrete and masonry bridges. NDT E Int. **36**(4), 265–275 (2003)
11. Aggelis, D.G., et al.: Combined use of thermography and ultrasound for the characterization of subsurface cracks in concrete. Constr. Build. Mater. **24**(10), 1888–1897 (2010)
12. Li, Z., et al.: Application of infrared thermography technique in building finish evaluation. J. Nondestr. Eval. **19**(1), 11–19 (2000)
13. Sebesta, S., Scullion, T.: Application of infrared imaging and ground-penetrating radar to detect segregation in hot-mix asphalt overlays. Transp. Res. Rec. J. Transp. Res. Board **1861** (1), 37–43 (2003)
14. Fangfang, H.E., Jialiang, X.U.: Studies on precipitation resource change in Shanghai since the 1990's. J. Nat. Resour. **21**(2), 210–216 (2006)
15. García, A., Castro-Fresno, D., Polanco, J.A.: Evolution of penetration resistance in fresh concrete. Cem. Concr. Res. **38**(5), 649–659 (2008)
16. Shokri, N., Or, D.: What determines drying rates at the onset of diffusion controlled stage-2 evaporation from porous media? Water Resour. Res. **47**(9), 1900–1904 (2011)

Thermal Cracking and Fatigue Analysis of Recycled Asphalt Mixture Using DCT Test and S-VECD Model

Yuefeng Zhu[1(✉)], Xingju Wang[1], Yaning Qiao[2], and Jiang Shu[1]

[1] Shijiazhuang Tiedao University, Shijiazhuang, China
yuefengzhu@stdu.edu.cn
[2] University of New Hampshire, Durham, USA

Abstract. In order to analyze the impact of reclaimed asphalt pavement (RAP) on cracking property of hot mix asphalt (HMA), different RAP contents (10%, 20% and 30%) and gradation (12.5 mm and 19 mm NMAS) of HMA were designed. This study evaluated thermal and fatigue cracking performances for six asphalt mixtures using two testing methods: Disk-Shaped Compact Tensile (DCT) test and Simplified Viscoelastic Continuum Damage (S-VECD) approach using uniaxial tensile fatigue test. Firstly, the DCT fracture test was conducted, and the toughness index (*TI*) was used to evaluate the low temperature cracking ability. Secondly, the fatigue test was conducted on different asphalt mixtures, and using S-VECD model to determine the Damage Characteristic Curve (*C-S*) and Energy-based failure criterion with experimental cycles to failure (G^R-N_f). Two tests show that with RAP content increase, the low temperature and fatigue cracking property of asphalt mixtures decrease, which means that RAP content should be controlled in the future application.

Keywords: Reclaimed asphalt pavement (RAP) · Cracking property
Disk-Shaped Compact Tensile (DCT) test · Toughness index (*TI*)
Fatigue test

1 Introduction

Currently, lots of money is spent on repair and maintenance of asphalt pavements all over the world. Cracking, includes thermal cracking and fatigue cracking, becomes the most common issues in asphalt pavement. With the increase of cracking, water penetration increase from surface to underlying soil layers, and also affects ride quality. Using recycled materials in asphalt mixtures become more and more popular which can save a lot of money. However, with reclaimed asphalt pavement (RAP) content increase, asphalt mixtures become stiffer which are less workable and more susceptible to cracking (Al-Qadi et al. 2015). This issue is the same with when virgin mixtures age in asphalt pavement which makes the cracking happen in the field. Vargas-Nordcbeck and Timm (2013) mentioned that using high RAP contents may cause the blending issue between the residual and virgin asphalt binders which can affect the performance of the mixtures. Some researchers applied warm-mix asphalt (WMA) technologies for

X. Shi et al. (Eds.): GSIC 2018, *Proceedings of GeoShanghai 2018 International Conference: Transportation Geotechnics and Pavement Engineering*, pp. 483–492, 2018.
https://doi.org/10.1007/978-981-13-0011-0_52

using RAP, the results showed that WMA with high percentage of RAP exhibited better performances than HMA (Xie et al. 2017).

Many researchers have been working on the cracking resistance of asphalt mixture using both numerical modeling and testing. The Disk-Shaped Compact Tension (DCT) test and Simplified Viscoelastic Continuum Damage (S-VECD) approach using uniaxial tensile fatigue test are two lab test methods that have drawn a lot of attention in this ten years for assessment of thermal cracking and fatigue, respectively. The DCT test method was first proposed by Wagoner et al. (2005) for asphalt mixture. The DCT test method has been specified as ASTM D7313 (2007) and was extensively assessed by the other researchers (Marasteanu et al. 2007; Dave et al. 2016). However, some studies found that only using fracture energy for evaluating low temperature cracking of asphalt mixture was inadequate. Toughness index was proposed by Pérez-Jiménez et al. (2013) for assessing the thermal cracking of asphalt mixture, which achieved good results. In addition, the fatigue crack resistance of asphalt mixture has always been a typical and difficult problem for assessing the fatigue performance of asphalt mixture. The S-VECD approach is based on three principles: elastic-viscoelastic correspondence principle, continuum damage mechanics, and time-temperature superposition with growing damage (Cao et al. 2016). However, this method has not been widely used.

The objective of this study is to evaluate thermal and fatigue cracking performance of asphalt mixture representing different RAP contents and aggregate gradation using DCT test and uniaxial tensile fatigue test, respectively. The research results of this paper can provide a more accurate method for evaluating the cracking resistance of asphalt mixtures, and obtain some suggestions on the proper content of RAP.

2 Materials and Testing Method

2.1 Materials

This study includes testing on 6 different asphalt mixtures that were produced in the lab. The virgin binder used in this study is PG 64-22. The virgin aggregate used in this study originated from Lebanon, NH. The aggregate included natural sand, FM 02, limestone CM 16, FM 20 and mineral filler. The surface mixture from State Route 12 near Westmoreland was the RAP source in this study, and the performance grade of RAP binder is 81.6–19.4 °C.

2.2 Mixture Designs

Superpave mixture design was conducted for the hot mix asphalt (HMA) mixtures at the three RAP levels (10, 20, and 30%) and two nominal maximum size of aggregate (NMAS) (12.5 and 19 mm). The six gradations used in this study are provided in Table 1.

Mixture prepared met the Superpave requirements for both 12.5 mm and 19 mm NMAS. Additionally, the design number of gyrations was set to 70 for meeting the 3 to 10 million Equivalent Single Axle Loads. Compaction and mixing temperatures follow

Table 1. Aggregate gradation of different asphalt mixtures

Sieve size /mm	12.5 mm 10% RAP	12.5 mm 20% RAP	12.5 mm 30% RAP	19 mm 10% RAP	19 mm 20% RAP	19 mm 30% RAP
	Passing rate/%					
25	100	100	100	100	100	100
19	100	100	100	99	99	98
12.5	98.6	98.3	98.9	83.2	83.3	83.5
9.5	86.4	86.8	86.1	70	70.5	69.8
4.75	58.9	59.2	60	47.2	46.3	46.2
2.36	41.8	41.5	42.3	32.4	32	31.8
1.18	30.3	30.9	31.5	23.5	23.3	23.6
0.6	21.1	21.2	22.3	16.2	16	16.1
0.3	11.6	11.6	12.3	9	9.3	9.2
0.15	6.0	5.9	6.2	4.7	5.2	5.1
0.075	3.8	3.8	4.2	3.1	3.3	3.1

the Asphalt Institute method for asphalt mixture design. Also, the cylindrical molds and plates should be heated to the compaction temperature. Table 2 shows mix design volumetric information. It should be noted that the mixtures were prepared using 10, 20, and 30% RAP by weight of mixture.

Table 2. Mixtures types and volumetric properties

NMAS/mm	Total binder designed/%	RAP content/%	VMA/ %	VFA/ %	Design Air void%
12.5	5.2	10	15.2	75.3	4.0
	5.2	20	14.9	74.1	4.0
	5.2	30	14.8	73.2	4.0
19	5.0	10	15.8	72.3	4.0
	5.0	20	15.3	71.6	4.0
	5.0	30	15.5	74.3	4.0

2.3 Mixture Testing

DCT Test

At present, disk-shaped compact tension (DCT) fracture test is specified as ASTM D7313 (2007) for low temperature fracture characterization of asphalt mixtures. The mixture specimens were fabricated to be 50 mm thick and 150 mm in diameter. Bore holes 25 mm in diameter were fabricated into which the loading rods were inserted. In the test procedure a notched disk shaped specimen is initially pre-loaded with 0.1 kN of seating load, thereafter the test control is switched over to achieve a constant crack mouth opening displacement (CMOD) rate of 0.167 mm/s. Based on the ASTM D7313, the DCT test temperature selected for our study was −12 °C (−22 + 10 °C).

Specimens were conditioned prior to testing in refrigerated chambers. Temperature was verified by temperature gauges, and specimens were tested within 1 °C of the desired temperature. Note: each mixture at least had three duplicate specimens to exclude the effect of accidental error.

(1) Fracture energy

Fracture energy (G_F) is an engineering property that can be determined from the DCT testing. The fracture work is defined as the area under the load versus crack mouth opening displacement (CMOD) curve. Fracture energy is determined by normalizing the fracture work for specimen thickness and ligament length. The fracture energy is considered to be the amount of energy required to develop a unit surface fracture of the asphalt mixture. The fracture energy can be calculated by Eq. (1).

$$G_F = \frac{W_D}{h \cdot l} \tag{1}$$

where G_F is the fracture energy, J/m^2; W_D is the work of fracture calculated by finding the area under the load-displacement curve, kN·mm; h is the specimen thickness, m; l is the initial ligament length, m.

$$W_D = \int_0^{\Delta R} F du \tag{2}$$

where F is the load, kN; u is the displacement, mm; ΔR is the displacement at F is the 0.1 kN post-peak load, mm.

(2) Toughness index

Fracture energy is calculated as the area under load-CMOD curve. Accordingly, for different mixtures with similar fracture energies, this parameter may not capture the ductility of mixture in the post peak region if the mix exhibits a very high peak load. For better understanding of asphalt mixtures' behavior at low temperature, Also with the intent to capture the post-peak fracture behavior of asphalt mixtures, Perez-Jimenez et al. (2013) proposed a normalized fracture energy parameter called toughness index (*TI*). Instead of using slope as normalizing parameter, *TI* is calculated by multiplying post-peak fracture energy (G_F Post-peak) with displacement between peak load and 50% peak load. The choice of 50% peak load to select displacements is arbitrary in this index. Using the CMOD displacement thresholds for DCT test as shown in Fig. 1, *TI* can be calculated as shown in Eq. 3.

$$TI = (G_F - G_{F\max}) \cdot (\Delta_{mdp} - \Delta_{F\max}) \cdot 10^{-3} \tag{3}$$

where *TI* is the toughness index, J/m; $G_{F\max}$ the fracture energy until $\Delta_{F\max}$, J/m^2; Δ_{mdp} the displacement at 50% of post-peak load, mm; $\Delta_{F\max}$ is the displacement at maximum load, mm.

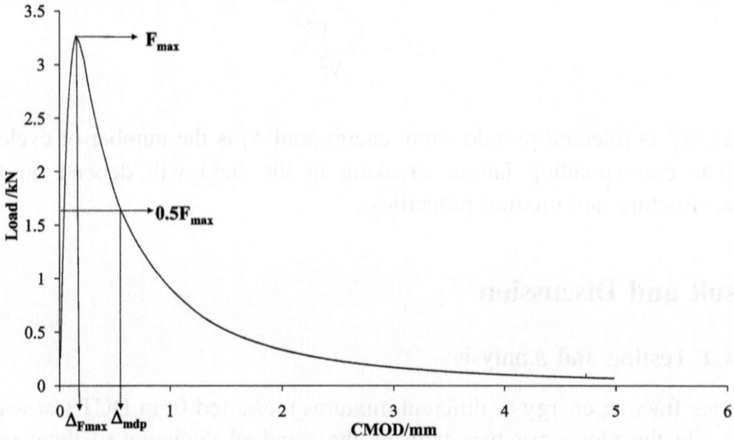

Fig. 1. Load-CMOD curve of DCT test

S-VECD Fatigue Testing

To evaluate the fatigue behavior of asphalt mixtures, uniaxial fatigue testing was conducted on different mixtures following AASHTO TP-107 (2014) procedure. The simplified viscoelastic continuum damage (S-VECD) approach to cracking, developed by Hou et al. (2010), models the constitutive response of asphalt concrete over cyclic loading in uniaxial tensile mode.

Fatigue testing was performed in uniaxial tension on the AMPT. Specimens were cut to dimensions of 100 mm in diameter by 130 mm tall and glued to end platens that were fixed in the AMPT. Testing was performed at 20.0 °C and 10 Hz. The mixtures were tested with three or four replicate specimens at varying microstrain levels ranging from 300 to 450 microstrain to cover a range of numbers of cycles to failure. More details of the test method can be found in AASHTO TP 107.

All cyclic tests were performed at a minimum of three different amplitudes to cover a range of numbers of cycles to failure (N_f). Once the fatigue tests are conducted, the damage characteristic curves are developed by calculating the secant pseudo-stiffness (C) and the damage parameter (S) at each cycle of loading. These values are cross-plotted to form the damage characteristic curve. For all the mixtures, the exponential form shown in Eq. 4 was used to fit the C versus S characteristic curves.

$$C = e^{aS^b} \tag{4}$$

where a, b is the fitting coefficients; C is the secant pseudo-stiffness; S is damage parameter.

And also, an energy-based failure criterion curve of G^R versus number of cycles to failure (N_f) are developed in this approach. G^R (Eq. 5) is defined as the rate of change of averaged released pseudo strain energy throughout the test. Sabouri and Kim (2014) developed G^R and showed it is strongly correlated with the number of cycles to failure (N_f).

$$G^R = \frac{\int_0^{N_f} W_C^R}{N_f^2} \tag{5}$$

Where W_C^R is released pseudo strain energy and N_f is the number of cycles before failure. The corresponding fatigue cracking in the field will depend on both the pavement structure and mixture properties.

3 Result and Discussion

3.1 DCT Testing and Analysis

The average fracture energy of different mixtures measured from DCT test was shown in Fig. 2. On the plot, error bars indicate the standard deviation of three replicates. Overall, the coefficient of variation varied from 5 to 15% throughout the study which is acceptable range for DCT tests. In general, all mixtures show high fracture energy. Another result can be seen from Fig. 2 is that 10% RAP content mixtures did not show evidently higher fracture energy compared with 30% RAP content. Similarly, two kinds of asphalt mixtures with different gradation also did not exhibit different fracture energies.

The result was the evidence that fracture energy may be unsuitable for evaluating the low temperature cracking ability of asphalt mixture. The reason can be explained as fracture energy may not capture the ductility of mixture in the post peak region if the mix exhibits a very high peak load (Zhu et al. 2017).

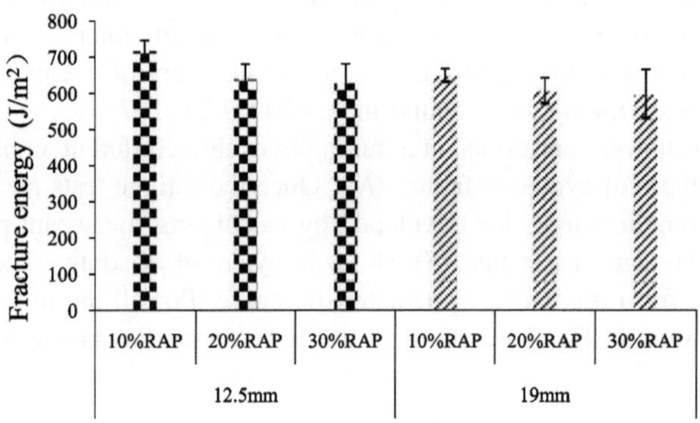

Fig. 2. Fracture energy (DCT test) for different mixture

Another parameter to evaluate mixtures behavior against thermal cracking is Toughness Index (*TI*) proposed by Perez-Jimenez et al. (2013). This parameter could better describe the whole fracture process during the DCT test. The results of toughness index for different mixtures are shown in Fig. 3. Overall, the 19 mm 30% RAP mixture shows the worst performance and the 12.5 mm 10% RAP mixtures the best. For 19 mm asphalt mixtures, when the RAP content increased from 10% to 30%, the

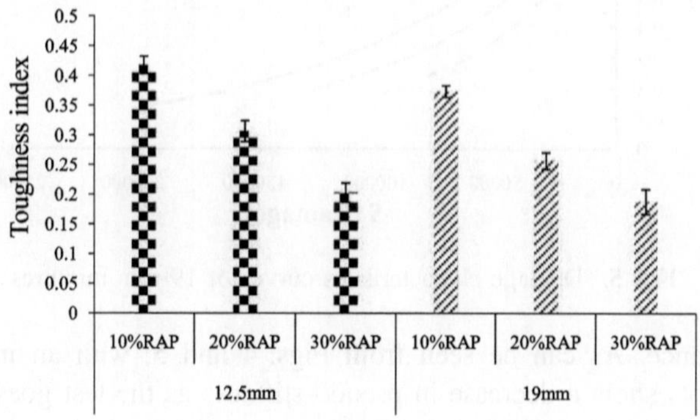

Fig. 3. Toughness index (DCT test) for different mixture

toughness reduced by half. The same situation can be found from 12.5 mm asphalt mixtures. The reason is that adding RAP in virgin mixtures would affect the relaxation ability and ductility which could lead to more prone to cracking at low temperature.

3.2 S-VECD Fatigue Cracking

The results from the uniaxial fatigue testing and analysis on different mixtures are shown in Figs. 4, 5, 6 and 7. Figures 4 and 5 show the average damage characteristic curves (DCC) of the replicate specimens. In the DCC curve, *S* represents damage along with the test progresses while *C* represents normalized pseudo-stiffness. Generally, DCC curve shows how the material integrity of the specimen decreases as damage is growing. Typically, the mixtures which have DCC curves further up and right will have

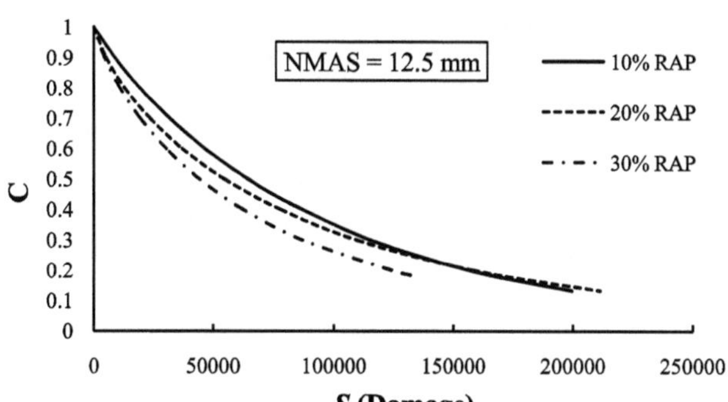

Fig. 4. Damage characteristics curves of 12.5 mm mixtures

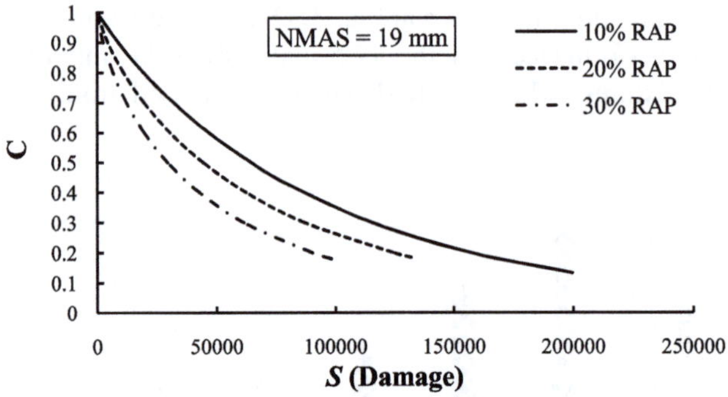

Fig. 5. Damage characteristics curves of 19 mm mixtures

better performance. As can be seen from Figs. 4 and 5, with an increase in RAP content, the plots show a decrease in pseudo-stiffness as the test goes on. This especially indicates that with the increase of RAP content, the cracking resistance of asphalt mixture decrease.

Figures 6 and 7 show the number of cycles (N_f) versus fatigue failure criterion (G^R) of different asphalt mixtures. In general, the higher G^R values at the same number of cycles (N_f) indicates better fatigue performance. As can be seen from two plots, G^R and N_f have a very strong relationship (high coefficient of determination, R^2). The 10% RAP content mixtures have better fatigue performance than the 30% RAP content mixtures, which means the lower RAP content mixtures show better fatigue life. The reason also can be explained as adding more RAP in asphalt mixture might result in a more brittle mixture.

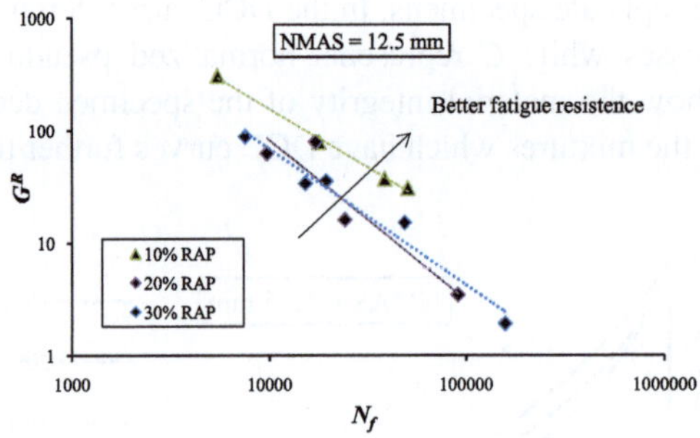

Fig. 6. G^R-N_f for 12.5 mm mixtures

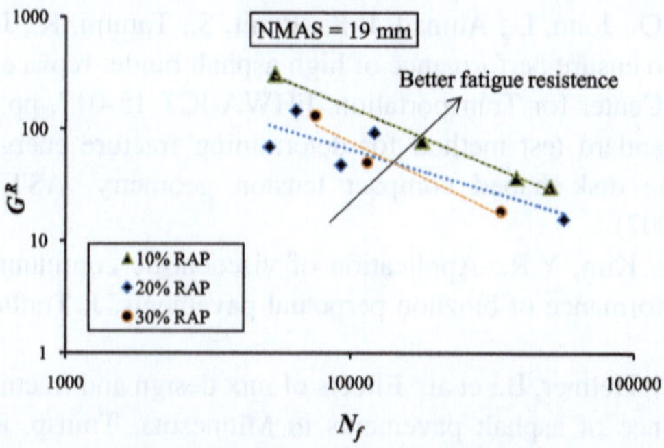

Fig. 7. G^R-N_f for 19 mm mixtures

4 Summary

The objective of this study was to evaluate fatigue and thermal cracking performance of different RAP content mixtures using compaction-tension fatigue test and DCT test, respectively. The following conclusions can be drawn on basis of the results and discussions presented in this paper:

(1) Fracture energy and toughness index were calculated from DCT test for evaluating the cracking property of mixtures. The fracture energy was found to be difficult to distinguish the fracture resistance of different asphalt mixtures, while the toughness index appeared to be a useful parameter for characterizing the low temperature property of asphalt mixtures. Mixtures with lower RAP contents and 12.5 mm NMAS had better fracture resistance.

(2) S-VECD fatigue testing was performed for six asphalt mixtures. An increase in RAP content resulted in an increase in pseudo-stiffness using the damage characteristic curves, which showed that the 30% RAP content mixtures were much more susceptible to fatigue cracking than the 10% RAP content mixtures. G^R-N_f plots showed little distinction between RAP contents, but identified 30% RAP contents as the most susceptible to fatigue cracking.

(3) Future research is planned to include more mixtures to achieve a larger result base and also study the effect of aging on cracking performance as well. At the same time, the cracking performance of actual field mixture is being investigated to be compared with the results of lab produced mixture. Also, mixtures without RAP are not included, which should be considered in the future study.

References

AASHTO TP 107-14: Standard method of test for determining the damage characteristic curve of asphalt mixtures from direct tension cyclic fatigue tests. American Association of State and Highway Transportation Officials, Washington, D.C. (2014)

Al-Qadi, I.L., Hasan, O., John, L., Ahmad, E.K., Punit, S., Tamim, K., José, R., Berangere, D.: Testing protocols to ensure performance of high asphalt binder replacement mixes using RAP and RAS. Illinois Center for Transportation. FHWA-ICT-15-017, pp. 60–69 (2015)

ASTM D7313-07: Standard test method for determining fracture energy of asphalt-aggregate mixtures using the disk-shaped compact tension geometry. ASTM International, West Conshohocken (2007)

Cao, W., Norouzi, A., Kim, Y.R.: Application of viscoelastic continuum damage approach to predict fatigue performance of binzhou perpetual pavements. J. Traffic Transp. Eng. **3**, 104–115 (2016)

Dave, E.V., Hoplin, C., Helmer, B., et al.: Effects of mix design and fracture energy on transverse cracking performance of asphalt pavements in Minnesota. Transp. Res. Rec. **2576**, 40–50 (2016)

Hou, T., Underwood, B.S., Kim, Y.R.: Fatigue performance prediction of north carolina mixtures using simplified viscoelastic continuum damage model. J. Assoc. Asphalt Paving Technol. **79** (2010), 35–80 (2010)

Marasteanu, M., Zofka, A., Turos, M., et al.: Investigation of low temperature cracking in asphalt pavements. Report no. 776, Minnesota Department of Transportation, St. Paul (2007)

Pérez-Jiménez, F., Ramón, B., Adriana, H.M., Rodrigo, M.: Analysis of the mechanical behaviour of bituminous mixtures at low temperatures. Constr. Build. Mater. **46**, 193–202 (2013)

Sabouri, M., Kim, Y.R.: Development of failure criterion for asphalt mixtures under different modes of fatigue loading. Transp. Res. Rec.: J. Transp. Res. Board **2447**, 117–125 (2014)

Vargas-Nordcbeck, A., Timm, D.: Physical and structural characterization of sustainable asphalt pavement sections at the NCAT test track. National Center for Asphalt Technology (NCAT) Report 13-02 (2013)

Wagoner, M.P., Buttlar, W.G., Paulino, G.H.: Disk-shaped compact tension test for asphalt concrete fracture. Exp. Mech. **45**, 270–277 (2005)

Xie, Z., Tran, N., Taylor, A., et al.: Evaluation of foamed warm mix asphalt with reclaimed asphalt pavement: field and laboratory experiments. Road Mater. Pavement Des. (2017). https://doi.org/10.1080/14680629.2017.1389077

Zhu, Y., Dave, E., Reyhaneh, R., Daniel, J., Zofka, A.: Comprehensive evaluation of low-temperature fracture indices for asphalt mixtures. Road Mater. Pavement Des. (2017). https://doi.org/10.1080/14680629.2017.1389085

Author Index

X. Shi et al. (Eds.): GSIC 2018, *Proceedings of GeoShanghai 2018 International Conference: Transportation Geotechnics and Pavement Engineering*, pp. 493–495, 2018.
https://doi.org/10.1007/978-981-13-0011-0

MIX
Papier aus verantwortungsvollen Quellen
Paper from responsible sources
FSC® C105338

If you have any concerns about our products,
you can contact us on
ProductSafety@springernature.com

In case Publisher is established outside the EU,
the EU authorized representative is:
Springer Nature Customer Service Center GmbH
Europaplatz 3, 69115 Heidelberg, Germany

Printed by Libri Plureos GmbH
in Hamburg, Germany